MODERN
INORGANIC
CHEMISTRY

MODERN INORGANIC CHEMISTRY

Second Edition

William L. Jolly

Professor of Chemistry
University of California, Berkeley

McGraw-Hill, Inc.

New York   St. Louis   San Francisco   Auckland   Bogotá   Caracas
Hamburg   Lisbon   London   Madrid   Mexico   Milan   Montreal
New Delhi   Paris   San Juan   São Paulo   Singapore   Sydney
Tokyo   Toronto

This book was set in Times Roman by Publication Services.
The editors were Kirk Emry and John M. Morriss;
the production supervisor was Kathryn Porzio.
The cover was designed by Jo Jones.
Project supervision was done by Publication Services.
R. R. Donnelley & Sons Company was printer and binder.

MODERN INORGANIC CHEMISTRY

1 2 3 4 5 6 7 8 9 0 DOC DOC 9 0 9 8 7 6 5 4 3 2 1

ISBN 0-07-032768-8

Library of Congress Cataloging-in-Publication Data

Jolly, William L.
   Modern inorganic chemistry / William L. Jolly. — 2nd ed.
     p.   cm.
   Includes bibliographical references and indexes.
   ISBN 0-07-032768-8
   1. Chemistry, Inorganic.   I. Title.
QD151.2.J637   1991   90-25206
546—dc20

CONTENTS

PREFACE

The purpose of this text is to describe and explain the intellectual tools used in the systematization of inorganic chemistry. This aim cannot be achieved in a vacuum, and therefore the text includes a large amount of descriptive material. It is assumed that the reader has completed a one-year course in general chemistry and an introductory course in organic chemistry. Thus familiarity with the rudiments of thermodynamics, kinetics, atomic and molecular orbital theory, orbital hybridization, the electronegativity concept, and simple bonding theory is assumed.

This new edition contains many changes. In many college chemistry curricula, students receive little exposure to applications of thermodynamics and kinetics in inorganic chemistry. Therefore the chapter that formerly emphasized reduction potentials has been broadened to include the use of thermodynamics in a variety of systems. And the chapter that formerly covered only the kinetics of gas phase reactions now also introduces the principles of kinetics in aqueous solutions. The chapter on the solid state has been expanded, reflecting changes in modern research interests. I deem the calculational aspects of molecular orbital theory to be of relatively little importance, and therefore the material in the chapter on molecular orbitals has been abbreviated. Almost every chapter has new examples and, I hope, improved discussions.

I believe that the text covers all the important aspects of modern inorganic chemistry. Although the descriptive material is organized topically rather than in periodic table arrangement, the reader has easy access to information through the index.

Too many chemistry students graduate with only a hazy notion of the applied side of chemistry. Therefore I offer no apologies for occasionally describing practical applications (and potential applications) of inorganic systems. It is important for students to recognize that much of the chemistry described in this book closely affects their daily lives and that soon they may be called upon to use inorganic chemistry to solve important problems.

Problems are provided at the end of each chapter, and answers to about half of these problems are given after Appendix G. The problems range from simple to difficult (the latter indicated by asterisks) and should be attempted by all students who seriously want to master the subject matter.

McGraw-Hill and the author would like to thank the following reviewers for their many helpful comments and suggestions: Kenneth R. Magnell, Central Michigan University; Edward A. Mottel, Rose-Hulman Institute of Technology; and Hans J. Mueh, United States Air Force Academy.

William L. Jolly

MODERN
INORGANIC
CHEMISTRY

ELECTRON CONFIGURATIONS OF ATOMS AND THE PERIODIC TABLE

Many properties of compounds are predictable or rationalizable from the electron configurations of the atoms that make up the compounds. Thus it is very important for chemists to be familiar with the electronic structures and related properties of atoms. In this chapter we briefly discuss these topics and show how the periodic table may be used to correlate properties with electronic structure.

Although some of the material covered in the chapter is discussed in elementary chemistry texts, you should not assume that this material is easily understood. It is included here because of its fundamental importance in the study of inorganic chemistry and to give you the opportunity to review. Throughout the remainder of the text, it will be assumed that you are familiar with the principles discussed in this chapter. You may find it helpful occasionally to refer back to parts of this chapter.

QUANTUM NUMBERS

In 1926 Schrödinger[1] proposed a differential equation (which now bears his name) for relating the energy of a system to the space coordinates of its constituent particles. For a particle in three dimensions, the equation may be written

$$\frac{\partial^2 \psi}{\partial x^2} + \frac{\partial^2 \psi}{\partial y^2} + \frac{\partial^2 \psi}{\partial z^2} + \frac{8\pi^2 m}{h^2}(E - V)\psi = 0$$

---

[1]   E. Schrödinger, *Ann. Phys.*, **79**, 361, 489, **81**, 109 (1926).

where ψ is the "wave function," x, y, and z are the cartesian coordinates of the particle, m is its mass, E is the total energy, and V is the potential energy. This equation incorporates both the wave character of the particle and the probability character of the measurements. The wave function ψ has properties analogous to the amplitude of a wave; its square, ψ^2, is proportional to the probability of finding a particle at the coordinates x, y, z.

When the Schrödinger equation is applied to the hydrogen atom or to any system with one electron and one nucleus, the solution includes three "integration constants." These are the familiar quantum numbers n, l, and m_l. The "principal" quantum number n may have any integral value from 1 to infinity:

$$n = 1, 2, 3,\ldots$$

The "azimuthal" or "orbital angular momentum" quantum number l may have any integral value from zero to $n - 1$:

$$l = 0, 1, 2,\ldots, (n - 1)$$

However, for historical reasons, l is usually not specified by these integers, but rather by the letters s, p, d, f, g, $\ldots$ (continuing alphabetically), which correspond to $l = 0, 1, 2, 3, 4, \ldots$, respectively. The n and l values of an electron are often designated by the notation nl, in which the value of l is indicated by the appropriate letter. Thus a $2p$ electron has $n = 2$ and $l = 1$. The "magnetic" quantum number m_l may have any integral value from $-l$ to $+l$:

$$m_l = -l, -(l - 1),\ldots, -1, 0, +1, +2,\ldots, l - 1, +l$$

Because an electron has spin, and consequently a magnetic moment which can be oriented either up or down, yet a fourth quantum number must be specified, the "spin" quantum number m_s. The permissible values of m_s are $\pm\frac{1}{2}$.

As a consequence of the restrictions on the quantum numbers, the electron of a hydrogen atom may be assigned only certain combinations of quantum numbers. These permissible combinations for $n = 1, 2, 3$, and 4 are indicated in Table 1.1.

TABLE 1.1
Some allowed values of the hydrogen atom quantum numbers

n	l	m_l		m_s	No. of combinations
1	0	0		$\pm\frac{1}{2}$	2
2	0	0		$\pm\frac{1}{2}$	2
2	1	$-1, 0, +1$		$\pm\frac{1}{2}$	6
3	0	0		$\pm\frac{1}{2}$	2
3	1	$-1, 0, +1$		$\pm\frac{1}{2}$	6
3	2	$-2, -1, 0, +1, +2$		$\pm\frac{1}{2}$	10
4	0	0		$\pm\frac{1}{2}$	2
4	1	$-1, 0, +1$		$\pm\frac{1}{2}$	6
4	2	$-2, -1, 0, +1, +2$		$\pm\frac{1}{2}$	10
4	3	$-3, -2, -1, 0, +1, +2, +3$		$\pm\frac{1}{2}$	14

Each allowed combination of n, l, and m_l corresponds to an atomic "orbital." We say that the electron may be "put into" or "assigned to" a particular orbital. Of course, in any orbital, the m_s quantum number may be either $+\frac{1}{2}$ or $-\frac{1}{2}$.

ORBITAL SHAPES AND ENERGIES FOR THE HYDROGEN ATOM

It is convenient to express the wave function of the hydrogen atom in terms of the polar coordinates r, θ, and ϕ and to factor the function into three separate parts, each of which is a function of only one coordinate:

$$\psi(r, \theta, \phi) = R(r) \cdot \Theta(\theta) \cdot \Phi(\phi)$$

In Tables 1.2 to 1.4, the functions $R(r)$, $\Theta(\theta)$, and $\Phi(\phi)$ are given for various values of the quantum number n, l, and m_l. Obviously the values of the three functions, and hence also the spatial distribution of the electron of a hydrogen atom, are markedly affected by the values of n and l. The spatial distributions can be indicated graphically in several ways. Let us first consider the case of $l = 0$, that is, s electrons. In Fig. 1.1, the radial wave function R is plotted as a function of r, the distance from the nucleus, for $n = 1$, 2, and 3. Three facts should be noted. First, in each case, the magnitude of the wave function has its maximum value at the nucleus. Second, for $n > 1$, the wave function is zero in certain regions called "nodes." (As a general rule, there are $n - 1$ nodes in an atomic wave function.) Third, the sign of R changes as it passes through a node.

Although R is a function which is not directly related to any experimentally measurable quantity, the function R^2 is proportional to electron density and therefore has considerable physical significance. In Fig. 1.2, R^2 is plotted versus

TABLE 1.2
Radial wave function $R(r)$ for hydrogen atom

n	l	$R(r)$†
1	0	$2\left(\dfrac{Z}{a_0}\right)^{3/2} e^{-Zr/a_0}$
2	0	$\left(\dfrac{1}{2\sqrt{2}}\right)\left(\dfrac{Z}{a_0}\right)^{3/2}\left(2 - \dfrac{Zr}{a_0}\right)e^{-Zr/2a_0}$
	1	$\left(\dfrac{1}{2\sqrt{6}}\right)\left(\dfrac{Z}{a_0}\right)^{3/2}\left(\dfrac{Zr}{a_0}\right)e^{-Zr/2a_0}$
3	0	$\left(\dfrac{2}{81\sqrt{3}}\right)\left(\dfrac{Z}{a_0}\right)^{3/2}\left(27 - \dfrac{18Zr}{a_0} + \dfrac{2Z^2r^2}{a_0^2}\right)e^{-Zr/3a_0}$
	1	$\left(\dfrac{4}{81\sqrt{6}}\right)\left(\dfrac{Z}{a_0}\right)^{3/2}\left(\dfrac{6Zr}{a_0} - \dfrac{Z^2r^2}{a_0^2}\right)e^{-Zr/3a_0}$
	2	$\left(\dfrac{4}{81\sqrt{30}}\right)\left(\dfrac{Z}{a_0}\right)^{3/2}\left(\dfrac{Zr}{a_0}\right)^2 e^{-Zr/3a_0}$

† a_0 is the "first Bohr radius," $h^2/(4\pi^2 me^2)$, or 0.529 Å.

TABLE 1.3
Angular wave function $\Theta(\theta)$ for hydrogen atom

l	m_l	$\Theta(\theta)$
0	0	$\frac{\sqrt{2}}{2}$
1	0	$\frac{\sqrt{6}}{2}\cos\theta$
	± 1	$\frac{\sqrt{3}}{2}\sin\theta$
2	0	$\frac{\sqrt{10}}{4}(3\cos^2\theta - 1)$
	± 1	$\frac{\sqrt{15}}{2}(\sin\theta\cos\theta)$
	± 2	$\frac{\sqrt{15}}{4}\sin^2\theta$

r for $1s$, $2s$, and $3s$ electrons. Note that R^2, like R, has its maximum value at the nucleus and that nodes appear for $n > 1$. Of course, R^2 is never negative; a negative electron density is physically meaningless. One important feature of any s-electron distribution that is not obvious from Figs. 1.1 and 1.2 is that the distribution is independent of θ and ϕ; that is, it is spherically symmetric. To show this feature in a graph or picture would be exceedingly difficult; however, we have done the next best thing in Fig. 1.3, where we have plotted electron density as a function of position on a plane passing through the nucleus for $1s$, $2s$, and $3s$ electrons. The electron density is indicated by the black dots; dark regions correspond to high electron density, and light regions correspond to low electron density. Figure 1.3 makes it easier to understand that the nodes in $2s$ and $3s$ wave functions are actually spherical *surfaces*.

Analogous plots can be drawn for $2p$ and $3p$ electrons. In Fig. 1.4, R is plotted versus r, and in Fig. 1.5, R^2 is plotted versus r. The angular distribution of a p electron is very different from that of an s electron. A nodal plane passes through the nucleus, and the electron density is concentrated in lobes on both sides of the nodal plane. Electron density contour maps for $2p$ and $3p$ electrons are shown in

TABLE 1.4
Angular wave function $\Phi(\phi)$ for hydrogen atom

m_l	$\Phi(\phi)$
0	$\frac{1}{\sqrt{2\pi}}$
1	$\frac{1}{\sqrt{\pi}}\cos\phi$
-1	$\frac{1}{\sqrt{\pi}}\sin\phi$
2	$\frac{1}{\sqrt{\pi}}\cos 2\phi$
-2	$\frac{1}{\sqrt{\pi}}\sin 2\phi$

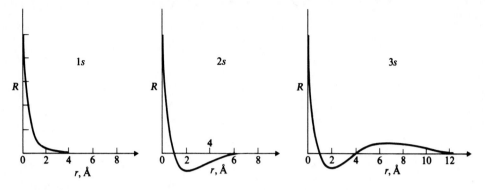

FIGURE 1.1

Plots of R versus r for $1s$, $2s$, and $3s$ orbitals of the hydrogen atom. The radius scale is the same throughout, but the scale for R is changed for the various orbitals.

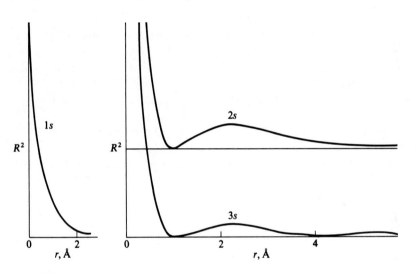

FIGURE 1.2

Plots of R^2 versus r for $1s$, $2s$, and $3s$ orbitals of the hydrogen atom. The radius scale is the same throughout, but the density scale is changed for the various orbitals.

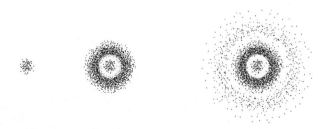

FIGURE 1.3

Electron density in a plane passing through the hydrogen atom for $1s$, $2s$, and $3s$ electrons.

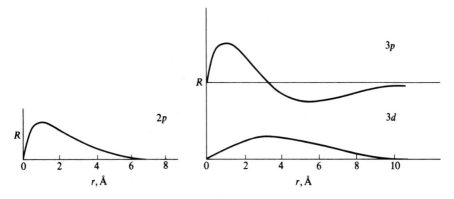

FIGURE 1.4
Plots of R versus r for $2p$, $3p$, and $3d$ orbitals of the hydrogen atom. The radius scale is the same throughout, but the scale for R is changed for the various orbitals.

Fig. 1.6. These maps show the distribution of electron density in planes which pass through the regions of maximum electron density and the nucleus. The p-electron distributions are cylindrically symmetric; a three-dimensional representation of the contour maps could be generated by rotating the diagrams in Fig. 1.6 about the z axes (the lines which pass through the nuclei and the points of maximum electron density).

Chemists commonly represent the spatial configuration of a p orbital by a figure which looks somewhat like a dumbbell or a sausage tightly tied at the middle, as shown in Fig. 1.7. The three p orbitals of a given n value can be

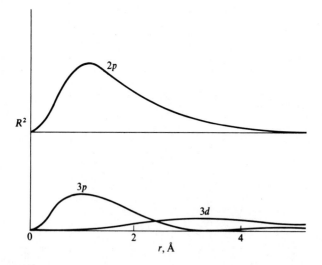

FIGURE 1.5
Plots of R^2 versus r for $2p$, $3p$, and $3d$ orbitals of the hydrogen atom. The density scale is different for the two plots.

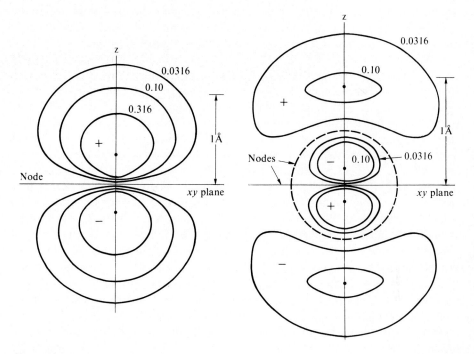

FIGURE 1.6
Electron density contours for $2p_z$ and $3p_z$ orbitals. The contour values are relative to the electron density maxima (indicated as dots). In the case of the $3p_z$ orbital, the xy plane and a sphere of radius 0.52 Å are nodal surfaces. [*Adapted from E. A. Ogryzlo and G. B. Porter, J. Chem. Educ.,* **40**, *256 (1963), with permission.*]

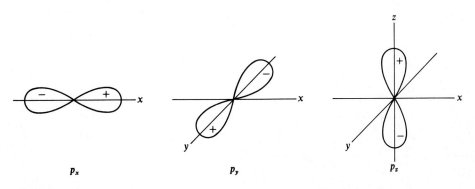

FIGURE 1.7
Sausage representations of p orbitals. Such diagrams are merely crude representations of orbital boundaries within which most of the electron density resides. The shapes have only qualitative significance.

designated as p_x, p_y, and p_z, the subscripts indicating the axes along which the orbitals lie. Although it is not obvious from the figures, the sum of the electron densities of such a set of p orbitals is spherically symmetric.

Plots of R and R^2 for $3d$ electrons are given in Figs. 1.4 and 1.5. There are six d functions, identical except for their spatial orientations, that satisfy the solution of the Schrödinger equation for the hydrogen-like atom for $l = 2$. An electron density contour map for one of the six d functions is shown in Fig. 1.8. Each of these functions has four lobes directed toward the corners of a square. Two such functions lie in each of the cartesian planes, with the lobes either directed along the coordinate axes or bisecting the right angles between the coordinate axes. Of course, there can be only five d orbitals, and therefore the orbitals must be constructed as linear combinations of the six d functions. It is conventional to make four of the d orbitals identical to corresponding d functions. These are the d_{xy}, d_{xz}, and d_{yz} orbitals (which lie in the xy, xz, and yz planes, respectively, with the lobes bisecting the angles between the axes) and the $d_{x^2-y^2}$ orbital (which lies in the xy plane, with lobes directed along the axes). The fifth d orbital (the d_{z^2} orbital) is a hybrid, or combination of the $d_{z^2-x^2}$ and $d_{z^2-y^2}$ functions, which lie in the xz and yz planes, with lobes along the axes. The d_{z^2} orbital (which might better be called the $d_{2z^2-x^2-y^2}$ orbital) consists of two main lobes directed along the z axis and a kind of torus or band lying in the xy plane. "Sausage" representations of the five $3d$ orbitals are shown in Fig. 1.9. Note that each $3d$ orbital has two nodal surfaces which pass through the nucleus. In the case of the $3d_{z^2}$ orbital, the nodal surfaces are cones with a common apex at the nucleus. The sum of the electron density distributions of the five d orbitals of a given n value is spherically symmetric.

The arbitrariness of the conventional choice of d orbitals is emphasized by the fact that it is possible to combine the six d functions to obtain five *equivalent*

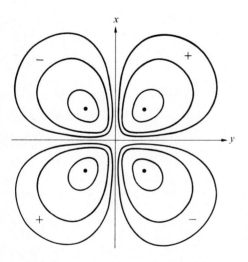

FIGURE 1.8
Electron density contours for a $3d_{xy}$ function. Except for differences in axis orientation, all six d functions, of which the five d orbitals are linear combinations, are identical.

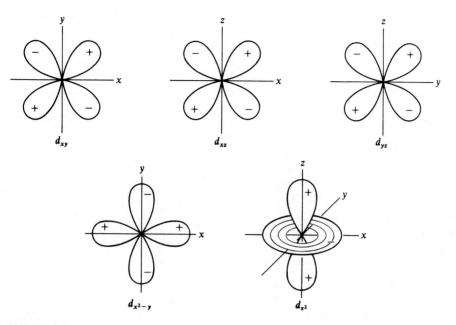

FIGURE 1.9
Sausage representations of the five conventional d orbitals. Notice that the axes are labeled differently for the d_{xy}, d_{xz}, and d_{yz} orbitals.

d orbitals (i.e., orbitals having the same shape) with lobes directed toward the corners of a pentagonal antiprism.[2]

In Figs. 1.7 and 1.9, the various lobes of the sausagelike representations of the p and d orbitals are marked with $+$ or $-$ signs. These refer to the sign of the original wave function, ψ, and do not indicate positive or negative charges. Orbitals may be classified in terms of their symmetry with respect to inversion through the center (nucleus). When inversion (which simply involves changing the signs of the x, y, and z coordinates) causes no change in the sign of ψ, the orbital and its wave function are described as *gerade* (German for "even"); when inversion causes a change in sign, the orbital and wave function are described as *ungerade* (German for "odd"). Thus s and d orbitals are gerade, and p orbitals are ungerade. Later, when we consider the formation of molecular orbitals by the combination of atomic orbitals, we shall appreciate the importance of these symmetry designations and the signs of the wave functions.

[2] R. E. Powell, *J. Chem. Educ.*, **45**, 45 (1968). The five equivalent d orbitals are generally not as convenient to use as the conventional d orbitals. However, they may have a slight advantage when considering the bonding in a molecule such as ferrocene, $Fe(C_5H_5)_2$, in the staggered conformation.

If one solves the Schrödinger equation for the energy levels of the hydrogen atom, or of a one-electron ion, one obtains[3]

$$E = \frac{2\pi^2 m Z^2 e^4}{n^2 h^2} = -13.60 \frac{Z^2}{n^2} \text{ eV}$$

where E is the energy relative to the infinitely separated electron and nucleus, m is the mass of the electron, Z is the charge on the nucleus, e is the electronic charge, and h is Planck's constant. It can be seen that the energy is independent of the values of the l and m_l quantum numbers but is dependent on the principal quantum number n. The energy of the atom increases with increasing n, and the atom is in its lowest energy state (the ground state) when $n = 1$, that is, when the electron is in the $1s$ orbital. Because electrostatic interaction energy decreases with increasing distance, it is clear that the average distance of the electron from the nucleus increases with increasing n. Calculation shows that the average value of $1/r$ is given by

$$\left\langle \frac{1}{r} \right\rangle = \frac{4\pi^2 m e^2 Z}{n^2 h^2}$$

ELECTRON CONFIGURATIONS OF ATOMS AND IONS

The Schrödinger equation has never been solved in closed form[4] except for the hydrogen atom and one-electron ions such as He^+, Li^{2+}, etc. However, spectroscopic data and accurate calculations based on successive approximations (such as the Hartree-Fock self-consistent field method) have shown that many-electron atoms possess atomic orbitals completely analogous to those of the hydrogen atom.[5] Thus we may speak of $1s$, $2p$, $4f$, etc., orbitals for any of the known elements. The Pauli exclusion principle states that in a given atom no two electrons can have the same values for all four quantum numbers. Consequently, each atomic orbital can contain no more than two electrons, of opposite spin. We can describe the building of a set of elements by a sequence of imaginary steps in each of which we simultaneously increase the nuclear charge by one unit and add one electron to an atomic orbital. If the electrons are added to the atomic orbitals in the proper order, the elements are generated in their ground states. In Table 1.5 we have listed, in order of atomic number, the elements and their electron configurations. By examination of Table 1.5, the reader can verify that, to a fair

---

[3] In 1913 Niels Bohr derived this relation by combining classical physics with some of the ideas of quantum theory. See G. Herzberg, "Atomic Spectra and Atomic Structure," 2d ed., pp. 13–21, Dover, New York, 1944.

[4] That is, by a method not involving successive approximations.

[5] H. F. Schaefer, "The Electronic Structure of Atoms and Molecules," Addison-Wesley, Reading, Mass., 1972.

TABLE 1.5
Electron configurations of the elements

Z	Element	Symbol	Electron configuration†	Z	Element	Symbol	Electron configuration†
1	Hydrogen	H	$1s$	54	Xenon	Xe	$[Kr]4d^{10}5s^25p^6$
2	Helium	He	$1s^2$	55	Cesium	Cs	$[Xe]6s$
3	Lithium	Li	$[He]2s$	56	Barium	Ba	$[Xe]6s^2$
4	Beryllium	Be	$[He]2s^2$	57	Lanthanum	La	$*[Xe]5d6s^2$
5	Boron	B	$[He]2s^22p$	58	Cerium	Ce	$*[Xe]4f5d6s^2$
6	Carbon	C	$[He]2s^22p^2$	59	Praseodymium	Pr	$[Xe]4f^36s^2$
7	Nitrogen	N	$[He]2s^22p^3$	60	Neodymium	Nd	$[Xe]4f^46s^2$
8	Oxygen	O	$[He]2s^22p^4$	61	Promethium	Pm	$[Xe]4f^56s^2$
9	Fluorine	F	$[He]2s^22p^5$	62	Samarium	Sm	$[Xe]4f^66s^2$
10	Neon	Ne	$[He]2s^22p^6$	63	Europium	Eu	$[Xe]4f^76s^2$
11	Sodium	Na	$[Ne]3s$	64	Gadolinium	Gd	$*[Xe]4f^75d6s^2$
12	Magnesium	Mg	$[Ne]3s^2$	65	Terbium	Tb	$[Xe]4f^96s^2$
13	Aluminum	Al	$[Ne]3s^23p$	66	Dysprosium	Dy	$[Xe]4f^{10}6s^2$
14	Silicon	Si	$[Ne]3s^23p^2$	67	Holmium	Ho	$[Xe]4f^{11}6s^2$
15	Phosphorus	P	$[Ne]3s^23p^3$	68	Erbium	Er	$[Xe]4f^{12}6s^2$
16	Sulfur	S	$[Ne]3s^23p^4$	69	Thulium	Tm	$[Xe]4f^{13}6s^2$
17	Chlorine	Cl	$[Ne]3s^23p^5$	70	Ytterbium	Yb	$[Xe]4f^{14}6s^2$
18	Argon	Ar	$[Ne]3s^23p^6$	71	Lutetium	Lu	$[Xe]4f^{14}5d6s^2$
19	Potassium	K	$[Ar]4s$	72	Hafnium	Hf	$[Xe]4f^{14}5d^26s^2$
20	Calcium	Ca	$[Ar]4s^2$	73	Tantalum	Ta	$[Xe]4f^{14}5d^36s^2$
21	Scandium	Sc	$[Ar]3d4s^2$	74	Tungsten	W	$[Xe]4f^{14}5d^46s^2$
22	Titanium	Ti	$[Ar]3d^24s^2$	75	Rhenium	Re	$[Xe]4f^{14}5d^56s^2$
23	Vanadium	V	$[Ar]3d^34s^2$	76	Osmium	Os	$[Xe]4f^{14}5d^66s^2$
24	Chromium	Cr	$*[Ar]3d^54s$	77	Iridium	Ir	$[Xe]4f^{14}5d^76s^2$
25	Manganese	Mn	$[Ar]3d^54s^2$	78	Platinum	Pt	$*[Xe]4f^{14}5d^96s$
26	Iron	Fe	$[Ar]3d^64s^2$	79	Gold	Au	$*[Xe]4f^{14}5d^{10}6s$
27	Cobalt	Co	$[Ar]3d^74s^2$	80	Mercury	Hg	$[Xe]4f^{14}5d^{10}6s^2$
28	Nickel	Ni	$[Ar]3d^84s^2$	81	Thallium	Tl	$[Xe]4f^{14}5d^{10}6s^26p$
29	Copper	Cu	$*[Ar]3d^{10}4s$	82	Lead	Pb	$[Xe]4f^{14}5d^{10}6s^26p^2$
30	Zinc	Zn	$[Ar]3d^{10}4s^2$	83	Bismuth	Bi	$[Xe]4f^{14}5d^{10}6s^26p^3$
31	Gallium	Ga	$[Ar]3d^{10}4s^24p$	84	Polonium	Po	$[Xe]4f^{14}5d^{10}6s^26p^4$
32	Germanium	Ge	$[Ar]3d^{10}4s^24p^2$	85	Astatine	At	$[Xe]4f^{14}5d^{10}6s^26p^5$
33	Arsenic	As	$[Ar]3d^{10}4s^24p^3$	86	Radon	Rn	$[Xe]4f^{14}5d^{10}6s^26p^6$
34	Selenium	Se	$[Ar]3d^{10}4s^24p^4$	87	Francium	Fr	$[Rn]7s$
35	Bromine	Br	$[Ar]3d^{10}4s^24p^5$	88	Radium	Ra	$[Rn]7s^2$
36	Krypton	Kr	$[Ar]3d^{10}4s^24p^6$	89	Actinium	Ac	$*[Rn]6d7s^2$
37	Rubidium	Rb	$[Kr]5s$	90	Thorium	Th	$*[Rn]6d^27s^2$
38	Strontium	Sr	$[Kr]5s^2$	91	Protactinium	Pa	$*[Rn]5f^26d7s^2$
39	Yttrium	Y	$[Kr]4d5s^2$	92	Uranium	U	$*[Rn]5f^36d7s^2$
40	Zirconium	Zr	$[Kr]4d^25s^2$	93	Neptunium	Np	$*[Rn]5f^46d7s^2$
41	Niobium	Nb	$*[Kr]4d^45s$	94	Plutonium	Pu	$[Rn]5f^67s^2$
42	Molybdenum	Mo	$*[Kr]4d^55s$	95	Americium	Am	$[Rn]5f^77s^2$
43	Technetium	Tc	$[Kr]4d^55s^2$	96	Curium	Cm	$*[Rn]5f^76d7s^2$
44	Ruthenium	Ru	$*[Kr]4d^75s$	97	Berkelium	Bk	$[Rn]5f^97s^2$
45	Rhodium	Rh	$*[Kr]4d^85s$	98	Californium	Cf	$[Rn]5f^{10}7s^2$
46	Palladium	Pd	$*[Kr]4d^{10}$	99	Einsteinium	Es	$[Rn]5f^{11}7s^2$
47	Silver	Ag	$*[Kr]4d^{10}5s$	100	Fermium	Fm	$[Rn]5f^{12}7s^2$
48	Cadmium	Cd	$[Kr]4d^{10}5s^2$	101	Mendelevium	Md	$[Rn]5f^{13}7s^2$
49	Indium	In	$[Kr]4d^{10}5s^25p$	102	Nobelium	No	$[Rn]5f^{14}7s^2$
50	Tin	Sn	$[Kr]4d^{10}5s^25p^2$	103	Lawrencium	Lr	$[Rn]5f^{14}6d7s^2$
51	Antimony	Sb	$[Kr]4d^{10}5s^25p^3$	104	Rutherfordium	Rf	$[Rn]5f^{14}6d^27s^2$
52	Tellurium	Te	$[Kr]4d^{10}5s^25p^4$	105	Hahnium	Ha	$[Rn]5f^{14}6d^37s^2$
53	Iodine	I	$[Kr]4d^{10}5s^25p^5$	106		...	$[Rn]5f^{14}6d^47s^2$

† Asterisks indicate exceptions to Madelung's rule for adding electrons.

approximation, the order of filling atomic orbitals is the order of increasing $n + l$, and, when there are two or more orbitals with the same value of $n + l$ (for example $5s$, $4p$, and $3d$), these orbitals are filled in order of increasing n. Although this approximate rule, sometimes called Madelung's rule,[6] has some theoretical justification,[7] it is essentially an empirical rule, useful as a mnemonic for the electron configurations of the elements. The exceptions to this simple rule for adding electrons are indicated by asterisks in Table 1.5. Most of the exceptions are due to the extra stability associated with electron subshells which are half-filled or completely filled.[8] For example, Cr and Cu have the configurations $3d^5 4s$ and $3d^{10} 4s$ instead of the predicted configurations $3d^4 4s^2$ and $3d^9 4s^2$, respectively. The exceptions, although chemically significant, can be ignored when considering the general construction of the periodic table.

We have already noted that the energy of a hydrogen atom is essentially independent of the l quantum number of the single electron. Therefore we might have expected, in building up the elements, that it would make no difference whether an electron was added to an empty $3s$ orbital, $3p$ orbital, or $3d$ orbital. The fact that it does make a difference and that the energy of an electron in a many-electron atom increases with increasing l as well as increasing n is a consequence of the fact that the shapes of the various orbitals are different. The lower the l value for a given n value, the more effectively the electron shields the nuclear charge for other electrons of the same n value. Thus an s electron (which has a high electron density in the region of the nucleus) more effectively shields the nucleus from a p electron (which has essentially no electron density at the nucleus) than vice versa. Consequently, both the effective nuclear charge felt by an electron and its binding energy decrease in the order $s, p, d, f, \ldots$.

The symbolic representations of electron configurations in Table 1.5 give no details regarding the filling of orbitals in partly filled shells of p, d, and f electrons. For example, consider the carbon atom, which has two $2p$ electrons. In this case, if we do not distinguish between the three $2p$ orbitals, the orbitals can be occupied in three different ways:

According to a rule propounded by Hund, the ground state of a set of electrons of given n and l values corresponds to an orbital occupancy such that the multiplicity is maximized; i.e., the electrons are unpaired as far as possible. Application of this rule tells us that the first of the three indicated configurations is the ground state and that the other configurations are excited states. The fact that the third configuration is not the ground state is easy to rationalize: When two electrons are

[6] E. Madelung, "Mathematische Hilfsmittel des Physikers," 3d ed., p 359, Springer, Berlin, 1936.

[7] D. P. Wong, *J. Chem. Educ.*, **56**, 714 (1979).

[8] This extra stability is a consequence of electron repulsion, exchange energy, and nuclear shielding effects, to be discussed later in this chapter.

in the same orbital, the electrostatic repulsion between them is greater than when they are in separate orbitals. The fact that the first configuration is more stable than the second is more difficult to rationalize: it is a consequence of so-called *exchange energy*[9] which is lost when two electrons of the same spin are forced to have their spins opposed.

One might suppose that, in the successive removal of electrons from an atom to form a series of stable positive ions, the electrons would leave the atomic orbitals in the reverse order in which they were filled in the hypothetical buildup of the elements. To some extent this is true; however, there are many exceptions. For example, the Ti^{2+} ion has the configuration $1s^2 2s^2 2p^6 3s^2 3p^6 3d^2$, whereas the Ca atom (which has the same number of electrons) has the configuration $1s^2 2s^2 2p^6 3s^2 3p^6 4s^2$. Exceptions of this type are not surprising when one recognizes that the nuclear charge of titanium is two units more positive than that of calcium and that the process of removing electrons from an atom is not the same are removing electrons while simultaneously removing an equal number of protons from the nucleus. In the case of TI^{2+}, all the atomic orbitals are smaller and closer to the nucleus than the corresponding orbitals of Ca, but the difference is more marked for d orbitals than for s and p orbitals. This effect is so great that in Ti^{2+} the $3d$ electrons are not shielded from the nucleus as well as $4s$ electrons would be.

As a general rule, ionization of an atom causes the relative energies of the orbitals to approach those of a hydrogen-like atom; that is, the orbital energies become relatively independent of l and mainly dependent on n. Usually the easiest electrons to remove from an atom or an ion are those with the maximum value of n; of this set of electrons, the easiest to remove are those with the maximum value of l.

The valence electrons of an atom are those which are relatively easily removed. Generally, removal of all the valence electrons of an atom yields an ion with an outer electron configuration of the type $ns^2 np^6$ or $ns^2 np^6 nd^{10}$. Such ions are called atomic cores or kernels.

THE PERIODIC TABLE

A periodic table is a listing of the elements in an array such that atomic number increases from left to right in any row and the elements in any column have outer electronic structures which are similar except for principal quantum number. A common form of the periodic table is shown in Fig. 1.10. Such a table is a very useful device for recalling groups of related atoms and for correlating properties with systematic changes in electronic structure. Every inorganic chemist should have a periodic table displayed near his or her desk or workbench. Even a chemist who has memorized the periodic table will find that it often helps to look at the table when thinking about inorganic problems.

---

[9] See, for discussion, L. E. Orgel, "An Introduction to Transition-Metal Chemistry: Ligand Field Theory," 2d ed., pp. 42–45, Wiley, New York, 1966.

PERIODIC TABLE OF THE ELEMENTS

Alkali metals

Alkaline earths

Chalcogens

Halogens

Noble gases

Transition metals

Coinage metals

Platinum metals

I																	VIII
1.008 **H** 1																	4.003 **He** 2

I	II											III	IV	V	VI	VII	VIII
6.939 **Li** 3	9.012 **Be** 4											10.811 **B** 5	12.011 **C** 6	14.007 **N** 7	15.999 **O** 8	18.998 **F** 9	20.183 **Ne** 10
22.990 **Na** 11	24.312 **Mg** 12	3 III	4 IV	5 V	6 VI	7 VII	8	9 VIII	10	11 I	12 II	26.982 **Al** 13	28.086 **Si** 14	30.974 **P** 15	32.064 **S** 16	35.453 **Cl** 17	39.948 **Ar** 18
39.102 **K** 19	40.08 **Ca** 20	44.956 **Sc** 21	47.90 **Ti** 22	50.942 **V** 23	51.996 **Cr** 24	54.938 **Mn** 25	55.847 **Fe** 26	58.933 **Co** 27	58.71 **Ni** 28	63.546 **Cu** 29	65.37 **Zn** 30	69.72 **Ga** 31	72.59 **Ge** 32	74.922 **As** 33	78.96 **Se** 34	79.904 **Br** 35	83.80 **Kr** 36
85.47 **Rb** 37	87.62 **Sr** 38	88.905 **Y** 39	91.22 **Zr** 40	92.906 **Nb** 41	95.94 **Mo** 42	(98) **Tc** 43	101.07 **Ru** 44	102.905 **Rh** 45	106.4 **Pd** 46	107.868 **Ag** 47	112.40 **Cd** 48	114.82 **In** 49	118.69 **Sn** 50	121.75 **Sb** 51	127.60 **Te** 52	126.904 **I** 53	131.30 **Xe** 54
132.905 **Cs** 55	137.34 **Ba** 56	138.91 ★**La** 57	178.49 **Hf** 72	180.948 **Ta** 73	183.85 **W** 74	186.2 **Re** 75	190.2 **Os** 76	192.2 **Ir** 77	195.09 **Pt** 78	196.967 **Au** 79	200.59 **Hg** 80	204.37 **Tl** 81	207.19 **Pb** 82	208.980 **Bi** 83	(209) **Po** 84	(210) **At** 85	(222) **Rn** 86
(223) **Fr** 87	(226.025) **Ra** 88	(227.028) ▲**Ac** 89	(261) **Rf** 104	(262) **Ha** 105	(263) 106												

★ Lanthanide series

140.12 **Ce** 58	140.907 **Pr** 59	144.24 **Nd** 60	(145) **Pm** 61	150.35 **Sm** 62	151.96 **Eu** 63	157.25 **Gd** 64	158.924 **Tb** 65	162.50 **Dy** 66	164.930 **Ho** 67	167.26 **Er** 68	168.934 **Tm** 69	173.04 **Yb** 70	174.97 **Lu** 71

▲ Actinide series

232.038 **Th** 90	(231.036) **Pa** 91	238.029 **U** 92	(237.048) **Np** 93	(244) **Pu** 94	(243) **Am** 95	(247) **Cm** 96	(247) **Bk** 97	(251) **Cf** 98	(252) **Es** 99	(257) **Fm** 100	(258) **Md** 101	(259) **No** 102	(260) **Lr** 103

FIGURE 1.10
Periodic table of the elements.

The elements may be classified in terms of their valence electrons. Those whose atoms contain d electrons in the valence shell are called "transition elements." These elements appear near the middle of the periodic table, as shown in Fig. 1.10. Special groups of transition elements, the lanthanides and the actinides, contain f electrons in their valence shells and appear in separate rows below the main body of the table. All the other elements of the periodic table are called "nontransition elements," "main-group elements," or "representative elements." Special names are given to certain "families" of elements which appear in particular columns or blocks of the periodic table. These names are indicated in Fig. 1.10.

The periodic table derives its name from the fact that the properties of the elements, when listed in order of atomic number, are periodic; that is, certain properties recur at regular intervals of atomic number. The periodic table is simply a device for listing the elements so that elements with similar properties are grouped together.

Often elements are classified in terms of their "group numbers," shown at the tops of the columns in the periodic table. Note that two sets of group numbers are given—those in Arabic numerals and those in Roman numerals. The Arabic numerals correspond to the new system recently recommended by the International Union of Pure and Applied Chemistry and the American Chemical Society.[10] The Roman numerals correspond to the traditional system, although with the usual A and B designations omitted. The A and B designations became problematical because of disagreement as to which columns corresponded to A elements and which columns corresponded to B elements. The problem was presumably solved by invention of the new 18-column format. The new proposal is not as radical as it first appears: Note that in the two-digit numbers, the units digit corresponds to the traditional group number. Nevertheless, there has been great resistance to the acceptance of the new scheme. Perhaps the simplest solution is to use the traditional scheme with a verbal indication as to whether one is referring to a main-group element or a transition element. For example, chromium can be referred to as a member of *transition group* VI, and oxygen as a member of *main group* VI.

IONIZATION-ENERGY TRENDS

The energy required to remove an electron from an atom, such that the removed electron has zero kinetic energy and is infinitely distant from the resulting ion, is called an "ionization energy," "ionization potential," or "electron binding energy." In the case of a neutral hydrogen or helium atom, there is only one kind of electron which can be removed, and hence only one ionization energy. However, in the case of a more complicated atom such as argon, there are separate ionization energies corresponding to each occupied set of atomic orbitals. For example, in the case

---

[10] K. L. Loening, *J. Chem. Educ.*, **61**, 136 (1984); *Chem. & Eng. News*, Feb. 4, 1985, p. 27.

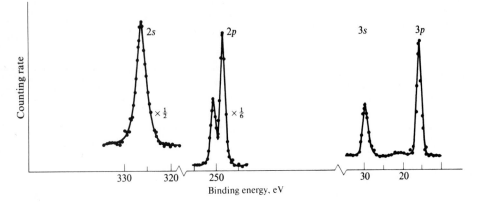

FIGURE 1.11

X-ray photoelectron spectrum of argon excited by Mg Kα x-radiation. The x-ray has insufficient energy to ionize the $1s$ electrons and is not sufficiently monochromatic to permit resolution of the $3p_{1/2}$ and $3p_{3/2}$ peaks. *(Reproduced with permission from K. Siegbahn et al., "ESCA Applied to Free Molecules." North-Holland, Amsterdam, 1969.)*

of argon, one can separately ionize a $1s$, $2s$, $2p$, $3s$, or $3p$ electron. In fact, in the case of the $2p$ or $3p$ electrons, there are two kinds of ionization processes (and corresponding ionization energies). One can remove a p electron such that the spin magnetic moment of the remaining odd electron and the orbital magnetic moment of the set of five remaining p electrons are either aligned (yielding a $^2P_{3/2}$ ion) or opposed (yielding a $^2P_{1/2}$ ion). (See Appendix B for a discussion of the significance of term symbols such as $^2P_{3/2}$.). One method of measuring ionization energies is photoelectron spectroscopy.[11] In this technique one irradiates a sample with monochromatic photons and measures the kinetic energy spectrum of the ejected photoelectrons. Electrons are observed at kinetic energies corresponding to the various electronic levels of the sample. It is then a simple matter to calculate the corresponding ionization energies by using the Einstein relation

$$h\nu = I + E_k$$

where $h\nu$ is the photon energy, I is the ionization energy of a given level, and E_k is the electron kinetic energy. An x-ray photoelectron spectrum of argon is shown in Fig. 1.11. The various ionization energies of argon are listed in Table 1.6.

The lowest ionization energy of an atom (corresponding to ionization of the highest-energy occupied level and formation of the ground-state ion) is often

---

[11] For an introduction to photoelectron spectroscopy, see D. N. Hendrickson, in R. S. Drago, "Physical Methods in Chemistry," pp. 566–584, Saunders, Philadelphia, 1977. For further material on this topic, see T. A. Carlson, "Photoelectron and Auger Spectroscopy," Plenum, New York, 1975; C. R. Brundle and A. D. Baker, "Electron Spectroscopy: Theory, Techniques and Applications," vols. 1 and 2, Academic, London, 1977, 1979; or K. Siegbahn, *Science*, **217**, 111 (1982).

TABLE 1.6
Ionization energies of argon†

Level	I, eV
$3p_{3/2}$	15.759
$3p_{1/2}$	15.937
$3s$	29.24
$2p_{3/2}$	248.52
$2p_{1/2}$	250.55
$2s$	326.3
$1s$	3205.9

† K. Siegbahn et al., "ESCA Applied to Free
Molecules," North-Holland, Amsterdam, 1969.

referred to simply as "the ionization energy" of the atom. The corresponding process may be represented by the following chemical equation:

$$M(g) \quad \rightarrow \quad M^+(g) + e^-(g)$$

To be precise, the energy of this process should be called the "first ionization energy" of the atom. The "second," "third," etc., ionization energies correspond to the processes

$$M^+(g) \quad \rightarrow \quad M^{2+}(g) + e^-(g)$$

$$M^{2+}(g) \quad \rightarrow \quad M^{3+}(g) + e^-(g)$$

etc.

Such ionization energies, corresponding to the formation of ground-state ions, are listed for the elements in Table 1.7. The ionization energies vary in a fairly systematic way throughout the periodic table. From left to right in a given row (corresponding to the filling of a particular shell of electrons), the first ionization energy generally increases. This increase is due to the fact that the valence electrons do not efficiently shield one another from the nucleus. Therefore as we go from one element to the next, the increase in nuclear charge causes an increased positive field in the valence-electron region.

Slater[12] devised a very approximate method for estimating the extent to which various electrons in an atom or ion shield a particular electron from the nucleus. He assumed that each electron is situated in a field of an "effective nuclear charge," $Z - S$, where Z is the true nuclear charge and S is a screening constant. The value of S is given by the following rules.

1. Orbitals are considered to be grouped in the following order, from inside out;
(1s) (2s, 2p) (3s, 3p)(3d)(4s, 4p)(4d)(4f). . . .

---

[12] J. C. Slater, *Phys. Rev.*, **36**, 57 (1930).

TABLE 1.7
Ionization energies of the elements (in electronvolts)†‡

Z	Element	I	II	III	IV	V	VI	VII	VIII
1	H	13.598							
2	He	24.587	54.416						
3	Li	5.392	75.638	122.451					
4	Be	9.322	18.211	153.893	217.713				
5	B	8.298	25.154	37.930	259.368	340.217			
6	C	11.260	24.383	47.887	64.492	392.077	489.981		
7	N	14.534	29.601	47.448	77.472	97.888	552.057	667.029	
8	O	13.618	35.116	54.934	77.412	113.896	138.116	739.315	871.387
9	F	17.422	34.970	62.707	87.138	114.240	157.161	185.182	953.886
10	Ne	21.564	40.962	63.45	97.11	126.21	157.93	207.27	239.09
11	Na	5.139	47.286	71.64	98.91	138.39	172.15	208.47	264.18
12	Mg	7.646	15.035	80.143	109.24	141.26	186.50	224.94	265.90
13	Al	5.986	18.828	28.447	119.99	153.71	190.47	241.43	284.59
14	Si	8.151	16.345	33.492	45.141	166.77	205.05	246.52	303.17
15	P	10.486	19.725	30.18	51.37	65.023	220.43	263.22	309.41
16	S	10.360	23.33	34.83	47.30	*72.68*	88.049	280.93	328.23
17	Cl	12.967	23.81	39.61	53.46	67.8	*97.03*	114.193	348.28
18	Ar	15.759	27.629	40.74	59.81	75.02	91.007	124.319	143.456
19	K	4.341	31.625	45.72	60.91	*82.66*	*100.0*	117.56	154.86
20	Ca	6.113	11.871	50.908	67.10	84.41	*108.78*	127.7	147.24
21	Sc	6.54	12.80	24.76	73.47	91.66	111.1	*138.0*	158.7
22	Ti	6.82	13.58	27.491	43.266	99.22	119.36	140.8	168.5
23	V	6.74	14.65	29.310	46.707	*65.23*	128.12	150.17	173.7
24	Cr	6.766	16.50	30.96	*49.1*	*69.3*	90.56	161.1	*184.7*
25	Mn	7.435	15.640	33.667	*51.2*	*72.4*	*95*	119.27	*196.46*
26	Fe	7.870	16.18	30.651	*54.8*	*75.0*	*99*	*125*	151.06
27	Co	7.86	17.06	33.50	*51.3*	*79.5*	*102*	*129*	*157*
28	Ni	7.635	18.168	35.17	*54.9*	*75.5*	*108*	*133*	*162*
29	Cu	7.726	20.292	36.83	*55.2*	*79.9*	*103*	*139*	*166*
30	Zn	9.394	17.964	39.722	*59.4*	*82.6*	*108*	*134*	*174*
31	Ga	5.999	20.51	30.71	64				
32	Ge	7.899	15.934	34.22	45.71	93.5			
33	As	9.81	18.633	28.351	50.13	62.63	127.6		
34	Se	9.752	21.19	30.820	42.944	*68.3*	81.70	155.4	
35	Br	11.814	21.8	36	*47.3*	59.7	88.6	*103.0*	192.8
36	Kr	13.999	24.359	36.95	*52.5*	64.7	78.5	*111.0*	126
37	Rb	4.177	27.28	40	*52.6*	71.0	84.4	*99.2*	136
38	Sr	5.695	11.030	*43.6*	57	71.6	90.8	106	122.3
39	Y	6.38	12.24	20.52	*61.8*	77.0	93.0	116	129
40	Zr	6.84	13.13	22.99	34.34	81.5			
41	Nb	6.88	14.32	25.04	38.3	50.55	102.6	125	
42	Mo	7.099	16.15	27.16	46.4	61.2	68	126.8	153
43	Te	7.28	15.26	29.54					
44	Ru	7.37	16.76	28.47					
45	Rh	7.46	18.08	31.06					
46	Pd	8.34	19.43	32.93					
47	Ag	7.576	21.49	34.83					
48	Cd	8.993	16.908	37.48					
49	In	5.786	18.869	28.03	54				

(continued)

TABLE 1.7 (*continued*)

Z	Element	I	II	III	IV	V	VI	VII	VIII
50	Sn	7.344	14.632	30.502	40.734	72.28			
51	Sb	8.641	16.53	25.3	44.2	56	108		
52	Te	9.009	18.6	27.96	37.41	58.75	70.7	137	
53	I	10.451	19.131	*33*					
54	Xe	12.130	21.21	32.1					
55	Cs	3.894	25.1						
56	Ba	5.212	10.004						
57	La	5.577	11.06	19.175	49.95				
58	Ce	5.47	10.85	20.20	36.72				
59	Pr	5.42	10.55	21.62	38.95	57.45			
60	Nd	5.49	10.72	22.1	40.41				
61	Pm	5.55	10.90	22.3	41.1				
62	Sm	5.63	11.07	23.4	41.4				
63	Eu	5.67	11.25	24.9	42.6				
64	Gd	6.14	12.1	20.63	44.0				
65	Tb	5.85	11.52	21.91	39.8				
66	Dy	5.93	11.67	22.8	41.5				
67	Ho	6.02	11.80	22.84	42.5				
68	Er	6.10	11.93	22.74	42.6				
69	Tm	6.18	12.05	23.71	42.7				
70	Yb	6.254	12.17	25.03	43.7				
71	Lu	5.426	13.9	20.96	45.19				
72	Hf	7.0	14.9	23.3	33.3				
73	Ta	7.89							
74	W	7.98							
75	Re	7.88							
76	Os	8.7							
77	Ir	9.1							
78	Pt	9.0	18.563						
79	Au	9.225	20.5						
80	Hg	10.437	18.756	34.2					
81	Tl	6.108	20.428	29.83					
82	Pb	7.416	15.032	31.937	42.32	68.8	88.3		
83	Bi	7.289	16.69	25.56	45.3	56.0			
84	Po	8.48							
85	At								
86	Rn	10.748							
87	Fr								
88	Ra	5.279	10.147						
89	Ac	5.2	12.1						
90	Th	6.1	11.5	20.0	28.8				
91	Pa	5.9							
92	U	6.05							
93	Np	6.2							
94	Pu	6.06							
95	Am	6.0							

† Most of the data are from C. E. Moore, "Ionization Potentials and Ionization Limits Derived from the Analyses of Optical Spectra," NSRDS-NBS 34, National Bureau of Standards, Washington, D.C., 1970.

‡ Italicized entries are estimates.

2. For an electron in a group of s, p electrons, the value of S is the sum of the following contributions.

   a. Nothing from electrons in groups outside the one considered.

   b. The amount 0.35 from each other electron in the group considered (except in the 1s group, where 0.30 is used instead).

   c. The amount 0.85 from each electron with quantum number $(n - 1)$.

   d. The amount 1.00 from all other inner electrons.

3. For an electron in a group of d and f electrons, parts a and b apply, as above. However, parts c and d are replaced with the rule that *all* inner electrons contribute 1.00 to S.

Thus, for a valence electron in aluminum $(1s^2 2s^2 2p^6 3s^2 3p)$, $S = 2 \times 1.00 + 8 \times 0.85 + 2 \times 0.35 = 9.5$, and $Z - S = 13 - 9.5 = 3.5$. As we move to the right, element by element, the quantity $Z - S$ gradually increases until it reaches a maximum value of $18 - 11.25 = 6.75$ at argon. By increasing the atomic number one more unit, we come to potassium $(1s^2 2s^2 2p^6 3s^2 3p^6 4s)$. The value of S for the valence electron in potassium is $2 \times 1.00 + 8 \times 1.00 + 8 \times 0.85 = 16.8$, and the effective nuclear charge is $19 - 16.8 = 2.2$. Thus Slater's simple rules correctly predict that the effective nuclear charge (and hence the ionization energy) reaches a maximum at the noble gas argon. In the same way, we can account for the maxima in first ionization energies observed for all the noble gases, which, except for helium, are characterized by valence shells containing complete sets of p electrons. Less pronounced maxima are observed for elements whose valence electrons consist of complete shells of d electrons, e.g. Zn and Cd. These maxima can also be rationalized by Slater's rules. However, Slater's rules are too simplified to account for the maxima at Be and Mg, corresponding to complete s shells, or the even less pronounced maxima at elements with half-filled valence p shells. The maxima at nitrogen and phosphorus occur because the next element in the periodic table has an np^4 valence-shell configuration, in which two electrons are paired in the same atomic orbital, with consequent loss of exchange energy and a relatively great increase in electron repulsive energy.[13]

The stability of completely filled shells of p, d, and f electrons is very important in controlling the chemistry of most of the elements in the periodic table. As a general rule, atoms tend to acquire filled electron configurations of these types. For example, the $2s^2 2p^6$ configuration is achieved in C^{4-}, N^{3-}, O^{2-}, F^-, Ne, Na^+, Mg^{2+}, and Al^{3+}. The $3s^2 3p^6 3d^{10}$ configuration is achieved in Cu^+, Zn^{2+}, and Ga^{3+}. The $4s^2 4p^6 4d^{10} 4f^{14} 5s^2 5p^6$ configuration (with a just-completed $4f$ shell) is achieved in Yb^{2+}, Lu^{3+}, Hf^{4+}, and Ta^{5+}. Even the stability of a filled s shell is important in species such as Tl^+, Pb^{2+}, and Bi^{3+}, which have the $5s^2 5p^6 5d^{10} 6s^2$ configuration. A tendency for atoms to achieve *half*-filled shells is shown by Eu^{2+}, Gd^{3+}, and Tb^{4+}, all of which have the $4s^2 4p^6 4d^{10} 4f^7 5s^2 5p^6$ configuration (with a half-filled $4f$ shell).

---

[13]   A. B. Blake, *J. Chem. Educ.*, **58**, 393 (1981).

It should be noted that the ionization energy of a negative species such as F^- is seldom referred to as such, but rather as the "electron affinity" of the corresponding neutral species F. Thus, the electron affinity of the fluorine atom is equivalent to the "zeroth" ionization energy of fluorine. The periodic-table trends of electron affinities are similar to those of ionization energies. Tables of electron affinities are given in Appendix C.

In conjunction with his scheme for estimating electron screening constants, Slater also proposed that a given electron should be assigned an "effective" quantum number n^*, which is the same as the true principal quantum number for $n \leq 3$, but which has the values 3.7 when $n = 4$, 4.0 when $n = 5$, and 4.2 when $n = 6$. The total binding energy of an atom or ion (the energy required to remove all the electrons from the nucleus) is given approximately by the sum of the quantities

$$13.6 \left(\frac{Z - S}{n^*} \right)^2$$

calculated for all the electrons of the atom or ion.

As an example of the use of this relation, we shall estimate the first ionization energy of fluorine. This energy is equal to the difference between the total binding energy of the normal atom and that of the ion. Because the $1s$ shell is complete in both the atom and the ion, and because the valence electrons are assumed not to contribute to the $1s$ shielding constant, the $(Z - S)/n^*$ values for the $1s$ electrons will cancel when the difference is taken and hence can be ignored. For the $2s$, $2p$ electrons of the *atom*, $S = 2(0.85) + 6(0.35) = 3.8$. For the $2s$, $2p$ electrons of the *ion*, $S = 2(0.85) + 5(0.35) = 3.45$. The total $2s$, $2p$ electron bonding energies are $7(13.6)[(9 - 3.8)/2]^2 = 643.6$ eV and $6(13.6)[(9 - 3.45)/2]^2 = 628.4$ eV, respectively, corresponding to an ionization energy of $643.6 - 628.4 = 15.2$ eV. This value may be compared with the actual value, 17.4 eV.

As a second example, we shall estimate the first ionization energy of iron. Again, we can neglect the $(Z - S)/n^*$ values for all the electrons but those in the outer shell. For the atom, $(Z - S)/n^* = 3.75/3.7 = 1.014$; for the ion, $(Z - S)/n^* = 4.1/3.7 = 1.108$. The ionization energy is given by $13.6[2(1.014)^2 - (1.108)^2] = 11.3$ eV, which may be compared with the actual value, 7.9 eV. Obviously ionization energies calculated by this method are only rough estimates whose principal value is in qualitatively predicting *trends* or *changes* in ionization energy.

ATOMIC-SIZE TRENDS

Another atomic property which varies systematically throughout the periodic table is atomic size. Unfortunately, atomic size, unlike ionization energy, is an ill-defined quantity. We shall later discuss several methods for defining and measuring effective radii for atoms and ions, but at this point it will suffice to discuss qualitative trends in atomic size. When the principal quantum number of the valence electrons is increased while the valence configuration is kept constant, atomic size increases. Thus the bromine atom is bigger than the chlorine atom.

This effect is due to the fact that an increase in principal quantum number corresponds to a decrease in electron binding energy and an increased average distance from the nucleus. Thus atomic size increases as one descends any column of the periodic table. On the other hand, when we move from left to right in the periodic table, i.e., when an increase in atomic number is matched by an increase in the number of valence electrons in the valence shell, atomic size decreases. This effect is a consequence of the increase in the effective nuclear charge and the contraction of the electron cloud. Thus atomic size decreases from left to right in any row of the periodic table, as long as the n or l values of the electrons added are not changed.

The contraction in size which occurs in the first series of transition metals (scandium through zinc) has interesting consequences in the succeeding elements, gallium, germanium, etc. The contraction causes gallium and germanium to be almost the same size as aluminum and silicon, respectively. Therefore, because chemical properties are highly dependent on atomic size, the chemical properties of aluminum and gallium are very similar, as are those of silicon and germanium. Similarly, the so-called lanthanide contraction which occurs from La to Lu, corresponding to the filling of the $4f$ shell, has important consequences in the series from Hf to Pt. The elements of the latter series are remarkably similar to the corresponding members of the second transition series, from Zr to Pd. Thus zirconium and hafnium are so similar that it is very difficult to separate these elements from each other. The block of six elements Ru, Rh, Pd, Os, Ir, and Pt have many chemical similarities and are called the "platinum metals." Beyond platinum, however, the effect of the lanthanide contraction markedly peters out; thus cadmium in some ways is more like zinc than like mercury, and tin in some ways is more like germanium than like lead.

RELATIVISTIC EFFECTS[14]

Probably the most familiar result of relativity theory is the increase in a particle's mass as its velocity approaches that of light. In the heavier elements of the periodic table, electronic velocities in the s orbitals, and to some extent in the p orbitals, are high enough to increase the electronic mass significantly—as much as 20 percent for the $1s$ electrons of mercury, for example. Because of the inverse proportionality between an electron's average distance from the nucleus and its mass (see p. 10), the relativistic effects cause a contraction in atomic radius. Thus, although the main cause of the lanthanide contraction is the incomplete shielding of the nucleus by the $4f$ electrons, calculations have shown that about 14 percent of the contraction is due to relativistic effects.

Because electronic velocities are highest in the region near the nucleus, it is not surprising that relativistic effects are greatest for s orbitals, which have the greatest density near the nucleus. The relativistic stabilization of the $6s^2$

---

[14] K. S. Pitzer, *Acc. Chem. Res.*, **12**, 271 (1979); P. Pyykkö and J. P. Desclaux, ibid., **12**, 276 (1979).

electron pair in the elements Hg, Tl, Pb, and Bi is believed to be responsible for the so-called "inert pair" effect in these elements. This effect manifests itself in remarkable stability of the oxidation states two units lower than the group number. Thus mercury is difficult to oxidize (corresponding to a very stable zero oxidation state), thallium readily forms the Tl^+ ion, lead is found mainly in the +2 oxidation state, and bismuth mainly in the +3 oxidation state. The contraction of the $6s^2$ shell is also responsible for the unexpected volatility of mercury: mercury acts as if it were a pseudo inert gas.

OXIDATION STATES OF THE LANTHANIDES AND ACTINIDES

The known oxidation states of the lanthanides and actinides are given in Table 1.8. In the lanthanides, the only exceptions to the +3 oxidation state are those cases in which the ion *corresponds to* (in Ce^{4+}, Eu^{2+}, Tb^{4+}, and Yb^{2+}) or *approaches* (in Pr^{4+}, Nd^{4+}, Sm^{2+}, and Dy^{4+}) a $4f^0$, $4f^7$, or $4f^{14}$ electron configuration. The first six actinide elements after actinium are not good analogs of the corresponding lanthanides. Thorium shows only the +4 oxidation state and is a better analog of hafnium than of cerium. The multiplicity of oxidation states shown by Pa, U, Np, Pu, and Am is very analogous to the situation in the elements Ta, W, Re, Os, and Ir. However, starting with curium, the heavier actinides show only two or three oxidation states, with the +3 state being the most important.

In the early actinide elements the $5f$ and $6d$ orbitals are of comparable energy, and for practical purposes these elements may be considered as members of a fourth transition series, in which electrons are added to the $6d$ orbitals as the atomic number is increased. However, at about curium, the stabilization of the $5f$ orbitals has reached the point where they are definitely more stable than the $6d$ orbitals, and thereafter electrons are added to the $5f$ shell as atomic number increases. Hence the heavier actinides are good analogs of the lanthanides.

TABLE 1.8
Oxidation states of the lanthanides and actinides

La	Ce	Pr	Nd	Pm	Sm	Eu	Gd	Tb	Dy	Ho	Er	Tm	Yb	Lu
					2	2							2	
3	3	3	3	3	3	3	3	3	3	3	3	3	3	3
	4	4	4?					4	4?					

Ac	Th	Pa	U	Np	Pu	Am	Cm	Bk	Cf	Es	Fm	Md	No	Lr
				2					2	2	2	2	2	
3		3	3	3	3	3	3	3	3	3	3	3	3	3
	4	4	4	4	4	4	4	4	4					
		5	5	5	5	5								
			6	6	6	6								
				7	7									

PROBLEMS[15]

(Many of these problems should be answerable without referring to the text or a periodic table.)

*1.1 The probability of finding the electron of a hydrogen atom at a distance r from the proton is at a maximum for $r = 0.529$Å. Does this statement contradict the fact that the electron density is greatest at the proton? Explain.

1.2 Describe the nodal surfaces of the $3p$ orbital.

*1.3 Using the angular parts of the hydrogen-atom wave function for p orbitals (see Tables 1.3 and 1.4), show that an atom with three equally occupied p orbitals, such as the nitrogen atom, is spherically symmetric.

1.4 How many unpaired electrons does each of the following atoms or ions have in its ground state: Al, S, Sc^{3+}, Cr^{3+}, Ir^{3+}, Dy^{3+}?

1.5 What $+2$ ion has just six $4d$ electrons in its ground state?

1.6 Write out the electron configuration of arsenic.

1.7 List, in order of atomic numbers, the symbols of the elements between calcium and gallium.

1.8 Which element has the following electronic configuration:

$$1s^2 2s^2 2p^6 3s^2 3p^6 3d^{10} 4s^2 4p^6 4d^4 5s^1?$$

1.9 Name the two elements whose chemical properties are very similar to those of chromium.

1.10 Which element has atomic number 25?

1.11 Which of the following gas-phase reactions can proceed spontaneously?

$$Kr + He^+ \quad \rightarrow \quad Kr^+ + He$$

$$Si + Cl^+ \quad \rightarrow \quad Si^+ + Cl$$

$$Cl^- + I \quad \rightarrow \quad I^- + Cl$$

1.12 Why does potassium have a lower first ionization energy than lithium?

1.13 Why does copper have an outer electron configuration $3d^{10} 4s^1$ instead of $3d^9 4s^2$?

1.14 Argue for the inclusion of B and Al in the same family with Sc, Y, La, and Ac.

1.15 What are the two common oxidation states of In, Sn, Sb, and Te?

1.16 Calculate the ninth ionization energy of fluorine.

1.17 Using Slater's rules, estimate the second ionization energies of Na and Mg, and the $1s$ binding energy of Ne.

1.18 Using Slater's rules, estimate the first ionization energies of S, Cl, Ar, K, and Ca. Show that the estimates are similar in magnitude and trend to the actual values.

1.19 On the basis of analogy with the lanthanides, which two actinides would you expect to show a stable $+2$ oxidation state?

---

[15] In this and subsequent problem sections, asterisks indicate more difficult problems.

*1.20 Arrange the ions of each group in order of increasing size:

(a) Y^{3+}, Ba^{2+}, Al^{3+}, Co^{3+}, Cs^+, La^{3+}, Ir^{3+}, Fe^{3+}

(b) Cl^-, I^-, Te^{2-}, Ar^+

*1.21 The $+4$ oxidation state of lead is much more strongly oxidizing than that of tin. Show that this fact can be explained by the first four ionization energies of tin and lead. Rationalize the trends in these four ionization energies in the series Ge, Sn, Pb.

*1.22 In the separation of nuclear fission products by elution from a column of ion-exchange resin, the following sequence of ions is observed: . . . , Yb^{3+}, Tm^{3+}, Er^{3+}, Ho^{3+}, Y^{3+}, Dy^{3+}, Tb^{3+}, Gd^{+3}, ... Explain the appearance of Y^{3+} in this series.

1.23 Rationalize the fact that solutions containing the Au^- ion can be prepared.

CHAPTER
2

SYMMETRY AND THE ELEMENTS OF GROUP THEORY

Symmetry is one of the most pervasive features of the universe, and we almost subconsciously become familiar with the effects of symmetry through daily experience with symmetrical things. Even a beginner in chemistry recognizes that the structures of many compounds are "symmetric," and thus it should be no surprise that the principles of symmetry play an important part in chemistry. The simple, intuitive grasp of symmetry that most observant people naturally acquire is adequate for understanding chemistry up to a point. However, a systematic study of symmetry and the ways of specifying it with mathematical precision are important in the advanced study of inorganic chemistry because of the wide variety of symmetric structures encountered. The principles of symmetry and its close cousin, group theory, aid in the following (and many other) activities:

1. Classifying molecular structures
2. Classifying molecular orbitals
3. Predicting splitting of electronic levels in electric fields of various symmetry
4. Constructing hybrid orbitals
5. Classifying electronic states of molecules
6. Classifying normal modes of vibration
7. Predicting allowed transitions in spectra

In this chapter, we shall briefly introduce the important concepts of symmetry and shall describe a few applications of group theory to chemical problems. In later chapters we shall apply the methods of this chapter to the first four listed activities. If the material in this chapter whets your appetite for more extensive discussion, various texts can be consulted.[1]

You are encouraged to make structural models of the various compounds discussed in this chapter. The symmetry features of some molecules are difficult to visualize, and the discussions will be much more easily understood if you have the appropriate models in hand. Indeed, you will find that a simple set of molecular model-making equipment is valuable in studying all the discussions of structure found throughout this text.

SYMMETRY OPERATIONS AND SYMMETRY ELEMENTS

If a molecule has two or more orientations in space that are indistinguishable, the molecule possesses symmetry. The process of going from one orientation to the other (a reflection, rotation, or a combination of these) is called a "symmetry operation," and the molecule possesses a corresponding "symmetry element." A symmetry element is a point, line, or plane about which the symmetry operation is carried out. For single molecules, there are five symmetry elements and corresponding symmetry operations.

1. *Identity, E.* The identity operation consists of doing nothing to the molecule. All molecules possess this symmetry element, inasmuch as the aspect of a molecule is unchanged after doing nothing to it. Although this may appear to be a trivial symmetry element, it is included because it is involved in group theoretical calculations.

2. *Center of symmetry (inversion center), i.* A molecule possesses a center of symmetry if every atom, when moved in a straight line through this center and an equal distance on the other side of the center, encounters a similar atom. Note that the square planar $PtCl_4^{2-}$ ion (Fig. 2.1a) has a center of symmetry, whereas the tetrahedral SiF_4 molecule (Fig. 2.1b) does not.

3. *n-fold axis of symmetry (rotational axis), C_n.* If rotation of a molecule by an angle of $2\pi/n$ about an axis gives an indistinguishable configuration, the axis is an axis of symmetry. Thus the water molecule possesses one twofold axis of symmetry (C_2), and the planar BF_3 molecule (Fig. 2.2) possesses one threefold axis (C_3) and three twofold axes (C_2). The highest-fold rotation axis of a molecule is conventionally taken as the vertical z axis.

[1] F. A. Cotton, "Chemical Applications of Group Theory," 2d ed., Wiley-Interscience, New York, 1971; L. H. Hall, "Group Theory and Symmetry in Chemistry," McGraw-Hill, New York, 1969; H. H. Jaffe and M. Orchin, "Symmetry in Chemistry," Academic, New York, 1965; J. P. Fackler, Jr., "Symmetry in Coordination Chemistry," Academic, New York, 1971; D. S. Schonland, "Molecular Symmetry," Van Nostrand, Princeton, N.J., 1965.

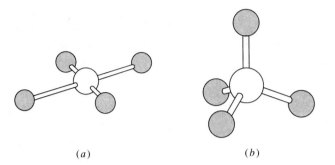

(a)   (b)

FIGURE 2.1
The planar $PtCl_4^{2-}$ ion (a) and the tetrahedral SiF_4 molecule (b).

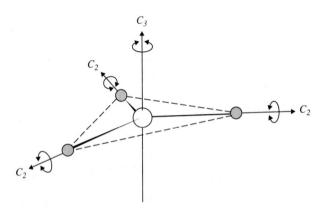

FIGURE 2.2
The BF_3 molecule, showing the four symmetry axes.

4. *Plane of symmetry (mirror plane), σ.* If reflection of all parts of a molecule through a plane yields an indistinguishable orientation of the structure, the plane is a plane of symmetry. For example, the water molecule (Fig. 2.3) possesses two planes of symmetry: the plane in which the three atoms lie, and the plane perpendicular to that plane and bisecting the H—O—H angle.

Mirror planes containing the vertical axis are "vertical planes," σ_v, and a mirror plane perpendicular to the vertical axis is the "horizontal plane," σ_h.

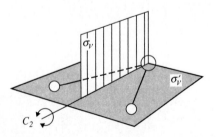

FIGURE 2.3
The water molecule, showing the symmetry elements.

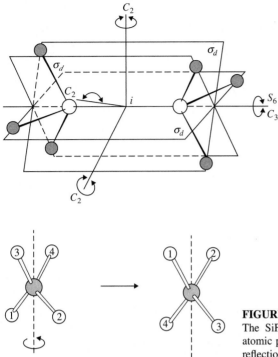

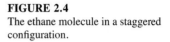

FIGURE 2.4
The ethane molecule in a staggered configuration.

FIGURE 2.5
The SiF$_4$ molecule, showing the changes in atomic positions accompanying an S$_4$ rotation-reflection.

"Diagonal planes," σ_d, are vertical planes that bisect the angles between successive twofold axes.

5. *n-fold rotation-reflection axis (improper rotational axis), S_n.* If rotation by an angle $2\pi/n$ about an axis, followed by reflection through a plane perpendicular to the axis, yields an indistinguishable configuration, the axis is a rotation-reflection axis. For example, the staggered configuration of ethane has an S_6 axis coincident with the C_3 axis (Fig 2.4), and SiF$_4$ has three S_4 axes coincident with the three C_2 axes which bisect the F—Si—F angles (Fig. 2.5). These symmetry elements are seen best with the aid of molecular models.

POINT GROUPS

The complete set of symmetry operations that can be carried out on a molecule constitutes a "point group." Although there are an infinite number of possible point groups, about 40 suffice to classify all known molecules. Point groups have labels, such as C_2, C_{2v}, D_{3h}, O_h, and T_d. A yes-no system for determining the point group of a given molecule is outlined in Fig. 2.6. We shall demonstrate the use of this diagram with several examples.

1. **HCl.** The HCl molecule possesses an ∞-fold axis of rotation along the H—Cl bond, but it possesses no inversion center; therefore, the molecule belongs to the $C_{\infty v}$ point group.

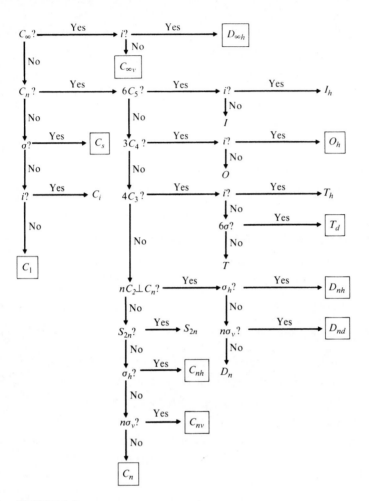

FIGURE 2.6
Procedure for determining the point group of a molecule.

2. **BFClBr.** This planar triangular molecule possesses only one symmetry element—the plane of symmetry in which the atoms lie. Consequently, the point group is C_s.

3. **trans-N$_2$O$_2^{2-}$,** $\overset{\text{O}^-}{\underset{\text{O}}{\overset{\displaystyle\nearrow}{\underset{\displaystyle\swarrow}{\ddot{\text{N}}=\text{N}}}}}$. This planar ion possesses only one symmetry axis, the twofold axis of symmetry perpendicular to the plane of the ion. There is no S_4 axis, but there is a horizontal plane. Consequently, the point group is C_{2h}.

4. **S$_8$.** The S_8 molecule consists of a zigzag closed chain of sulfur atoms, as pictured in Fig. 2.7. Notice that the molecule has one fourfold axis and four

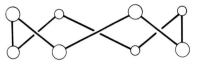

FIGURE 2.7
Side view of an S_4 ring.

twofold axes perpendicular to the fourfold axis. It has no horizontal plane, but it has four vertical planes. Thus, the point group is D_{4d}.

5. **NH₃.** The pyramidal ammonia molecule possesses a threefold rotational axis. It does not have a horizontal mirror plane, but it does have a set of three vertical planes. Hence the point group is C_{3v}.

6. **CH₄.** This regular tetrahedral molecule has four threefold axes, no center of symmetry, and six mirror planes, corresponding to the T_d point group.

Further examples are given in Table 2.1. The student should become adept at recognizing symmetry elements and at determining the point groups of molecules. A list of compounds and objects is given in the problems at the end of the chapter for practice.

Even if we pursued the subject of symmetry no further, you would find that familiarity with point groups is an asset. For example, consider how much structural information is condensed in the simple statement that the $Ni(CN)_4{}^{2-}$ ion has D_{4h} symmetry. This statement implies that (1) the ion is completely planar, (2) the Ni—C—N bond angles are all 180°, (3) the C—Ni—C bond angles are all 90°, (4) the four Ni—C bond lengths are all equal, and (5) the four C—N

TABLE 2.1
Examples of some common point groups

Point group	Examples
C_1	SiFClBrI, SOFCl
C_s	ONCl, HOCl, SOCl₂
C_2	Nonplanar H₂O₂
C_{2h}	Trans-planar H₂O₂, *trans*-C₂H₂Cl₂
C_{2v}	H₂O, SO₂F₂, SCl₂, ClO₂⁻
C_{3v}	NH₃, SiH₃Cl, PF₃
C_{4v}	XeOF₄, SF₅Cl
D_{2h}	N₂O₄, C₂O₄²⁻
D_{3h}	BCl₃, PCl₅, SO₃
D_{4h}	PtCl₄²⁻, Ni(CN)₄²⁻, *trans*-SF₄Cl₂
D_{5h}	Eclipsed Fe(C₅H₅)₂
D_{6h}	Cr(C₆H₆)₂
D_{2d}	Staggered B₂Cl₄, H₂C=C=CH₂
D_{3d}	Staggered Si₂H₆
D_{4d}	Mn₂(CO)₁₀
D_{5d}	Staggered Fe(C₅H₅)₂
T_d	GeCl₄, ClO₄⁻
O_h	UF₆, SF₆, PF₆⁻
$D_{\infty h}$	H₂, N₃⁻, CO₂
$C_{\infty v}$	HCl, CO, OCS

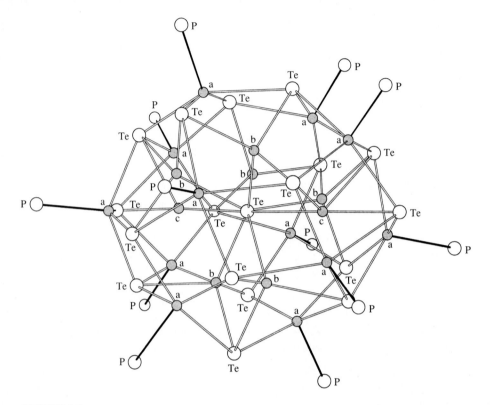

FIGURE 2.8
Structure of $Ni_{20}Te_{18}(PEt_3)_{12}$, with the ethyl groups omitted for clarity. Atoms labeled a, b, and c are nickel atoms in square pyramidal, tetrahedral, and distorted trigonal bipyramidal environments, respectively. The overall structure has no symmetry element other than the identity element. *(Reproduced with permission from ref. 2.)*

bond lengths are all equal. You will find that the inorganic chemical literature, in which point group symbols are commonly used to describe structures concisely, is much more easily understood when you are familiar with point groups.

Most of the molecules that chemists have isolated have more than one symmetry element. The molecule $Ni_{20}Te_{18}[P(C_2H_5)_3]_{12}$. pictured in Fig. 2.8, is unusual because it is completely lacking in symmetry—except for the identity element.[2]

CHARACTER TABLES AND IRREDUCIBLE REPRESENTATIONS

For each point group there is a corresponding "character table." Some of the more important character tables are given in Table 2.2. At the top of each character table are listed horizontally the various classes of symmetry operations of the

---

[2] J. G. Brennan, T. Siegrist, S. M. Stuczynski, and M. L. Steigerwald, *J. Am. Chem. Soc.*, **111**, 9240 (1990).

TABLE 2.2
Character tables for some important point groups

C_1	E
A	1

C_s	E	σ_h	
A'	1	1	x, y, x^2, y^2, z^2, xy
A''	1	-1	z, yz, xz

C_2	E	C_2	
A	1	1	z, x^2, y^2, z^2, xy
B	1	-1	x, y, xy, xz

C_{2h}	E	C_2	i	σ_h	
A_g	1	1	1	1	x^2, y^2, z^2, xy
B_g	1	-1	1	-1	xz, yz
A_u	1	1	-1	-1	z
B_u	1	-1	-1	1	x, y

C_{2v}	E	C_2	$\sigma_v(xz)$	$\sigma_v'(yz)$	
A_1	1	1	1	1	z, x^2, y^2, z^2
A_2	1	1	-1	-1	xy
B_1	1	-1	1	-1	x, xz
B_2	1	-1	-1	1	y, yz

C_{3v}	E	$2C_3$	$3\sigma_v$	
A_1	1	1	1	$z, x^2 + y^2, z^2$
A_2	1	1	-1	
E	2	-1	0	$(x, y), (x^2 - y^2, xy), (xz, yz)$

C_{4v}	E	$2C_4$	C_2	$2\sigma_v$	$2\sigma_d$	
A_1	1	1	1	1	1	$z, x^2 + y^2, z^2$
A_2	1	1	1	-1	-1	
B_1	1	-1	1	1	-1	$x^2 - y^2$
B_2	1	-1	1	-1	1	xy
E	2	0	-2	0	0	$(x, y), (xz, yz)$

D_{2h}	E	$C_2(z)$	$C_2(y)$	$C_2(x)$	i	$\sigma(xy)$	$\sigma(xz)$	$\sigma(yz)$	
A_g	1	1	1	1	1	1	1	1	x^2, y^2, z^2
B_{1g}	1	1	-1	-1	1	1	-1	-1	xy
B_{2g}	1	-1	1	-1	1	-1	1	-1	xz
B_{3g}	1	-1	-1	1	1	-1	-1	1	yz
A_u	1	1	1	1	-1	-1	-1	-1	
B_{1u}	1	1	-1	-1	-1	-1	1	1	z
B_{2u}	1	-1	1	-1	-1	1	-1	1	y
B_{3u}	1	-1	-1	1	-1	1	1	-1	x

(*continued*)

TABLE 2.2 (*continued*)

D_{3h}	E	$2C_3$	$3C_2$	σ_h	$2S_3$	$3\sigma_v$	
A_1'	1	1	1	1	1	1	$x^2 + y^2, z^2$
A_2'	1	1	-1	1	1	-1	
E'	2	-1	0	2	-1	0	$(x, y), (x^2 - y^2, xy)$
A_1''	1	1	1	-1	-1	-1	
A_2''	1	1	-1	-1	-1	1	z
E''	2	-1	0	-2	1	0	(xz, yz)

D_{4h}	E	$2C_4$	C_2	$2C_2'$	$2C_2''$	i	$2S_4$	σ_h	$2\sigma_v$	$2\sigma_d$	
A_{1g}	1	1	1	1	1	1	1	1	1	1	$x^2 + y^2, z^2$
A_{2g}	1	1	1	-1	-1	1	1	1	-1	-1	
B_{1g}	1	-1	1	1	-1	1	-1	1	1	-1	$x^2 - y^2$
B_{2g}	1	-1	1	-1	1	1	-1	1	-1	1	xy
E_g	2	0	-2	0	0	2	0	-2	0	0	(xz, yz)
A_{1u}	1	1	1	1	1	-1	-1	-1	-1	-1	
A_{2u}	1	1	1	-1	-1	-1	-1	-1	1	1	z
B_{1u}	1	-1	1	1	-1	-1	1	-1	-1	1	
B_{2u}	1	-1	1	-1	1	-1	1	-1	1	-1	
E_u	2	0	-2	0	0	-2	0	2	0	0	(x, y)

D_{5h}	E	$2C_5$	$2C_5^2$	$5C_2$	σ_h	$2S_5$	$2S_5^3$	$5\sigma_v$	
A_1'	1	1	1	1	1	1	1	1	$x^2 + y^2, z^2$
A_2'	1	1	1	-1	1	1	1	-1	
E_1'	2	$2\cos 72°$	$2\cos 144°$	0	2	$2\cos 72°$	$2\cos 144°$	0	(x, y)
E_2'	2	$2\cos 144°$	$2\cos 72°$	0	2	$2\cos 144°$	$2\cos 72°$	0	$(x^2 - y^2, xy)$
A_1''	1	1	1	1	-1	-1	-1	-1	
A_2''	1	1	1	-1	-1	-1	-1	1	z
E_1''	2	$2\cos 72°$	$2\cos 144°$	0	-2	$-2\cos 72°$	$-2\cos 144°$	0	(xz, yz)
E_2''	2	$2\cos 144°$	$2\cos 72°$	0	-2	$-2\cos 144°$	$-2\cos 72°$	0	

D_{6h}	E	$2C_6$	$2C_3$	C_2	$3C_2'$	$3C_2''$	i	$2S_3$	$2S_6$	σ_h	$3\sigma_d$	$3\sigma_v$	
A_{1g}	1	1	1	1	1	1	1	1	1	1	1	1	$x^2 + y^2, z^2$
A_{2g}	1	1	1	1	-1	-1	1	1	1	1	-1	-1	
B_{1g}	1	-1	1	-1	1	-1	1	-1	1	-1	1	-1	
B_{2g}	1	-1	1	-1	-1	1	1	-1	1	-1	-1	1	
E_{1g}	2	1	-1	-2	0	0	2	1	-1	-2	0	0	(xz, yz)
E_{2g}	2	-1	-1	2	0	0	2	-1	-1	2	0	0	$(x^2 - y^2, xy)$
A_{1u}	1	1	1	1	1	1	-1	-1	-1	-1	-1	-1	
A_{2u}	1	1	1	1	-1	-1	-1	-1	-1	-1	1	1	z
B_{1u}	1	-1	1	-1	1	-1	-1	1	-1	1	-1	1	
B_{2u}	1	-1	1	-1	-1	1	-1	1	-1	1	1	-1	
E_{1u}	2	1	-1	-2	0	0	-2	-1	1	2	0	0	(x, y)
E_{2u}	2	-1	-1	2	0	0	-2	1	1	-2	0	0	

(*continued*)

TABLE 2.2 (*continued*)

D_{2d}	E	$2S_4$	C_2	$2C_2'$	$2\sigma_d$	
A_1	1	1	1	1	1	x^2+y^2, z^2
A_2	1	1	1	-1	-1	
B_1	1	-1	1	1	-1	x^2-y^2
B_2	1	-1	1	-1	1	z, xy
E	2	0	-2	0	0	$(x,y),(xz,yz)$

D_{3d}	E	$2C_3$	$3C_2$	i	$2S_6$	$3\sigma_d$	
A_{1g}	1	1	1	1	1	1	x^2+y^2, z^2
A_{2g}	1	1	-1	1	1	-1	
E_g	2	-1	0	2	-1	0	$(x^2-y^2, xy),(xz,yz)$
A_{1u}	1	1	1	-1	-1	-1	
A_{2u}	1	1	-1	-1	-1	1	z
E_u	2	-1	0	-2	1	0	(x,y)

D_{4d}	E	$2S_8$	$2C_4$	$2S_8^{\ 3}$	C_2	$4C_2'$	$4\sigma_d$	
A_1	1	1	1	1	1	1	1	x^2+y^2, z^2
A_2	1	1	1	1	1	-1	-1	
B_1	1	-1	1	-1	1	1	-1	
B_2	1	-1	1	-1	1	-1	1	z
E_1	2	$\sqrt{2}$	0	$-\sqrt{2}$	-2	0	0	(x,y)
E_2	2	0	-2	0	2	0	0	(x^2-y^2, xy)
E_3	2	$-\sqrt{2}$	0	$\sqrt{2}$	-2	0	0	(xz,yz)

D_{5d}	E	$2C_5$	$2C_5^{\ 2}$	$5C_2$	i	$2S_{10}^{3}$	$2S_{10}$	$5\sigma_d$	
A_{1g}	1	1	1	1	1	1	1	1	x^2+y^2, z^2
A_{2g}	1	1	1	-1	1	1	1	-1	
E_{1g}	2	$2\cos 72°$	$2\cos 144°$	0	2	$\cdot\ 2\cos 72°$	$2\cos 144°$	0	(xz,yz)
E_{2g}	2	$2\cos 144°$	$2\cos 72°$	0	2	$2\cos 144°$	$2\cos 72°$	0	(x^2-y^2, xy)
A_{1u}	1	1	1	1	-1	-1	-1	-1	
A_{2u}	1	1	1	-1	-1	-1	-1	1	z
E_{1u}	2	$2\cos 72°$	$2\cos 144°$	0	-2	$-2\cos 72°$	$-2\cos 144°$	0	(x,y)
E_{2u}	2	$2\cos 144°$	$2\cos 72°$	0	-2	$-2\cos 144°$	$-2\cos 72°$	0	

T_d	E	$8C_3$	$3C_2$	$6S_4$	$6\sigma_d$	
A_1	1	1	1	1	1	$x^2+y^2+z^2$
A_2	1	1	1	-1	-1	
E	2	-1	2	0	0	$(2z^2-x^2-y^2, x^2-y^2)$
T_1	3	0	-1	1	-1	
T_2	3	0	-1	-1	1	$(x,y,z),(xy,xz,yz)$

(*continued*)

TABLE 2.2 (*continued*)

O_h	E	$8C_3$	$6C_2$	$6C_4$	$3C_2(=C_4{}^2)$	i	$6S_4$	$8S_6$	$3\sigma_h$	$6\sigma_d$	
A_{1g}	1	1	1	1	1	1	1	1	1	1	$x^2+y^2+z^2$
A_{2g}	1	1	-1	-1	1	1	-1	1	1	-1	
E_g	2	-1	0	0	2	2	0	-1	2	0	$(2z^2-x^2-y^2, x^2-y^2)$
T_{1g}	3	0	-1	1	-1	3	1	0	-1	-1	
T_{2g}	3	0	1	-1	-1	3	-1	0	-1	1	(xz, yz, xy)
A_{1u}	1	1	1	1	1	-1	-1	-1	-1	-1	
A_{2u}	1	1	-1	-1	1	-1	1	-1	-1	1	
E_u	2	-1	0	0	2	-2	0	1	-2	0	
T_{1u}	3	0	-1	1	-1	-3	-1	0	1	1	(x, y, z)
T_{2u}	3	0	1	-1	-1	-3	1	0	1	-1	

$C_{\infty v}$	E	$2C_\infty{}^\Phi$	$\ldots$	$\infty\sigma_v$	
$A_1\equiv\Sigma^+$	1	1	$\ldots$	1	z, x^2+y^2, z^2
$A_2\equiv\Sigma^-$	1	1	$\ldots$	-1	
$E_1\equiv\Pi$	2	$2\cos\Phi$	$\ldots$	0	$(x, y), (xz, yz)$
$E_2\equiv\Delta$	2	$2\cos 2\Phi$	$\ldots$	0	(x^2-y^2, xy)
$E_3\equiv\Phi$	2	$2\cos 3\Phi$	$\ldots$	0	
....	....				

$D_{\infty h}$	E	$2C_\infty{}^\Phi$	$\ldots$	$\infty\sigma_i$	i	$2S_\infty{}^\Phi$	$\ldots$	∞C_2	
$\Sigma_g{}^+$	1	1	$\ldots$	1	1	1	$\ldots$	1	x^2+y^2, z^2
$\Sigma_g{}^-$	1	1	$\ldots$	-1	1	1	$\ldots$	-1	
Π_g	2	$2\cos\Phi$	$\ldots$	0	2	$-2\cos\Phi$	$\ldots$	0	(zx, yz)
Δ_g	2	$2\cos 2\Phi$	$\ldots$	0	2	$2\cos 2\Phi$	$\ldots$	0	(x^2-y^2, xy)
..									
$\Sigma_u{}^+$	1	1	$\ldots$	1	-1	-1	$\ldots$	-1	z
$\Sigma_u{}^-$	1	1	$\ldots$	-1	-1	-1	$\ldots$	1	
Π_u	2	$2\cos\Phi$	$\ldots$	0	-2	$2\cos\Phi$	$\ldots$	0	(x, y)
Δ_u	2	$2\cos 2\Phi$	$\ldots$	0	-2	$-2\cos 2\Phi$	$\ldots$	0	
..									

point group. These classes are indicated by the symbols for the operations of the point group. For example, the classes of symmetry operations of the C_{2v} point group are E, C_2, $\sigma_v(xz)$, and $\sigma_v{}'(yz)$.[3] In this case, there are four classes of symmetry operations, and there is one operation for each class. In some cases, a point group may possess more than one symmetry operation of a given class. For example, the C_{3v} point group has three different σ_v planes, and one finds $3\sigma_v$ in the top row of the character table. For the same point group, there are two possible C_3 rotations (through 120 and 240°) other than rotation through 360° (the identity operation); thus, one finds $2C_3$ in the table.[4] The number preceding the symmetry

---

[3] For planar C_{2v} molecules, the x axis is generally chosen perpendicular to the plane.

[4] The effect of carrying out a rotation m times may be written $C_n{}^m$. Thus, the operation C_3 or $C_3{}^1$ corresponds to rotation by 120°; $C_3{}^2$ corresponds to rotation by 240°; and $C_3{}^3$ corresponds to rotation by 360° ($C_3{}^3 = E$). Often, there are equivalent ways of representing an operation—for example, $C_6{}^3 = C_2$; $C_6{}^2 = C_3{}^1$. Similar remarks apply to the other symmetry operations, σ, i, and S.

operator symbol is the number of distinct operations of that particular class and is represented by the general symbol g_R, where R refers to the Rth column in the table (absence of such a number implies that $g_R = 1$). The "order" of the point group, g, is the total number of operations in the group, or simply the sum of the g_R terms, $g = \sum_R g_R$.

Below the listing of the symmetry operations, there is a matrix of terms $(1, -1, 0, 2,$ and so forth). Each term is a "character," and each horizontal row contains the characters of an "irreducible representation" of the point group. Each irreducible representation is labeled by a symbol (A_1, B_2, and so forth) which is the "symmetry species" of the irreducible representation. These symbols are presented in the column on the left side of the table. It can be seen, for example, that in the C_{2v} point group the irreducible representation $1, 1, -1, -1$ corresponds to the symmetry species A_2. The symmetry species symbols are used for labeling normal vibrational modes, electronic states, molecular orbitals, etc. It will be noted from the character tables that the number of irreducible representations of a group is always equal to the number of classes of symmetry operations of the group. That is, each character table contains a square matrix of characters. On the right side of the table are listed various functions, such as x, y, and $x^2 - y^2$. The symbols x, y, and z are the cartesian coordinates (which can stand for vectors of translation or dipole moments). Each of these functions can serve as a "basis" for the particular irreducible representation whose characters are given on the same line. This statement will be clarified in the following paragraphs.

For most symmetric molecules it is possible to describe an entity (e.g., an atom, a bond vector, an atomic orbital, etc.) which lies on all the symmetry elements of the molecule. When any symmetry operation of the molecule is carried out, the position of this entity remains unchanged. If the entity is a property having *sign* (e.g., a vector or p orbital), the entity will have its sign either unchanged or changed. We say that the entity is either "symmetric" or "antisymmetric," respectively, with respect to the symmetry operation. To each operation we assign a "character," which is the factor by which the magnitude of the entity is changed by the operation. Thus the character 1 corresponds to a symmetric transformation, and the character -1 corresponds to an antisymmetric transformation.

For example, consider the p_x orbital of the oxygen atom of the water molecule, which is parallel to the x axis of the molecule and perpendicular to the plane of the molecule. The oxygen atom lies on all the symmetry elements of the molecule, and therefore all its orbitals obviously have symmetry properties characteristic of the symmetry of the molecule. The identity operation leaves the molecule and the p_x orbital unchanged; hence the orbital is clearly symmetric with respect to this operation. Rotation about the C_2 axis inverts the orbital, corresponding to a change in sign. Hence the orbital is antisymmetric with respect to this operation. Reflection through the vertical (xz) plane causes no change in the orbital; the orbital is symmetric with respect to this operation. However, the orbital changes sign on reflection through the plane of the molecule (yz) and therefore is antisymmetric with respect to this operation. Thus the four characters corresponding to the oxygen p_x orbital are, respectively, $1, -1, 1, -1$. These characters constitute the irreducible representation of the oxygen p_x orbital of water. From the C_{2v} character table we see that the symmetry label for this irreducible representation, and hence for this orbital, is B_1. The reader should verify, as a useful exercise, that the oxygen $2p_z$, $3d_{xy}$, and $2p_y$ orbitals[5] are of A_1, A_2, and B_2 symmetry, respectively.

For a slightly more complicated case, consider the nitrogen p_x and p_y orbitals of the ammonia molecule. Rotation by 120 or 240° about the C_3 axis converts p_x and p_y into two new p orbitals both of which have p_x and p_y character. (It should be remembered that a p orbital in the xy plane can be written as a linear combination of p_x and p_y.) That is p_x and p_y are mixed by symmetry and *must be considered as an inseparable degenerate pair*. The character corresponding to C_3 rotation for the nitrogen (p_x, p_y) pair of ammonia is calculated as the sum of the separate factors for p_x and p_y:

$$p_x' = -\tfrac{1}{2}p_x - \tfrac{1}{2}\sqrt{3}p_y$$
$$p_y' = +\tfrac{1}{2}\sqrt{3}p_x - \tfrac{1}{2}p_y$$

Hence $-\tfrac{1}{2}$ from p_x and $-\tfrac{1}{2}$ from p_y gives the character -1.

The irreducible representation of the (p_x, p_y) pair of orbitals of the nitrogen atom of ammonia is an example of a "doubly degenerate" representation.[6] The symmetry species of any doubly degenerate representation is E.[7] Triply degenerate representations are always designated by T. You will perhaps have already noticed that nondegenerate representations correspond to the symmetry species A and B.

In point groups with centers of inversion, the subscript g (from German *gerade*, even) is attached to the symbols of irreducible representations which are symmetric (no change in sign) with respect to inversion, and the subscript u (from German *ungerade*, uneven) is attached in the case of representations which are antisymmetric (change in sign) with respect to inversion.

REDUCIBLE REPRESENTATIONS

In chemistry, it is often necessary to determine the representation of a set of entities (e.g., atoms, bonds, atomic orbitals, etc.) located at stereochemically equivalent points in a molecule. The following procedure is followed. For a particular member of the set, carry out one operation for each class of symmetry operation in the point group. Assign a number to each of these operations. If the operation does not move or change the sign of the entity, the number is 1. In the case of an entity having sign, if the operation leaves the entity unmoved but with a change of sign, the number is -1. For any other change the number is 0. Determine a similar series of numbers for each member of the set (always taking care to use the same operation for a given class of operation). Then add the numbers corresponding to

---

[5] The oxygen $3d$ orbital would not be importantly involved in the bonding of H_2O. However, sulfur $3d$ orbitals would perhaps be significant in H_2S, an analogous molecule.

[6] Some authors refer to the degeneracy of a representation as the "dimension" of the representation.

[7] This E should not be confused with the identity operator.

the same class of symmetry operation. The resulting sums are the characters of the reducible representation of the entities.

As an example, we shall determine the reducible representation of the group of three N—H bonds of ammonia. For convenience, we make a table, using the headings of the C_{3v} point group. The sets of numbers corresponding to the three equivalent bonds are indicated.

C_{3v}

Bond	E	$2C_3$	$3\sigma_v$
a	1	0	1
b	1	0	0
c	1	0	0
Reducible representation	3	0	1

It is obvious from an inspection of the C_{3v} character table that the representation 3, 0, 1 is not one of the irreducible representations. However, it can be *reduced* to a set of irreducible representations. That is, there is a set of irreducible representations which, when the corresponding characters are added, yields the reducible representation 3, 0, 1. By inspection of the character table, it can be readily seen that these irreducible representations are those of symmetry species A_1 and E.

A_1	1	1	1
E	2	-1	0
	3	0	1

Incidentally, this result tells us that the N—H stretching vibrations of NH_3 are of two types: A_1 (nondegenerate) and E (doubly degenerate). The sum of the degeneracies of the irreducible representations is the number of bonds.

Let us consider a more complicated set of entities, the four chlorine $3p$ orbitals of $PtCl_4^{2-}$ (Fig. 2.1a) which are perpendicular to the plane of the molecule. The reducible representation of these four equivalent orbitals is obtained as follows:

D_{4h}

E	$2C_4$	C_2	$2C_2'$	$2C_2''$	i	$2S_4$	σ_h	$2\sigma_v$	$2\sigma_d$
1	0	0	-1	0	0	0	-1	1	0
1	0	0	0	0	0	0	-1	0	0
1	0	0	-1	0	0	0	-1	1	0
1	0	0	0	0	0	0	-1	0	0
4	0	0	-2	0	0	0	-4	2	0

In this case is it relatively difficult to reduce the reducible representation by mere inspection of the character table. Fortunately there is a foolproof analytical method for doing this. The number of times a particular irreducible representation contributes to a reducible representation may be calculated from the expression

$$n = \frac{1}{g} \sum_R g_R \chi(R) \cdot \gamma(R)$$

where $\chi(R)$ is the character of the symmetry operation R in the irreducible representation (taken from the character table) and $\gamma(R)$ is the character of the symmetry operation R in the reducible representation. Thus, the reducible representation of the four chlorine $3p$ orbitals of $PtCl_4^{2-}$ which are perpendicular to the molecular plane may be reduced as follows:

$$N_{A_{1g}} = \frac{1}{16}[1 \cdot 1 \cdot 4 + 2 \cdot 1 \cdot 0 + 1 \cdot 1 \cdot 0 + 2 \cdot 1 \cdot (-2) + 2 \cdot 1 \cdot 0 + 1 \cdot 1 \cdot 0 \\ + 2 \cdot 1 \cdot 0 + 1 \cdot 1 \cdot (-4) + 2 \cdot 1 \cdot 2 + 2 \cdot 1 \cdot 0] = 0$$

$$N_{A_{2g}} = \frac{1}{16}[1 \cdot 1 \cdot 4 + 2 \cdot 1 \cdot 0 + 1 \cdot 1 \cdot 0 + 2 \cdot (-1) \cdot (-2) + 2 \cdot (-1) \cdot 0 \\ + 1 \cdot 1 \cdot 0 + 2 \cdot 1 \cdot 0 + 1 \cdot 1 \cdot (-4) + 2 \cdot (-1) \cdot 2 + 2 \cdot (-1) \cdot 0] = 0$$

$$N_{B_{1g}} = \frac{1}{16}[1 \cdot 1 \cdot 4 + 2 \cdot (-1) \cdot 0 + 1 \cdot 1 \cdot 0 + 2 \cdot 1 \cdot (-2) + 2 \cdot (-1) \cdot 0 \\ + 1 \cdot 1 \cdot 0 + 2 \cdot (-1) \cdot 0 + 1 \cdot 1 \cdot (-4) + 2 \cdot 1 \cdot 2 + 2 \cdot (-1) \cdot 0] = 0$$

$$N_{B_{2g}} = \frac{1}{16}[1 \cdot 1 \cdot 4 + 2 \cdot (-1) \cdot 0 + 1 \cdot 1 \cdot 0 + 2 \cdot (-1) \cdot (-2) + 2 \cdot 1 \cdot 0 \\ + 1 \cdot 1 \cdot 0 + 2 \cdot (-1) \cdot 0 + 1 \cdot 1 \cdot (-4) + 2 \cdot (-1) \cdot 2 + 2 \cdot 1 \cdot 0] = 0$$

$$N_{E_g} = \frac{1}{16}[1 \cdot 2 \cdot 4 + 2 \cdot 0 \cdot 0 + 1 \cdot (-2) \cdot 0 + 2 \cdot 0 \cdot (-2) + 2 \cdot 0 \cdot 0 + 1 \cdot 2 \cdot 0 \\ + 2 \cdot 0 \cdot 0 + 1 \cdot (-2) \cdot (-4) + 2 \cdot 0 \cdot 2 + 2 \cdot 0 \cdot 0] = 1$$

$$N_{A_{1u}} = \frac{1}{16}[1 \cdot 1 \cdot 4 + 2 \cdot 1 \cdot 0 + 1 \cdot 1 \cdot 0 + 2 \cdot 1 \cdot (-2) + 2 \cdot 1 \cdot 0 + 1 \cdot (-1) \cdot 0 \\ + 2 \cdot (-1) \cdot 0 + 1 \cdot (-1) \cdot (-4) + 2 \cdot (-1) \cdot 2 + 2 \cdot (-1) \cdot 0] = 0$$

$$N_{A_{2u}} = \frac{1}{16}[1 \cdot 1 \cdot 4 + 2 \cdot 1 \cdot 0 + 1 \cdot 1 \cdot 0 + 2 \cdot (-1)(-2) + 2 \cdot (-1) \cdot 0 \\ + 1 \cdot (-1) \cdot 0 + 2 \cdot (-1) \cdot 0 + 1 \cdot (-1) \cdot (-4) + 2 \cdot 1 \cdot 2 + 2 \cdot 1 \cdot 0] = 1$$

$$N_{B_{1u}} = \frac{1}{16}[1 \cdot 1 \cdot 4 + 2 \cdot (-1) \cdot 0 + 1 \cdot 1 \cdot 0 + 2 \cdot 1 \cdot (-2) + 2 \cdot (-1) \cdot 0 \\ + 1 \cdot (-1) \cdot 0 + 2 \cdot 1 \cdot 0 + 1 \cdot (-1) \cdot (-4) + 2 \cdot (-1) \cdot 2 + 2 \cdot 1 \cdot 0] = 0$$

$$N_{B_{2u}} = \frac{1}{16}[1 \cdot 1 \cdot 4 + 2 \cdot (-1) \cdot 0 + 1 \cdot 1 \cdot 0 + 2 \cdot (-1) \cdot (-2) + 2 \cdot 1 \cdot 0 \\ + 1 \cdot (-1) \cdot 0 + 2 \cdot 1 \cdot 0 + 1 \cdot (-1) \cdot (-4) + 2 \cdot 1 \cdot 2 + 2 \cdot (-1) \cdot 0] = 1$$

$$N_{E_u} = \frac{1}{16}[1 \cdot 2 \cdot 4 + 2 \cdot 0 \cdot 0 + 1 \cdot (-2) \cdot 0 + 2 \cdot 0 \cdot (-2) + 2 \cdot 0 \cdot 0 \\ + 1 \cdot (-2) \cdot 0 + 2 \cdot 0 \cdot 0 + 1 \cdot 2 \cdot (-4) + 2 \cdot 0 \cdot 2 + 2 \cdot 0 \cdot 0] = 0$$

Thus we find that the symmetry species of the irreducible representations of the orbitals are E_g, A_{2u}, and B_{2u}, with a total degeneracy of 4. As we shall see later (in Chap. 4), these symmetry species correspond to "group orbitals" which are suitable linear combinations of the four p orbitals.

PROBLEMS

2.1 Determine the point groups of the following molecules or objects:

(a) SiHDBr₂ ("tetrahedral")

(b) SiFClBrI ("tetrahedral")

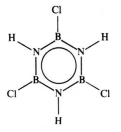

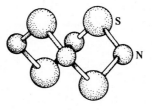

FIGURE 2.9
The structures of $B_3N_3H_3Cl_3$ and S_4N_4.

(c) SiH_2Br_2 ("tetrahedral")

(d) PCl_3 (pyramidal)

(e) $OPCl_3$ (pyramidal)

(f) CO_2 (linear)

(g) P_4O_6 (tetrahedral)

(h) $B, B, B,$-Trichloroborazine, $B_3N_3H_3Cl_3$ (planar six-membered ring of alternating B's and N's, shown in Fig. 2.9)

(i) S_4N_4 (N's square planar; S's approximately tetrahedral, as shown in Fig. 2.9)

(j) $Mn(CO)_5I$ ("octahedral")

(k) The letter "T"

(l) The letter "Z"

(m) The numeral 8 (idealized)

(n) A tennis ball (including seam)

*2.2 Show that the character corresponding to $3\sigma_v$ for the (p_x, p_y) pair of NH_3 is 0.

2.3 What are the symmetry species of the $3d$ orbitals of sulfur in SF_6 (O_h point group)?

2.4 What are the symmetry species of the symmetry-allowed combinations of the following atomic orbitals?

(a) The six carbon $2p$ orbitals of benzene which are perpendicular to the plane of the molecule

(b) The four hydrogen $1s$ orbitals of CH_4

(c) The oxygen $2p$ orbitals of the terminal atoms of O_3 (C_{2v} point group) which are perpendicular to the plane of the molecule

(d) The six hydrogen $1s$ orbitals of hypothetical SH_6 (O_h point group)

CHAPTER
3

COVALENT BONDING

We envision two extremes of chemical bonding: covalent and ionic. We say that a bond is purely covalent when the bonding valence electrons are equally shared between the bonded atoms, and that a bond is purely ionic when the atoms have charges of opposite sign and there are no valence electrons shared between the atoms. Although the bonding in a homonuclear diatomic molecule such as Cl_2 obviously fits the criterion for pure covalency, there is no known example of purely ionic bonding. (In Chap. 11 we will show that the bonding in certain crystals only *approximates* the ionic extreme.) In other words, all chemical bonds are at least partially covalent, with some valence electron density shared between the atoms. In this chapter we discuss a simple theory for describing covalent bonds (the Lewis octet theory) and a method for predicting the geometries of molecules (the valence-shell electron repulsion theory). We also discuss methods used for systematizing properties of chemical bonds, such as bond distances, bond strengths, and bond polarities. All these topics are important in inorganic chemistry, particularly in the study of compounds of the nontransition elements. As was the case in Chap. 1, you may already be familiar with some of the material in this chapter; to that extent, this chapter will be a review.

THE LEWIS OCTET THEORY

The Lewis octet theory[1] is extremely useful for describing the bonding in practically all compounds of the nontransition elements. The theory is based on the

---

[1] The modern concept of the Lewis octet theory is an evolutionary product of the theory originally outlined by G. N. Lewis in "Valence and the Structure of Atoms and Molecules," Chemical Catalog, New York, 1923.

stability of valence shells containing eight electrons in the s and p orbitals (s^2p^6). According to the theory, each atom in a compound achieves an octet of valence electrons by sharing electrons with the other atoms to which it is bonded. Of course, hydrogen atoms do not achieve octets; their complete valence shells contain only two electrons.

Lewis octet structures for molecules can be represented in several ways. For example, the structure of ammonia can be written in the following ways:

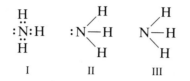

The first way, in which all the valence electrons are indicated by dots, is somewhat cumbersome and is usually abandoned in the introductory chemistry course. The second and third ways are commonly used by practicing chemists; the choice between these methods is determined by whether or not one wishes to emphasize the presence of nonbonding electron pairs.

In writing octet structures, it is common practice to indicate the charges which atoms would have if each pair of bonding electrons were equally divided between the bonded atoms. These charges are called "formal charges" and should not be confused with the actual charges on the atoms. In NH_3, none of the atoms has a formal charge. However, many other molecules, and of course all ions, have formally charged atoms. The structures of a few of these are shown below.

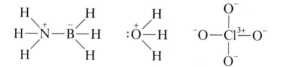

All chemists should be proficient in the assignment of formal charges to the atoms of a Lewis octet structure. One simple method for making this assignment is based on the "total bond order" of an atom, i.e., the sum of the bond orders of the bonds to an atom. The total bond order has an easily remembered characteristic value for a given atom with zero formal charge. Thus it is 1 for a halogen atom, 2 for an atom of the oxygen family, 3 for an atom of the nitrogen family, etc. When the total bond order is 1 greater than the characteristic value, the formal charge is $+1$; when it is 2 greater, the formal charge is $+2$, etc. Similarly, when the total bond order is 1 less than the characteristic value, the formal charge is -1, etc. These rules[2] are summarized in Table 3.1.

---

[2]   There is some difficulty in extending the rules to elements to the left of group IV. Thus it is impossible for boron to have both a complete octet and a zero formal charge.

TABLE 3.1
Formal charges on atoms with complete octets

Total bond order	Periodic table group					
	III	**IV**	**V**	**VI**	**VII**	**VIII**
0	...	...	−3	−2	−1	0
1	...	−3	−2	−1	0	+1
2	...	−2	−1	0	+1	+2
3	−2	−1	0	+1	+2	+3
4	−1	0	+1	+2	+3	+4

Proficiency in the assignment of the formal charges can be very helpful in writing Lewis structures. For example, if the structure of nitrosyl chloride were mistakenly written

the structure would immediately be recognized as wrong, because the oxygen and nitrogen atoms would each have −1 formal charges, corresponding to a −2 ion. By introducing the N=O double bond,

a satisfactory structure for the neutral molecule is achieved.

Sometimes more than one structure can be written for a species with multiple bonds. In such a case the multiple bonding is distributed throughout the molecule, a situation often referred to as resonance[3] or delocalized π bonding. (A detailed discussion of π bonding is presented in Chap. 4.) Thus for ozone we may write either

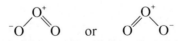 or

Neither structure is a satisfactory representation of the bonding because the two terminal oxygen atoms in ozone are known to be equivalent. The actual bonding is midway between these extreme "resonance structures." We may represent the actual molecule, which is a "resonance hybrid," by the structure

[3]   L. Pauling, "The Nature of the Chemical Bond," 3rd ed., Cornell University Press, Ithaca, N.Y., 1960.

Here each bond has a bond order of $1\frac{1}{2}$, and each terminal oxygen atom has a formal charge of $-\frac{1}{2}$, corresponding to the average values of these quantities in the two resonance structures. Similarly, we may represent the bonding in the nitrate ion by the hybrid structure

where each bond has a bond order of $1\frac{1}{3}$.

The two examples of resonance which we have just discussed involve resonance structures which are equivalent, i.e., structures which are indistinguishable upon the execution of a symmetry operation such as rotation about an axis of symmetry or reflection through a mirror plane. However, sometimes nonequivalent resonance structures can be written, as in the case of the cyanate ion:

$$^-O-C\equiv N \qquad O=C=N^- \qquad {}^+O\equiv C-N^{2-}$$

In such cases the best representation of the bonding is usually not just a simple average of the individual resonance structures but rather some weighted average of these structures. The third cyanate structure, with a negative formal charge on the nitrogen atom and a positive formal charge on the oxygen atom, has a charge distribution which contradicts the fact that oxygen is more electronegative (electron-attracting) than nitrogen. Therefore it is reasonable to suppose that the actual bonding is a resonance hybrid mainly of the first two structures, with negligible contribution from the third structure.

Let us consider a more complicated example, the $S_3N_2Cl^+$ ion, which is found in the salt $(S_3N_2Cl)Cl$.[4]

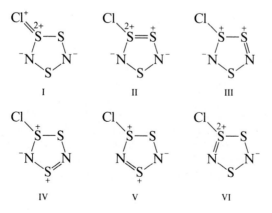

Of the six different resonance structures, structures I to III are very unimportant because they involve adjacent atoms with formal charges of the same sign—an

[4] A. Zalkin, T. E. Hopkins, and D. H. Templeton, *Inorg. Chem.*, **5**, 1767 (1966).

electrostatically unfavorable situation. Structure IV is the most favorable in terms of the electrostatic interaction of the formal charges. Indeed, the measured bond distances, given in the following diagram, are in accord with the main contribution from the structure IV. The shortest S—N bond corresponds to the double bond of resonance structure IV.

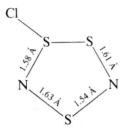

A special kind of resonance, called "no-bond" resonance or "hyperconjugation," is occasionally employed to rationalize extraordinary bonding features. This special resonance involves an increase in the order of one bond at the expense of the order of another bond:

$$X—Y—Z \quad \leftrightarrow \quad {}^+X{=}Y \quad Z^-$$

For example, in ONF_3, the N—F bonds appear to be abnormally long, as determined by electron diffraction.[5] Resonance of the following type can explain this "abnormality."

$$^-O{-}\overset{+}{N}{-}F \quad \longleftrightarrow \quad O{=}\overset{+}{N} \quad F^-$$
(with F substituents above and below N)

Similarly, the unexpectedly strong acid-base interaction between BH_3 and CO has been explained in terms of hyperconjugation of the following type[6]:

$$H{-}\overset{-}{B}{-}C{\equiv}O^+ \quad \longleftrightarrow \quad {}^+H \quad B{=}C{=}O$$
(with H substituents above and below B)

Although many structural rationalizations based on hyperconjugation are somewhat ad hoc, they are nevertheless reasonable insofar as the hyperconjugation causes a change in formal charge distribution consistent with electronegativities.[7] Thus in ONF_3 hyperconjugation causes a shift in negative formal charge from oxygen to fluorine, and in H_3BCO it causes a shift in positive formal charge from oxygen to hydrogen.

---

[5]   V. Plato, W. D. Hartford, and K. Hedberg, *J. Chem. Phys.*, **53**, 3488 (1970).

[6]   B. E. Douglas, D. H. McDaniel, and J. J. Alexander, "Concepts and Models of Inorganic Chemistry," 2d ed., pp. 535–536. Wiley, New York, 1983.

[7]   Electronegativities are discussed on pp. 62–63 and 71–76.

Some species have more valence electrons than can be accommodated by a Lewis octet structure involving only single bonds. For example, if we write single-bonded structures for I_3^- and SF_6, we find that the central atoms are required to have 10 and 12 valence electrons, respectively:

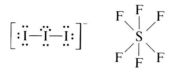

Such species are "electron-rich," and the central atoms are referred to as "hypervalent." There are two common ways of handling this problem. One way is to invoke "no-bond" resonance structures or, what is equivalent, fractional bonds. Thus we may write the following Lewis structures for I_3^-:

$$I—I \quad I^- \qquad I^- \quad I—I$$

These correspond to a symmetric resonance hybrid which can be represented as follows:

$$^{-1/2}I \text{ --- } I \text{ --- } I^{-1/2}$$

In the latter structure each bond has a bond order of $\frac{1}{2}$. The molecule SF_6 may be similarly treated; the structure involves six $\frac{2}{3}$ bonds:

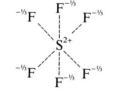

Another way of handling hypervalent atoms is to invoke the use of valence-shell d orbitals. Then, instead of restricting atoms like iodine and sulfur to an s^2p^6 valence shell, we can permit electrons to be accepted by the empty d orbitals of the same principal quantum number. In principle, we can thereby accommodate hypervalency for any atom heavier than neon. Such d-orbital participation has been postulated even in species for which satisfactory Lewis octet structures can be written. Thus, in recognition of the fact that the central atoms in SO_4^{2-} and SiF_4 use $3d$ orbitals, resonance structures of the following type have been written:

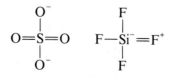

Although d orbitals unquestionably participate to some extent in the bonding of compounds having atoms with empty valence-shell d orbitals,[8] most of the properties of such compounds can be qualitatively accounted for using octet structures.

Straightforward octet structures cannot be written for molecules with an odd number of electrons, such as NO. However, Linnett[9] devised a procedure in which the valence electrons of a molecule are split into two parts (corresponding to electrons of opposite spin) so that each atom achieves a *quartet* of each type of electron. By this "double quartet" theory, one indicates the separate quartets for NO as follows:

$$\overset{\times}{\underset{\times}{N}}\overset{\times}{\underset{\times}{O}}\overset{\times}{} \qquad \overset{\circ}{}N\overset{\circ}{\underset{\circ}{}}O\overset{\circ}{}$$

Notice that we have accounted for all the valence electrons of NO: six electrons of one spin and five of the opposite spin. Five electrons are situated in the bonding region between the atoms, corresponding to 2.5 electron-pair bonds. Pairs of electrons having opposed spins in the same region of the molecule can be represented by lines; thus the molecule can be represented as follows:

$$\overset{\circ}{\underset{}{}}\overset{\times}{\underset{\times}{N}}\overset{\circ}{\underset{\circ}{}}O\overset{\times}{\underset{\times}{}}\circ \qquad \text{or} \qquad \overset{\times}{\underline{}}N\!\!=\!\!O\overset{\times}{\underline{}}$$

Because nitric oxide has one more electron of one spin than of the other, it has a net electronic spin of $\frac{1}{2}$.

In the case of the molecule O_2, we can combine electron quartets in two ways:

(1) $\overset{\times}{\underset{\times}{O}}\overset{\times}{\underset{\times}{O}}\times$ and $\overset{\circ}{\underset{\circ}{O}}\overset{\circ}{\underset{\circ}{O}}\overset{\circ}{\underset{\circ}{}}$ yielding $\overset{\diagdown}{\underset{\diagup}{O}}\!\!=\!\!\overset{\diagup}{\underset{\diagdown}{O}}$

(2) $\overset{\times}{\underset{\times}{O}}\times\overset{\times}{\underset{\times}{O}}$ and $\circ O\overset{\circ}{\underset{\circ}{}}O\circ$ yielding $\overset{\times}{\underset{\times}{}}O\overset{\circ}{\underset{\circ}{}}O\overset{\times}{\underset{\times}{}}$

Both combinations give double-bonded molecular oxygen. However, the first gives oxygen with no unpaired electrons (an excited state of O_2), and the second gives oxygen with two unpaired electrons (the ground state). The greater stability of the second structure can be rationalized in terms of reduced repulsion between the electrons of opposite spin. For further applications of this method, the book or article by Linnett[9] should be consulted. Probably the main significance of Linnett's method is that it reaffirms the fundamental importance of the octet theory and the strong tendency for atoms to achieve rare-gas electron configurations.

THE ISOELECTRONIC PRINCIPLE

Molecules that have the same number of electrons and the same number of heavy atoms (i.e., atoms heavier than hydrogen) usually have similar electronic structures, similar heavy-atom geometries, and similar chemical properties. This state-

---

[8] T. B. Brill, *J. Chem. Educ.*, **50**, 392 (1973). Also see pp. 117–118 and 130 in this text for a discussion of molecular orbitals involving d orbitals in nonmetal compounds. See A. E. Reed and F. Weinhold [*J. Am. Chem. Soc.*, **108**, 3586 (1986)] for a discussion of π donor bonding from the fluorine atoms to the sulfur d orbitals in SF_6.

[9] J. W. Linnett, "The Electronic Structure of Molecules," Wiley, New York, 1964; *J. Am. Chem. Soc.*, **83**, 2643 (1961).

TABLE 3.2
Vibrational frequencies for linear triatomic species†

Molecule	Bond bending frequency, cm^{-1}	Symmetric stretching frequency, cm^{-1}	Asymmetric stretching frequency, cm^{-1}
$NO_2{}^+$	538	1400	2375
H_2CN_2	564	1170	2102
HNCO	572	1327	2274
H_2CCO	588	1120	2152
N_2O	589	1285	2223
$BO_2{}^-$	610	1070	1970
NCO^-	629	1205	2170
$N_3{}^-$	630	1348	2080
CO_2	667	1388	2349

† H. A. Bent., *J. Chem. Educ.*, **43**, 170 (1966).

ment is the "isoelectronic principle," first discussed by Langmuir.[10] As an example of a set of isoelectronic species, the molecules and ions consisting of 3 heavy atoms and 22 electrons may be considered:

$H_2C\!=\!C\!=\!CH_2$
Allene

$H_2C\!=\!C\!=\!O$
Ketene

$HN\!=\!C\!=\!O$
Isocyanic acid

$O\!=\!C\!=\!O$
Carbon dioxide

$^-N\!=\!\overset{+}{N}\!=\!O$
Nitrous oxide

$^-N\!=\!\overset{+}{N}\!=\!N^-$
Azide ion

$H_2C\!=\!\overset{+}{N}\!=\!N^-$
Diazomethane

$O\!=\!\overset{+}{N}\!=\!O$
Nitryl ion

$F\!-\!C\!\equiv\!N$
Cyanogen fluoride

$H_3\bar{B}\!-\!C\!\equiv\!N$
Cyanoborohydride ion

$H_3C\!-\!C\!\equiv\!N$
Acetonitrile

$^-N\!=\!C\!=\!N^-$
Cyanamide ion

$H_3\bar{B}\!-\!C\!\equiv\!O^+$
Borane carbonyl

$O\!=\!\bar{B}\!=\!O$
Metaborate ion

$CH_3\!-\!C\!\equiv\!CH$
Methyl acetylene

The important common feature of these species is not that they have the same total number of electrons, but rather that they have the same number of *valence* electrons (in this case, 16). Indeed, the word "isoelectronic" is usually loosely interpreted as meaning *having the same number of valence electrons*. Thus we may include the following species in the preceding group of 3-heavy-atom, 16-valence-electron species:

$O\!=\!C\!=\!S$
Carbonyl sulfide

$S\!=\!C\!=\!S$
Carbon disulfide

$^-N\!=\!C\!=\!S$
Thiocyanate ion

$Br\!-\!C\!\equiv\!N$
Cyanogen bromide

etc.

One feature which all these species have in common is a linear heavy-atom skeleton. They also have similar values for their corresponding vibrational frequencies, as shown by the data in Table 3.2. Many of the species undergo analogous reactions. Thus Lewis bases such as OH$^-$ and H_2O react by attacking the middle atom:

[10] I. Langmuir, *J. Am. Chem. Soc.*, **41**, 868, 1543 (1919).

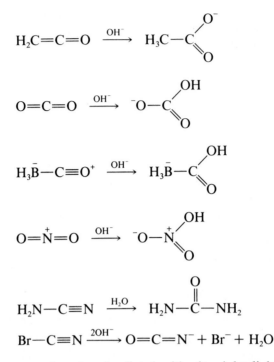

Many of the compounds, when irradiated with ultraviolet light, break into two fragments:

$$H_2CCO \xrightarrow{h\nu} CH_2 + CO$$
$$H_2CNN \xrightarrow{h\nu} CH_2 + N_2$$
$$HNNN \xrightarrow{h\nu} NH + N_2$$
$$OCO \xrightarrow{h\nu} O + CO$$
$$OCS \xrightarrow{h\nu} S + CO$$

etc.

The existence of isoelectronic analogs of an unknown compound has often served as the impetus for its first synthesis. For example, in 1971, the following isoelectronic compounds were known: $Ni(CO)_4$, $Co(CO)_3NO$, $Fe(CO)_2(NO)_2$, and $Mn(NO)_3CO$. The last member of this series, $Cr(NO)_4$, was unknown. However, in 1972, several chemists[11] had sufficient faith in the isoelectronic principle to photolyze a solution of $Cr(CO)_6$ in the presence of NO, and thus they prepared the elusive compound.

For many years chemists unsuccessfully tried to prepare the perbromate ion BrO_4^-, which is analogous to the well-known perchlorate and periodate ions.

---

[11] B. I. Swanson and S. K. Satija, *J. Chem. Soc. Chem. Commun.*, 40 (1973); M. Herberhold and A. Razavi, *Angew. Chem. Int. Ed.*, **11**, 1092 (1972).

The first successful synthesis of perbromate involved an isoelectronic species as the starting material. The synthesis involved the β decay of radioactive ^{83}Se incorporated in a selenate[12]:

$$^{83}SeO_4{}^{2-} \quad \rightarrow \quad ^{83}BrO_4{}^- + \beta^-$$

The ^{83}Br is itself radioactive, and its coprecipitation with $RbClO_4$ was taken as evidence that it was in the form of $BrO_4{}^-$. We now know that the difficulty in the synthesis of perbromate by conventional methods is due to the fact that perbromate is an extremely powerful oxidizing agent, much more powerful than either perchlorate or periodate. The most efficient method now known for the preparation of perbromate involves the oxidation of a strongly alkaline $BrO_3{}^-$ solution by elemental fluorine.[13]

One must use reasonable caution in applying the isoelectronic principle. For example, because of the large differences in the properties of the fluorine and nitrogen atoms, one should not expect FCN to be a close analog of CO_2. Indeed, the trimerization reaction of FCN has no known parallel with CO_2.

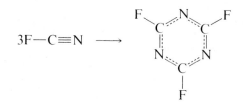

Similarly, borazine, $B_3N_3H_6$, is an imperfect analog of benzene, C_6H_6:

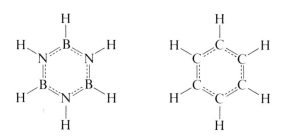

Borazine forms an adduct with 3 mol of HCl in which the ring nitrogen atoms are protonated,[14] whereas benzene is inert toward HCl under ordinary conditions. Although hexamethylborazine forms transition-metal complexes such as $B_3N_3(CH_3)_6Cr(CO)_3$ (analogous to complexes formed by the corresponding benzene derivative), the B_3N_3 ring in these complexes is believed to be puckered, in contrast to the planar C_6 rings of the benzene derivative complexes.[15]

[12]   E. H. Appelman, *J. Am. Chem. Soc.*, **90**, 1900 (1968).
[13]   E. H. Appelman, *Inorg. Syn.*, **13**, 1 (1972).
[14]   A. W. Laubengayer, O. T. Beachley, Jr., and R. F. Porter, *Inorg. Chem.*, **4**, 578 (1965).
[15]   J. L. Adcock and J. J. Lagowski, *Inorg. Chem.*, **12**, 2533 (1973).

BOND DISTANCES

Interatomic distances are now known for many molecules with probable errors of ± 0.03 Å or less.[16] One finds that the distance between two bonded atoms (say Si and Cl) is very nearly the same in different molecules, as long as the general nature of the bonding is the same in the various molecules. Some idea of the constancy of covalent bond distances can be obtained from Table 3.3, where the experimental Si—Cl bond distances for 10 different compounds are listed. The average value, 2.02 Å, is taken as the characteristic Si—Cl bond distance and may be used as an estimate of the Si—Cl bond distance in any other molecule containing such a bond. A brief listing of other characteristic covalent single-bond distances is given in Table 3.4.

The Si—Cl bond distances listed in Table 3.3 have practically the same value because the bonds not only have the same order but they also involve silicon atoms which have the same number of bonds. A bond distance changes significantly as the total number of bonds to the bonded atoms changes. The effect is quite obvious in the C—C bonds of hydrocarbons, as can be seen from the data in Table 3.5. As the number of secondary bonds to a C—C single bond is reduced from six (in ethane) to two (in a diacetylene), the bond distance decreases. This effect is undoubtedly related to the correlation of bond strength with orbital hybridization,[17] a topic which will be discussed in more detail in Chap. 4. At this point we will merely point out that, as the fractional s character

---

[16]   L. E. Sutton (ed.), "Tables of Interatomic Distances and Configuration in Molecules and Ions," Special Publ. 11, The Chemical Society, London, 1958; Supplement, Special Publ. 18, 1965.
[17]   H. A. Bent, *Chem. Rev.*, **61**, 275 (1961).

TABLE 3.3
Silicon-chlorine bond distances in various molecules†

Compound	Si—Cl
$SiClF_3$	2.00
$SiClH_3$	2.05
$SiCl_2H_2$	2.02
$SiCl_3H$	2.02
$SiCl_3SH$	2.02
$SiCl_4$	2.01
Si_2Cl_6	2.01
Si_2Cl_6O	2.02
Cl_3CSiCl_3	2.01
$C_6H_5SiCl_3$	2.00

†L. E. Sutton (ed.), "Tables of Interatomic Distances and Configuration in Molecules and Ions," Special Publ. 11, The Chemical Society, London, 1958; Supplement, Special Publ. 18, 1965.

TABLE 3.4
Covalent single-bond distances for the nonmetals (in angstroms)†

	As	B	Br	C	Cl	F	Ge	H	I	N	O	P	S	Sb	Se	Si	Sn	Te
As	2.44	2.33	2.28	1.96	2.16	1.71		1.52	2.54		1.78		2.24	2.20	1.98	2.16	2.46	2.51
B		1.72	1.93	1.56	(1.74)	1.42		1.12		(1.42)	(1.36)							
Br			2.28	1.94	2.14	1.76	2.30	1.41		[2.14]			2.27	2.51				
C				1.54	1.77	1.35	1.94	1.09	2.14	1.47	1.43	1.84	1.82	2.20	1.98	1.87	2.15	
Cl					1.99	1.63		1.27	2.32	1.75	[1.70]	2.01	2.03	2.32		2.02	2.33	2.33
F						[1.42]	1.68	0.92		1.36	[1.42]	1.52	1.53			1.56		
Ge							2.44	1.53										
H								0.74	1.61	1.01	0.96	1.44	1.34	1.71	1.46	1.48	1.70	
I									2.67							2.44	2.69	
N										[1.45]	[1.41]	1.66	1.67			1.74		
O											[1.48]	2.24	1.5			1.63		
P												2.24	1.86		2.24			
S													2.05		2.32	2.15	2.05	
Sb														2.80				
Se															2.32			
Si																2.34		
Sn																	2.81	
Te																		2.86

†Bracketed values are probably high because of lone-pair repulsions or hyperconjugation, and parenthesized values are probably low because of multiple bonding.

TABLE 3.5
Carbon-carbon single-bond lengths and hybridizations in various hydrocarbons

Bond type	Hybridization	r, Å
$\underset{/}{\overset{\backslash}{>}}C-C\underset{\backslash}{\overset{/}{<}}$	$sp^3\text{-}sp^3$	1.54
$\underset{/}{\overset{\backslash}{>}}C-C\underset{\Bbb{\backslash}}{\overset{/}{<}}$	$sp^3\text{-}sp^2$	1.51
$\underset{/}{\overset{\backslash}{>}}C-C\equiv$	$sp^3\text{-}sp$	1.46
$\overset{\backslash}{\underset{\backslash}{>}}C-C\underset{\Bbb{\backslash}}{\overset{/}{<}}$	$sp^2\text{-}sp^2$	1.46
$\overset{\backslash}{\underset{\backslash}{>}}C-C\equiv$	$sp^2\text{-}sp$	1.44
$\equiv C-C\equiv$	$sp\text{-}sp$	1.37

of the bonding orbitals increases on going from sp^3 to sp hybridization, the bond distance decreases and the bond strength increases.[18]

By use of the data in Table 3.4, it can be shown that, in most cases, the distance of a heteronuclear covalent bond, A—B, is approximately the arithmetic average of the distances of the corresponding homonuclear covalent bonds, A—A and B—B. Therefore it is possible to assign covalent radii to atoms and to use these to estimate unknown covalent bond distances. The covalent radii listed in Table 3.6 are, in most cases, simply one-half of the homonuclear single-bond distances. However, in some cases the values were chosen so as to give as good agreement as possible with the heteronuclear bond distances, especially with the bonds to carbon. Notice the correlation of covalent radius with position in the periodic table. The data bear out the trends discussed in Chap. 1.

As one might expect, bond distance decreases as bond order increases. This can be seen from the covalent radii for carbon, nitrogen, and oxygen in Table 3.7.

---

[18] C. A. Coulson, "Valence," p. 199. Oxford University Press, New York, 1952.

TABLE 3.6
Single-bond covalent radii (in angstroms)

				H 0.30
B 0.86	C 0.77	N 0.70	O 0.66	F 0.58
	Si 1.17	P 1.12	S 1.02	Cl 1.00
	Ge 1.22	As 1.22	Se 1.16	Br 1.14
	Sn 1.40	Sb 1.40	Te 1.43	I 1.34

TABLE 3.7
Covalent radii for carbon, nitrogen, and oxygen

Atom	r, Å		
	Single bond	Double bond	Triple bond
C	0.77	0.67	0.60
N	0.70	0.60	0.55
O	0.66	0.55	0.53

This general correlation can be used to detect partial multiple-bond character, or delocalized π bonding, in molecules. For example, in SO_2 the S—O bond order is 1.5, and the observed bond distance, 1.43 Å, is 0.25 Å shorter than the sum of the single-bond covalent radii $(1.02 + 0.66 = 1.68$ Å$)$. In BF_3, the B—F bond order is 1.33, and the observed bond distance, 1.30 Å, is 0.14 Å shorter than the sum of the single-bond covalent radii $(0.86 + 0.58 = 1.44$ Å$)$. The correlation of N—O bond distance with bond order can be seen from the data in Table 3.8.

In SiF_4, the Si—F bond distance, 1.56 Å, is 0.19 Å shorter than the sum of the single-bond radii $(1.17 + 0.58 = 1.75$ Å$)$. The cause of this apparent bond shortening is a subject of some dispute. Schomaker and Stevenson[19] proposed that the shortening is a result of the relatively high ionic character of the bond. That is, they proposed that the bond is extraordinarily strong and short because of the electrostatic attraction between the very positive silicon atom and the very negative

---

[19] V. Schomaker and D. P. Stevenson, *J. Am. Chem. Soc.*, **63**, 37 (1941).

TABLE 3.8
Nitrogen-oxygen bond distances

Compound	Formal bond order	Bond distance, Å
OṄ—OH, $H_2\ddot{N}$—OH	1	1.46
O_2N—OH	1	1.41
$(CH_3)_3\overset{+}{N}$—O^-	1	~1.4
NO_3^- { N_2O_5	1.33	1.24
NO_3^- { $NaNO_3$	1.33	1.21
NO_2^-	1.5	1.24
$ClNO_2$	1.5	1.24
CH_3NO_2	1.5	1.22
$HONO_2$ (nitro group)	1.5	1.22
$N\equiv\overset{+}{N}—\overset{-}{O} \leftrightarrow \overset{-}{N}=\overset{+}{N}=O$	~1.5	1.19
NO_2	1.75	1.19
$ClN\equiv O$	2	1.14
NO_2^+	2	1.15
NO	2.5	1.15
NO^+	3	1.06

fluorine atom. Others have proposed that the shortening of such bonds is a result of $p\pi \rightarrow d\pi$ bonding,[20] which is not feasible in the case of bonds involving only first-row atoms, such as the C—F bond. Another possibility is that the Si—F bond distance is normal but that the C—F bond distance is extraordinarily long because of repulsion between the lone-pair electrons of the fluorine and the electrons of the other three bonds to the carbon atom. (We shall later say more about the repulsive interactions of nonbonding and bonding electrons.) Perhaps the apparent shortening of the Si—F bond is due to a combination of the effects that we have discussed. In any event, the phenomenon serves as an indication of the errors which one can make by incautious use of covalent atomic radii.

The covalent radius for hydrogen, which is consistent with the other covalent radii and the various X—H bond distances, is much less than one-half the H—H bond distance. Probably the H—H bond is abnormally long (and weak) because the $1s$ orbitals of the atoms cannot overlap adequately without undue proton-proton repulsion. On the other hand, when a hydrogen atom forms a bond to an atom which offers a highly directional bonding orbital (e.g., an orbital with appreciable p character), the hydrogen atom can *immerse itself* in the other atom's bonding orbital, and thus very strong orbital overlap can be achieved.

The covalent single-bond radii for fluorine, oxygen, and nitrogen in Table 3.6, which were calculated by subtracting the covalent radius of carbon from the C—F, C—O, and C—N bond distances, respectively, are considerably less than one-half the F—F, O—O, and N—N single-bond distances, respectively. It is generally supposed that the latter distances are abnormally large because of strong repulsive interaction between the lone-pair electrons on adjacent atoms. Because the fluorine, oxygen, and nitrogen atoms are relatively small, lone pairs on adjacent atoms are close together and repel each other more strongly than they do in bonds such as Cl—Cl and S—S. The N—N bond distance in hydrazine ($H_2\overset{..}{N}$—$\overset{..}{N}H_2$) is reduced from 1.47 to 1.40 Å by protonation of the two lone pairs and formation of the hydrazinium ion, $H_3\overset{+}{N}$—$\overset{+}{N}H_3$. The same sort of effect is observed on going from hydroxylamine, $H_2\overset{..}{N}$—$\overset{..}{O}H$ ($r_{N—O} = 1.46$ Å), to trimethylamine oxide, $(CH_3)_3\overset{+}{N}$—O^- ($r_{N—O} = 1.36$ Å).

It is rather interesting that the O—O bond distance in O_2F_2 is only 1.22 Å, almost as short as that in O_2 (1.21 Å).[21] This short bond has been explained in terms of hyperconjugation, which shifts nonbonding electrons from the oxygen atoms to the fluorine atoms.

$$F—O—O—F \quad \leftrightarrow \quad F^-\; O{=}O^+{—}F \quad \leftrightarrow \quad F—O^+{=}O \;\; F^-$$

As would be expected on the basis of such resonance, the O—F bond length is exceptionally long, 1.58 Å.

---

[20]   T. B. Brill, *J. Chem. Educ.*, **50**, 392 (1973).

[21]   R. H. Jackson, *J. Chem. Soc.*, 4285 (1962); J. K. Burdett, D. J. Gardiner, J. J. Turner, R. D. Spratley, and P. Tchir, *J. Chem. Soc. Dalton*, 1928 (1973).

It should come as no surprise that bonds of order less than 1 are longer than single bonds. Thus in I_3^-, which has two bonds of order 0.5, the I—I bond distance is 2.91 Å, or 0.24 Å longer than that in I_2. The molecule PF_5 has a trigonal bipyramidal structure, with two axial fluorines, each 1.58 Å from the phosphorus, and three equatorial fluorines, each 1.53 Å from the phosphorus. These bond distances are slightly longer than the P—F bond distance in PF_3, 1.52 Å, and reflect the fact that the average bond order in PF_5 is 0.8, if we ignore the participation of phosphorus $3d$ orbitals in the bonding.

BOND STRENGTH

Dissociation Energies

The most straightforward measure of the strength of the bond in a diatomic molecule is the "dissociation energy," i.e., the energy required to break the gaseous molecule into its constituent atoms. A listing of dissociation energies for some diatomic molecules is given in Table 3.9.[22] From these data it can be seen that, within families of molecules such as HF, HCl, HBr, HI or C_2, Si_2, Pb_2, the heavier analogs generally have lower dissociation energies. Comparison of the data for N_2, O_2, and F_2 and for P_2, S_2, and Cl_2 shows that, for similar molecules, the dissociation energy increases with increasing bond order. The data as a whole indicate that, for related molecules, there is an inverse relationship between dissociation energy and bond length.

In the case of a polyatomic molecule, the energy required to break just one bond, with formation of two molecular fragments, is called a "bond dissociation energy." A brief selection of bond dissociation energies is given in Tables 3.10 and 3.11.[22] Bond dissociation energies are valuable data for thermochemical calculations, but their interpretation as a measure of bond strength is somewhat hazardous because of the structural and electronic rearrangements which occur in the molecular fragments that form when a bond in a molecule is broken. Unfortunately, the sum of the separate dissociation energies of the bonds of a molecule is not equal to the energy of dissociating the molecule completely into atoms. For example, consider the molecule OCS:

$$OCS(g) \rightarrow CO(g) + S(g) \qquad D(OC{=}S) = 74 \, \text{kcal mol}^{-1}$$
$$OCS(g) \rightarrow O(g) + CS(g) \qquad D(O{=}CS) = 149 \, \text{kcal mol}^{-1}$$
$$OCS(g) \rightarrow O(g) + C(g) + S(g) \qquad \underset{\text{atom}}{\circ} \qquad {}^{-1}$$

---

[22] The dissociation energies, bond energies, and other thermodynamic data given in this chapter are taken from S. W. Benson, "Thermochemical Kinetics," Wiley-Interscience, New York, 1976; Rosenstock et al., *J. Phys. Chem. Ref. Data*, **6**, suppl. 1 (1977); D. A. Johnson, "Some Thermodynamic Aspects of Inorganic Chemistry," Cambridge University Press, London, 1968; S. R. Gunn, *Inorg. Chem.*, **11**, 796 (1972); L. Pauling, "The Nature of the Chemical Bond," 3d ed., Cornell University Press, Ithaca, N.Y., 1960; L. Brewer, Lawrence Berkeley Laboratory Report LBL 3720 Rev., May 4, 1977.

TABLE 3.9
Dissociation energies of some gaseous diatomic molecules at 298 K

Molecule	D, kcal mol^{-1}†	Molecule	D, kcal mol^{-1}†
H_2	104.2	LiCl	114
D_2	106.0	BeCl	93
		BCl	128
HF	135.8	CCl	80
HCl	103.2	OCl	64.3
HBr	87.5	NaCl	98
HI	71.3	MgCl	74
LiH	58	AlCl	118.1
BH	79	KCl	101.5
CH	81	CuCl	78
NH	75	AgCl	75
OH	102.3	HgCl	25
NaH	48	O_2	119.1
AlH	68	S_2	102.6
PH	73	SO	124.7
SH	85	Se_2	73.6
KH	44	Te_2	53.8
HgH	10	BeO	107
F_2	37.8	BO	182
Cl_2	58.2	CO	256.9
Br_2	46.1	NO	151.0
I_2	36.1	MgO	91
ClF	61.0	AlO	116
BrF	59.4	SiO	190
IF	66.2	PO	141
BrCl	51.5	BrO	56.3
ICl	49.6	C_2	142.4
IBr	41.9	Si_2	81
LiF	137	Pb_2	13
BeF	147	N_2	226.0
BF	180	P_2	117
CF	129	Sb_2	70
NF	72	PN	168
OF	52	Li_2	26
NaF	115	Na_2	18
MgF	107.3	K_2	13
AlF	159	Cu_2	46
KF	118.1	Ag_2	37
		CN	183
		CS	183
		B_2	71

†1 kcal = 4.1840 kJ

TABLE 3.10
Some bond dissociation energies of gaseous molecules and radicals

Molecule	D, kcal mol^{-1}†	Molecule	D, kcal mol^{-1}†
HO—H	119	CH_2=NH	154
CH_3COO—H	112	H_2N—NH_2	58
H_3CO—H	102	HO—OH	51
HOO—H	90	NH_2—Cl	60
C_6H_5O—H	85	HO—Cl	60
ClO—H	78	HO—I	56
NC—H	130	OO—H	47
Cl_3C—H	96	CH_2O—H	31
HC≡CH	230	COO—H	12
HC≡N	224	CH_2—H	106
CH_2=CH_2	163	CH—H	106
CH_2=O	175	CH_2CH_2—H	39
HN=O	115		

† 1 kcal = 4.1840 kJ

The sum of $D(OC$=$S)$ and $D(O$=$CS)$ is much less than the atomization energy of OCS. However, each of the sums $D(OC$=$S) + D(C$≡$O)$ and $D(O$=$CS) + D(C$≡$S)$ must equal the atomization energy of OCS.

At least a few remarks should be made regarding the methods used to obtain dissociation energies. In most cases, the dissociation energies were calculated from heats of formation such as those given in Appendix E and Table 3.12. Thus from the heats of formation of HCl(g), H(g), and Cl(g), one can readily calculate $D(H$—$Cl)$. The heat of formation of HCl(g) was obtained by directly measuring the heat of reaction of H_2 and Cl_2 in a calorimeter; the heats of formation of H(g) and Cl(g) are calculated from the dissociation energies of H_2 and Cl_2, which in turn were obtained from electronic absorption spectral data. The heats of formation of molecules such as NaCl(g), which at room temperature are unstable with respect to crystalline phases, obviously cannot be directly measured in a calorimeter. One method for obtaining the heat of formation of such gaseous molecules is by measuring the vapor pressure of the diatomic molecule over the solid as a function of temperature. From the derived heat of vaporization and the known heat of formation of the solid (obtained calorimetrically), one readily calculates the heat of formation of the gaseous molecule. These and other techniques for determining dissociation energies are discussed by Cottrell.[23]

Bond Energies

Because of the nonadditivity of bond dissociation energies of molecules, chemists commonly measure bond strengths in terms of *bond energies* which, for poly-

[23] T. L. Cottrell, "The Strengths of Chemical Bonds," 2d ed., Butterworth, London, 1958.

TABLE 3.11
Bond-dissociation energies for some gaseous organic molecules R—X (in kilocalories per mole)†

	(52.1) H	(19.8) F	(28.9) Cl	(26.7) Br	(25.5) I	(9.5) OH	(46 ± 1) NH_2	(4 ± 1) OCH_3	(34 ± 1) CH_3	(26 ± 1) C_2H_5	(18 ± 1) $i\text{-}C_3H_7$	(8.0 ± 1) $t\text{-}Bu$	(78.5 ± 1) C_6H_5
(34 ± 1)‡ CH_3	104	109	83.5	70	56	91.5	85	82	88	85	84	81	100
(26 ± 1) C_2H_5	98	108	81.5	68	53.5	91.5	84	82	85	82	80	78	97
(21 ± 1) $n\text{-}C_3H_7$	98	108	81.5	68	53.5	91.5	84	82	85	82	80	78	97
(18 ± 1) $i\text{-}C_3H_7$	94.5	105	81	68	53	92	84	82	84	80	77.5	74	95
(8.0 ± 1) $t\text{-}Bu$	92	.....	80	64	51	92	84	81	81	78	74	70	92
(78.5 ± 1) C_6H_5	110.5	125	.....	78	64	112	105	100	100	97	95	92	116
(45 ± 1) $C_6H_5CH_2$	85	.....	.....	.....	40	77	.....	70	72	69	67.5	65	77
(40 ± 1) Allyl	87	.....	.....	.....	43.5	80	.....	(68)§	74.5	71.5	69.5	66	78
(−5) CH_3CO	87	118	82.5	.....	51.6	108	101	97	81	78	80	.....	95
(−4.5 ± 1) CH_3CH_2O	104	.....	.....	.....	.....	43	.....	38	81	81	82	82	99
(−1.5) $CH_3CH_2O_2$	90	.....	.....	.....	.....	(28)	.....	21	72	72	(71)	(69)	91
(67.5 ± 2) $CH_2{=}CH$	108	.....	91	.....	60	95	.....	95	99	96	94	92	112

† Data taken with permission from S. W. Benson, "Thermochemical Kinetics," Wiley-Interscience, New York, 1976. 1 kcal = 4.1840 kJ
‡ Values in parentheses near radicals and atoms are $\Delta H^\circ_{f\,300}$.
§ Values in parentheses are estimates.

TABLE 3.12
Heats of formation of the gaseous atoms at 25°C (in kilocalories per mole)†

Aluminum	78.7	Hydrogen	52.1	Radon	0
Americium	63	Indium	58.0	Rhenium	185.4
Antimony	63.5	Iodine	25.52	Rhodium	133
Argon	0	Iridium	160.3	Rubidium	19.33
Arsenic	68.4	Iron	99.3	Ruthenium	155.8
Barium	43.5	Krypton	0	Samarium	49.4
Beryllium	77.5	Lanthanum	103.1	Scandium	90.3
Bismuth	50.1	Lead	46.6	Selenium	52.0
Boron	135	Lithium	38.1	Silicon	107.7
Bromine	26.7	Lutecium	102.2	Silver	68.1
Cadmium	26.7	Magnesium	35.0	Sodium	25.6
Calcium	42.6	Manganese	67.7	Strontium	39.5
Carbon	171.29	Mercury	14.5	Sulfur	66.29
Cerium	99.5	Molybdenum	157.6	Tantalum	187
Cesium	18.18	Neodymium	78.3	Technetium	158
Chlorine	29.1	Neon	0	Tellurium	51.4
Chromium	95	Neptunium	109	Terbium	93
Cobalt	101.7	Nickel	102.9	Thallium	43.2
Copper	80.7	Niobium	175.2	Thorium	142.8
Dysprosium	68.5	Nitrogen	113.0	Thulium	55.5
Erbium	75.5	Osmium	188.6	Tin	72.5
Europium	42.4	Oxygen	59.2	Titanium	112.5
Fluorine	18.9	Palladium	90.0	Tungsten	205.5
Gadolinium	95.2	Phosphorus	79.8	Uranium	128
Gallium	65.0	Platinum	134.9	Vanadium	123
Germanium	89.5	Plutonium	83.1	Xenon	0
Gold	88.0	Polonium	34.5	Ytterbium	37.0
Hafnium	148.5	Potassium	21.33	Yttrium	101.0
Helium	0	Praeseodynium	85.0	Zinc	31.2
Holmium	71.9	Protactinium	146	Zirconium	144.5

† 1 kcal = 4.1840 kJ

atomic molecules, are hypothetical because they are calculated on the *assumption* that the sum of the bond energies of a molecule equals the atomization energy of the molecule. The usefulness of this concept lies in the fact that the calculated bond energy for a particular type of bond, say an S—H bond, is found to be approximately constant in different molecules containing that type of bond. We may demonstrate the approximate validity of the method by a simple calculation. Consider the atomization energy of cyclooctasulfur:

$$S_8(g) \quad \rightarrow \quad 8S(g) \qquad \Delta H^\circ_{atom}(S_8) = 508.4 \text{ kcal mol}^{-1}$$

We assume that the atomization energy is eight times the S—S bond energy, $E(S—S)$; hence we calculate $E(S—S) = 508.4/8 = 63.6$ kcal mol^{-1}. Similarly, for the atomization of hydrogen sulfide,

$$H_2S(g) \quad \rightarrow \quad 2H(g) + S(g) \qquad \Delta H^\circ_{atom}(H_2S) = 175.7 \text{ kcal mol}^{-1}$$

We assume that the atomization energy is two times the S—H bond energy, $E(S—H)$. Thus we calculate $E(S—H) = 175.7/2 = 87.8$ kcal mol^{-1}. Using these bond energies, we estimate $63.6 + 2(87.8) = 239.3$ kcal mol^{-1} for the atomization of disulfane,

$$H_2S_2(g) \quad \rightarrow \quad 2H(g) + 2S(g)$$

The actual atomization energy for this molecule is 235 kcal mol^{-1}.

A wide variety of bond energies are listed in Table 3.13.[22] These approximate bond energies, in combination with the relatively accurately known heats of formation of the gaseous atoms in Table 3.12,[22] may be used to estimate the heat of formation of any nonresonant molecule containing these bonds.

From the data in Table 3.13 and some of the data in Table 3.9, it can be shown that the bond energies for heteronuclear bonds are considerably greater than the average of those for the corresponding homonuclear bonds. For example, the As—Cl bond energy (74 kcal mol^{-1}) is much greater than either as As—As bond energy (40 kcal mol^{-1}) or the Cl—Cl bond energy (58.2 kcal mol^{-1}). Pauling[3] ascribed this enhancement of a heteronuclear bond energy to the ionic character of the bond and showed that the bond energy in kilocalories per mole can be represented by the equation

$$E(A\text{—}B) = \tfrac{1}{2}[E(A\text{—}A) + E(B\text{—}B)] + 23(x_A - x_B)^2 \qquad (3.1)$$

TABLE 3.13
Thermochemical (average) bond energies (in kilocalories per mole)†

As—As	40		Ge—Ge	38		S—S	64
As—F	116		Ge—F	113		S—Cl	65
As—Cl	74		Ge—Cl	81		S—Br	51
As—Br	61		Ge—Br	67		S—H	88
As—H	71		Ge—I	51		Sb—Sb	34
As—O	79		Ge—H	69		Sb—Cl	75
B—B	79		Ge—O	86		Sb—Br	63
B—F	154		Ge—N	61		Sb—H	61
B—Cl	106		N—N	38		Se—Se	38
B—Br	88		N=N	100		Se—Cl	58
B—I	65		N≡N	226		Se—H	73
B—H	91		N—F	67		Si—Si	46
B—O	125		N—Cl	45		Si—Ge	42
C—C	83		N—H	93		Si—F	143
C=C	147		N—O	39		Si—Cl	96
C≡C	194		N=O	142		Si—Br	79
C—F	117		O—O	34		Si—H	76
C—Cl	78		O=O	119		Si—O	111
C—Br	65		O—F	51		Si—N	80
C—I	57		O—Cl	49		Si—S	54
C—H	99		O—H	111		Si—C	73
C—O	86		P—P	47		Sn—Sn	36
C=O	174		P—F	119		Sn—Cl	75
C—N	73		P—Cl	79		Sn—Br	64
C=N	147		P—Br	64		Sn—H	60
C≡N	213		P—H	77		Sn—C	50
C—S	114		P—O	88			

† 1 kcal = 4.1840 kJ

TABLE 3.14
The original Pauling scale of electronegativities showing periodic table trends†

H																
2.1																

Li	Be	B										C	N	O	F
1.0	1.5	2.0										2.5	3.0	3.5	4.0

Na	Mg	Al										Si	P	S	Cl
0.9	1.2	1.5										1.8	2.1	2.5	3.0

K	Ca	Sc	Ti	V	Cr	Mn	Fe	Co	Ni	Cu	Zn	Ga	Ge	As	Se	Br
0.8	1.0	1.3	1.5	1.6	1.6	1.5	1.8	1.8	1.8	1.9	1.6	1.6	1.8	2.0	2.4	2.8

Rb	Sr	Y	Zr	Nb	Mo	Tc	Ru	Rh	Pd	Ag	Cd	In	Sn	Sb	Te	I
0.8	1.0	1.2	1.4	1.6	1.8	1.9	2.2	2.2	2.2	1.9	1.7	1.7	1.8	1.9	2.1	2.5

Cs	Ba	La–Lu	Hf	Ta	W	Re	Os	Ir	Pt	Au	Hg	Tl	Pb	Bi	Po	At
0.7	0.9	1.1–1.2	1.3	1.5	1.7	1.9	2.2	2.2	2.2	2.4	1.9	1.8	1.8	1.9	2.0	2.2

Fr	Ra	Ac	Th	Pa	U	Np–No
0.7	0.9	1.1	1.3	1.5	1.7	1.3

† For revised Pauling values and values calculated by other methods, see Table 3.18.

where x_A and x_B refer to the empirically evaluated electronegativity values listed in Table 3.14. It is obvious from Eq. 3.1 that the bond energy for a bond between two atoms of equal electronegativity would be expected to be the arithmetic average of the homonuclear bond energies.[24] It should be emphasized that Eq. 3.1 is a very approximate relation and should be used only when a very crude estimate will suffice. Nevertheless, it serves as the basis for a very useful generalization: The most stable arrangement of covalent bonds connecting a group of atoms is that arrangement in which the atom with the highest electronegativity is bonded to the atom with the lowest electronegativity. There are many examples of spontaneous reactions, such as the following, which bear out this rule.

$$BBr_3 + PCl_3 \quad \rightarrow \quad BCl_3 + PBr_3$$

$$2HI + Cl_2 \quad \rightarrow \quad 2HCl + I_2$$

$$TiCl_4 + 4EtOH \quad \rightarrow \quad Ti(OEt)_4 + 4HCl$$

Force Constants

Another useful measure of the strength of a bond is the bond-stretching force constant. This is the constant of proportionality between the force tending to restore the equilibrium bond distance and an infinitesimal displacement from that

---

[24] Pauling[3] has pointed out that it is more accurate to use the geometric mean, $[E(A-A) \cdot E(B-B)]^{1/2}$. However, the arithmetic mean is easier to apply and in most cases is of similar magnitude.

distance. For small displacements, most bonds act as harmonic oscillators, and we can write the following relationship between the vibrational frequency ν and the force constant k of a bond A—B:

$$\nu = \frac{1}{2\pi c}\sqrt{\frac{k}{\mu}}$$

Here μ is the reduced mass of A and B, $\mu = m_A m_B/(m_A + m_B)$. If we express ν in wave numbers (reciprocal centimeters), k in millidynes per angstrom, and μ in the usual atomic-weight units, we obtain the relation $\nu^2 = (1.70 \times 10^6)k/\mu$. A diatomic molecule has only one vibrational frequency, which obviously corresponds to the bond stretching. However, polyatomic molecules have a number of vibrational frequencies, and these are not necessarily directly related to the stretching of individual bonds. The calculation of bond force constants from the vibrational frequencies of a polyatomic molecule is a relatively complicated procedure, involving assumptions regarding the dependence of the energy of the molecule on the atomic coordinates. The calculated force constants are strongly dependent on

TABLE 3.15
Some bond-stretching force constants†

Bond	Force constant, mdyn Å^{-1}	Bond	Force constant, mdyn Å^{-1}
N_2	22.6	F_2	4.5
CO	18.7	Cl_2	3.2
$HC\equiv CH$	15.7	Br_2	2.4
P_2	5.5	I_2	1.7
As_2	4.0	$HgCl_2$	2.11
O_2	11.4	$HgCl_3^-$	1.27
$H_2C\equiv CH_2$	9.6	$HgCl_4^{2-}$	0.82
$OC\equiv S$	7.5	$GeCl_4$	3.27
H_2	5.1	$GaCl_4^-$	2.50
CH_4	5.0	ZnI_4^{2-}	1.64
NH_3	6.4	$MnCl_4^{2-}$	1.39
PH_3	3.2	$MnBr_4^{2-}$	1.16
KH	0.53	MnI_4^{2-}	0.92
LiI	0.77	$TiCl_6^{2-}$	1.13
PCl_3	2.1	$ZrCl_6^{2-}$	0.88
PBr_3	1.6	$HfCl_6^{2-}$	0.86
Na_2	1.7	$PtCl_6^{2-}$	1.86
NO	15.5		
HF	8.8		
HCl	4.8		
HBr	3.8		
HI	2.9		

† Data from T. L. Cottrell, "The Strengths of Chemical Bonds," 2d ed., Butterworth, London, 1958; and A. Finch, P. N. Gates, K. Radcliffe, F. N. Dickson, and F. F. Bentley, "Chemical Applications of Far Infrared Spectroscopy," Academic, London, 1970.

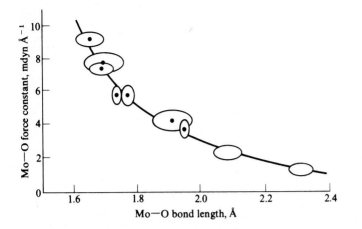

FIGURE 3.1
Plot of Mo—O force constants versus the corresponding bond lengths. [*Reproduced with permission from F. A. Cotton and R. M. Wing, Inorg. Chem.,* **4**, *867 (1965). Copyright 1965 American Chemical Society.*]

the nature of these assumptions. Some bond force constants for both diatomic and polyatomic molecules, calculated using very simple assumptions,[25] are given in Table 3.15.

For a series of related bonds, the force constant increases with increasing bond energy, with increasing bond order, and with decreasing bond length. The relationship between force constant and bond length for Mo—O bonds is shown in Fig. 3.1.

BOND POLARITY

The electrons of a heteronuclear bond are generally not shared equally by the two bonded atoms. The bonding electron density is almost always greater near one atom (the more electronegative atom) than the other. This polarization is depicted graphically for the gaseous LiH molecule in Fig. 3.2. This figure is a plot of the electron density of LiH minus the electron density of the hypothetical pair of neutral, noninteracting atoms, for a plane passing through the two nuclei.[26] The plot shows that bond formation causes electron density to leave the outer regions of the lithium atom, including a region in the direction of the hydrogen atom, and to enter the region around the hydrogen atom (especially in the bonding region)

---

[25] Force constants for the diatomics were calculated using the harmonic oscillator (Hooke's law) approximation. For polyatomics, the "valence force field" was assumed; that is, a force constant was associated with every valence bond, representing its resistance to stretching, and to every valence angle, representing its resistance to bending.

[26] A. Streitwieser, Jr., and P. H. Owens, "Orbital and Electron Density Diagrams," Macmillan, New York, 1973.

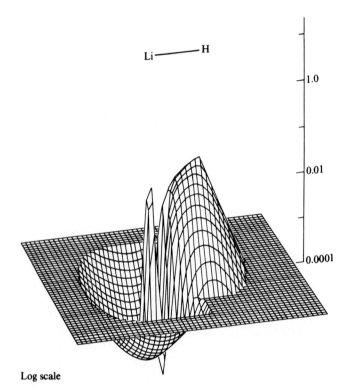

Log scale

FIGURE 3.2
Plot of electron density in LiH less electron density of neutral Li + H. *(Reproduced with permission from A. Streitwieser, Jr., and P. H. Owens, "Orbital and Electron Density Diagrams," Macmillan, New York, 1973.)*

and the region near the lithium nucleus. This polarization is qualitatively what one would expect for a bond between the electropositive lithium atom and the relatively electronegative hydrogen atom. The electron flow is not complete, in the sense that the compound does not correspond to Li^+ and H^-, but the covalent bond between the atoms does have a lot of ionic character. The effective nuclear charge of the lithium atom is increased because the reduced electron density gives poorer shielding; hence some electron density crowds around the nucleus, and the atom is effectively smaller. Conversely, the hydrogen atom has greater electron density, and is effectively larger, than a neutral H atom.

Dipole Moments

One way of quantitatively expressing the ionic character of a bond is by assigning fractional charges to the atoms. The dipole moment μ of LiH is 5.88×10^{-18} esu cm = 5.88 Debye units; if we assume that this moment is due to two point

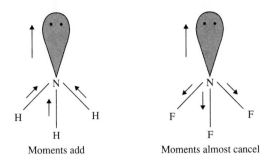

FIGURE 3.3
Explanation of the low dipole moment of NF_3. Arrows indicate the direction from + to −.

charges, $+q$ and $-q$, separated a distance r equal to the Li—H bond distance (1.595Å), it is an easy matter to calculate q from the relation $\mu = qer$.

$$q = \frac{\mu}{er} = \frac{5.88 \times 10^{-18}}{(4.8 \times 10^{-10})(1.595 \times 10^{-8})} = 0.768$$

Thus the dipole moment may be accounted for by assigning a charge of $+0.768$ to the lithium atom and a charge of -0.768 to the hydrogen atom.

   The treatment of LiH as a pair of point charges, or even as a pair of charged spherically symmetric atoms, is an approximation. From Fig. 3.2 it can be seen that the charge distribution of the molecule would be better represented by a pair of charged atoms, each with a dipole moment. The atomic moments would be relatively small, however, and the representation of the overall dipole moment with simple atomic charges is fairly reasonable in the case of LiH. However, in the case of a molecule with nonbonding valence electrons, such as HCl or NH_3, such a representation is extremely misleading because of the relatively large moments due to the nonbonding electrons. It is instructive to compare the dipole moments of the pyramidal molecules $NF_3(\mu = 0.23$ Debye unit) and NH_3 ($\mu = 1.47$ Debye units). Although the bond angle of NF_3 is more acute than that of NH_3 and although the N—F bond would be expected to be at least as polar as the N—H bond (but with opposite sign), the dipole moment of NF_3 is much smaller than that of NH_3. The apparent anomaly is caused by the large moment of the lone pair of electrons on the nitrogen atom. In NH_3, the lone-pair–nitrogen moment adds to the resultant of the N—H moments. Thus NH_3 has a large dipole moment. However, in NF_3, the lone-pair–nitrogen moment partly cancels the resultant of the N—F moments. Thus NF_3 has a small dipole moment. Reference to Fig. 3.3 will make this clear.

   Simple electrostatic molecular models which provide both for the polarity of bonds and for the existence of lone-pair electrons have been devised. Benson[27] has described a molecular model in which atoms are point charges and in which lone pairs are point dipole moments. Halogen atom lone pairs are represented by

27   S. W. Benson, *Angew. Chem. Int. Ed.*, **17**, 812 (1978).

single point dipoles centered at the nuclei and oriented along the bond axes. Thus a methyl halide is represented by this type of model as follows:

Benson has shown that the heats of formation and dipole moments of molecules can be used to calculate consistent sets of atomic charges and lone-pair moments. In the CH_nF_{4-n} series a lone-pair moment for fluorine of about 0.7 Debye unit and a charge separation across each C—F bond of about 0.15 comes close to reproducing the thermochemical and dipole data. Scheraga et al.[28] have used a different model, in which molecules are made up of atomic cores (with point integral charges) and both bonding and "lone" electron pairs (-2 point charges). Thus the water molecule is represented by this type of model as follows:

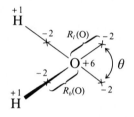

By assuming that $R_l(O)$ is a constant for all 2-coordinate oxygen compounds and by making certain approximations about the transferability of $R_b(O)$ to other compounds, it has been shown that the dipole moment data for a variety of molecules are consistent with the following parameters: $R_b(O) = 0.570$ Å, $R_l(O) = 0.297$ Å, $\theta = 120°$. These methods of accounting for lone-pair electron density are very promising, and perhaps one of them, or a related method, will find general applicability to both organic and inorganic molecules.

Nuclear Magnetic Resonance

Other physical properties of molecules are related to molecular charge distribution. Chemical shifts and spin-spin coupling constants from nuclear magnetic resonance (nmr) spectra[29] are sensitive to electron density at the nucleus. Thus in Fig. 3.4

---

[28] L. L. Shipman, A. W. Burgess, and H. A. Scheraga, *Proc. Nat. Acad. Sci. USA*, **72**, 543 (1975); A. W. Burgess, L. L. Shipman, and H. A. Scheraga, ibid., **72**, 854 (1975); A. W. Burgess, L. L. Shipman, R. A. Nemenoff, and H. A. Scheraga, *J. Am. Chem. Soc.*, **98**, 23 (1976)

[29] Nuclear magnetic resonance spectroscopy, nuclear quadrupole resonance spectroscopy, and Mössbauer spectroscopy are discussed in W. L. Jolly, "The Synthesis and Characterization of Inorganic Compounds," Prentice-Hall, Englewood Cliffs, N.J., 1970; R. S. Drago, "Physical Methods in Chemistry," Saunders, Philadelphia, 1977; and I. S. Butler and J. F. Harrod, "Inorganic Chemistry," Benjamin/Cummings, Redwood City, Calif., 1989.

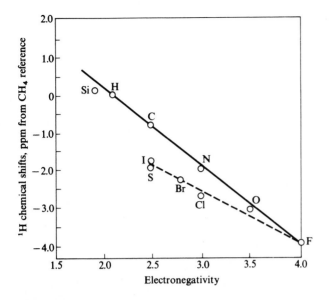

FIGURE 3.4
Plot of pmr chemical shifts for methane derivatives versus the Pauling electronegativity of the substituent atoms. [*Reproduced with permission from H. Spieseke and W. G. Schneider, J. Chem. Phys., 35, 722 (1961).*]

is shown a correlation of proton chemical shift in methyl compounds with the electronegativity of the atom bonded to the methyl group. One can rationalize the data by arguing that the more electronegative atoms pull electrons away from the protons, thus deshielding them and allowing the protons to resonate at a lower applied magnetic field. Unfortunately, however, nmr parameters are affected by other features besides electron density, and these effects are not understood well enough to permit the use of nmr parameters as reliable measures of atomic charge.

Nuclear Quadrupole Resonance

The nuclear quadrupole (nqr) coupling constant[29] of an atom is a measure of the asymmetry of the electric field at the nucleus. A free chloride ion, for example, should exhibit no quadrupole coupling. The greater the electronegativity of the atom bonded to a chlorine atom, the more asymmetric is the electric field at the chlorine nucleus, and the greater is the absolute magnitude of the nqr coupling constant. The trend of the chlorine-35 coupling constant with the electronegativity of the substituent is shown by the data in Table 3.16. Unfortunately the interpretation of these data in terms of chlorine's atomic charge is complicated by the fact that the electric field gradient is a function of the hybridization of the bonding orbital as well as the atomic charge. Because of the difficulty of estimating the *s* character of a chlorine bonding orbital, data such as those in Table 3.16 are restricted to qualitative interpretation.

TABLE 3.16
Chlorine-35 nuclear quadrupole coupling constants†

Compound	Coupling constant, MHz
Free Cl^-	0.0
LiCl	−6.1
$(CH_3)_3CCl$	−62.1
$(CH_3)_2CHCl$	−64.1
C_2H_5Cl	−65.8
CH_3Cl	−68.1
CH_2Cl_2	−72.0
$CHCl_3$	−76.7
CCl_4	−81.2
ICl	−82.5
BrCl	−103.6
Cl_2	−109.0
ClF	−146.0

† J. E. Huheey, "Inorganic Chemistry," 2d ed.,
p. 178, Harper & Row, New York, 1978.

Mössbauer Spectroscopy

In Mössbauer spectroscopy[29] one can measure shifts in the energy of a nuclear transition caused by changes in the electronic environment of the nucleus. These "isomer shifts" are caused by changes in s electron density at the nucleus and are essentially independent of the population of other atomic orbitals. Thus isomer shifts can be meaningfully interpreted in terms of changes in atomic charge only for compounds having similar structures and orbital hybridizations. For example, shifts for the tin nucleus in a series of hexahalostannate salts, listed in Table 3.17, can readily be correlated with the electron-withdrawing abilities of the coordinated halide ions.

X-ray Photoelectron Spectroscopy

X-ray photoelectron spectroscopy (often referred to as ESCA)[30] is a useful experimental technique for estimating relative magnitudes of atomic charges. We have already alluded to this technique in Chap. 1 in connection with the determination of the ionization energies (binding energies) of the various kinds of electrons in argon. With respect to the estimation of atomic charges, *core*-electron binding energies are of principal interest. It has been shown that, for a given core level of an element in a set of compounds, the binding energies E_B may be approximately represented by the following equation:

[30] K. Siegbahn et al., "ESCA; Atomic, Molecular and Solid State Structure by Means of Electron Spectroscopy," Almqvist and Wiksells, Uppsala, Sweden. 1967; K. Siegbahn et al., "ESCA Applied to Free Molecules." North-Holland, Amsterdam, 1969; J. M. Hollander and W. L. Jolly, *Acc. Chem. Res.*, **3**, 193 (170).

TABLE 3.17
Mössbauer isomer shifts for some hexahalostannates†

Compound	Isomer shift, mm s^{-1}‡
K_2SnF_6	-0.36
$(Et_4N)_2SnCl_4F_2$	$+0.29$
$(Et_4N)_2SnCl_6$	$+0.52$
$(Et_4N)_2SnBr_4F_2$	$+0.53$
$(Et_4N)_2SnCl_4Br_2$	$+0.67$
$(Et_4N)_2SnBr_4Cl_2$	$+0.77$
$(Et_4N)_2SnCl_4I_2$	$+0.78$
$(Et_4N)_2SnBr_6$	$+0.84$
$(Et_4N)_2SnBr_4I_2$	$+0.96$
$(Et_4N)_2SnI_4Br_2$	$+1.09$
$(Et_4N)_2SnI_6$	$+1.23$

† C. A. Clasen and M. L. Good, *Inorg. Chem.*, **9**, 817 (1970).

‡ Shifts are customarily expressed as the experimentally measured relative velocities of the sample and a gamma-ray source
required to give resonance absorption.

$$E_B = kQ + l \qquad\qquad (3.2)$$

In this expression, Q is the charge of the atom which has the core binding energy E_B, and k and l are constants characteristic of the element. Inasmuch as the constant k is always positive, the first term in this expression shows that the core binding energy increases with increasing atomic charge. This relation is qualitatively reasonable; one would expect the difficulty of removing an electron from an atom to increase with the positive charge of the atom.

The xenon $3d_{5/2}$, xenon $3d_{3/2}$, and fluorine $1s$ spectra for elementary xenon, molecular fluorine, and a series of xenon fluorides are shown in Fig. 3.5. It can be seen that, as expected, the xenon core binding energies increase as the oxidation state of the xenon increases and that the fluorine $1s$ binding energy decreases on going from the free element to the fluorides.

Electronegativities

Pauling defined electronegativity as the power of an atom in a molecule to attract electrons to itself. Clearly this property is closely related to bond polarity and deserves discussion here. As shown on pages 62–63, Pauling calculated electronegativities from thermodynamic data; his calculated values are given in Table 3.14. Allred[31] has recalculated these values by the Pauling method, using more recent thermochemical data; his "revised Pauling" values, for the nontransition elements, are given in Table 3.18. Ever since Pauling's pioneering work in this area, many other methods of evaluating electronegativities have been proposed, and we shall now consider a few of these.

[31] A. L. Allred, *J. Inorg. Nucl. Chem.*, **17**, 215 (1961).

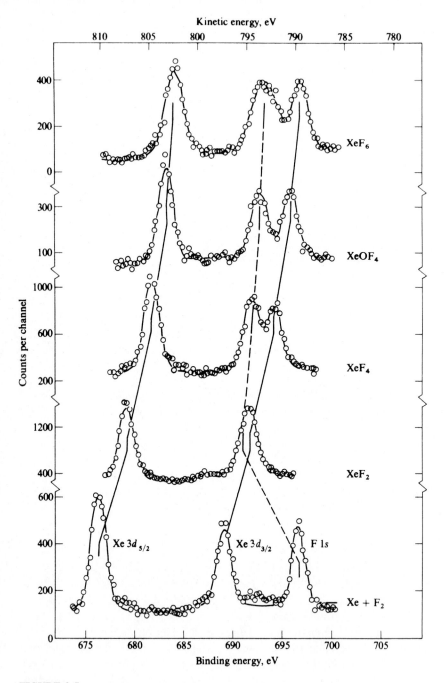

FIGURE 3.5
X-ray photoelectron spectra of Xe, F_2, XeF_2, XeF_4, $XeOF_4$, and XeF_6. [*Reproduced with permission from T. X. Carroll et al., J. Am. Chem. Soc.,*
Society.]

TABLE 3.18
Electronegativities of the nontransition elements

Element	Revised Pauling	Mulliken	Allred-Rochow	New Sanderson	Allen
H	2.2	2.8	2.20	2.59	2.30
Li	0.98	1.3	0.97	0.67	0.91
Be	1.57	.....	1.47	1.81	1.58
B	2.04	1.8	2.01	2.28	2.05
C	2.55	2.5	2.50	2.75	2.54
N	3.04	2.9	3.07	3.19	3.07
O	3.44	3.0	3.50	3.65	3.61
F	3.98	4.1	4.10	4.00	4.19
Na	0.93	1.2	1.01	0.56	0.87
Mg	1.31	.....	1.23	1.32	1.29
Al	1.61	1.4	1.47	1.71	1.61
Si	1.90	2.0	1.74	2.14	1.92
P	2.19	2.3	2.06	2.52	2.25
S	2.58	2.5	2.44	2.96	2.59
Cl	3.16	3.3	2.83	3.48	2.87
K	0.82	1.1	0.91	0.45	0.73
Ca	1.00	.....	1.04	0.95	1.03
Sc	1.36	.....	1.20	.....	.....
Zn	1.65	.....	1.66	2.22	.....
Ga	1.81	1.4	1.82	2.42	1.76
Ge	2.01	1.9	2.02	2.62	1.99
As	2.18	2.2	2.20	2.82	2.21
Se	2.55	2.4	2.48	3.01	2.42
Br	2.96	3.0	2.74	3.22	2.68
Rb	0.82	1.0	0.89	0.31	0.71
Sr	0.95	.....	0.99	0.72	0.96
Y	1.22	1.4	1.11	.....	.....
Cd	1.69	.....	1.46	1.98	.....
In	1.78	1.3	1.49	2.14	1.66
Sn	1.96	1.8	1.72	2.30	1.82
Sb	2.05	2.0	1.82	2.46	1.98
Te	2.1	2.2	2.01	2.62	2.16
I	2.66	2.7	2.21	2.78	2.36
Cs	0.79	1.0	0.86	0.22	.....
Ba	0.89	.....	0.97	0.65	.....
La	1.10	.....	1.08	.....	.....
Hg	2.00	.....	1.44	2.20	.....
Tl	2.04	.....	1.44	2.25	.....
Pb	2.33	.....	1.55	2.29	.....
Bi	2.02	.....	1.67	2.34	.....

It was pointed out by Mulliken[32] that the average of the first ionization energy (IE) and electron affinity (EA) of an atom should be a measure of the electron attraction of the neutral atom and hence of its electronegativity. Indeed, Pearson[33] calls this average the "absolute electronegativity." If one defines Mulliken elec-

[32] R. S. Mulliken, *J. Chem. Phys.*, **2**, 782 (1934); **3**, 573 (1935).

[33] R. G. Pearson, *J. Am. Chem. Soc.*, **107**, 6801 (1985).

tronegativities by the linear relation[34]

$$x_M = 0.187(\text{IE} + \text{EA}) + 0.17$$

(where IE and EA are in electronvolts), the calculated x_M values are similar to the corresponding Pauling electronegativities, as can be seen by comparing these values in Table 3.18.

Allred and Rochow[35] defined electronegativity as the electrostatic force exerted by the nucleus of an atom on the valence electrons. Using effective nuclear charges calculated by Slater's rules (see p. 17), they proposed the equation

$$x_{AR} = 0.359\frac{Z - S}{r^2} + 0.744$$

(where r is the covalent radius in angstroms). This equation yields x_{AR} values that are remarkably similar to the corresponding Pauling values, as can be seen from Table 3.18.

Sanderson[36] reasoned that, because atomic size decreases and electronegativity increases on going from left to right in the periodic table, it should be possible to define electronegativity in terms of the average electron density of an atom. He assumed electronegativity values to be a linear function of the reciprocal of the nonpolar covalent radius cubed within each period of the periodic table. On this basis, using a procedure too complicated to describe here, he calculated the electronegativity values given in the fifth column of Table 3.18. It can be seen that the values are again similar to the Pauling values.

Allen[37] defined "spectroscopic electronegativity" on an energy-per-electron basis as

$$\chi = \frac{m\epsilon_p + n\epsilon_s}{m + n}$$

where m and n are the number of p and s valence electrons, respectively, and ϵ_p and ϵ_s are the corresponding one-electron energies, for the free gaseous atom. These values are given in the last column of Table 3.18. The fact that comparable electronegativities can be calculated from a wide variety of atomic properties is simply due to the fact that, on going from one atom to another, the changes in various properties (ionization potential, effective nuclear charge, size, etc.) are all fundamentally due to the same thing, viz., the change in electronic structure.

Electronegativities have many uses in chemistry. Of course the Pauling values can be used, together with Eq. 3.1, to estimate rough values of bond energies. Sanderson has devised a somewhat more accurate, although more complicated,

---

[34] J. E. Huheey, "Inorganic Chemistry," 2d ed., p. 167, Harper & Row, New York, 1978.

[35] A. L. Allred and E. G. Rochow, *J. Inorg. Nucl. Chem,*, **5**, 264 (1958).

[36] R. T. Sanderson, *J. Am. Chem. Soc.*, **105**, 2259 (1983); "Polar Covalence," Academic Press, New York, 1983.

[37] L. C. Allen, *J. Am. Chem. Soc.*, **111**, 9003 (1989).

procedure for estimating bond energies in a wide variety of compounds, including nonmolecular solids as well as gaseous molecules. The procedure requires
a knowledge of bond lengths and his electronegativity values, and the reader is
referred to his publications for details.

Any set of electronegativities can be used to obtain a qualitative, if not a
quantitative, measure of the relative charges on atoms in molecules. And because
the charge distribution in a molecule affects the interactions of the molecule with
other polar groups and ions, electronegativities give information regarding relative
chemical reactivity. For example, the relative electronegativities of carbon and
silicon indicate that the hydrogen atoms in silicon hydrides should be considerably
more negatively charged than those in hydrocarbons:

$$\overset{\delta+}{\text{Si}}\text{——}\overset{\delta-}{\text{H}} \qquad \overset{\delta-}{\text{C}}\text{——}\overset{\delta+}{\text{H}}$$
$$1.8 \qquad 2.1 \qquad 2.5 \qquad 2.1$$

Thus we can rationalize the fact that silicon hydrides react with sources of protons
to give molecular hydrogen, whereas hydrocarbons do not.

Many quantifiable chemical and physical properties are closely correlated
with charge distribution, and one generally finds that a plot of such properties
versus appropriate electronegativity values is a smooth curve or straight line.
By interpolation or extrapolation in such plots, one can predict the properties of
molecules for which there are no data. Examples of such plots are shown in Figs.
3.4, 3.6, and 3.7.

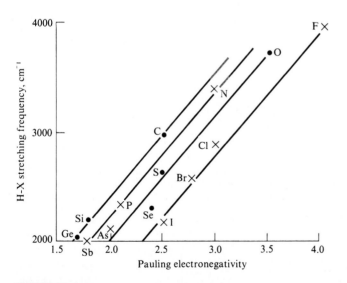

FIGURE 3.6
Plot of typical X—H stretching frequency (from infrared spectra) against the electronegativity of X.
*(Reproduced with permission from L. J. Bellamy, "The Infra-Red Spectra of Complex Molecules,"
2d ed., p. 392, Wiley, New York, 1958.)*

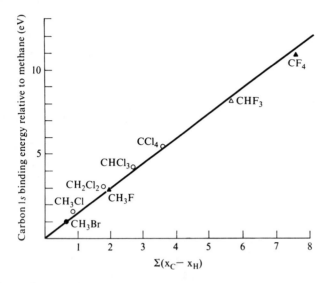

FIGURE 3.7
Plot of carbon $1s$ binding energy (relative to that of methane) for halomethanes against the sum of electronegativity differences over all the ligands. [*Reproduced with permission from T. D. Thomas, J. Am. Chem. Soc.*, **92**, 4184 (1970). Copyright 1970 American Chemical Society.]

PREDICTION OF MOLECULAR TOPOLOGY

The empirical formula of a polyatomic molecule tells nothing about the structure of the molecule. For example, the formula S_2O could correspond to a three-membered ring,

or a chain with either an oxygen atom or a sulfur atom in the middle,

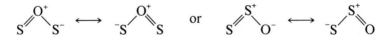

The chain structures could be either linear or bent. From this and other examples we see that a description of molecular structure consists of two stages. The first stage is simply a statement of which atoms are bonded to each other, i.e., a topological description of the bonding. The second stage is the specification of bond angles and bond distances. In this section we shall discuss methods for predicting bonding topologies; methods for predicting structural parameters will be discussed in the next section.

The following procedure may be used to predict the most stable of several possible bonding topologies for a molecule or ion. First, write valence structures for the various possible topologies. Then discard the less stable structures according to the following rules, in the order given, until only one structure remains.

1. Discard structures which do not satisfy the Lewis octet rule. Thus for CH_3N discard the structure in which a central carbon atom is attached to each of the other four atoms in favor of the $HN{=}CH_2$ structure.

2. Discard structures with three-membered rings. Thus the ring structure for S_2O is discarded in favor of the chain structures.

3. Discard structures that require adjacent atoms to have formal charges of the same sign. For example, for S_2N_2 discard the $^{-}N{=}\overset{+}{S}{-}\overset{+}{S}{=}N^{-}$ structure in favor of the following ring structure:

$$^{-}N{-}\overset{+}{S}$$
$$|\quad\parallel$$
$$S{-}N$$

4. Discard structures in which nearby atoms have unlike formal charges contradicting electronegativities, or in which an atom has an unlikely formal charge (according to electronegativities). Thus for N_2O discard the $^{-}N{=}\overset{2+}{O}{=}N^{-}$ structure in favor of the $N{\equiv}N^{+}{-}O^{-}$ structure.

5. Discard structures for which relatively few stable resonance forms can be written. For example, we predict nitramide, $H_2N{-}NO_2$ (for which two equivalent stable resonance structures can be written), to be stable with respect to hyponitrous acid, $HON{=}NOH$ (for which only one stable structure can be written). This example illustrates the utility of the procedure for estimating the relative stabilities of structural isomers.

6. The most stable structure of those remaining is generally that for which the sum of the electronegativity differences of adjacent atoms is the greatest. Discard all the other structures. Thus the structure $H{-}O{-}Cl$ is predicted to be more stable than the structure $H{-}\overset{+}{Cl}{-}O^{-}$.

VALENCE-SHELL ELECTRON REPULSION

The concept of valence-shell electron repulsion is the basis for a simple, yet remarkably accurate, scheme for predicting the shapes of nontransition element molecules. The essentials of this valuable concept were outlined by Sidgwick and Powell in 1940[38]; the method has been refined and popularized by Gillespie and Nyholm.[39] Although some of the assumptions and rules of the method may not be theoretically justifiable, the method nevertheless has practical value for predicting and rationalizing structures.

The spatial arrangement of the bonds to an atom is strongly correlated with a quantity which we shall call the "effective coordination number" of the atom. We

---

[38] N. V. Sidgwick and H. M. Powell, *Proc. Roy. Soc. A*, **176**, 153 (1940).

[39] R. J. Gillespie and R. S. Nyholm, *Q. Rev.*, **11**, 339 (1957); R. J. Gillespie, *J. Chem. Educ.*, **47**, 18 (1970); "Molecular Geometry," Van Nostrand Reinhold, Princeton, N.J., 1972. The discussion of the concept in the following pages includes some ideas quite different from those of previous authors.

define the effective coordination number as the sum of the number of ligands and valence lone pairs. Thus the oxygen atom of H_2O has an effective coordination number of 4 (two ligands and two lone pairs), and the boron atom of BF_3 has an effective coordination number of 3 (three ligands and no lone pairs). The ligands and lone pairs of an atom are generally located at or near the vertices of a regular polyhedron which has as many vertices as the effective coordination number of the atom. Such an arrangement is energetically favorable because it keeps all the valence-shell electron pairs as far apart as possible and minimizes the electron-electron repulsions. When more than one arrangement of ligands and lone pairs is possible at the vertices of a polyhedron, the observed stable geometry is generally consistent with the following empirical rules: (1) Electron repulsions decrease in the order lone-pair–lone-pair repulsion > lone-pair–bond-pair repulsion > bond-pair–bond-pair repulsion, and (2) repulsions between electrons at vertices which subtend an angle at the central atom greater than about 115° can be neglected. A summary of the stereochemistries of a wide variety of molecules which can be rationalized by valence-shell electron repulsion is presented in Table 3.19.

Obviously the ability to quickly determine the number of lone pairs on an atom is an asset when using the valence-shell electron repulsion theory to predict molecular structure. It is helpful to recognize that the number of nonbonding valence electrons on an atom is the number of valence electrons on the free neutral atom minus the number of electrons engaged in bonding. Thus, if we are willing to assume that coordinated hydrogen atoms and alkyl groups are negatively charged, we may write the equation

TABLE 3.19
Predictions of the valence-shell electron repulsion theory of directed valency

Effective coordination no.	Arrangement of lone pairs and ligands	No. of ligands	No. of lone pairs	Shape of molecule	Examples
2	Linear	2	0	Linear	$BeCl_2$, $HgCl_2$, ZnI_2, CO_2
3	Equilateral triangular	3	0	Planar triangular	BCl_3, NO_3^-
		2	1	V-shaped	O_3, NO_2^-, $SnCl_2$
4	Tetrahedral	4	0	Tetrahedral	CH_4, Al_2Cl_6, ClO_4^-
		3	1	Pyramidal	NF_3, H_3O^+, $(TIOR)_4$, ClO_3^-
		2	2	V-shaped	H_2O, SCl_2, ClO_2^-
5	Trigonal bipyramidal	5	0	Trigonal bipyramidal	PCl_5, $PF_3(CH_3)_2$
		4	1	Irregular tetrahedral	SF_4, R_2TeCl_2
		3	2	T-shaped	ClF_3, $C_6H_5ICl_2$
		2	3	Linear	ICl_2^-, XeF_2
6	Octahedral	6	0	Octahedral	SF_6, PCl_6^-, S_2F_{10}
		5	1	Square pyramidal	BrF_5, $XeOF_4$
		4	2	Square planar	ICl_4^-, XeF_4

$$\text{Number of lone pairs} = \frac{1}{2}\left[\left(\begin{array}{c}\text{Number of valence electrons}\\ \text{on free atom}\end{array}\right) - \left(\begin{array}{c}\text{Oxidation}\\ \text{state}\end{array}\right)\right]$$

We shall discuss specific applications of the theory in the following paragraphs.

Effective Coordination Number 2

When only two ligands are coordinated to a central atom, with no nonbonding electrons on the central atom, the repulsions between the two ligands and between the bonding electrons are minimized when the ligands form a 180° bond angle with the central atom. Therefore a linear structure is generally observed in such cases. It is convenient to think of an s orbital and a p orbital in terms of their linear combinations, i.e., as two hybrid orbitals[40] directed 180° from each other. If we wish the hybrid orbitals to be equivalent, they are formed as the sum and difference of the atomic orbitals, as shown schematically in Fig. 3.8. In linear molecules such as CO_2 and the high-temperature gaseous molecules $BeCl_2$ and $HgCl_2$, we may think of the central atoms as using such hybrid sp orbitals in the formation of the σ bonds between the central atoms and the ligands.

When the central atom is large and the ligand atoms are small, as in the high-temperature gaseous molecule BaF_2, the molecule is V-shaped. This result

---

[40] Hybridization is discussed in Chap. 4.

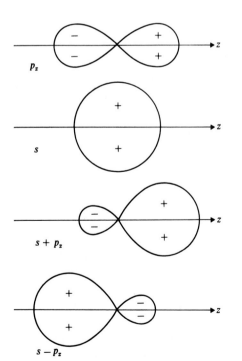

FIGURE 3.8
Formation of sp hybrid orbitals by combination of an s and a p orbital. Notice that, for simplicity, a $1s$ orbital is combined with a $2p$ orbital.

may be rationalized by considering the molecule as a combination of a polarizable central cation and two anions. The two anions are not bound at an angle of 180°, because then the induced dipole moments in the central cation would cancel and there would be no extra bonding energy attributable to interaction of the negative anions with a polarized cation. However, if the angle between the anions is less than 180°, the central cation undergoes a net polarization and the system is stabilized by the interaction between the resultant induced dipole and the anions. As the bond angle decreases, the polarization stabilization increases and the anion-anion electrostatic repulsion increases. The observed bond angle corresponds to a balance of these effects.

Effective Coordination Number 3

When three ligands are coordinated to a central atom, with no nonbonding electrons on the central atom, repulsions are minimized when the ligands are disposed at the corners of an equilateral triangle, with 120° bond angles. In such cases it is often convenient to consider the σ bonds as being formed by the overlap of sp^2 hybrid orbitals on the central atom with appropriate orbitals of the ligands.

If one of the ligands is replaced by a lone pair, the lone pair occupies the site of the replaced ligand, and the resulting molecule is V-shaped. Because of the relative stability of s orbitals compared with p orbitals, a lone pair is more stable the greater the s character of its orbital. This effect is somewhat counterbalanced by the fact that bond strength increases with increasing s character of the atomic orbitals involved (at least up to a hybridization of sp). Nevertheless, replacement of a ligand by a lone pair almost always causes rehybridization such as to shift more s character to the lone-pair orbital. Thus, in the case of species such as O_3 and NO_2^-, the σ bonds are formed using orbitals from the central atoms which have more p character than corresponds to sp^2 hybridization. Therefore the bond angles are less than 120°. For O_3 and NO_2^-, the bond angles are 117 and 115°, respectively.

The singlet carbenes, $:CX_2$, represent an interesting class of compound in which the carbon atom can have an effective coordination number of either 2 or 3. The singlet carbenes in which X = F, OH, OMe, NH_2, BH_2, BeH, Li, and H are believed to have the geometries given in Table 3.20. Schoeller[41] has rationalized these data by claiming that singlet carbenes are linear when X is less electronegative (Li to B) than carbon and are bent when X is more electronegative (N to F) than carbon, and he and Pauling[42] have offered simple explanations of the structures based on orbitals. However, the data may be equally well rationalized using valence-shell electron repulsion theory. In the case of X = F, OH, OMe, or NH_2, Lewis octet structures can be written for the singlet carbenes; thus

[41]   W. W. Schoeller, *J. Chem. Soc. Chem. Commun.*, 124 (1980).

[42]   L. Pauling, ibid., 688 (1980).

TABLE 3.20
Geometries of the singlet carbenes, :CX$_2$†

X	r_{C-X}, Å	∠XCX, degrees
F	1.323	102.8
OH	1.354	101.9
OMe	1.358	101.0
NH$_2$	1.356	109.2
H	1.127	100.2
BH$_2$	1.419	180.0
BeH	1.612	180.0
Li	1.842	180.0

† W. W. Schoeller, *J. Chem. Soc. Chem. Commun.*, 124 (1980). The geometries are not experimental, but are based on high-quality *ab initio* quantum-mechanical calculations.

Clearly in these compounds the carbon atom has an effective coordination number of 3, and the molecules should be bent. In the case of X = BH$_2$, BeH, or Li, Lewis octet structures cannot be written, but the lone pair on the carbon atom would be expected to engage in bonding to the electron-deficient atoms; thus,

$$\text{Li—}\overset{+}{\text{C}}\text{=Li}^- \qquad ^-\text{Li=}\overset{+}{\text{C}}\text{—Li}$$

In these compounds the carbon atom has an effective coordination number of 2, and the molecules should be linear. In the case of :CH$_2$, the hydrogen atoms have no lone-pair electrons which can be used to double-bond to the carbon atom, and they have no empty valence orbitals which can share the lone pair of the carbon atom. Hence the carbon atom has an effective coordination number of 3, and the molecule is bent:

$$\begin{array}{c} \text{H} \\ | \\ :\text{C—H} \end{array}$$

Effective Coordination Number 4

In the case of an atom with four identical ligands, repulsions are minimized for a regular tetrahedral geometry with sp^3 hybridization and bond angles of 109.47°. Replacement of ligands by lone pairs causes a shift of s character to the lone pairs and a decrease in the bond angles. It is interesting to compare the bond angles of NH$_3$ and H$_2$O with those for the corresponding hydrides of the second-, third-, and fourth-row elements; the data are given in Table 3.21. The bond angles for the first-

TABLE 3.21
Bond angles of some hydrides

NH_3	106.8°	H_2O	104.5°
PH_3	93.3°	H_2S	92.2°
AsH_3	91.5°	H_2Se	91.0°
SbH_3	91.3°	H_2Te	89.5°

row hydrides are slightly less than the tetrahedral value; the bond angles for the hydrides of the heavier elements are remarkably close to 90°, which corresponds to the use of pure p orbitals by the central atom. These results can be rationalized as follows: In the heavier hydrides, the central atoms are so big that there is very little repulsion between valence electrons. Thus bonding is accomplished using the p orbitals of the central atoms, without the need of the promotional energy required by hybridization. In the first-row hydrides, 90° bond angles would cause excessive crowding of the bonding electron pairs. Hybridization, with the introduction of s character into the bonding orbitals almost to the extent of sp^3 hybridization, permits the lone pairs and bonding pairs to separate, with reduction of the repulsion.

The bond angle of OF_2 is 102°, significantly less than that of H_2O. Presumably the bond angle is reduced because the highly electronegative fluorine atoms withdraw much more of the bonding electron density than the hydrogen atoms do. Such electron withdrawal is energetically favored by increasing the p character of the oxygen orbitals used in the bonding. On the other hand, the bond angle of OCl_2 is $\sim$ 111°, appreciably greater than that of H_2O. Apparently the chlorine atoms are so big that repulsion between nonbonding electrons on separate chlorine atoms overwhelms the electronegativity effect.

The lone-pair electrons of a molecule such as water are often illustrated as separate lobes or "rabbit ears," as shown in the following structure:

Although such structures are helpful for illustrating the stereochemical influence of lone pairs, they are very misleading representations of valence-electron density. The electron density plot for H_2O, given in Fig. 3.9, shows that the lone pairs overlap and that the separate lobes are not easily discerned.[26]

Effective Coordination Number 5

When an atom with no lone pairs is bonded to five ligands, the ligands are almost always located at the vertices of a trigonal bipyramid. Such a structure is obtained in the case of PF_5, which is illustrated in Fig. 3.10. There are two stereochemically distinct types of ligands in a trigonal bipyramidal MX_5 molecule;

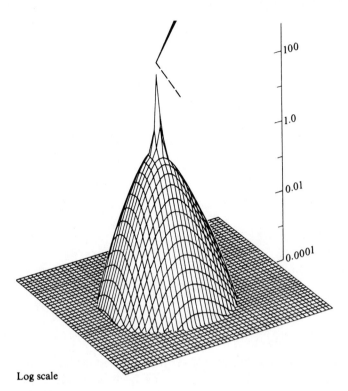

Log scale

FIGURE 3.9
Electron density plot for H_2O in the plane of the lone pairs. *(Reproduced with permission from A. Streitwieser, Jr., and P. H. Owens, "Orbital and Electron Density Diagrams," Macmillan, New York, 1973.)*

two axial ligands and three equatorial ligands. For convenience, let us define the threefold axis of the molecule as the z axis of the central atom. Then the central atoms's p_z orbital lies along the threefold axis and the p_x and p_y orbitals lie in the equatorial plane. Because the axial bonds must share the p_z orbital, each axial bond has available one-half of a p orbital, and because the equatorial bonds must share the p_x and p_y orbitals, each of these bonds has available two-thirds of a p orbital. The spherically symmetric s orbital could, in principle, be distributed in any proportion between the axial and equatorial bonding orbitals, although the two axial orbitals must have the same s character, and the three equatorial orbitals must have the same s character. At one extreme, there could be a full sp hybrid orbital for each axial bond and a $\frac{2}{3}p$ orbital for each equatorial bond. At the other extreme, there could be a $\frac{1}{2}p$ orbital for each axial bond and a full sp^2 hybrid orbital for each equatorial bond. We shall assume, as a reasonable working hypothesis, that the s orbital is equally distributed among all five bonds. Thus we obtain bond orders of 0.7 for the axial bonds and 0.867 for the equatorial bonds, and formal charges of -0.3 for the axial ligand atoms and -0.133 for the equatorial ligand atoms. Because the equatorial bond order is greater than the axial bond order, the equatorial bond lengths are shorter than the axial bond

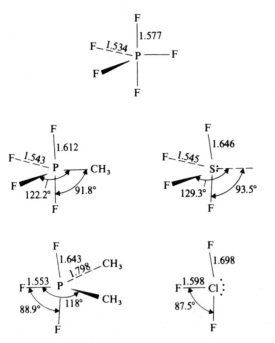

FIGURE 3.10
Structures of PF_5, PF_4CH_3, $PF_3(CH_3)_2$, SF_4, and ClF_3. [*Reproduced with permission from R. J. Gillespie, J. Chem. Educ.*, **47**, 18 (1970).]

lengths. (For PF_5, $r_{ax} = 1.58$ Å and $r_{eq} = 1.53$ Å; for PCl_5, $r_{ax} = 2.19$ Å and $r_{eq} = 2.04$ Å.) Because, on the basis of our structural hypothesis, the formal charge on the axial ligand atoms is more negative than that on the equatorial ligand atoms, we would expect that, when more than one type of ligand is bonded to the same central atom, the more electronegative ligands would preferentially occupy the axial positions. This effect is clearly seen in the structures of PF_4CH_3 and $PF_3(CH_3)_2$, illustrated in Fig. 3.10. Because CH_3 groups are more electropositive than the F atoms, the CH_3 groups occupy equatorial positions.

The trigonal bipyramidal configuration for molecules with five ligands and no lone pairs on the central atom is not exceedingly stable with respect to other configurations. Distortions of the bond angles readily occur, and it is believed that many pentacoordinate molecules undergo rapid interconversion between a trigonal bipyramidal configuration and a square pyramidal configuration, even though the former configuration may be the stable, equilibrium configuration. For example, the ^{19}F nmr spectrum of PF_5 shows a signal apparently due to only one kind of fluorine atom, even though one would expect separate signals due to the axial and equatorial fluorine atoms. Apparently the axial and equatorial atoms are scrambled at a rate greater than the difference in their nmr frequencies, and therefore the nmr signal corresponds to a weighted average signal of the axial and equatorial atoms. The mechanism shown in Fig. 3.11 has been postulated to explain the scrambling.[43] It will be noted that the molecule passes through an intermediate square pyramidal configuration in the scrambling process. It should

---

[43] S. Berry, *J. Chem. Phys.*, **32**, 933 (1960); R. J. Gillespie, *Angew. Chem. Int. Ed.*, **6**, 819 (1967).

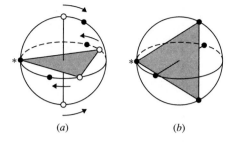

FIGURE 3.11
The mechanism postulated to account for fluorine scrambling in PF_5. (*a*) Original positions of F atoms, o; final positions of F atoms, ● (*b*) Pseudorotated trigonal bipyramid. [*Reproduced with permission from R. J. Gillespie, Angew. Chem. Int. Ed., 6, 819 (1967).*]

also be noted that one fluorine atom (marked with an asterisk in Fig. 3.11) is essentially unmoved from its equatorial position during the illustrated process. In effect, the molecule undergoes a rotation about the P—F* axis.

The ^{19}F nmr spectrum of CH_3PF_4 shows only one kind of fluorine on the nmr time scale,[44] and again the pseudorotation scrambling mechanism can be invoked to explain the interchange of fluorine atoms.

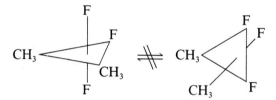

However, although the proton nmr spectrum of $(CH_3)_2PF_3$ shows only one kind of methyl group, the ^{19}F nmr spectrum shows two different kinds of fluorine atoms, in the ratio 2:1. The absence of scrambling in $(CH_3)_2PF_3$ can be explained by the pseudorotation mechanism and the tendency for relatively electropositive substituents to remain at equatorial positions.[45] Pseudorotation of $(CH_3)_2PF_3$ would force one of the methyl groups into an apical position.

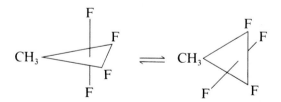

Apparently the configuration with an apical methyl group is of such high energy that the pseudorotation does not occur and both methyl groups remain at equatorial positions.

Another kinetic manifestation of the tendency for relatively electropositive groups to occupy equatorial positions is found in the hydrolysis of five-membered cyclic esters of phosphoric acid.[45] These hydrolyze rapidly, both with ring opening

---

[44]   E. L. Muetterties, W. Mahler, and R. Schmutzler, *Inorg. Chem.*, **2**, 613 (1963).
[45]   F. H. Westheimer, *Acc. Chem. Res.*, **1**, 70 (1968).

and with retention of the ring, as shown by the example of methyl ethylene phosphate:

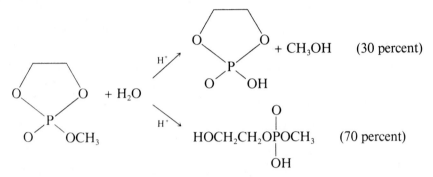

The hydrolysis of the methyl ester of propylphostonic acid also occurs rapidly, but almost exclusively with ring opening:

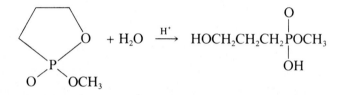

These results are explained by assuming the formation of a 5-coordinate intermediate which can undergo pseudorotation. The five-membered rings are assumed to span one equatorial and one apical position. It is also assumed that the incoming group is initially attached to an apical position and that the leaving group comes from an apical position. Thus the formation of two products in the hydrolysis of methyl ethylene phosphate is rationalized by the following mechanism:

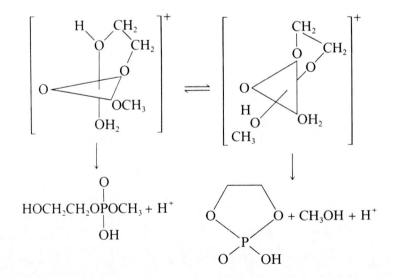

In the case of the phostonate, pseudorotation would give an unstable intermediate, and so essentially only one product is observed:

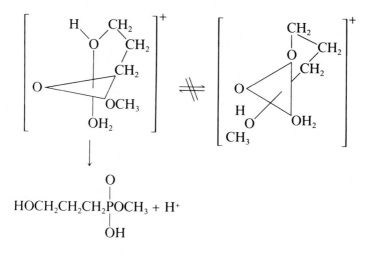

$$HOCH_2CH_2CH_2POCH_3 + H^+$$

with O double-bonded above P and OH below.

The rate of hydrolysis of cyclic phosphinates such as

is low because there is no way to form a trigonal bipyramidal intermediate of low energy; either the ring angle must be expanded to 120° or an alkyl group must occupy an apical position.

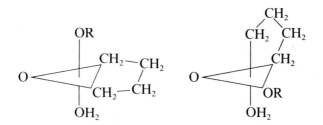

A few molecules, including pentaphenylantimony[46] and the phosphorus compound[47] pictured on the following page,

[46] A. L. Beauchamp, M. J. Bennett, and F. A. Cotton, *J. Am. Chem. Soc.*, **90**, 6675 (1968).
[47] J. A. Howard, D. R. Russell, and S. Trippett, *J. Chem. Soc. Chem. Commun.*, 856 (1973).

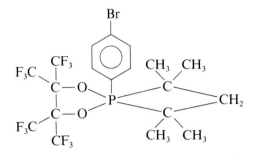

are so distorted in the solid state that their configurations are closer to square pyramidal than to trigonal bipyramidal. These distortions are not understood. They may be a consequence of solid-state crystal packing forces; perhaps the free molecules would have trigonal bipyramidal configurations.

When one, two, or three of the ligands in a trigonal bipyramidal molecule are replaced by lone pairs, the lone pairs always occupy equatorial positions, presumably to minimize lone-pair–bonding-pair repulsions. Thus molecules of this type always have two ligands in the axial positions. The structures of SF_4 and ClF_3 are shown in Fig. 3.10. The deviations of the bond angles from those of a perfect trigonal bipyramid are significant. In both structures the fluorine atoms are "bent away" from the lone-pair electrons, in accord with the rule that lone-pair–bonding-pair repulsion is greater than bonding-pair–bonding-pair repulsion. Thus in SF_4, the angle subtended by the lone pair and an apical fluorine is considerably greater than 90° (93.5°), and the angle subtended by the lone pair and an equatorial fluorine is considerably greater than 120° (129.3°). Similarly, in ClF_3, the F—Cl—F bond angle is less than 90° (87.5°).

Effective Coordination Number 6

In a 6-coordinate molecule such as SF_6 the ligand atoms are situated at the vertices of a regular octahedron. If we arbitrarily place the ligand atoms on the x, y, and z axes of the central atom, it is easy to see that each bonding orbital of the central atom consists of a combination of $\frac{1}{2}$ p orbital and $\frac{1}{6}$ s orbital. Neglecting any contributions of the valence-shell d orbitals, each bond has a bond order of $\frac{2}{3}$. If we replace one of the ligands with a lone pair, the resulting molecule has a square pyramidal configuration, with the axial ligand trans to the lone pair. As an illustration of this type of bonding, the structure of BrF_5 is shown in Fig. 3.12. Because of the relatively high lone-pair–bonding-pair repulsions, the bromine atom lies below the basal plane of the four equivalent fluorine atoms, and the bond angles are less than 90°. In order to replace a ligand of a 6-coordinate molecule by a lone pair, it is necessary to extract electron density from the other bonds. (The average bond order drops from $\frac{2}{3}$ to $\frac{3}{5}$.) Because of the repulsion of the lone pair by the bonding electrons of the four basal bonds, it is to be expected that more electron density will be extracted from these basal bonds than from the axial bond. Hence the axial bond order is greater than the basal bond

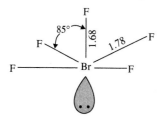 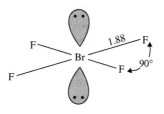

FIGURE 3.12
Structures of BrF_5 and BrF_4^-.

order, in qualitative agreement with the corresponding bond distances (1.68 and 1.79 Å, respectively, in BrF_5). If we replace two of the ligands in a 6-coordinate molecule with lone pairs, it is reasonable to place these in trans positions to minimize the strong lone-pair–lone-pair repulsion. Thus we expect the ligands to occupy a square planar configuration. Indeed such a configuration is found for all tetracoordinate species with two lone pairs on the central atom (for example, BrF_4^-, shown in Fig. 3.12, and XeF_4). In such species the bond order is $\frac{1}{2}$, with one-half of a p orbital from the central atom involved in each bond.

The molecule SF_4CH_2 has a trigonal bipyramidal arrangement of ligands around the sulfur atom, with the methylene group occupying an equatorial position.[48] These structural features are readily predicted using the rules we have previously discussed for five-coordination. However, the fact that the hydrogen atoms of the CH_2 group lie in a plane containing the axial fluorines, rather than in the equatorial plane, requires further explanation. In this case it is helpful to consider the carbon atom as 4-coordinate (tetrahedral) and the sulfur atom as 6-coordinate (octahedral). The observed structure corresponds to linking the octahedron with the tetrahedron, as follows:

Obviously the S—C bond should be considered a double bond.

Effective Coordination Number 7

When the effective coordination number exceeds 6, there are usually various symmetric, polyhedral configurations which have very similar energies, and one

---

[48] G. Kleemann and K. Seppelt, *Angew. Chem. Int. Ed.*, **17**, 516 (1978).

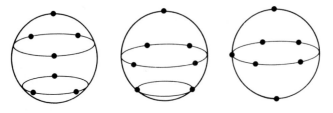

FIGURE 3.13
Three ways of achieving sevenfold coordination. [*Reproduced with permission from R. J. Gillespie, Angew. Chem. Int. Ed., **6**, 819 (1967).*]

often finds examples of each of these configurations in actual compounds. It does not seem to be possible to predict structures with certainty in such cases, particularly if the ligands are not equivalent or if they are chelated. For example, let us consider effective coordination number 7. Three plausible configurations are shown in Fig. 3.13. The first structure is a monocapped octahedron, the second is a monocapped trigonal prism, and the third is a pentagonal bipyramid. The first structure has been observed in the case of $NbOF_6^{3-}$ and ZrF_7^{3-}, the second structure in the case of NbF_7^{2-} and TaF_7^{2-}, and the third structure in the case of $UO_2F_5^{3-}$ and IF_7. If one of the ligands is replaced with a lone pair, one might expect to obtain the structure in which the lone pair has the minimum number of nearest neighbors, i.e., the first structure, with the lone pair in the unique axial position. However, the structure of the trisoxalatoantimonate(III) ion, $Sb(C_2O_4)_3^{3-}$, in which the antimony atom has one lone pair and six oxygen atoms in its coordination sphere, is based on a pentagonal bipyramid with the lone pair presumably occupying an axial position.[49] The structure of this ion is shown in Fig. 3.14. Two oxalate ions lie in the equatorial plane, and one bridges an equatorial site and an axial site. The structure of XeF_6, another molecule in which the central atom has one lone pair and six ligands, is unknown.[50] Nevertheless, on the basis of electron diffraction data it is clear that the XeF_6 molecule does not have O_h symmetry and that the molecule is fluxional, i.e., that it is very flexible, with rapid scrambling of the fluorine atoms, as in the case of PF_5.

When the ligands are large, repulsions between ligands can be of primary importance in determining stereochemistry. For example, $TeCl_6^{2-}$, $TeBr_6^{2-}$, and $SbBr_6^{3-}$ have regular octahedral structures even though the central atoms have nonbonding electrons. Apparently the lone pair is forced inside the shell of ligands into a spherical s orbital.

Higher coordination numbers are known, but for none of these is the geometry known to be affected by the presence of lone pairs on the central atom. Therefore we shall defer consideration of such structures until we discuss transition-metal complexes.

---

[49]   M. C. Poore and D. R. Russell, *Chem. Commun.*, 18 (1971).

[50]   N. Bartlett and F. O. Sladky, "Comprehensive Inorganic Chemistry," p. 213, Pergamon Press, Oxford, 1973.

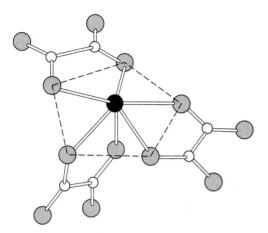

FIGURE 3.14
Perspective view of the $Sb(C_2O_4)_3{}^{3-}$ ion.
[*Reproduced with permission from M. C. Poore and D. R. Russell, Chem. Commun., 18 (1971).*]

A Caveat

There is no question that the valence-shell electron repulsion theory works; it fairly reliably predicts the structures of a wide variety of molecules, and it has been shown to be consistent with some theoretical studies.[51] However, it has been claimed that some of the fundamental postulates of the theory are incorrect.[52,53] In particular, the rule that repulsions decrease in the order lone-pair−lone-pair > lone-pair−bond-pair > bond-pair−bond-pair is said to be wrong; the repulsions between bond pairs are in fact the most important. Hall's papers[53] should be consulted for a discussion of the relation between bond angles, bond-pair−bond-pair repulsions, and the *p-s* energy level separation of the central atom. Undoubtedly, further theoretical studies will have to be carried out before we have a simple, yet theoretically sound, scheme for predicting the structures of nontransition-metal molecules.

CONFORMATIONAL ISOMERISM
BASED ON BOND ROTATIONS

Many molecules have groups of atoms joined by a bond such that rotation of the groups with respect to one another around the bond axis causes a continuous change in the energy of the molecule. Rotation about a double bond causes the molecule to pass through conformations of such high energy that rotation is strongly resisted. However, in the case of a single bond, the rotational barriers

[51] G. W. Schnuelle and R. G. Parr, *J. Am. Chem. Soc.*, **94**, 8974 (1972); C. A. Naleway and M. E. Schwartz, *J. Am. Chem. Soc.*, **95**, 8235 (1973); L. S. Bartell and Y. Z. Barshad, *J. Am. Chem. Soc.*, **106**, 7700 (1984).

[52] W. E. Palke and B. Kirtman, *J. Am. Chem. Soc.*, **100**, 5717 (1978); C. Edmiston, J. Bartleson, and J. Jarvie, *J. Am. Chem. Soc.*, **108**, 3593 (1986).

[53] M. B. Hall, *J. Am. Chem. Soc.*, **100**, 6333 (1978); *Inorg. Chem.*, **17**, 2261 (1978); also see J. K. Burdett, "Molecular Shapes," Chaps. 3–7, Wiley, New York, 1980.

TABLE 3.22
Stable conformations of several molecules†

Compound	Projected conformation	Dihedral angle (ϕ), degrees
N_2H_4 P_2H_4		90-95 90-100
H_2O_2		111
H_2S_2 O_2F_2 S_2F_2		91 88 88
FCH_2OH		60

† S. Wolfe, *Acc. Chem. Res.*, **5**, 102 (1972).

are low (usually less than 4 kcal mol^{-1}), and rotation is relatively unrestricted at ordinary temperatures. Nevertheless the stable conformation for a single bond is of considerable interest both because of its relation to the dipole moment of the molecule and because of the theoretical problems posed. A conformation is generally described in terms of the dihedral angle between two atoms or lone pairs on the adjacent atoms, i.e., the angles projected by the atoms or lone pairs on a plane perpendicular to the bond axis. Conformational data for several molecules are given in Table 3.22. Such data are difficult to rationalize. Probably one must consider nucleus-electron attractions, nucleus-nucleus repulsions, and electron-electron repulsions; and no simple theory seems to be adequate. For a discussion of the theoretical aspects of bond conformations, the reader is referred to a paper by Wolfe.[54]

It is interesting to compare the structures, shown in Fig. 3.15, of the cyclo-octasulfur molecule, S_8, its two-electron oxidation product, S_8^{2+}, and an isoelec-

54 S. Wolfe, *Acc. Chem. Res.*, **5**, 102 (1972).

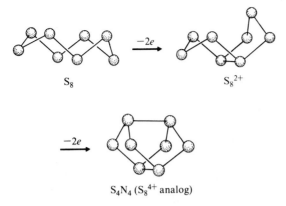

S_8

$\xrightarrow{-2e}$

$S_8{}^{2+}$

$\xrightarrow{-2e}$

S_4N_4 ($S_8{}^{4+}$ analog)

FIGURE 3.15

Relationships between the structures of S_8, $S_8{}^{2+}$, and S_4N_4. [*Reproduced with permission from R. J. Gillespie, Chem. Soc. Rev., 8, 315 (1979).*]

tronic analog of its four-electron oxidation product, S_4N_4. The eight-membered S_8 ring is puckered; the sulfur atoms lie in two parallel planes, giving the ring a crown shape, with an S—S distance of 2.05 Å and an S—S—S angle of 108°.[55] The ion $S_8{}^{2+}$ consists of an eight-membered ring in which one of the sulfur atoms has flipped up into an exo position and in which a trans-annular bond has formed.[56] The S—S bonds around the $S_8{}^{2+}$ ring have an average length of 2.04 Å, essentially the same as in S_8. However, the trans-annular bond has a length of 2.86 Å, corresponding to a rather weak bond. The reason for the endo-exo conformation is not clear, but the conformation presumably minimizes nonbonding electron repulsions between the atoms adjacent to the trans-annularly bonded atoms. The $S_8{}^{4+}$ cation is unknown; it appears to be unstable with respect to $S_4{}^{2+}$. However, the isoelectronic S_4N_4 is well known; it has a structure in which a second sulfur atom has flipped up and a second trans-annular S—S bond has formed. The four coplanar atoms are nitrogen atoms.

The molecule $S_4N_4F_4$ contains a highly puckered eight-membered ring of alternating sulfur and nitrogen atoms, as shown in Fig. 3.16.[57] The remarkable feature of this structure is that there are two S—N bond distances (1.66 and 1.54 Å), which occur alternately in the ring. In Fig. 3.16, the S_1—N_1 distance is 1.66 Å, the N_1—S_2 distance is 1.54 Å, and so on around the ring. These alternating bond lengths are often interpreted in terms of localized single and double S—N bonds. However, a simpler, a priori, explanation is available.[58] By examination of Fig. 3.16, it is seen that the S_1—N_1 and N_1—S_2 bonds have quite different stereochemical environments. The S_1—N_1 bond is longer, probably because of repulsive interaction of the nonbonding electrons of fluorine atom F_1 with the nonbonding electrons of nitrogen atom N_1. The N_1—S_2 bond is shorter, probably

55   B. Meyer, *Chem. Rev.*, **76**, 367 (1976).
56   C. G. Davies, R. J. Gillespie, J. J. Park, and J. Passmore, *Inorg. Chem.*, **10**, 2781 (1971).
57   G. A. Wiegers and A. Vos, *Acta Crystallogr.*, **14**, 562 (1961); **16**, 152 (1963).
58   W. L. Jolly, *Adv. Chem. Ser.*, **110**, 92 (1972).

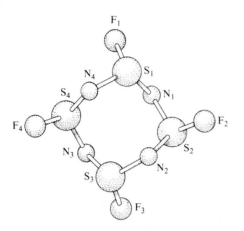

FIGURE 3.16
The structure of the $S_4N_4F_4$ molecule. [*Reproduced with permission from G. A. Wiegers and A. Vos, Acta Crystallogr, **14**, 562 (1961); **16**, 152 (1963).*]

because there is much less interaction between the nonbonding electrons of atoms F_2 and N_1. If atoms F_1 and F_2 were flipped to the upper side of the ring, there would be no resultant net advantage in terms of nonbonding electron interactions. In such a conformation, the S—N bonds would be expected to occur in the sequence s, s, 1, 1, s, s, 1, 1 (s = short, l = long).

PROBLEMS

3.1 Write Lewis octet structures for the following species. (Show all non-bonding valence electrons and indicate formal charges.) (*a*) Al_2Cl_6, (*b*) $SnCl_3^-$, (*c*) XeF_2, (*d*) BrF_4^-, (*e*) NS^+, (*f*) SO_3F^-, (*g*) HOClO, (*h*) $S_4N_3^+$ (S—S—N—S—N—S—N).

3.2 The S—O bond distance in SO_2F_2 is 1.37 Å, whereas that in SO_2 is 1.43 Å. Explain in terms of octet structures.

3.3 How can you account for CH_5^+ in terms of an octet structure?

3.4 For each of the following, name three well-known isoelectronic species containing the same number of atoms: (*a*) O_3, (*b*) OH^-, (*c*) BH_4^-, (*d*) NO_3^-.

3.5 How might you estimate or predict, by analogy with known compounds, (*a*) the ultraviolet spectrum of CO_2^-, (*b*) the bond distance in BF, (*c*) the dissociation energy of F_2, (*d*) the symmetric stretching frequency of BF_2^+, (*e*) the bond angles in NH_3^+, (*f*) the structure of solid $AlPO_4$, (*g*) the infrared spectrum of ClO_2^+, and (*h*) the bond distances and angles in B_2F_4?

3.6 What are the values of n and m in the anionic species $Fe(CO)_n^{-m}$? Hint: Consider $Ni(CO)_4$.

*3.7 The Mössbauer tin isomer shifts, relative to IS = 0.00 for grey tin metal, are positive for tin(II) salts such as $SnCl_2$, and negative for tin(IV) compounds such as $SnCl_4$. Explain.

*3.8 The Mn $2p_{3/2}$ binding energy of MnF_2 is greater than that of MnO_2, even though the manganese oxidation states in these compounds are +2 and +4, respectively. Explain.

3.9 Which of the following topologies for $N_2F_3^+$ is more stable? (The lines do not necessarily indicate *single* bonds; they merely indicate bonds.)

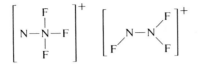

3.10 Draw valence-bond octet-satisfying structures for the following species, showing all nonbonding electron pairs and formal charges: HNO_3, NOF, $NSCl$, NOF_3, NSF_3, ClO_3^-, N_3^-, PH_2^-, $SbCl_5^{2-}$, $IO_2F_2^-$. Indicate when other resonance structures are important. Indicate the shapes of the molecules.

3.11 Which of the following molecules would you expect to have permanent dipole moments: SO_2, CO_2, BF_3, NF_3, SiF_4, XeF_4, SF_4, SF_2, C_2H_2, N_2F_4, *trans*-N_2F_2, SiH_3Cl, O_3, PF_5, BrF_5, S_2Cl_2?

3.12 Predict structures, including estimated bond distances and bond angles, for the molecules ClO_2F, N_2O_5, and N_4O.

3.13 Estimate the heats of formation of the gaseous molecules N_3H_5 and $(GeH_3)_2Se$.

3.14 Rationalize the following bond-angle data for alkaline-earth dihalide *molecules*.

MX_2	$\angle XMX$, degrees
MgF_2	158
CaF_2	140
SrF_2	108
BaF_2	~100
$CaCl_2$	~180
$SrCl_2$	120
$BaCl_2$	~100

3.15 From data in Table 3.6, estimate the N—F bond length in NF_3. Rationalize the marked difference between this value and the actual value, 1.37 Å.

3.16 Assuming octet structures, what are the bond orders in the species XeF_4, XeF_6, SO_2, CH_3^-, and NO_2?

3.17 Which S—N bonds in the seven-membered ring of $S_4N_3^+$ would you expect to be the shortest? [See Prob. 3.1(*h*).]

3.18 Explain why the sum of $D(OC{=}S)$ and $D(O{=}CS)$ is much less than the atomization energy of OCS.

*3.19 In diamond, each carbon atom is bonded to four other carbon atoms. Rationalize the relative magnitudes of the sublimation energy of diamond (170.8 kcal mol^{-1}) and the C—C bond energy.

3.20 From simple electronegativity considerations, predict whether or not the following gas-phase reaction proceeds as written:

$$ONCl + H_2O \rightleftharpoons HONO + HCl$$

Explain the fact that ONCl undergoes hydrolysis in liquid water.

3.21 Using Slater's rules, estimate the $1s$ binding energies of $C(g)$ and $C^+(g)$.

*3.22 The base-catalyzed hydrolysis of

$$CH_3\overset{\displaystyle O}{\overset{\displaystyle \|}{C}}-\overset{\displaystyle CH_3}{\overset{\displaystyle |}{CH}}-\overset{\displaystyle R}{\overset{\displaystyle |}{OPO(OCH_3)}}$$

proceeds through an intermediate anion

$$CH_3\overset{\displaystyle OH}{\overset{\displaystyle |}{\underset{\displaystyle \underset{\displaystyle O_-}{|}}{C}}}-\overset{\displaystyle CH_3}{\overset{\displaystyle |}{CH}}-\overset{\displaystyle R}{\overset{\displaystyle |}{OPO(OCH_3)}}$$

to give methanol and

$$CH_3\overset{\displaystyle O}{\overset{\displaystyle \|}{C}}-\overset{\displaystyle CH_3}{\overset{\displaystyle |}{CH}}-\overset{\displaystyle R}{\overset{\displaystyle |}{OPO(OH)}}$$

when R $=$ OCH_3, and to give

$$CH_3\overset{\displaystyle O}{\overset{\displaystyle \|}{C}}-\overset{\displaystyle OH}{\overset{\displaystyle |}{CHCH_3}}$$

and methyl methylphosphonate when R $=$ CH_3. Rationalize the results, assuming that the intermediate anion rearranges to form a species with a five-membered ring and a 5-coordinate phosphorus atom.

*3.23 The dissociation energy of F_2 seems out of line when compared with the values for the other halogens (see Table 3.9). Explain.

3.24 Estimate the P—P bond energy from the atomization energy of tetrahedral P_4, 305 kcal mol^{-1}.

CHAPTER
4

MOLECULAR ORBITAL THEORY

The Lewis octet theory and the valence-shell electron repulsion theory discussed in Chap. 3 are usually adequate for predicting the nature of the bonding and the structures of molecules of non-transition-element compounds. However, these simple theories have serious limitations. There are many properties of molecules (e.g., spectral properties) that cannot be explained by the theories. And, in the case of transition-metal compounds, many aspects of bonding and structure cannot be treated using these theories. In this chapter we will show that molecular orbital (MO) theory is a much broader concept which is, in principle, applicable to all substances and capable of predicting most of the phenomena in which chemists are interested.

SIMPLE LCAO THEORY

In Chap. 1, we discussed the electron orbitals associated with an atomic nucleus, that is, atomic orbitals. In this chapter we shall discuss the orbitals associated with a set of two or more atomic nuclei, i.e., MOs. We shall make the reasonable approximation that an MO is a linear combination of atomic orbitals (LCAO). The approximation is based on the assumption that, as a valence electron circulates among the nuclei, it will at any one time be much closer to one nucleus than any other and the MO wave function will be fairly well represented by a valence atomic orbital for that nucleus. To illustrate this idea we shall develop the MOs and their energies for the case of a single valence electron associated with two atoms. The results are applicable to molecule ions such as H_2^+ (in which two $1s$ atomic orbitals are involved in the bonding), Li_2^+ (two $2s$ orbitals), and LiH^+ (a

$1s$ orbital and a $2s$ orbital). (In the case of Li_2^+ and LiH^+ we make the further very good approximation that the $1s^2$ closed shells are not involved in bonding.)

The Schrödinger equation which was introduced in Chap. 1 may be rearranged as follows:

$$\left[V - \frac{h^2}{8\pi^2 m} \left(\frac{\partial^2 \psi}{\partial x^2} + \frac{\partial^2 \psi}{\partial y^2} + \frac{\partial^2 \psi}{\partial z^2} \right) \right] \psi = E\psi$$

The left side of the equation can be considered the action of an operator (called the "hamiltonian operator") on ψ, where ψ is now an MO. Therefore we can abbreviate the expression as

$$H\psi = E\psi \qquad (4.1)$$

where H is the hamiltonian operator. Just as in the case of atomic wave functions, the MO wave function ψ can be either positive or negative, and ψ^2 is a quantity proportional to electron density. If we multiply both sides of Eq. 4.1 by ψ and integrate over all space, we obtain

$$\int \psi H\psi \, d\tau = E \int \psi^2 \, d\tau \qquad (4.2)$$

In Eq. 4.2 we have allowed a single integral sign to stand for a triple integral sign and have made the substitution $d\tau = dx \, dy \, dz$. By simple rearrangement we obtain the following expression for the MO energy:

$$E = \frac{\int \psi H\psi \, d\tau}{\int \psi^2 \, d\tau} \qquad (4.3)$$

The wave function ψ is represented by the following linear function,

$$\psi = c_1 \phi_1 + c_2 \phi_2 \qquad (4.4)$$

where ϕ_1 and ϕ_2 are the atomic orbital wave functions of atoms 1 and 2, and c_1 and c_2 are coefficients to be determined. Combination of Eqs. 4.3 and 4.4 yields

$$E = \frac{\int (c_1\phi_1 + c_2\phi_2)H(c_1\phi_1 + c_2\phi_2) \, d\tau}{\int (c_1\phi_1 + c_2\phi_2)^2 \, d\tau}$$

$$= \frac{\int (c_1\phi_1 H c_1\phi_1 + c_1\phi_1 H c_2\phi_2 + c_2\phi_2 H c_1\phi_1 + c_2\phi_2 H c_2\phi_2) \, d\tau}{\int (c_1^2\phi_1^2 + 2c_1c_2\phi_1\phi_2 + c_2^2\phi_2^2) \, d\tau}$$

Now we shall accept, without proof,[1] the facts that

$$H c_1\phi_1 = c_1 H\phi_1$$

and

$$\int \phi_1 H\phi_2 \, d\tau = \int \phi_2 H\phi_1 \, d\tau$$

---

[1]   M. C. Day, Jr., and J. Selbin, "Theoretical Inorganic Chemistry," 2d ed., pp. 165–166, 571–574, Reinhold, New York, 1969.

Therefore we may write

$$E = \frac{c_1^2 \int \phi_1 H \phi_1 \, d\tau + 2c_1 c_2 \int \phi_1 H \phi_2 \, d\tau + c_2^2 \int \phi_2 H \phi_2 \, d\tau}{c_1^2 \int \phi_1^2 \, d\tau + 2c_1 c_2 \int \phi_1 \phi_2 \, d\tau + c_2^2 \int \phi_2^2 \, d\tau}$$

For simplification, we make the following substitutions:

$$H_{11} = \int \phi_1 H \phi_1 \, d\tau$$

$$H_{22} = \int \phi_2 H \phi_2 \, d\tau$$

$$H_{12} = \int \phi_1 H \phi_2 \, d\tau$$

$$S_{11} = \int \phi_1^2 \, d\tau$$

$$S_{22} = \int \phi_2^2 \, d\tau$$

$$S_{12} = \int \phi_1 \phi_2 \, d\tau$$

Hence

$$E = \frac{c_1^2 H_{11} + 2c_1 c_2 H_{12} + c_2^2 H_{22}}{c_1^2 S_{11} + 2c_1 c_2 S_{12} + c_2^2 S_{22}} \tag{4.5}$$

We are interested in determining the minimum value of E, corresponding to the equations

$$\left(\frac{\partial E}{\partial c_1} \right)_{c_2} = 0 \quad \text{and} \quad \left(\frac{\partial E}{\partial c_2} \right)_{c_1} = 0$$

Differentiation yields the equations

$$c_1(H_{11} - ES_{11}) + c_2(H_{12} - ES_{12}) = 0$$
$$c_1(H_{12} - ES_{12}) + c_2(H_{22} - ES_{22}) = 0$$

These are called the "secular equations." A nontrivial solution to these equations can be expressed in terms of the "secular determinant":

$$\begin{vmatrix} H_{11} - ES_{11} & H_{12} - ES_{12} \\ H_{12} - ES_{12} & H_{22} - ES_{22} \end{vmatrix} = 0 \tag{4.6}$$

The terms H_{11} and H_{22} are called "coulomb integrals." From our previous definition and Eq. 4.3, we see that a coulomb integral is approximately the energy of an electron in the valence atomic orbital, α. At least this approximation is reasonable for a neutral molecule, in which electron-electron and nucleus-nucleus repulsions somewhat compensate. Hence we may write $H_{11} = \alpha_1$ and $H_{22} = \alpha_2$.

The term H_{12} is called the "exchange integral"[2] and is essentially the interaction energy of the two atomic orbitals, β. Both α and β have negative values.

If we assume that the atomic orbital wave functions ϕ_1 and ϕ_2 of Eq. 4.4 are "normalized," then

$$S_{11} = \int \phi_1{}^2\, d\tau = S_{22} = \int \phi_2{}^2\, d\tau = 1 \tag{4.7}$$

Equation 4.7 simply states that the probability of finding an electron in the orbital is exactly unity. The term S_{12} is called the "overlap integral" because it is a measure of the extent to which orbitals 1 and 2 overlap. For simplification we shall omit the subscripts and write S for the overlap integral. The secular determinant reduces to

$$\begin{vmatrix} \alpha_1 - E & \beta - ES \\ \beta - ES & \alpha_2 - E \end{vmatrix} = 0 \tag{4.8}$$

For a homonuclear species such as $H_2{}^+$, we may substitute $\alpha_1 = \alpha_2 = \alpha$. The determinantal equation then corresponds to

$$(\alpha - E)^2 = (\beta - ES)^2$$

The two solutions of this equation are

$$\alpha - E = -(\beta - ES) \quad \text{or} \quad E = \frac{\alpha + \beta}{1 + S} \tag{4.9}$$

and

$$\alpha - E = (\beta - ES) \quad \text{or} \quad E = \frac{\alpha - \beta}{1 - S} \tag{4.10}$$

By appropriate substitution in the first secular equation, we obtain

$$c_1(\alpha - E) + c_2(\beta - ES) = 0$$

or

$$c_1 = -\frac{\beta - ES}{\alpha - E}c_2$$

From this relation it can be seen that, when $E = (\alpha + \beta)/(1 + S)$, then $c_1 = c_2$, and when $E = (\alpha - \beta)/(1 - S)$, then $c_1 = -c_2$. Thus the MO wave function can be written as follows:

$$\psi = c_1\phi_1 \pm c_1\phi_2$$

To evaluate c_1, we must normalize the wave function:

$$\int \psi^2\, d\tau = c_1{}^2 \int \phi_1{}^2\, d\tau \pm 2c_1{}^2 \int \phi_1\phi_2\, d\tau + c_1{}^2 \int \phi_2{}^2\, d\tau = 1$$
$$= c_1{}^2 S_{11} \pm 2c_1{}^2 S + c_1{}^2 S_{22} = 1$$

---

[2]   Sometimes referred to as the "resonance integral" or "bond integral."

Hence

$$c_1{}^2(2 \pm 2S) = 1$$

and

$$c_1 = \pm \frac{1}{\sqrt{2 \pm 2S}}$$

The + sign under the radical sign corresponds to $c_1 = c_2$, and the − sign under the radical sign corresponds to $c_1 = -c_2$. Obviously the following wave functions are normalized.

$$\psi_B = \frac{1}{\sqrt{2 + 2S}}(\phi_1 + \phi_2) \tag{4.11}$$

$$\psi_A = \frac{1}{\sqrt{2 - 2S}}(\phi_1 - \phi_2) \tag{4.12}$$

The valence-electron density is obtained by squaring these functions:

$$\psi_B{}^2 = \frac{1}{2 + 2S}(\phi_1{}^2 + \phi_2{}^2 + 2\phi_1\phi_2)$$

$$\psi_A{}^2 = \frac{1}{2 - 2S}(\phi_1{}^2 + \phi_2{}^2 - 2\phi_1\phi_2)$$

$\psi_B{}^2$ shows an increase in electron density in the region of overlap between the atoms over that of the individual atoms. Such an electron distribution stabilizes the system, and we refer to ψ_B as the "bonding" MO. The energy level of ψ_B is given by $E = (\alpha + \beta)/(1 + S)$. $\psi_A{}^2$ shows a decrease in electron density in the overlap region, and the system is unstable relative to the separate atoms. We refer to ψ_A as the "antibonding" MO, for which $E = (\alpha - \beta)/(1 - S)$.

In Fig. 4.1 is shown a plot of $\phi_1{}^2$, $\phi_2{}^2$, $\psi_B{}^2$, and $\psi_A{}^2$ along the internuclear line. The dashed lines indicate $\phi_1{}^2$ and $\phi_2{}^2$, i.e., the electron density of the individual atomic orbitals. The lower solid line indicates $\psi_A{}^2$, the electron density of the antibonding MO, and the upper solid line indicates $\psi_B{}^2$, the electron density of the bonding MO. Figure 4.2 is an energy level diagram which graphically indicates the energies of the two MOs which arise from the interaction of two atomic orbitals. Overlap integrals are generally fairly small (often in the range 0.2 to 0.3); hence the antibonding MO is destabilized approximately the same amount that the bonding MO is stabilized. As a matter of fact, in simple LCAO theory, it is often assumed that $S = 0$. This assumption simplifies the calculations and is actually not as drastic an approximation as it might appear.[3] With this approximation, the energy levels of ψ_B and ψ_A are $\alpha + \beta$ and $\alpha - \beta$, respectively.

---

[3] A. Streitwieser, Jr., "Molecular Orbital Theory for Organic Chemists," pp. 101–103, Wiley, New York, 1961.

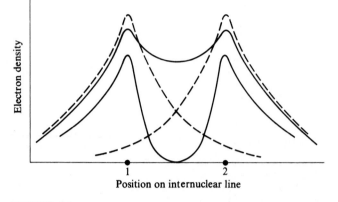

FIGURE 4.1
Plot of electron densities for the orbitals ϕ_1 and ϕ_2 (dashed lines), ψ_B (upper solid line), and ψ_A (lower solid line) along the internuclear axis of H_2^+.

In the case of a heteronuclear bond (such as in LiH^+), the secular determinant yields, if we neglect S,

$$(\alpha_1 - E)(\alpha_2 - E) = \beta^2$$

Solving for E gives

$$E = \frac{\alpha_1 + \alpha_2}{2} \pm \frac{1}{2}\sqrt{\alpha_1^2 + \alpha_2^2 + 2\alpha_1\alpha_2 - 4\alpha_1\alpha_2 + 4\beta^2}$$

$$= \frac{\alpha_1 + \alpha_2}{2} \pm \frac{1}{2}\sqrt{(\Delta\alpha)^2 + 4\beta^2} \qquad (4.13)$$

The corresponding energy level diagram is given in Fig. 4.3. Notice that, according to the approximate LCAO method that we are now employing, the energy of

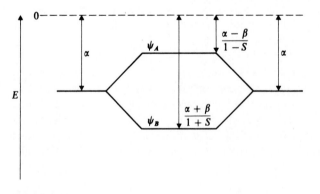

FIGURE 4.2
Energy level diagram for the molecular orbitals formed from similar atomic orbitals in a homonuclear molecule.

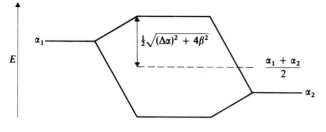

FIGURE 4.3
Energy level diagram for the molecular orbitals formed from dissimilar atomic orbitals in a heteronuclear molecule. The overlap integral has been neglected.

the bonding MO is depressed from that of the more stable atomic orbital by the same amount that the energy of the antibonding MO is raised from that of the less stable atomic orbital. If $|\beta|$ is very small, the energy spread between the bonding and antibonding levels is very little more than the separation between α_1 and α_2, and the MOs are essentially slightly perturbed atomic orbitals.

It should be remembered that the MOs which we have been discussing are "one-electron" MOs, for which the calculated energies are valid only for one-electron occupancy. However, the results we have obtained are at least qualitatively applicable to many-electron systems. Hence the energy level diagrams of Figs. 4.2 and 4.3 can be used to predict the electron configurations of systems containing one to four valence electrons. The MOs are filled in the order of increasing energy. Table 4.1 gives the electron configurations for four species with $1s$ valence atomic orbitals to illustrate the filling of the bonding MOs, represented by the symbol $(1s\sigma_B)$, and the antibonding MOs, represented by the symbol $(1s\sigma^*)$. Asterisks are used to designate antibonding MOs, except in the case of ψ_A.[4] In MO theory, bond order is defined as one-half the number of bonding electrons minus one-half the number of antibonding electrons. Hence the species in Table 4.1 have bond orders as indicated in the third column. Note that the species He_2 is predicted to have a bond order of zero, corresponding to no net bonding between the atoms.

---

[4]   The symbol ψ^* is reserved for the complex conjugate of ψ.

TABLE 4.1

Species	Electron configuration	Bond order
H_2^+	$(1s\sigma_B)^1$	$\frac{1}{2}$
H_2	$(1s\sigma_B)^2$	1
HHe	$(1s\sigma_B)^2(1s\sigma^*)^1$	$\frac{1}{2}$
He_2	$(1s\sigma_B)^2(1s\sigma^*)^2$	0

CRITERIA FOR STABLE MOLECULAR ORBITALS

The energy $-\beta$ may be taken as a rough measure of the covalent bond energy. It is commonly assumed in simple MO theory that $-\beta$ is proportional to the corresponding overlap integral S and to the average energy of the atomic orbitals. Thus stable covalent bonds can form whenever the overlap integral is large. The following three criteria must be met to ensure large values of S.

1. The energies of the atomic orbitals must be comparable. A small negative value for α corresponds to a large, diffuse atomic orbital, and a large negative value for α corresponds to a small, compact atomic orbital. Two orbitals with dissimilar values of α overlap poorly, as indicated by the following diagram:

Poor overlap; low S Good overlap; high S

Examples of diatomic molecules which have low overlap integrals because of a mismatch of atomic orbital energies are high-temperature gaseous KF and the hypothetical molecule NeS. In spite of a low β value, the KF molecule is stable because of the large amount of energy released on transferring a valence electron from potassium to the bonding MO, which has an energy slightly lower than that of the fluorine atomic orbital. The bonding in KF is principally ionic. The NeS molecule has very low stability because very little energy is released on transferring two electrons from the neon atomic orbital to the bonding MO.

2. The atomic orbitals must be positioned so that good overlap can occur. For example, consider the C—C bonds in cyclopropane and cyclohexane. The C—C—C bond angles in cyclopropane are 60°, which is much lower than the tetrahedral angle or even the 90° angle corresponding to use of pure p orbitals. Obviously the C—C bonds in cyclopropane must be "bent"; i.e., the overlapping atomic orbitals meet at an angle less than 180°. Such overlap is poorer (i.e., S is lower) than in the case of normal "straight" bonds such as probably exist in cyclohexane, in which the C—C—C bond angles are essentially tetrahedral angles. Evidence for the strain in the cyclopropane configuration is found in the high exothermicity of the hypothetical conversion of cyclopropane into cyclohexane:

$$2C_3H_6 \quad \rightarrow \quad C_6H_{12} \quad \Delta H° = -54.9 \text{ kcal mol}^{-1}$$

3. The atomic orbitals must have the same, or approximately the same, symmetry with respect to the bond axis. (Symmetry is briefly discussed in Chap. 2.) In the following examples, we shall consider the overlap between orbitals on atoms whose z axes lie on the bond axis. In the case of an s orbital and p_y orbital, the overlapping volume element on one side of the p_y orbital has the same magnitude but opposite sign as that on the other side of the p_y orbital.

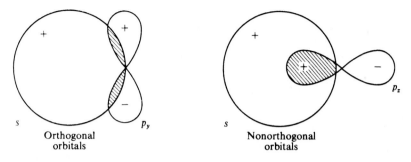

FIGURE 4.4

Overlap between s and p orbitals. In the case of the p_y orbital, $S = 0$; in the case of the p_z orbital, $S > 0$.

This fact is obvious from Fig. 4.4. Thus the integration over all space is exactly zero; i.e., $S = 0$. The atomic orbitals are said to be "orthogonal." On the other hand, the overlap of an s orbital and a p_z orbital is finite, and $S \neq 0$. Such orbitals are "nonorthogonal." As another example, consider the overlap of a p_y orbital on one atom with a d orbital on another atom (again with the atomic z axes coincident with the bond axis). In the case of d_{xz}, d_{xy}, $d_{x^2-y^2}$, and d_{z^2} orbitals, the overlap integral is exactly zero. However, in the case of the d_{yz} orbital, S is finite. The reader is invited to ponder the orthogonalities of all the other possible combinations of s, p, and d orbitals.

SIGMA, PI, AND DELTA MOLECULAR ORBITALS

An MO can be classified in terms of the number of nodal surfaces which pass through at least one atomic nucleus.

A nodal surface is a surface where the MO wave function has zero amplitude as a result of its changing sign on going from one side of the surface to the other. In simple or highly symmetric molecules (such as water and benzene), the pertinent nodal surfaces are usually planes. A sigma (σ) MO has *no* nodal surface passing through a nucleus. A pi (π) MO has *one* nodal surface passing through at least one nucleus. A delta (δ) MO has *two* nodal surfaces passing through at least one nucleus.

Consider the bonding and antibonding MOs resulting from the interaction of two atoms. If we take the internuclear axis as the z axis, then σ orbitals are formed from atomic orbital overlaps of the type $s + s$, $s + p_z$, $p_z + p_z$, $p_z + d_{z^2}$, etc. Pi MOs are formed from overlaps of the type $p_x + p_x$, $p_y + p_y$, $p_y + d_{yz}$, etc. The relatively rare δ MOs are formed from overlaps of the type $d_{xy} + d_{xy}$ and $d_{x^2-y^2} + d_{x^2-y^2}$. Some examples of the formation of σ and π MOs are schematically illustrated in Fig. 4.5. It is not possible to have more than two orthogonal $p\pi$ bonds between two atoms. These two π bonds have nodal planes which are perpendicular and which pass through the bond axis, as shown in Fig. 4.6. It can be seen that the positive overlap between lobes of the same sign exactly cancels the negative overlap between lobes of opposite sign.

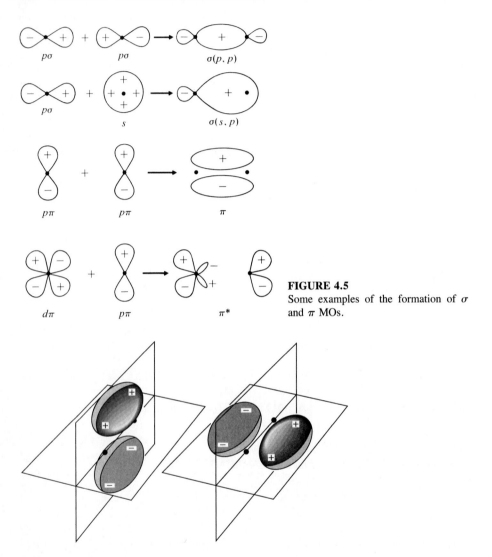

FIGURE 4.5
Some examples of the formation of σ and π MOs.

FIGURE 4.6
Schematic diagram showing two π bonds between a pair of atoms and the mutually perpendicular nodal planes. *(Reproduced with permission from F. A. Cotton, "Chemical Applications of Group Theory," 2d ed., Wiley-Interscience, New York, 1971.)*

DIATOMIC MOLECULES

Let us now consider all the MOs which form by the overlap of the valence-shell s and p orbitals in a homonuclear diatomic molecule in which the internuclear axis is taken as the z axis. The two s orbitals yield an $s\sigma_B$ and an $s\sigma^*$ MO, and the six p orbitals yield a $p\sigma_B$ and a $p\sigma^*$ MO, and two equivalent $p\pi_B$ and $p\pi^*$ MOs. Because the $p\pi$ overlap is poorer than the $p\sigma$ overlap, $|\beta_{p\sigma}| > |\beta_{p\pi}|$. If there were no interaction between the s and p_z orbitals, the energy level diagram would

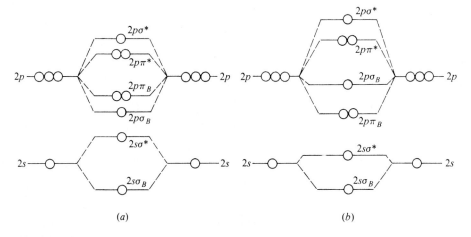

FIGURE 4.7
Valence-shell energy level diagrams for a homonuclear diatomic molecule containing first-row atoms. (*a*) corresponds to no interaction between $2s$ and $2p$ levels; (*b*) corresponds to substantial $2s$-$2p$ interaction.

look like that shown in Fig. 4.7*a*. However, in the molecules of the first-row elements, a significant amount of $2s$-$2p_z$ overlap occurs, causing the energies of the $2s\sigma$ orbitals to be depressed and the energies of the $2p\sigma$ orbitals to be raised. Consequently, for the molecules lighter than O_2, the $2p\pi_B$ levels lie below the $2p\sigma_B$ level, as shown in Fig. 4.7*b*.

Just as the electron configurations of the elements can be deduced by appropriate filling of the atomic orbitals, the electron configurations of the first-row diatomic molecules can be deduced by filling the MOs shown in Fig. 4.7. The configurations for the homonuclear molecules from Li_2 to Ne_2, as well as that of O_2^-, are given in Table 4.2. It should be noted that, as expected, dissociation energy increases and bond distance decreases with increasing bond order. The series O_2^+ ($r = 1.122$ Å), O_2 ($r = 1.21$ Å), O_2^- ($r = 1.28$ Å), and O_2^{2-}

TABLE 4.2
Diatomic species of first-row elements

Species	Valence electron configuration	Unpaired electrons	Bond order	D, kcal mol^{-1}†	r, Å
Li_2	$(s\sigma_B)^2$	0	1	26	2.67
Be_2	$(s\sigma_B)^2(s\sigma^*)^2$	0	0		
B_2	$(s\sigma_B)^2(s\sigma^*)^2(p\pi_B)^2$	2	1	71	1.59
C_2	$(s\sigma_B)^2(s\sigma^*)^2(p\pi_B)^4$	0	2	142	1.31
N_2	$(s\sigma_B)^2(s\sigma^*)^2(p\pi_B)^4(p\sigma_B)^2$	0	3	226	1.10
O_2	$(s\sigma_B)^2(s\sigma^*)^2(p\sigma_B)^2(p\pi_B)^4(p\pi^*)^2$	2	2	119	1.21
O_2^-	$(s\sigma_B)^2(s\sigma^*)^2(p\sigma_B)^2(p\pi_B)^4(p\pi^*)^3$	1	1.5	. . . .	1.33
F_2	$(s\sigma_B)^2(s\sigma^*)^2(p\sigma_B)^2(p\pi_B)^4(p\pi^*)^4$	0	1	38	1.42
Ne_2	$(s\sigma_B)^2(s\sigma^*)^2(p\sigma_B)^2(p\pi_B)^4(p\pi^*)^4(p\sigma^*)^2$	0	0		

† 1 kcal = 4.1840 kJ

$(r = 1.49 \text{ Å})$ also shows the expected inverse correlation of bond order and bond distance.

Probably the most significant feature of the simple MO treatment of the diatomic molecules is the ease with which the paramagnetism of B_2 and O_2 is accounted for. In each of these molecules, a pair of degenerate $p\pi$ MOs contain two electrons. According to Hund's rule, these electrons occupy separate orbitals and have parallel spins. Thus, although each of these molecules has an even number of electrons, each has a permanent magnetic moment and is paramagnetic. Another valuable achievement of the simple MO treatment is the prediction of zero bond order (i.e., no net bonding) in Be_2. This result is not at all obvious from simple valence pictures.

In heteronuclear molecules such as CO, NO, CS, and SiC, the corresponding valence atomic orbitals of the two atoms do not have the same energy. Nevertheless, if the electronegativities of the two atoms are not too dissimilar, the qualitative ordering of the MO energy levels in such molecules is similar to that in Fig. 4.7b, and the bonding can be discussed in terms of analogous MOs. For example, consider carbon monoxide, which is isoelectronic with molecular nitrogen. The MOs of CO are like those of N_2, with perturbations due to the difference in symmetry and the electronegativity difference between carbon and oxygen. A schematic representation of the MO energy level diagram is given in Fig. 4.8. The 1σ and 2σ orbitals are the oxygen 1s and carbon 1s atomic orbitals, respec-

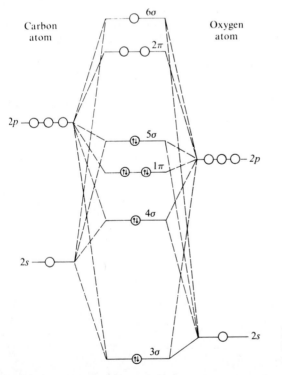

FIGURE 4.8
MO energy level diagram for carbon monoxide.

tively, and are not shown. Notice that the g and u subscripts are omitted from the symmetry labels because the molecule lacks a center of symmetry. The MO nomenclature scheme used in Fig. 4.8 is described below.

ELECTRON DENSITY CONTOUR MAPS

With modern computing facilities, it is a relatively easy problem to calculate accurate wave functions for the first-row diatomic molecules. Electron density contour maps for the MOs of the O_2 molecule, based on such calculations,[5] are shown in Fig. 4.9. The contours correspond to the electron densities in a plane passing through the two atoms. The MOs are labeled according to a scheme which is now common in the literature: A symbol such as σ or π is used to indicate the symmetry of the orbital; the subscript g (for "gerade") is used if the orbital wave function is symmetric with respect to inversion, and the subscript u (for "ungerade") is used if the orbital wave function is antisymmetric with respect to inversion.[6] The initial arabic numeral indicates the relative position of that particular type of orbital on an absolute energy scale; the lowest energy orbital of a particular symmetry is given the number 1, the next higher in energy the number 2, etc. This MO nomenclature scheme is superior to that used in Table 4.2 and Fig. 4.7 because it avoids the incorrect implication that the MOs are constructed from pure atomic orbitals, without any atomic orbital interaction or "mixing." It should be noted that the $1s$ core shells are so contracted because of the higher effective nuclear charge in the core region that they are essentially spherical with no significant overlap. Therefore the core electrons contribute very little to chemical bonding. It can be seen that the bonding orbitals ($2\sigma_g$, $3\sigma_g$, and $1\pi_u$) show an increase in electron density in the region between the nuclei and that the antibonding orbitals ($2\sigma_u$ and $1\pi_g$) show a reduction in electron density between the nuclei.

A different graphical method for representing the valence MOs of the oxygen molecule, which gives a good indication of the three-dimensional nature of the orbitals,[7] is shown in Fig. 4.10. Here, for the sake of completeness, the unoccupied $3\sigma_u$ (or "$2p\sigma^*$") orbital is illustrated, as well as the various occupied MOs.

Wave function contours for some of the carbon monoxide valence MOs are shown in Fig. 4.11.[8] Orbital 4σ is related to the "$2s\sigma^*$" orbital of N_2, but the contours indicate that the oxygen $2p\sigma$ orbital on the bond axis makes a

---

[5] A. C. Wahl, *Science*, **151**, 961 (1966).

[6] Inversion corresponds to reversing the directions of the x, y, and z coordinates. See Chap. 2 for a discussion of symmetry operations. A function which is symmetric with respect to inversion does not change sign upon inversion; a function which is antisymmetric with respect to inversion changes sign upon inversion.

[7] W. L. Jorgensen and L. Salem, "The Organic Chemist's Book of Orbitals," Academic, New York, 1973.

[8] J. B. Johnson and W. G. Klemperer, *J. Am. Chem. Soc.*, **99**, 7132 (1977).

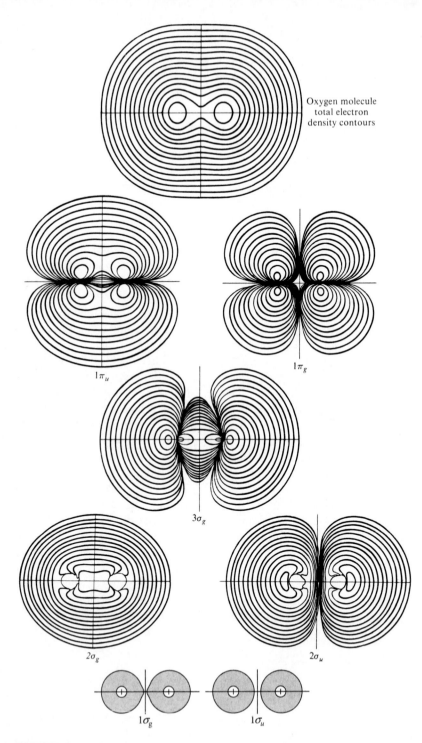

Oxygen molecule
total electron
density contours

$1\pi_u$

$1\pi_g$

$3\sigma_g$

$2\sigma_g$

$2\sigma_u$

$1\sigma_g$

$1\sigma_u$

FIGURE 4.9
Electron density contour maps for the MOs of O_2. [*Reproduced with permission from A. C. Wahl, Science, **151**, 961 (1966).*]

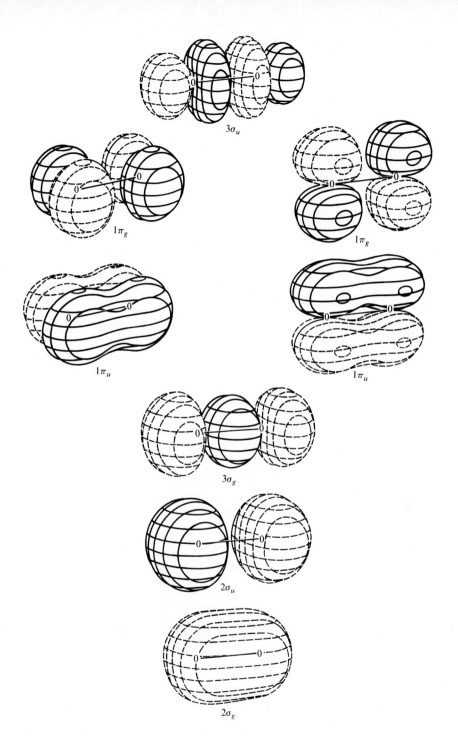

FIGURE 4.10
Three-dimensional representations of the O_2 MOs. The solid and dashed contours correspond to
$\psi > 0$ and $\psi < 0$, respectively. *(Reproduced with permission from W. L. Jorgensen and L. Salem,
"The Organic Chemist's Book of Orbitals," Academic, New York, 1973.)*

2π

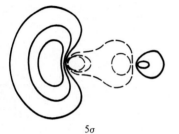

5σ

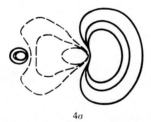

1π

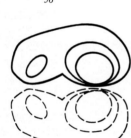

4σ

FIGURE 4.11
Wave function contour maps for selected carbon monoxide MOs. Solid and broken lines indicate contours of opposite sign having absolute values of 0.3, 0.2, and 0.1. The carbon atom is on the left. [*Reproduced with permission from J. B. Johnson and W. G. Klemperer, J. Am. Chem. Soc., **99**, 7132 (1977). Copyright 1977 American Chemical Society.*]

significant contribution. The MO is approximately derived from the carbon $2s$ orbital overlapping with a mixture of oxygen $2s$ and $2p$ orbitals. Some σ bonding results, but the MO is mainly a nonbonding lone-pair orbital.

The two 1π orbitals are bonding, but the contours show the dominance of the oxygen $p\pi$ orbitals; the bonding π electron density is concentrated on the more electronegative oxygen atom.

The highest occupied orbital, 5σ, is strongly polarized toward carbon. This MO has some σ-bonding character but is essentially a carbon lone-pair orbital.

The lowest unoccupied MOs, the 2π orbitals, are analogs of the $2p\pi^*$ orbitals of N_2 but are mainly located on the carbon atom. These orbitals figure importantly in the interaction of CO with transition-metal atoms. Notice that, if we consider the 3σ and 1π orbitals as bonding, and the 4σ and 5σ orbitals as nonbonding, we end up with a total of three bonds, just as in the $C{\equiv}O$: representation.

To illustrate the flow of electron density which occurs when atoms form chemical bonds, electron density "difference" plots can be used. One plots contours corresponding to the difference in electron density of the actual molecule and that of the hypothetical noninteracting spherical atoms. Figure 4.12 shows calculated electron density difference contours for molecular hydrogen and diatomic lithium.[9] The solid lines denote regions of increased electron density, and the dotted lines denote regions of decreased electron density. It is clear in each case that electron density moves from the far sides of the molecule into the overlap region between the nuclei. In both H_2 and Li_2, there is a distinct buildup of electron density near the atomic nuclei. This latter phenomenon is at first sight surprising but becomes reasonable upon recognizing that the large increase in electron density between the atoms allows the remaining electron density to become more concentrated in the region near the nuclei.

The amount of electron density which moves into the overlap region is quite small, typically about 0.1 to 0.3 of an electron. The smallness of this electron flow explains why, until recent years, experimental evidence for it was quite poor. However, now accurate x-ray and neutron diffraction data yield good electron density difference maps that show essentially the same sort of information as given by the theoretically calculated maps of Fig. 4.12. For example, accurate x-ray diffraction data for 2,5-di-N-chlorothioimino-3,4-dicyanothiophene,[10] yield

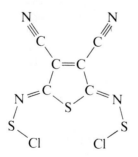

the electron density difference map shown in Fig. 4.13. This diagram reveals both

---

[9] Data of R. W. F. Bader, reported by C. A. Coulson, "The Shape and Structure of Molecules," pp. 36–37, Clarendon, Oxford, 1973.

[10] F. Wudl and E. T. Zellers, *J. Am. Chem. Soc.*, **102**, 4283 (1980).

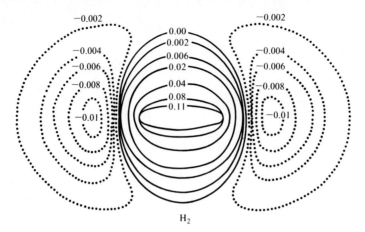

H$_2$

(a)

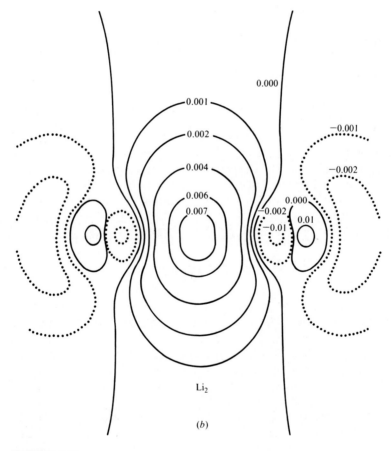

Li$_2$

(b)

FIGURE 4.12
Calculated electron density difference plots for H$_2$ (a) and Li$_2$ (b). The contours refer to electron density differences between the molecules and the atoms. Solid lines denote increased electron density, and dotted lines denote decreased electron density. *(Reproduced with permission from C. A. Coulson, "The Shape and Structure of Molecules." Clarendon, Oxford, 1973.)*

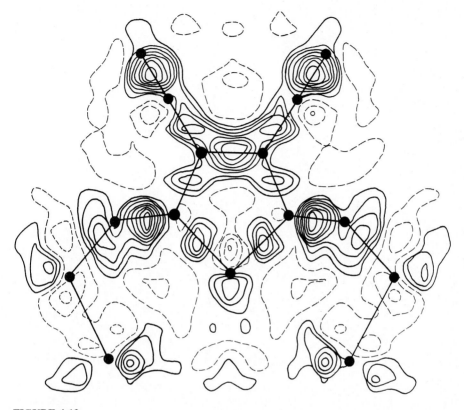

FIGURE 4.13

Electron density difference plot for 2,5-di-*N*-chlorothioimino-3,4-dicyanothiophene, obtained from x-ray diffraction data. Solid lines denote increased electron density. *(Reproduced with permission from Molecular Structure Corporation, College Station, Texas.)*

the expected buildup of density in the C—C, C—S, C—N, and N—S bond regions and the concentration of density in the lone-pair regions of the sulfur and chlorine atoms. Quantitative measurements by this method of the excess electron density in the bonding region of various carbon-carbon bonds have shown that the excess "bonding charge" is correlated with the C—C bond length.[11] Figure 4.14 shows a plot of measured bonding charge versus bond length. The bonding density falls off almost linearly with increasing bond length, in accord with the idea that a chemical bond is due to the accumulation of negative charge in the region between atoms to an extent sufficient to balance the nuclear repulsion forces.

---

[11] Z. Berkovitch-Yellin and L. Leiserowitz, *J. Am. Chem. Soc.*, **99,** 6106 (1977).

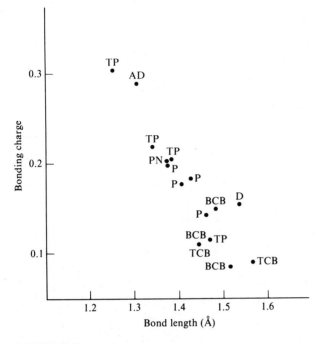

FIGURE 4.14

Plot of "bonding charge" (increased electron density in the bonding region) versus bond length for various C—C bonds. TP = tetraphenylbutatriene; AD = allenedicarboxylic acid–acetamide; PN = *p*-nitropyridine oxide; D = diamond; TCB = tetracyanocyclobutane; P = perylene; BCB = a bicyclobutane derivative. [*Reproduced with permission from Z. Berkovitch-Yellin and L. Leiserowitz, J. Am. Chem. Soc., 99, 6106 (1977). Copyright 1977 American Chemical Society.*]

SYMMETRY AND POLYATOMIC MOLECULES

In symmetric molecules, there are sets of equivalent atoms and atomic orbitals. When considering the MOs of such molecules, it is helpful to form linear combinations of the equivalent atomic orbitals, called "group orbitals." These group orbitals can then be treated just like individual atomic orbitals; they can be combined with other atomic orbitals or with other group orbitals to form MOs. However, no matter what types of orbitals are combined to form MOs, they must have the same symmetry properties and they must have satisfactory overlap.[12]

To illustrate the use of group orbitals, we shall discuss the MOs of the triiodide ion, I_3^-. From valence-shell electron repulsion theory (Chap. 3) we know that this is a linear species. The middle atom can be considered as having three nonbonding electron pairs in sp^2 hybrid orbitals in a plane normal to the

---

[12]   Group-theoretical labels such as σ_g^+, b_1, and t_2 are useful for designating the symmetry properties of orbitals and combinations of orbitals. A full understanding of their significance is not necessary to follow the discussion of this text. A brief introduction to the subject is given in Chap. 2.

bonds. Each of the terminal atoms also has a set of three nonbonding electron pairs, but the orientation of these pairs is not certain; they probably occupy orbitals having somewhat more p character than sp^2. Thus we have three collinear atomic orbitals which can be involved in σ bonding: a p orbital on the middle atom, and hybrid orbitals on the terminal atoms. The two equivalent hybrid orbitals on the terminal atoms, ϕ_1 and ϕ_3, can be combined into two group orbitals:

$$\phi_G = \frac{1}{\sqrt{2}}(\phi_1 \pm \phi_3)$$

These combinations are pictured below:

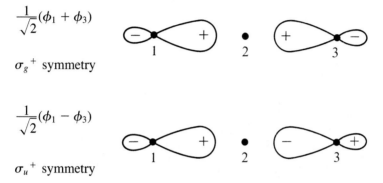

Note that the $\sigma_g{}^+$ group orbital[12] is symmetric with respect to inversion through the center of the $I_3{}^-$ ion and that the $\sigma_u{}^+$ group orbital is antisymmetric with respect to inversion. The p orbital on the middle atom is antisymmetric with respect to inversion; like the $\phi_1 - \phi_3$ group orbital, it has $\sigma_u{}^+$ symmetry.[12]

The overlap integral of the symmetric group orbital and the middle p orbital is exactly zero. Thus only the antisymmetric group orbital can interact with the middle p orbital to form a bonding and an antibonding MO. These MOs are represented as follows:

$$\psi = \frac{1}{\sqrt{2}}\phi_2 \pm \frac{1}{2}(\phi_1 - \phi_3)$$

The symmetric group orbital $(1/\sqrt{2})(\phi_1 + \phi_3)$ is essentially a nonbonding MO. The reader may wish, as an exercise, to prove that all three MOs, as written, are normalized (under the assumption that $S = 0$).

An energy level diagram for $I_3{}^-$ is shown in Fig. 4.15. Note that these orbitals are occupied by four valence electrons. Two are nonbonding and two are bonding. Obviously we should look upon the I—I bonds in $I_3{}^-$ as half-bonds, in agreement with simple octet structure considerations. The other 18 valence electrons occupy the 9 nonbonding orbitals that we referred to earlier.

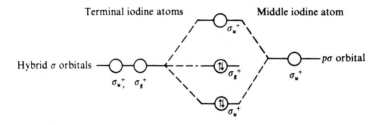

FIGURE 4.15
Energy level diagram for the σ molecular orbitals of I_3^- formed from the atomic orbitals lying on the bonding axis. Valence-shell d orbitals are ignored.

The nonbonding orbital of Fig. 4.15 is shared by the terminal atoms; thus $3\frac{1}{2}$ lone pairs are located on each terminal atom.

The model of I_3^- which we have just described completely ignores valence-shell d-orbital participation. If we choose to have the z axis of the middle atom lie along the bonds, we see that the d_{z^2} orbital of the middle atom has the same symmetry as that of the symmetric group orbital $(1/\sqrt{2})(\phi_1 + \phi_3)$. Now it is likely that the overlap between this d_{z^2} orbital and the terminal-atom hybrid orbitals is very poor; the orbital energies are probably quite far apart. However, if we are willing to admit a weak interaction between the middle-atom d_{z^2} orbital and the symmetric group orbital, we may draw a modified energy level diagram as shown in Fig. 4.16. Most evidence indicates that d-p bonding of the type shown is relatively unimportant; for most practical purposes the σ_g^+ orbital is nonbonding.[13]

---

[13]   T. B. Brill, *J. Chem. Educ.*, **50**, 392 (1973); C. D. Cornwell and R. S. Yamasaki, *J. Chem. Phys.*, **27**, 1060 (1957); A. E. Reed and F. Weinhold, *J. Am. Chem. Soc.*, **108**, 3586 (1986).

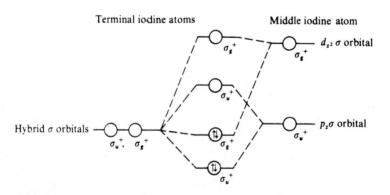

FIGURE 4.16
Energy level diagram for the σ molecular orbitals of I_3^- formed from the atomic orbitals lying on the bonding axis. A valence-shell d_{z^2} orbital is included.

Let us now consider the π bonding in the ozone molecule. Assume that this V-shaped molecule lies in the yz plane; we are then concerned with the overlap of the p_x orbitals on each oxygen atom. The two terminal p_x orbitals are equivalent; their corresponding group orbitals are

$$\phi_G = \frac{1}{\sqrt{2}}(\phi_1 \pm \phi_3)$$

These may be shown as the following projections:

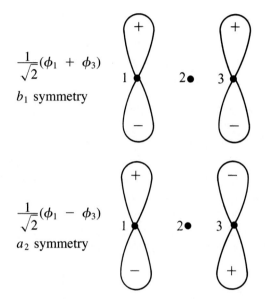

$$\frac{1}{\sqrt{2}}(\phi_1 + \phi_3)$$

b_1 symmetry

$$\frac{1}{\sqrt{2}}(\phi_1 - \phi_3)$$

a_2 symmetry

The p_x orbital on the middle oxygen atom has the same symmetry, b_1, as the $(1/\sqrt{2})(\phi_1 + \phi_3)$ group orbital.

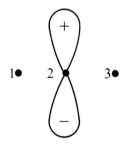

Therefore these orbitals combine to form antibonding and bonding MOs, and the group orbital of a_2 symmetry is essentially a nonbonding MO. The corresponding energy level diagram is shown in Fig. 4.17. Inasmuch as there are only two bonding π electrons, the π bond order in O_3 is $\frac{1}{2}$.

Finally we shall consider the σ bonding in a tetrahedral species such as BH_4^-, CH_4, NH_4^+, ClO_4^-, or PCl_4^+. The central atom has an s orbital and

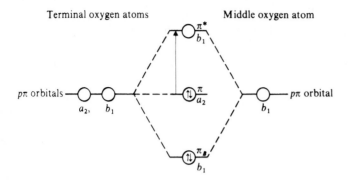

FIGURE 4.17
Energy level diagram for the π molecular orbitals of ozone, O_3. The vertical arrow indicates the $\pi \rightarrow \pi^*$ transition energy.

three p orbitals in its valence shell, which are involved in bonding. Each ligand atom has one atomic orbital involved in the bonding (a $1s$ orbital in the case of a hydrogen atom, and a hybrid orbital in the case of an oxygen or chlorine atom). The group orbital wave function which can combine with the central-atom s orbital is

$$\phi_G = \frac{1}{\sqrt{4}}(\phi_1 + \phi_2 + \phi_3 + \phi_4)$$

where ϕ_1, ϕ_2, ϕ_3, and ϕ_4 are the individual ligand orbital wave functions. The corresponding MO wave functions are

$$\psi = \frac{1}{\sqrt{2}}\phi_s \pm \frac{1}{2\sqrt{2}}(\phi_1 + \phi_2 + \phi_3 + \phi_4)$$

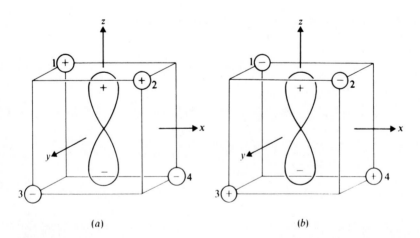

FIGURE 4.18
Bonding (a) and antibonding (b) combinations of the hydrogen s orbitals and a carbon $2p_z$ orbital in methane. Similar combinations can be drawn for the degenerate $2p_x$ and $2p_y$ orbitals.

In order to see which combinations of ligand orbitals can combine with the central-atom p orbitals, it is helpful to refer to Fig. 4.18. Figure 4.18a shows the bonding combination,

$$\psi_B(p_z) = \frac{1}{\sqrt{2}}\phi_{p_z} + \frac{1}{2\sqrt{2}}(\phi_1 + \phi_2 - \phi_3 - \phi_4)$$

and Fig. 4.18b shows the antibonding combination,

$$\psi_A(p_z) = \frac{1}{\sqrt{2}}\phi_{p_z} - \frac{1}{2\sqrt{2}}(\phi_1 + \phi_2 - \phi_3 - \phi_4)$$

Similarly, for the p_x and p_y orbitals, we write

$$\psi_B(p_x) = \frac{1}{\sqrt{2}}\phi_{p_x} + \frac{1}{2\sqrt{2}}(-\phi_1 + \phi_2 - \phi_3 + \phi_4)$$

$$\psi_A(p_x) = \frac{1}{\sqrt{2}}\phi_{p_x} - \frac{1}{2\sqrt{2}}(-\phi_1 + \phi_2 - \phi_3 + \phi_4)$$

$$\psi_B(p_y) = \frac{1}{\sqrt{2}}\phi_{p_y} + \frac{1}{2\sqrt{2}}(-\phi_1 + \phi_2 + \phi_3 - \phi_4)$$

$$\psi_A(p_y) = \frac{1}{\sqrt{2}}\phi_{p_y} - \frac{1}{2\sqrt{2}}(-\phi_1 + \phi_2 + \phi_3 - \phi_4)$$

Because of the symmetry of the molecule, the p_x, p_y, and p_z orbitals are equivalent. Therefore $\psi_B(p_x)$, $\psi_B(p_y)$, and $\psi_B(p_z)$ are degenerate; i.e., they have the same energy levels. The same is true of $\psi_A(p_x)$, $\psi_A(p_y)$, and $\psi_A(p_z)$. Thus the energy level diagram for this system, in Fig. 4.19, shows that the eight bonding electrons occupy a single orbital (of a_1 symmetry) and three degenerate orbitals (of t_2 symmetry). Although the four σ bonds are equivalent, there are essentially two kinds of bonding electrons, with entirely different ionization energies. The separate ionization energies corresponding to the two MOs are obvious in the ultraviolet photoelectron spectrum of Fig. 4.20.

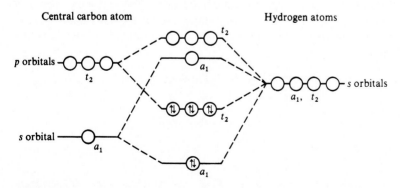

FIGURE 4.19
Energy level diagram for the valence molecular orbitals of methane.

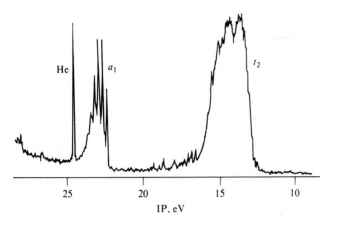

FIGURE 4.20
Ultraviolet photoelectron spectrum of methane. The photon energy was 41 eV; the signal at 24.6 eV
is due to helium. [*Reproduced with permission from G. Bieri and L. Asbrink, J. Electron Spectrosc.
Rel. Phen., 20, 149 (1980).*]

HYBRIDIZATION[14]

Although the concept of hybridization is strictly a part of valence bond theory, we
have postponed discussion of this topic until now, when the reader is familiar with
the concept of linear combinations of atomic orbitals and can directly compare
the application of hybridization theory and MO theory to a particular molecule.

We know from experiment that the methane molecule is tetrahedral and that
the four C—H bonds are equivalent. If we assume that the C—H bonds are single
bonds, formed by combination of the hydrogen $1s$ orbitals with the carbon $2s$ and
$2p$ orbitals, then each bond must involve, or "use," one-quarter of the carbon $2s$
orbital and one-quarter of the three carbon $2p$ orbitals. If we wish to describe the
bonding of methane in terms of localized two-electron bonds, we can assume that
the one $2s$ and three $2p$ orbitals of carbon "mix" to form four equivalent hybrid
orbitals directed toward the hydrogen atoms. Then each bond can be considered
the result of the overlap of an sp^3 hybrid orbital of carbon and a $1s$ orbital of
hydrogen. If we let the symbols s, p_x, p_y, and p_z stand for the valence atomic
orbitals of carbon, the four hybrid orbitals can be represented as the following
linear combinations:

$$\phi_1 = \tfrac{1}{2}(s + p_x + p_y + p_z)$$

$$\phi_2 = \tfrac{1}{2}(s + p_x - p_y - p_z)$$

$$\phi_3 = \tfrac{1}{2}(s - p_x - p_y + p_z)$$

$$\phi_4 = \tfrac{1}{2}(s - p_x + p_y - p_z)$$

[14] L. Pauling, "The Nature of the Chemical Bond," 3d ed., pp. 111–123, Cornell University Press,
Ithaca, N.Y., 1960; F. A. Cotton, "Chemical Applications of Group Theory," 2d ed., chap. 8,
Wiley-Interscience, New York, 1971.

TABLE 4.3
Some common hybridizations and their geometries

sp	Linear
sp^2	Trigonal (120° bond angles)
sp^3, sd^3	Tetrahedral (109.5° bond angles)
d^2sp^3	Octahedral
dsp^2	Square planar
dsp^3	Trigonal bipyramidal or square pyramidal

Similar sets of hybrid orbitals can be constructed from other combinations of atomic orbitals. Some of these types of hybridization are listed in Table 4.3. In each case, except dsp^3, it is possible to construct the hybrid orbitals so that they are equivalent. However, in the case of trigonal bipyramidal and square pyramidal geometries, such equivalence is obviously impossible.

The properties of a hybrid orbital are highly dependent on the relative proportions of the different kinds of atomic orbitals which are combined to form the hybrid orbital. For example, the overlap of a hybrid orbital with another orbital changes markedly as the fractional s (or p) content of the hybrid changes. The calculated overlap integrals for C—C and C—H bonds as a function of the percent s character of the carbon orbital[15] are plotted in Fig. 4.21. It can be seen that the overlap integrals have maximum values at hybridizations intermediate between pure s and pure p, and that in the hybridization range generally found in molecules (i.e, between pure p and sp hybridization), the overlap integral increases with s character. We have already pointed out that covalent bond strength is proportional to the overlap integral. Hence we expect that bond strength should

---

[15] A. Maccoll, *Trans. Faraday Soc.*, **46**, 369 (1950).

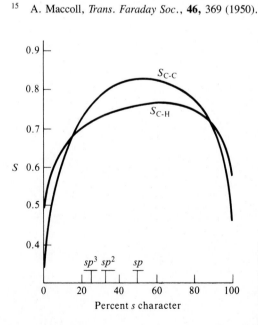

FIGURE 4.21
The calculated overlap integral, S, for C—C and C—H bonds as a function of percent s character. [*Adapted from a plot of A. Maccoll, Trans. Faraday Soc., 46, 369 (1950).*]

increase as the s character of a hybrid orbital increases. Stronger bonds correspond to shorter bonds, and thus we have an explanation of the data we saw in Table 3.5, which show a shortening of the C—C single bond on going from aliphatic hydrocarbons to compounds in which the other bonds to the carbon atoms are involved in π bonding. The π bonding uses only the p orbitals of the carbon atoms, and therefore the remaining σ bonds must be formed from hybrid orbitals possessing relatively high s character.

An s orbital is lower in energy, and has a higher electronegativity, than a p orbital. Therefore the effective electronegativity of an s, p hybrid orbital increases as the fractional s content (s character) increases.[16] Consider the hybridization of the lone-pair orbitals in ammonia, pyridine, and molecular nitrogen. In pyridine one nitrogen p orbital is involved in π bonding, and in molecular nitrogen two p orbitals of each nitrogen atom are involved in π bonding. Thus the average s character of the σ orbitals (the lone-pair and the σ-bonding orbitals) increases on going from NH_3 to NC_5H_5 to N_2. If the s character were divided equally among the lone-pair and σ-bonding orbitals in each of these cases, the lone pair would occupy an sp^3 hybrid in NH_3, an sp^2 hybrid in NC_5H_5, and an sp hybrid in N_2. The relative base strengths of the compounds are in accord with this simple picture. Ammonia is a stronger base than pyridine, which is a much stronger base than molecular nitrogen.

In the absence of steric effects, replacement of X in the structure X—A—Y by a more electronegative group causes the adjacent A—Y bond to become shorter. The effect is most pronounced when fluorine is introduced into a molecule (the C—F distances in CH_3F and CF_4, for example, are 1.391 and 1.323 Å, respectively) but generally can be seen whenever an atom is replaced by a more electronegative atom. Some examples follow.

Bond	Molecule	Bond length, Å
C—Cl	CH_3Cl	1.783
	CH_2Cl_2	1.772
	$CHCl_3$	1.764
	CCl_4	1.766
C—Br	CH_3Br	1.939
	$CHBr_3$	1.930
C—Cl	CH_3Cl	1.784
	CF_3Cl	1.751
C—C	C_2H_6	1.536
	C_2F_6	1.51
N—Cl	$N(CH_3)_2Cl$	1.77
	$N(CH_3)Cl_2$	1.74
C=O	H_2CO	1.225
	HFCO	1.192
	F_2CO	1.17

[16] H. A. Bent, *Chem. Rev.*, **61**, 275 (1961).

The data can be understood by recognizing that atomic p character concentrates in orbitals directed toward electronegative substituents (and, of course, that atomic s character concentrates in orbitals directed toward electropositive substituents). This rule follows from the fact that an electron in a p orbital is on the average farther from the nucleus than an electron in an s orbital; the p-orbital electron is more readily shared by an orbital centered on another atom. Replacement of a substituent by a more electronegative atom causes a rehybridization, i.e., a shift of p character to the orbital involved in bonding to the electronegative atom and a shift of s character to the other bonds. Hence the other bonds are strengthened and shortened.

It is instructive to discuss the bonding in the water molecule in terms of orbital hybridization and valence-shell electron repulsion. The ground-state electron configuration of the oxygen atom is $2s^2 2p^4$, with two unpaired $2p$ electrons. If this atom were to react with two hydrogen atoms without any hybridization, one would expect the resulting water molecule to have an H—O—H bond angle of 90°. In such a molecule the lone pairs would occupy the $2s$ and one $2p$ orbital or, what is precisely equivalent, two sp hybrid orbitals directed at right angles to the plane of the molecule. Obviously in such a molecule there would be a particularly strong repulsion between the lone-pair electrons and the bonding electrons. Now let us assume that we start with an oxygen atom in which the two lone pairs occupy hybrid orbitals which have more than 50 percent p character and in which the unpaired electrons occupy orbitals which have less than 100 percent p character. Energy is required to promote a ground-state oxygen atom to this hypothetical hybridized state, because the promotion corresponds to the excitation of a fraction of an electron from a $2s$ orbital to a $2p$ orbital. (If a complete electron were excited, one would obtain an atom with the configuration $2s^1 2p^5$.) In the hybridized oxygen atom, the angle between the lone pairs would be less than 180°, and the promotion energy can be partly ascribed to the resultant extra electron-electron repulsion. The hybridized atom would react with two hydrogen atoms to form a molecule in which the H—O—H bond angle is greater than 90° and in which the angle between either lone pair and either O—H bond is greater than 90°. The promotion energy would be compensated by the reduction in the electron-electron repulsions and by the fact that an increase in the s character of an orbital causes an increase in the bond energy. Obviously in the actual water molecule these energy terms are exactly compensating. Such balancing of promotion energy, electron repulsion energy, and bond energy is a common feature of orbital hybridization schemes.

It should be emphasized that the concept of hybridization is essentially a mathematical operation which we use mainly for convenience in the description of bonding. One must be careful not to draw incorrect conclusions regarding the excited states of molecules for which the ground states can be described in terms of hybridized orbitals. For example, in the case of methane, if we consider only the four equivalent bonding orbitals corresponding to the use of hybrid sp^3 orbitals, it is not apparent that there should be two different ionization potentials for the bonding electrons. However, evidence for these two ionizations is seen in the photoelectron spectrum of Fig. 4.20, which shows two separate photolines.

WALSH DIAGRAMS

When considering the structure of a molecule which could conceivably assume any conformation between, or including, two limiting conformations, it is instructive to study the correlation of MOs of one limiting conformation with those of the other. Walsh[17] was the first to apply this method extensively, and correlation diagrams of this type are now often called "Walsh diagrams." Consider the general case of an AH_2 molecule, for which we may consider the limiting values of the H—A—H bond angles to be 90 and 180°. We shall assume that the MOs are formed solely from valence s and p orbitals of atom A and $1s$ orbitals of the hydrogen atoms. Crude pictures of these MOs and a qualitative Walsh diagram are shown in Fig. 4.22.[18]

The lowest-energy valence MO for the linear geometry, $2\sigma_g$, is formed from the s orbital of A and the symmetric combination of hydrogen $1s$ orbitals. Upon

[17]   A. D. Walsh, *J. Chem. Soc.*, 2260, 2266, 2288, 2296, 2306 (1953). The method was first outlined by R. S. Mulliken, *Rev. Mod. Phys.*, **14**, 204 (1942).

[18]   B. M. Gimarc, *J. Am. Chem. Soc.*, **93**, 593 (1971).

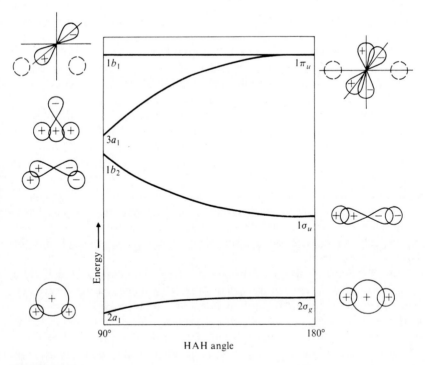

FIGURE 4.22
Walsh diagram for AH_2 molecule, with diagrams of the valence MOs.

bending the molecule, the $2a_1$ orbital which forms is slightly lower in energy than the original $2\sigma_g$ because the hydrogen $1s$ orbitals then overlap better with each other and with the p orbital of A on the twofold molecular axis, which becomes a minor component of $2a_1$. The $1\sigma_u$ MO of the linear molecule is formed from the p orbital of A on the molecular axis and the asymmetric combination of hydrogen $1s$ orbitals. Bending the molecule converts this MO to the $1b_2$ MO, which has higher energy because the hydrogen $1s$ orbitals then do not overlap as well with the p orbital. The highest occupied MOs of the linear molecule are the doubly degenerate $1\pi_u$ orbitals, which consist of the two p orbitals of A perpendicular to the molecular axis which have no net overlap with the hydrogen $1s$ orbitals. On bending, the p orbital which lies on the twofold molecular axis overlaps with the symmetric combination of hydrogen $1s$ orbitals and thus forms the $3a_1$ orbital, with lower energy than $1\pi_u$. In the bent molecule, a small amount of A-atom s orbital can also contribute to $3a_1$. In Fig. 4.22 it can be seen that the $3a_1$ level falls from the $1\pi_u$ level faster than the $1b_2$ level rises from the $1\sigma_u$ level. This difference can be rationalized by the fact that, on bending the molecule, the increase in the overlap integral of $3a_1$ is greater than the decrease in the overlap integral on going from $1\sigma_u$ to $1b_2$.[18] The p orbital perpendicular to the molecular plane (the $1b_1$ MO) has no overlap with the hydrogen $1s$ orbitals and therefore has the same energy as $1\pi_u$.

Because the diagram of Fig. 4.22 is based on approximate calculations, and because factors such as nucleus-nucleus repulsions and electron-electron repulsions have not been taken into account, it can only be used to predict approximate shapes of molecules—i.e., whether a molecule is linear or bent. Let us consider some examples. A molecule containing only four valence electrons, such as BeH_2, will have the lowest two orbitals filled. From the diagram, we see that a minimum in energy will be achieved by having a linear molecule, with the $2\sigma_g$ and $1\sigma_u$ orbitals filled. On the other hand, the first excited state of BeH_2 should be bent, corresponding to one electron each in the $3a_1$ and $1b_2$ orbitals.

The CH_2 molecule, with six valence electrons, is similarly expected to be bent. From the qualitative diagram, the bond angle cannot be predicted. However, very accurate calculations[19] have shown that the ground state has a bond angle of $134°$. At that bond angle, the $3a_1$ and $1b_1$ energy levels are so close together that the energy difference is less than the repulsion energy between two electrons in the $3a_1$ orbital. Hence the ground state is a triplet, and the electron configuration is

$$2a_1{}^2 \; 1b_2{}^2 \; 3a_1{}^1 \; 1b_1{}^1$$

The singlet state lies only 8 to 9 kcal above the ground state; the bond angle is $102.4°$, and the electron configuration is

$$2a_1{}^2 \; 1b_2{}^2 \; 3a_1{}^2$$

---

[19]   C. W. Bauschlicher, H. F. Schaefer, and P. S. Bagus, *J. Am. Chem. Soc.*, **99**, 7106 (1977).

The water molecule has two more electrons than CH_2, and all the orbitals in Fig. 4.22 are filled. Because of the steepness of the $3a_1-1\pi_u$ correlation line, we predict the water molecule, like CH_2, to be bent.

In general, we can rationalize why a given AH_2 molecule is bent or linear in terms of the tendency for the molecule to lower its total energy by keeping the lower-energy s orbital of atom A as fully occupied as possible. In the case of molecules containing fewer than eight valence electrons, this condition is usually achieved by making the empty or partially filled orbitals pure p orbitals. The particular bond angle assumed by a bent AH_2 molecule is determined by other considerations, the most important of which are electron-electron repulsions.[20]

PI БONDING BEYOND THE FIRST ROW
OF THE PERIODIC TABLE

There are many known compounds containing elements of the first row of the periodic table in which these elements are π-bonded either to themselves or to other first-row elements. However, there are few examples of compounds in which the heavier analogs of these elements are π-bonded. For example, carbon forms double bonds in CO_2, CH_3COOH, C_2H_4, etc., whereas silicon and germanium form no analogous molecules that are stable under ordinary conditions. Nitrogen forms N_2, NO_3^-, HCN, etc., whereas the phosphorus and arsenic analogs are extremely unstable species. Because of these facts, it was once believed that only elements of the first row are capable of forming stable π bonds. However, high-quality calculations[21] on compounds of heavier elements tend to refute this conclusion. For example, it has been calculated that $H_2Si{=}CH_2$ is stable with respect to the triplet state $H_2\dot{S}i-\dot{C}H_2$ by about 28 kcal mol^{-1} and that the bond strength of the Si—C $3p\pi$-$2p\pi$ bond as measured by the rotational barrier is about 46 kcal mol^{-1}. Therefore we conclude that the paucity of π-bonded compounds of silicon, phosphorus, etc., is not due to an intrinsic instability of the π bonds, but rather to a high reactivity which leads to conversion of the π-bonded compounds to more stable σ-bonded compounds. For example, the existence of short-lived intermediates of the type R_2SiCH_2 has been deduced from studies of the thermal decomposition of monosilacyclobutanes which yield 1,3-disilacyclobutanes and ethylene. Kinetic data support the mechanism

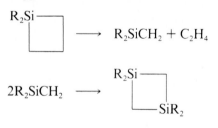

[20]   M. B. Hall, *J. Am. Chem. Soc.*, **100**, 6333 (1978); *Inorg. Chem.*, **17**, 2261 (1978).

[21]   R. Ahlrichs and R. Heinzmann, *J. Am. Chem. Soc.*, **99**, 7452 (1977). However, theoretical studies of HC≡SiH indicate that it is unstable with respect to the isomer $H_2C{=}Si$. See M. S. Gordon and J. A. Pople, *J. Am. Chem. Soc.*, **103**, 2945 (1981).

Kinetically stable compounds containing carbon-silicon double bonds have been prepared by surrounding the Si=C bond with extremely bulky groups. The large groups prevent cyclization and polymerization. For example, the substituted silaethylene with the following structure has been isolated as a crystalline solid, stable at room temperature.[22]

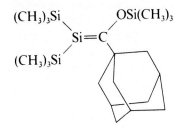

Similarly, tetramesityldisilene, one of several known compounds containing silicon-silicon double bonds, is a solid stable at room temperature.[23] This material is formed by the ultraviolet photolysis of 2,2-bis(2,4,6-trimethylphenyl)hexamethyltrisilane and undergoes addition reactions across the Si=Si double bond analogous to those of olefins:

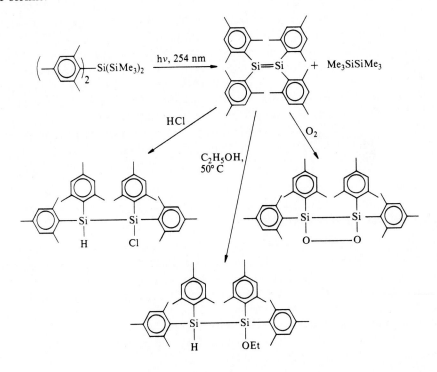

---

[22]   A. G. Brook et al., *J. Am. Chem. Soc.*, **104**, 5667 (1982).

[23]   R. West, M. J. Fink, and J. Michl, *Science*, **214**, 1343 (1981); R. West, *Science*, **225**, 1109 (1984).

The nonexistence of molecular SiO_2 under ordinary conditions is due to the fact that SiO_2 readily polymerizes to form solid silica, in which each silicon atom is σ-bonded to four oxygen atoms. Carbon dioxide undergoes no analogous polymerization probably because four oxygen atoms would be crowded if bonded to a single carbon atom.

Although trimethylamine has a pyramidal conformation (with a C—N—C bond angle of 110.9°), trisilylamine, $(SiH_3)_3N$, has a planar conformation for the Si_3N skeleton (with an Si—N—Si bond angle of 120°). This planarity has often been attributed to $p\pi \rightarrow d\pi$ bonding, in which the nitrogen lone-pair electrons expand onto the silicon atoms because of the overlap of the nitrogen $2p$ orbital (perpendicular to the Si_3N plane) with empty $3d$ orbitals of the silicon atoms.

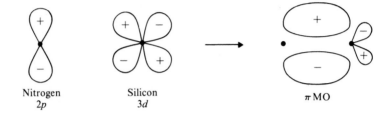

Nitrogen   Silicon π MO
$2p$ $3d$

This type of π bonding, or delocalization of the nitrogen lone pair, would necessarily be much reduced in a pyramidal conformation, because then the nitrogen $2p$ and silicon $3d$ orbitals would not be appropriately oriented for good overlap. Indeed, theoretical calculations show that there is a small but significant amount of $p\pi \rightarrow d\pi$ bonding in trisilylamine.[24] However, it is not clear whether this π bonding is primarily responsible for the planarity of the molecule. It has been argued that the planarity is caused by electrostatic repulsion between SiH_3 groups. The calculations indicate, as one might expect from the electronegativities, that the Si—N bond is much more polar than the C—N bond. The calculated carbon atom charge in $(CH_3)_3N$ is $+0.23$, and the calculated silicon atom charge in $(SiH_3)_3N$ is $+0.80$.

SIMPLE HÜCKEL THEORY

One of the best known and most important applications of *simple* MO theory is the study of π bonding in the cyclic aromatic hydrocarbons, C_nH_n, as originally carried out by Hückel.[25] Because aromatic hydrocarbons are planar, one can rigorously separate the MOs into σ MOs and π MOs. In the simple Hückel theory, one considers only the π MOs derived from the n orbitals of $p\pi$ symmetry and makes the approximations that *all* $S_{ij} = 0$ and that $H_{ij} = 0$ except when the ith and jth $p\pi$ orbitals are on adjacent atoms. Thus the general form of the secular determinant for a cyclic hydrocarbon, C_nH_n, is

[24]   L. Noodleman and N. L. Paddock, *Inorg. Chem.*, **18**, 354 (1979).
[25]   E. Hückel, *Z. Physik*, **70**, 204 (1931); **76**, 628 (1932); *Z. Elecktrochem.*, **43**, 752 (1937).

$$\begin{vmatrix} H_{11} - E & H_{21} & 0 & \cdots & H_{n1} \\ H_{12} & H_{22} - E & H_{32} & \cdots & 0 \\ 0 & H_{23} & H_{33} - E & \cdots & 0 \\ \cdots\cdots\cdots\cdots\cdots\cdots\cdots\cdots\cdots\cdots\cdots \\ H_{1n} & 0 & 0 & \cdots & H_{nn} - E \end{vmatrix} = 0$$

To illustrate the calculational methods, we shall apply the simple Hückel theory to the hypothetical square planar cyclobutadiene, C_4H_4. The p orbitals perpendicular to the plane of the molecule can be numbered as follows:

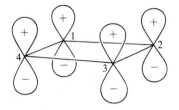

We write $\psi = c_1\phi_1 + c_2\phi_2 + c_3\phi_3 + c_4\phi_4$. We again make the substitutions $H_{ii} = \alpha$ and, when the ith and jth orbitals are adjacent, $H_{ij} = \beta$. The secular determinant for the π orbitals of C_4H_4 then becomes

$$\begin{vmatrix} \alpha - E & \beta & 0 & \beta \\ \beta & \alpha - E & \beta & 0 \\ 0 & \beta & \alpha - E & \beta \\ \beta & 0 & \beta & \alpha - E \end{vmatrix} = 0$$

For simplification, we divide through by β and replace $(\alpha - E)/\beta$ by x.

$$\begin{vmatrix} x & 1 & 0 & 1 \\ 1 & x & 1 & 0 \\ 0 & 1 & x & 1 \\ 1 & 0 & 1 & x \end{vmatrix} = 0$$

This equation is equivalent to $x^2(x^2 - 4) = 0$, whose roots are -2, 0, 0 and $+2$. Hence there are two nonbonding MOs of energy α, a bonding MO of energy $\alpha + 2\beta$, and an antibonding MO of energy $\alpha - 2\beta$.

When similar calculations are carried out for other C_nH_n molecules, it becomes apparent that the planar C_nH_n π energy level diagram always has a strongly bonding nondegenerate level with one or more doubly degenerate levels lying at higher energies and, when n is an even number, a strongly antibonding nondegenerate level. The energy level diagrams for $n = 3$, 4, 5, 6, 7, and 8, are shown in Fig. 4.23. A stable system results when all the MOs at or below the "center of gravity" of the levels are filled. Thus the species $C_3H_3^+$ (2 π electrons), $C_5H_5^-$ (6 π electrons), C_6H_6 (6 π electrons), $C_7H_7^+$ (6 π electrons), and $C_8H_8^{2-}$ (10 π electrons) are stable species. Hückel long ago pointed out that cyclic coplanar systems in which the number of π electrons is 2, 6, 10, 14, ..., are "aromatic" and possess relative electronic stability. That is, aromaticity is associated with $4n + 2$ π electrons, where n is zero or an integer and is not to be confused with

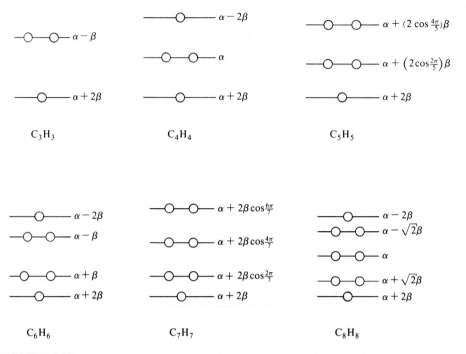

C_3H_3

—O—O— $\alpha - \beta$

—O— $\alpha + 2\beta$

C_4H_4

—O— $\alpha - 2\beta$

—O—O— α

—O— $\alpha + 2\beta$

C_5H_5

—O—O— $\alpha + (2\cos\frac{4\pi}{5})\beta$

—O—O— $\alpha + (2\cos\frac{2\pi}{5})\beta$

—O— $\alpha + 2\beta$

C_6H_6

—O— $\alpha - 2\beta$

—O—O— $\alpha - \beta$

—O—O— $\alpha + \beta$

—O— $\alpha + 2\beta$

C_7H_7

—O—O— $\alpha + 2\beta\cos\frac{6\pi}{7}$

—O—O— $\alpha + 2\beta\cos\frac{4\pi}{7}$

—O—O— $\alpha + 2\beta\cos\frac{2\pi}{7}$

—O— $\alpha + 2\beta$

C_8H_8

—O— $\alpha - 2\beta$

—O—O— $\alpha - \sqrt{2}\beta$

—O—O— α

—O—O— $\alpha + \sqrt{2}\beta$

—O— $\alpha + 2\beta$

FIGURE 4.23

MO energy level diagrams for planar cyclic C_nH_n molecules; n = 3, 4, 5, 6, 7, and 8.

the n of C_nH_n. This important relation is now known as the $4n + 2$ rule or the Hückel criterion of aromaticity.

In the case of C_4H_4, we would predict the square planar molecule to have a triplet ground state, with only two bonding π electrons, corresponding to an average π bond order of $\frac{1}{4}$. Cyclobutadiene has been observed only at very low temperatures or as an unstable transient species; however, both experimental and theoretical evidence indicate that it has a rectangular configuration, with two more-or-less localized double bonds:[26]

This distortion of the molecule from the square configuration causes the doubly degenerate level to split, and stabilization occurs by having the two unpaired electrons pair up in the lower of the split levels. However, the square form of C_4H_4 can be stabilized in the form of complexes with transition metals, as in

[26] S. Masamune, F. A. Souto-Bachiller, T. Machiguchi, and J. E. Bertie, *J. Am. Chem. Soc.*, **100**, 4889 (1978).

$C_4H_4Fe(CO)_3$. (See Chap. 16.) Cyclooctatetraene, C_8H_8, would also be expected to be a triplet molecule if it were planar with all the C—C bonds equivalent. However, it is in fact a conjugated but nonplanar tetraolefin. It can readily be reduced to the planar aromatic $C_8H_8^{2-}$ ion, which forms a "sandwich" complex with uranium(IV): $C_8H_8UC_8H_8$.[27]

It is interesting to apply simple Hückel theory to inorganic systems. The molecule P_4 contains just as many valence electrons as cyclobutadiene, C_4H_4, and if one ignores the phosphorus lone pairs just as one ignores the C—H bonds in simple Hückel theory, one can discuss the instability of the hypothetical square planar structure of P_4 just as one does for C_4H_4. If P_4 were analogous to C_4H_4, it would have a rectangular configuration, with two localized double bonds:

Instead P_4 exists in a tetrahedral configuration, with a total of six equivalent P—P bonds. Apparently these six σ bonds are more stable than the combination of four σ bonds and two π bonds of the rectangular configuration. Perhaps eventually somebody will stabilize square planar P_4 in the form of a transition-metal complex, just as square planar C_4H_4 has been stabilized.

A large number of cyclic sulfur-nitrogen species are known. Many of these sulfur-nitrogen ring systems, including S_4N_4, $S_4N_4H_4$, S_7NH, and $S_3N_3Cl_3$, are nonplanar, but at least four species are known which are planar[28]: S_2N_2, $S_3N_3^-$, $S_4N_3^+$, and $S_5N_5^+$. The species S_2N_2 and $S_3N_3^-$ have square and hexagonal geometries, respectively, with alternating sulfur and nitrogen atoms in the rings. The structures of the cations $S_4N_3^+$ and $S_5N_5^+$ are as follows:

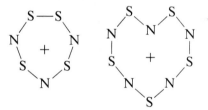

Each of the planar species is consistent with the $4n + 2$ rule if one assumes that each ring atom is σ-bonded to its neighboring atoms and that it possesses one σ nonbonding electron pair.[29] All the remaining valence electrons are assumed to occupy the π orbitals. On this basis we calculate that the species S_2N_2, $S_3N_3^-$,

[27] A. Streitwieser, Jr., and U. Muller-Westerhoff, *J. Am. Chem. Soc.*, **90**, 7363 (1968).

[28] J. Bojes, T. Chivers, W. G. Laidlaw, and M. Trsic, *J. Am. Chem. Soc.*, **101**, 4517 (1979); D. A. Johnson, G. D. Blyholder, and A. W. Cordes, *Inorg. Chem.*, **4**, 1790 (1965); A. J. Banister and H. G. Clark, *J. Chem. Soc. Dalton Trans.*, 2661 (1972).

[29] A. J. Banister, *Nature (London) Phys. Sci.*, **237**, 92 (1972); W. L. Jolly, *Adv. Chem. Ser.*, **110**, 92 (1972).

$S_4N_3^+$, and $S_5N_5^+$ have 6, 10, 10, and 14 π electrons, respectively. It is remarkable that the $S_4N_3^+$ cation "obeys" the $4n + 2$ rule even though there are two adjacent sulfur atoms in the ring. Apparently the overlap of the π orbitals on these adjacent sulfur atoms is comparable to that on adjacent sulfur and nitrogen atoms. In fact, it appears that it is possible to insert a group which provides no further π electrons into a Hückel ring without destroying the planarity of the system: Thus the insertion of $Cl-\overset{+}{S}$ into S_2N_2 gives $S_3N_2Cl^+$, which contains a planar five-membered ring[30]:

There are numerous inorganic compounds containing planar six-membered rings, analogous to the benzene ring. For instance, borazine ($B_3N_3H_6$) and its many derivatives have long been known.[31] Even rings containing second-row atoms, as in boraphosphabenzenes, $(RBPR')_3$, and aluminaazabenzenes, $(RAlNR')_3$, are now known.[32] The principal evidence for π-electron delocalization in the latter compounds is the planarity of the rings and the equivalence and shortness of the ring bonds (B—P 1.84 Å and Al—N 1.78 Å).

HYPERCONJUGATION

In Chap. 3, the phenomenon known as hyperconjugation was alluded to in a discussion of the bonding in molecules such as ONF_3. Hyperconjugation may be defined generally as the interaction within a molecule of a π system with a σ system. Consider the interaction of nonbonding $p\pi$ orbitals on the oxygen atom of ONF_3 with the bonding and antibonding MOs of the N—F bonds.[33] These hyperconjugative interactions are illustrated in Fig. 4.24, and an MO energy level diagram showing the effects of these interactions is shown in Fig. 4.25. The N—F σ^* levels are raised and the N—F σ levels are lowered as a consequence of interactions with the O $p\pi$ orbitals, but there is little net change in the energy of the O $p\pi$ orbitals, so that the latter orbitals remain essentially nonbonding as far as their energy is concerned. However, only the lower four energy levels

[30] A. Zalkin, T. E. Hopkins, and D. H. Templeton, *Inorg. Chem.*, **5**, 1767 (1966).

[31] *Gmelin Handbuch der Anorganischen Chemie*, Springer-Verlag, Berlin (New Supplement Series vol. 51). Also see p. 51, this text.

[32] H. V. R. Dias and P. P. Power, *Angew. Chem. Int. Ed. Engl.*, **26**, 1270 (1987); K. M. Waggoner, H. Hope, and P. P. Power, ibid., **27**, 1699 (1988).

[33] C. J. Eyermann, W. L. Jolly, S. A. Kinkead, J. M. Shreeve, and S. F. Xiang, *J. Fluorine Chem.* **23**, 389 (1983).

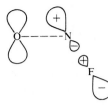

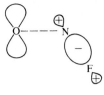

Oxygen $p\pi$ orbital and N—F σ^* orbital.

Oxygen $p\pi$ orbital and N—F σ orbital.

FIGURE 4.24
Schematic drawing showing the steric relation of an oxygen $p\pi$ orbital with σ and σ^* N—F orbitals in ONF_3.

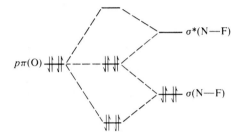

FIGURE 4.25
Energy level diagram illustrating hyperconjugation in ONF_3.

are occupied, and consequently there is a net stabilization of the system by the hyperconjugation. The lowest orbitals, derived from the O $p\pi$ and N—F σ orbitals, acquire considerable N—O π-bonding character and lose some of their N—F σ-bonding character. The oxygen lone-pair orbitals become quite delocalized; considerable nonbonding electron density shifts from the oxygen atom to the nitrogen and fluorine atoms. Thus qualitative MO theory allows us to rationalize the shortening of the N—O bond and the lengthening of the N—F bonds in ONF_3.

The data in Table 4.4 permit us to compare the changes in the structural parameters on going from NF_3 to ONF_3[34] with those on going from PF_3 to OPF_3.[35]

---

[34] V. Plato, W. D. Hartford, and K. Hedberg, *J. Chem. Phys.*, **53**, 3488 (1970).

[35] T. Moritani, K. Kuchitsu, and Y. Morino, *Inorg. Chem.*, **10**, 344 (1971).

TABLE 4.4
Structural parameters of MF_3 and OMF_3 molecules

Molecule	$r_{M—F}$, Å	$\angle F—M—F$, degree
NF_3	1.371	102.2
ONF_3	1.432	100.5
PF_3	1.570	97.8
OPF_3	1.524	101.3

It is remarkable that the changes in the P—F bond length and the F—P—F bond angle are in opposite directions from the changes in the N—F bond length and the F—N—F bond angle. The data suggest that, on going from PF_3 to OPF_3, the effects of rehybridization and an increase in phosphorus $3d$ orbital participation dominate, whereas, on going from NF_3 to ONF_3, the effects of hyperconjugation in ONF_3 dominate. In PF_3, the F—P—F bond angle is rather small, corresponding to considerable p character in the phosphorus bonding orbitals. In OPF_3 the bond angle approaches the tetrahedral value, corresponding to an increase in the s character of the σ-bonding orbitals of the phosphorus (an approach toward sp^3 hybridization).[36] This increase in s character strengthens and shortens the P—F bonds. It is also probable that, because of the greater positive charge on the phosphorus atom in OPF_3, the phosphorus $3d$ orbitals play a significant part in the P—F σ bonding and thus strengthen the P—F bonds. In the case of NF_3, because of the smallness of the nitrogen atom, there is considerable repulsion between the valence electrons, and the nitrogen bonding orbitals have much more s character than the phosphorus bonding orbitals in PF_3. Hence there is much less rehybridization on going from NF_3 to ONF_3 than on going from PF_3 to OPF_3, and hyperconjugation in ONF_3 is probably mainly responsible for the changes on going from NF_3 to ONF_3.

We have just considered the hyperconjugative interaction of a filled π orbital with a σ orbital. Another type of hyperconjugation involves the interaction of an empty π orbital with a σ orbital. This is exemplified by the $\sigma \rightarrow \pi^*$ delocalization in borane carbonyl:

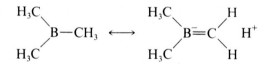

The planar $B(CH_3)_3$ molecule is another example:

H_3C\
\   B—CH_3 ⟷ H_3C\ /H
H_3C/   B=C H^+
                  H_3C/ \H

Indeed such hyperconjugation of methyl groups is probably responsible for the fact that the methyl group generally acts as a better electron donor than does a hydrogen atom.

The existence of hyperconjugation shows that the distinction between σ and π bonding is rather fuzzy in many molecules. The distinction is clean-cut and unambiguous only in the case of molecules in which all the atoms are coplanar, such as benzene, ethylene, formaldehyde, and water.

---

[36]   A. Serafini, J.-F. Labarre, A. Veillard, and G. Vinot, *Chem. Commun.*, 996 (1971). Also see H. Wallmeier and W. Kutzelnigg, *J. Am. Chem. Soc.*, **101**, 2804 (1979).

CALCULATIONS

Extended Hückel Theory

The simple Hückel theory is quite restricted in its applications. It only considers π MOs derived from $p\pi$ atomic orbitals of planar systems, and the values of the parameters α and β are usually determined empirically by fitting theoretical calculations to observed data. Because the parameters required to fit one property often differ from those required to fit another, the predictive value of the method is rather low.

Hückel theory was widely broadened with the development of extended Hückel theory by Hoffmann.[37] In this method, s, p, and d orbitals can be considered, all overlap integrals are evaluated, and definite procedures are followed for the evaluation of coulomb integrals and resonance integrals.

For the "basis set" of approximate atomic orbital wave functions, Slater[38] functions are used,

$$\phi = A r^{n-1} e^{-\zeta r} Y_{lm}(\theta, \phi)$$

where A is a normalization constant, r is the electron-nucleus distance in units of the Bohr radius of the hydrogen atom (0.529 Å), $\zeta = (Z - S)/n$ (the effective nuclear charge[39] divided by the principal quantum number), and $Y_{lm}(\theta, \phi)$ is a spherical harmonic introduced to give angular dependence to the function when $l > 0$. Values of ζ can be evaluated using the simple Slater recipe for the shielding constants outlined in Chap. 1, or theoretically calculated values (such as those given by Clementi and Raimondi[40] or Cusachs and Corrington[41] can be used. The diagonal H_{ii} values (coulomb integrals) are chosen as valence-state ionization potentials (VSIPs), which can be obtained from various tabulations.[42] The off-diagonal H_{ij} values (resonance integrals) are usually approximated by the following relation, first discussed by Mulliken,[43] and parameterized by Wolfsberg and Helmholtz[44].

$$H_{ij} = 1.75 \left(\frac{H_{ii} + H_{jj}}{2} \right) S_{ij} \tag{4.14}$$

[37] R. Hoffmann, *J. Chem. Phys.*, **39**, 1397 (1963); S. P. McGlynn et al., "Introduction to Applied Quantum Chemistry," pp. 97–156, Holt, Rinehart and Winston, New York, 1972.

[38] J. C. Slater, *Phys. Rev.*, **36**, 57 (1930).

[39] In this case, S represents the screening constant, not the overlap integral.

[40] E. Clementi and D. L. Raimondi, *J. Chem. Phys.*, **38**, 2686 (1963).

[41] L. C. Cusachs and J. H. Corrington, in "Sigma Molecular Orbital Theory," O. Sinanoglu and K. B. Wiberg, eds., chap. VI-4, Yale University Press, New Haven, 1970.

[42] Cusachs and Corrington tabulate VSIPs from various sources; also see G. Pilcher and H. A. Skinner, *J. Inorg. Nucl. Chem.*, **24**, 937 (1962), and J. Hinze and H. H. Jaffé, *J. Am. Chem. Soc.*, **84**, 540 (1962).

[43] R. S. Mulliken, *J. Phys. Chem.*, **56**, 295 (1952).

[44] M. Wolfsberg and L. Helmholtz, *J. Chem. Phys.*, **20**, 837 (1952).

TABLE 4.5
Extended Hückel study of phosphine. Matrix of the atomic orbital (AO) overlap integrals, S_{ij}

				AO			
AO	1	2	3	4	5	6	7
1	1.0000	0.0000	0.0000	0.0000	0.4118	0.4118	0.4118
2	0.0000	1.0000	0.0000	0.0000	−0.2809	−0.2809	−0.2809
3	0.0000	0.0000	1.0000	0.0000	0.4404	−0.2202	−0.2202
4	0.0000	0.0000	0.0000	1.0000	0.0000	0.3814	−0.3814
5	0.4118	−0.2809	0.4404	0.0000	1.0000	0.1185	0.1185
6	0.4118	−0.2809	−0.2202	0.3814	0.1185	1.0000	0.1185
7	0.4118	−0.2809	−0.2202	−0.3814	0.1185	0.1185	1.0000

The calculations are carried out using a computer program. The input data include the number of filled MOs, the number of half-filled MOs (if any), the cartesian coordinates of the atoms, and specification of the basis set, i.e., the atomic orbitals to be used. To illustrate the use of the method, we shall give the results of a calculation for the phosphine molecule, PH_3, which has a P—H bond length of 1.4206 Å and an H—P—H bond angle of 93.8°.[45] The x, y, z coordinates were chosen to make the threefold axis coincide with the z axis: P, 0.0, 0.0, 0.0; H, 1.19773, 0.0, −0.76390; H, −0.59887, 1.03727, −0.76390; H, −0.59887, −1.03727, −0.76390. The basis set was composed of the 3s orbital of phosphorus (atomic orbital 1), the phosphorus 3p orbitals (atomic orbitals, 2, 3, and 4), and the hydrogen 1s orbitals (atomic orbitals 5, 6, and 7).

A matrix of the overlap integrals, S_{ij}, is given in Table 4.5. The diagonal values of unity simply indicate that normalized wave functions were used. The

---

[45] C. A. Burrus, Jr., A. Jache, and W. Gordy, *Phys. Rev.*, **95,** 700 (1954).

TABLE 4.6
Extended Hückel study of phosphine. Matrix of the coulomb and resonance integrals, H_{ij}

				AO			
AO	1	2	3	4	5	6	7
1	−18.9430	0.0000	0.0000	0.0000	−11.7270	−11.7269	−11.7269
2	0.0000	−10.6540	0.0000	0.0000	5.9616	5.9615	5.9615
3	0.0000	0.0000	−10.6540	0.0000	−9.3473	4.6736	4.6736
4	0.0000	0.0000	0.0000	−10.6540	0.0000	−8.0950	8.0950
5	−11.7270	5.9616	−9.3473	0.0000	−13.6000	−2.8193	−2.8193
6	−11.7269	5.9615	4.6736	−8.0950	−2.8193	−13.6000	−2.8192
7	−11.7269	5.9615	4.6736	8.0950	−2.8193	−2.8192	−13.6000

TABLE 4.7
Extended Hückel study of phosphine.
MO energies

MO	Energy, eV
1	−21.83
2	−15.07
3	−15.07
4	−11.97
5	1.82
6	1.82
7	19.72

off-diagonal values were obtained from integrals[46] of the type

$$S_{ij} = \int \phi_i \phi_j \, d\tau$$

The zeros correspond to orbitals i and j which are orthogonal. Notice the small but finite overlap between the orbitals of the hydrogen atoms, which we ordinarily consider to be nonbonded.

A matrix of the coulomb and resonance integrals is given in Table 4.6. The diagonal elements are simply VSIPs, and the off-diagonal elements are the H_{ij} values calculated by Eq. 4.14.

Table 4.7 lists the MO energies, and Table 4.8 is a matrix of the atomic orbital coefficients, c_{ij}. From the magnitudes of the coefficients, one sees that the lowest-energy MO corresponds to a bonding MO formed by overlap of the phosphorus $3s$ orbital with the hydrogen $1s$ orbitals. The next two MOs are degenerate and correspond to bonds formed by overlap of the phosphorus $3p_x$

---

[46] Formulas for calculating overlap integrals, and extensive tables of values, are given by R. S. Mulliken, C. A. Rieke, D. Orloff, and H. Orloff, *J. Chem. Phys.*, **17**, 1248 (1949).

TABLE 4.8
Extended Hückel study of phosphine. Matrix of the AO coefficients, c_{ij}

AO	MO						
	1	2	3	4	5	6	7
1	0.6562	0.0000	0.0000	−0.4122	0.0000	0.0000	1.1979
2	0.0107	0.0000	0.0000	−0.8373	0.0000	0.0000	−0.8836
3	0.0000	0.5007	0.0038	0.0000	0.0000	−1.1145	0.0000
4	0.0000	0.0038	−0.5007	0.0000	1.1145	0.0000	0.0000
5	0.2313	0.5431	0.0041	0.1410	0.0000	0.9132	−0.7778
6	0.2313	−0.2680	−0.4723	0.1410	−0.7909	−0.4566	−0.7778
7	0.2313	−0.2751	0.4683	0.1410	0.7909	−0.4566	−0.7778

TABLE 4.9
Extended Hückel study of phosphine. Net and overlap orbital electronic populations

AO	AO						
	1	2	3	4	5	6	7
1	1.2010	0.0000	0.0000	0.0000	0.0771	0.0771	0.0771
2	0.0000	1.4022	0.0000	0.0000	0.0649	0.0649	0.0649
3	0.0000	0.0000	0.5014	0.0000	0.2395	0.0599	0.0559
4	0.0000	0.0000	0.0000	0.5014	0.0000	0.1797	0.1797
5	0.0771	0.0649	0.2395	0.0000	0.7365	−0.0176	−0.0176
6	0.0771	0.0649	0.0599	0.1797	−0.0176	0.7365	−0.0176
7	0.0771	0.0649	0.0599	0.1797	−0.0176	−0.0176	0.7365

and $3p_y$ orbitals with the hydrogen $1s$ orbitals. The fourth, highest occupied, MO is principally located on the phosphorus atom and is essentially a nonbonding (lone-pair) orbital comprising the phosphorus $3s$ and $3p_z$ orbitals.

Table 4.9 is a listing of the net and overlap electron populations of the atomic orbitals. Each element of the matrix is the sum of the $c_i c_j$ products for the four occupied MOs, multiplied by 2 times the overlap integral:

$$P_{ij} = 2S_{ij} \sum_{k=1}^{occ} c_i^k c_j^k$$

The diagonal values P_{ii} may be loosely interpreted as the "net" or "unshared" electron populations of the atomic orbitals. Each off-diagonal value P_{ij} may be interpreted as one-half of the electron population in the overlap region of the ith and jth atomic orbitals. In the process of assigning electrons to atomic orbitals, the overlap populations are split equally between the two atoms involved. Thus the sum of the numbers in any column or row is taken to be the total electron population of the corresponding atomic orbital. By adding up the electron populations of all the atomic orbitals of an atom, one obtains the total valence electron population of the atom, from which one can readily calculate the atomic charge. This procedure for calculating orbital electron populations and atomic charges is called a "Mulliken population analysis."[47] The reader can verify, from the data in Table 4.9, that the calculated phosphorus and hydrogen atom charges are +0.249 and −0.083, respectively.

Extended Hückel theory is useful for assessing the nature and relative importance of σ and π orbitals and for predicting the ground-state conformations of molecules. (Conformational prediction is accomplished by studying the total molecular energy as a function of systematic structural changes.) Atomic charges obtained from extended Hückel theory often appear to be too large (i.e., the ionic character of bonds appears to be exaggerated), and spectral energy predictions are usually quite poor.

---

[47]   R. S. Mulliken, *J. Chem. Phys.*, **23**, 1833, 1841 (1955).

Self-Consistent Field Methods

You may have noticed that, in the simple Hückel and extended Hückel methods, the hamiltonian is never used in an explicit form. In self-consistent field (SCF) methods, the full hamiltonian is actually calculated. The hamiltonian consists of the kinetic energy operator for an electron, the potential energy of attraction between an electron and all atomic cores of the molecule, and the potential energy of repulsion of the electrons. The algebraic expression for the hamiltonian involves the coefficients of the atomic orbitals, which of course are among the important quantities to be calculated. Hence the equations must be solved iteratively. A rough estimate is made of the coefficients (say, by a Hückel calculation), and then one solves the secular equations to obtain improved values of the coefficients. The cycle of the calculation is repeated until the coefficients do not change significantly, i.e., until the input and output coefficients are self-consistent.

In the "complete neglect of differential overlap" (CNDO) method,[48] various approximations are made to simplify the SCF calculations. The most important approximation is the neglect of all electron repulsion integrals involving the overlap distributions of different atomic orbitals, ϕ_i and ϕ_j. In addition, the corresponding overlap integrals S_{ij} are neglected in the normalization of the MOs. Some other integrals, instead of being evaluated rigorously, are chosen in a semiempirical manner to fit experimental data.

A version of CNDO known as CNDO/2 is most commonly used. There are various computer programs for making CNDO/2 calculations which differ merely in details of parameterization. The input data for these programs are very similar to those used in the extended Hückel program. A CNDO/2 calculation[49] was carried out for the molecule CIF, in which the basis set consisted of the following atomic orbitals, in order: chlorine $3s$, $3p_x$, $3p_y$, $3p_z$, $3d_{z^2}$, $3d_{xz}$, $3d_{yz}$, $3d_{x^2-y^2}$, $3d_{xy}$, and fluorine $2s$, $2p_x$, $2p_y$, $2p_z$. The Cl—F bond distance was taken as 1.628 Å.[50] In Table 4.10 are listed the calculated MO energy levels and experimental values of the ionization potentials from the three highest occupied MOs, as determined by ultraviolet photoelectron spectroscopy.[51] It is generally a fair approximation to assume, on the basis of Koopmans' theorem,[52] that an orbital ionization potential is equal to the negative of the orbital energy as obtained from a good SCF calculation. In the data we see the typical result of the use of CNDO/2 to calculate orbital energies: The calculated energy levels are in approximately the right order, but the calculated energies are too large. In general,

[48]   J. A. Pople and D. L. Beveridge, "Approximate Molecular Orbital Theory," McGraw-Hill, New York, 1970.

[49]   Using a CNDO program as modified by P. M. A. Sherwood, *J. Chem. Soc. Faraday Trans. 2*, **72**, 1791 (1976).

[50]   D. A. Gilbert, A. Roberts, and P. A. Griswold, *Phys. Rev.*, **76**, 1723 (1949).

[51]   C. P. Anderson, G. Mamantov, W. E. Bull, F. A. Grimm, J. C. Carver, and T. A. Carlson, *Chem. Phys. Lett.*, **12**, 137 (1971); R. L. DeKock, B. R. Higginson, D. R. Lloyd, A. Breeze, D. W. J. Cruickshank, and D. R. Armstrong, *Mol. Phys.*, **24**, 1059 (1972).

[52]   T. Koopmans, *Physica*, **1**, 104 (1934).

TABLE 4.10
CNDO/2 study of chlorine fluoride. MO energies

MO	Symmetry label	Calculated energy, eV	Experimental adiabatic ionization potential, eV
1	5σ	−44.77	
2	6σ	−29.50	
3,4	2π	−21.92	16.39
5	7σ	−19.95	17.80
6,7	3π	−16.16	12.66
8	....	−0.17	
9,10	....	5.32	
11,12	....	5.70	
13	....	7.99	

CNDO/2 calculations are useful for estimating charge distributions (including derived properties such as dipole moments) but are less successful for predicting molecular energies and ionization potentials. It has been found that the method tends to overestimate the extent of π bonding in molecules.

Various offshoots of the CNDO method have been devised, in which different approximations have been made. For example, we have "partial neglect of differential overlap"[53] (PNDO), "intermediate neglect of differential overlap"[54] (INDO), "modified intermediate neglect of differential overlap"[55] (MINDO), "neglect of diatomic differential overlap"[56] (NDDO), and "modified neglect of diatomic overlap"[57] (MNDO). Of these alternative methods, MINDO and MNDO appear to have been the most useful.[58] The reader interested in these methods is referred to the original references or to the comparative discussion by Murrell and Harget.[59]

Two other SCF MO methods which have found considerable application in inorganic chemistry are the parameter-free MO methods of Fenske and his coworkers[60] and the X-alpha scattered wave method of Slater.[61] The latter approach, in

[53] M. J. S. Dewar and G. Klopman, *J. Am. Chem. Soc.*, **89**, 3089 (1967).

[54] R. N. Dixon, *Mol. Phys.*, **12**, 83 (1967); J. A. Pople, D. L. Beveridge, and P. A. Dobosh, *J. Chem. Phys.*, **47**, 2026 (1967).

[55] N. C. Baird and M. J. S. Dewar, *J. Chem. Phys.*, **50**, 1262 (1969).

[56] J. A. Pople, D. P. Santry, and G. A. Segal, *J. Chem. Phys.*, **43**, S129 (1965).

[57] M. J. S. Dewar and W. Thiel, *J. Am. Chem. Soc.*, **99**, 4899 (1977); M. J. S. Dewar, M. L. McKee, and H. S. Rzepa, *J. Am. Chem. Soc.*, **100**, 3607 (1978).

[58] M. J. S. Dewar, *Chem. Brit.*, **11**, 97 (1975); "Further Perspectives in Organic Chemistry," Ciba Symposium 53, p. 107, ASP, Amsterdam, 1978.

[59] J. N. Murrell and A. J. Harget, "Semi-empirical Self-Consistent-Field Molecular Orbital Theory of Molecules," Wiley-Interscience, New York, 1972.

[60] R. F. Fenske and D. D. Radtke, *Inorg. Chem.*, **7**, 479 (1968); M. B. Hall and R. F. Fenske, *Inorg. Chem.*, **11**, 768 (1972).

[61] J. C. Slater and K. H. Johnson, *Phys. Today*, 34 (October 1974).

conjunction with the "transition state" method for accounting for relaxation energy, yields relatively accurate values for electronic transition energies.[62] For an interesting application of the X-alpha method to sulfur tetrafluoride and for a comparison of the results with those from several other MO studies, the reader is referred to a paper by Cowley, Lattman, and Walker.[63]

Finally we mention *ab initio* methods, in which no approximations (such as neglect of certain integrals) are made when the wave function is determined by SCF minimization of the total energy. These methods have risen in popularity in recent years because of the availability of high-speed digital computers and the development of sophisticated theoretical and computational methods. Some of them are easy to use because of the availability of appropriate computer programs, such as the STO-3G/Gaussian-70 program.[64] The main factor determining the accuracy of an *ab initio* calculation is the size of the basis set used. The so-called minimum basis set includes one function (such as a Slater function) for each occupied atomic orbital in the free atoms of the compound. Thus for H_2S, the minimum basis set consists of sulfur $1s$, $2s$, $3s$, $2p$, $3p$ and hydrogen $1s$ functions. A very popular basis set is the "double zeta" set, which includes two Slater functions for each occupied atomic orbital. Larger basis sets are called "extended basis" sets; these include functions for atomic orbitals higher than the occupied levels.

The speed and accuracy of various MO calculations, using the more popular semiempirical and minimum basis set *ab initio* methods, have been compared in two independent studies.[65] The interested reader is advised to consult these papers.

PROBLEMS

*4.1 Explain the fact that $(SiH_3)_3N$ is a much weaker base than $(CH_3)_3N$.

4.2 Calculate by simple MO theory whether the linear $[H \cdots H \cdots H]^+$ or triangular

$$\begin{bmatrix} H \\ H \cdots H \end{bmatrix}^+$$

form of H_3^+ is the more stable.

---

[62] J. C. Slater and K. H. Johnson, *Phys. Rev.*, **B5**, 844 (1972); J. C. Slater, in "Quantum Theory of Molecules and Solids," vol. 4, p. 583, McGraw-Hill, New York, 1974.

[63] A. H. Cowley, M. Lattman, and M. L. Walker, *J. Am. Chem. Soc.*, **101**, 4074 (1979).

[64] W. J. Hehre, R. F. Stewart, and J. A. Pople, *J. Chem. Phys.*, **51**, 2657 (1969); W. J. Hehre, W. A. Lathan, R. Ditchfield, M. D. Newton, and J. A. Pople, Program No. 236, QCPE, Indiana University, Bloomington, Ind. STO-3G stands for Slater-type orbital—approximated by a linear combination of Gaussian functions.

[65] T. A. Halgren, D. A Kleier, J. H. Hall, L. D. Brown, and W. N. Lipscomb, *J. Am. Chem. Soc.*, **100**, 6595 (1978); M. J. S. Dewar and G. P. Ford, *J. Am. Chem. Soc.*, **101**, 5558 (1979).

4.3 Construct a qualitative MO energy level diagram for the linear ion HF_2^-, involving only the hydrogen $1s$ orbital and the two fluorine hybrid orbitals that lie on the bonding axis.

4.4 For each of the following species, indicate the bond order and the number of unpaired electrons: NeO^+, O_2^+, CN^+, BN, SiF^+, NO^-, PCl, I_2^+, NeH^+.

4.5 Why is π bonding generally weaker than σ bonding?

4.6 Propose a structure for the P_4^{2+} ion.

*4.7 Explain why the $\pi \rightarrow \pi^*$ transition energy of N_3^- is lower than that of CO_2.

4.8 Predict the structures of Bi_4^{2-}, Te_4^{2-}, and $S_4N_4^{2+}$.

*4.9 If you have available an extended Hückel or CNDO/2 program and a suitable computer, use the program to calculate (*a*) a Walsh diagram for NH_3, in which the MO energy levels are plotted for H—N—H bond angles from 90 to 120°, or (*b*) an energy-versus-bond distance plot for a diatomic molecule such as N_2.

*4.10 Can you rationalize the fact that planar pseudoaromatic sulfur-nitrogen species (e.g., $S_4N_3^+$) are known, whereas analogous nitrogen-oxygen, phosphorus-oxygen, and phosphorus-sulfur species are unknown?

*4.11 The electron density difference plot (showing the difference in density between the molecule and the hypothetical noninteracting atoms) for ethane shows a buildup of electron density in the C—C bonding region, whereas that for molecular fluorine shows a diminution of electron density in the F—F bonding region. Explain. Hint: Consider the electron distribution for a pair of fluorine atoms which are "prepared" for bonding, i.e., with one electron in each of the p_z orbitals.

*4.12 The two lowest-energy bands in the photoelectron spectrum of F_2 correspond to removal of electrons from the $2p\pi^*$ and $2p\pi_B$ MOs (consistent with Fig. 4.7a), whereas the lowest ionization potential of N_2 corresponds to ionization of a σ MO (consistent with Fig. 4.7b). Explain the difference in terms of the effective nuclear charges and the $2s$-$2p$ energy separations.

4.13 Assuming the unlikelihood of ionic charges greater than ± 2, what charge, if any, would you expect to find on the planar S_3N_3 ring? How many π electrons would the species have?

*4.14 Show, that if the overlap integral is not ignored, an antibonding orbital is destabilized more than the corresponding bonding orbital is stabilized. Show how this result (*a*) explains lone-pair–lone-pair repulsion, such as in the F_2 molecule, and (*b*) explains why this repulsion is greater in F_2 than in Cl_2.

4.15 Predict the structure and describe the bonding of the molecule $C_2S_4N_4$, which has the topology shown below.

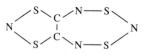

CHAPTER
5

THERMODYNAMIC ASPECTS OF INORGANIC CHEMISTRY

Thermodynamics is concerned with systems at equilibrium. Thermodynamic data give information regarding the driving forces of reactions and thus are a source of guidance for chemists. We assume that the reader is already familiar with the fundamental principles of thermodynamics; in this chapter we show how thermodynamics can be applied to a wide variety of inorganic problems and how it can be used to systematize reactions.

THERMODYNAMIC FUNCTIONS

The driving forces for chemical reactions are commonly expressed in the form of equilibrium constants or free-energy changes. However, the tabulation of such information for large numbers of related reactions is usually most efficiently accomplished not by tabulating the values of K or $\Delta G°$ for all the individual reactions, but by tabulating the free energies *of formation* of all the species involved in the various reactions. For example, one can write many thousands of reactions using as few as 100 different chemical species. Values of $\Delta G_f°$, as well as $\Delta H_f°$ and $S°$, are tabulated for various species in Appendix E.

It often happens that one or more of the $\Delta G_f°$ values required to calculate $\Delta G°$ for a reaction is unknown, but that the value of $\Delta H°$ for the reaction is known. In such cases an estimated value of $\Delta S°$ for the reaction can be very valuable. The ideal situation exists when the reaction is fairly symmetric, with species of similar complexity as reactants and products. For example, it would be a fairly good approximation to assume $\Delta S° = 0$ for the reaction

$$H_3PO_3 + H_3AsO_4 \quad \rightarrow \quad H_3PO_4 + H_3AsO_3$$

and therefore $\Delta H°$ is a good approximation for $\Delta G°$ for this reaction. Probably $\Delta S°$ for this reaction differs from zero by no more than ± 5 cal deg^{-1} mol^{-1}, and therefore at room temperature $\Delta H°$ and $\Delta G°$ probably differ by no more than 1.5 kcal mol^{-1}. On the other hand, it would be a very poor approximation to assume $\Delta S° = 0$ for the reaction

$$MnO_4^- + 5Fe^{2+} + 8H^+ \quad \rightarrow \quad Mn^{2+} + 5Fe^{3+} + 4H_2O$$

In this case, the highly charged Fe^{3+} ions have very negative entropies which are the principal cause of a very negative $\Delta S°$ for this reaction (-209.5 cal deg^{-1} mol^{-1}). Hence for this reaction at 25°C, $\Delta G°$ is 62.5 kcal mol^{-1} more positive than $\Delta H°$.

It should be remembered that enthalpy, or heat content, H, is defined such that

$$\Delta H = \Delta E + \Delta(PV)$$

where ΔE is the change in energy and $\Delta(PV)$ is the change in the pressure-volume product. For reactions at constant pressure in which only liquids and solids are involved, very little volume change occurs. The same is true for reactions involving gases when the number of moles of gaseous reactants equals the number of moles of gaseous products. Hence for many reactions $\Delta H \approx \Delta E$, and the heats and energies of such reactions can be used interchangeably.

When using chemical thermodynamic data, one must keep in mind the limitations of thermodynamics. Thermodynamic data give information only regarding the extent to which reactions can go, not regarding the rates at which they go. For example, although both of the following disproportionation reactions have equilibrium constants greater than 10^6, only the first proceeds at an appreciable rate under ordinary conditions:

$$2NO_2 + 2OH^- \rightarrow NO_2^- + NO_3^- + H_2O \qquad \text{(fast)}$$

$$2N_2O \rightarrow N_2 + 2NO \qquad \text{(slow)}$$

The science, or art, of predicting the rates of chemical reactions is in a developmental stage. At present, often the best that can be done is to classify roughly certain classes of compounds as fast-reacting and others as slow-reacting. However, sometimes thermodynamic data are useful even in the absence of kinetic information. Thus, when thermodynamics tells us that a reaction cannot go, we are saved the trouble of trying the reaction.

APPLICATIONS TO SYNTHESIS

Hydrazine

Let us suppose that we wish to know whether hydrazine can be prepared by the oxidation of aqueous ammonia by nitric oxide. From a table of free energies of formation such as that in Appendix E, we could obtain the following data:

Species	ΔG_f° (at 25°C), kcal mol^{-1}†
$H_2O(l)$	-56.69
$N_2H_4(aq)$	30.6
$NO(g)$	20.69
$NH_3(aq)$	-6.35

† 1 kcal = 4.1840 kJ

Thus we could calculate $\Delta G^\circ = +49.0$ kcal mol^{-1} and $K = 1.2 \times 10^{-36}$ for the reaction

$$8NH_3(aq) + 2NO(g) \rightarrow 5N_2H_4(aq) + 2H_2O(l)$$

We would then realize that, under ordinary conditions, the reaction cannot proceed sufficiently to give an appreciable concentration of hydrazine.

Fluorides from Oxides

Shreeve and her coworkers have shown that carbonyl fluoride (COF_2) is a useful reagent for displacing oxygen by fluorine from the oxides of V, Nb, Ta, Cr, Mo, W, B, Si, Ge, Sn, P, Se, Te, I, and U.[1] These results prompted Schack and Christe[2] to wonder whether the method could be extended to chlorine and nitrogen, specifically, to the synthesis of $FClO_2$ and FNO_2 from alkali metal chlorates and nitrates, respectively. The formation of Cl—F and N—F bonds usually requires extremely powerful fluorinating agents, and they were worried that COF_2 might not work in these cases. So the first thing they did was to determine the thermodyamic feasibility of the syntheses by calculating ΔH° values for the following reactions from available data.

$$NaClO_3 + COF_2 \rightarrow NaF + FClO_2 + CO_2$$

$$NaNO_3 + COF_2 \rightarrow NaF + FNO_2 + CO_2$$

They found that both of the ΔH° values are near zero (the reader can verify this with data from Appendix E), and, because the reactions probably have positive ΔS° values, it seemed likely that the ΔG° values are negative. Thus encouraged, Schack and Christe heated the appropriate reagents together and indeed obtained good yields of $FClO_2$ and FNO_2. Because of the ease of handling COF_2 and its demonstrated ability to function as a fluorinating agent, COF_2 will undoubtedly find many more applications in fluorine chemistry.

[1]   S. P. Mallela, O. D. Gupta, and J. M. Shreeve, *Inorg. Chem.*, **27**, 208 (1988).
[2]   C. J. Schack and K. O. Christe, *Inorg. Chem.*, **27**, 4771 (1988).

Gas-Solid Equilibria

However, one must be cautious in such applications of thermodynamic data. Consider the reaction of a nonvolatile phase (or phases) to yield a volatile product (or products) for which $\Delta G° \gg 0$. Reactions of this type often can be carried to completion by pumping on the reactants or by continuous sweeping of the nonvolatile phase with an inert carrier gas. An example of such a process is the dehydration of a solid compound in a desiccator. Even though the equilibrium dissociation pressure of water vapor over the compound may be far below 1 atm, it may be possible to dehydrate it completely in the desiccator, where the vapor is continuously removed from the gas phase by absorption by the desiccant. Similarly, consider a reaction, for which $\Delta G° \gg 0$, in which a gas reacts with a nonvolatile phase to yield a volatile product. Often such reactions can be carried to completion by sweeping the nonvolatile phase with the reactant gas, or by otherwise continuously removing the volatile product. We shall now discuss an example of such a process.

Every chemist should be familiar with the relation

$$\frac{d \ln K}{dT} = \frac{\Delta H°}{RT^2}$$

which says that the fractional increase in an equilibrium constant with temperature is proportional to $\Delta H°$. This relation played an important part in the development of a synthesis of platinum hexafluoride by the reaction of platinum with fluorine.[3] It was presumed that although PtF_6 is probably stable with respect to decomposition to the elements, it would be unstable with respect to dissociation into fluorine and a lower platinum fluoride. It was expected that $\Delta H°$ for the reaction

$$PtF_4 + F_2 \quad \rightarrow \quad PtF_6$$

would be positive (endothermic) and that the formation of PtF_6 would be more favorable at very high temperatures. The apparatus used for the synthesis is schematically illustrated in Fig. 5.1. A current was passed through the platinum wire to start the exothermic reaction, which then continued without the passage of current until the platinum was consumed. The steady-state temperature of the platinum during the reaction was approximately 1000°C. Although PtF_6 is more stable at high temperatures than at low temperatures, it is nevertheless unstable, and it was isolable only because convection currents in the reactor brought the PtF_6 in contact with the cold walls of the reactor (where it froze out) faster than it could decompose.

Consider a gaseous phase containing several types of molecules (say, monomers, dimers, etc.) in equilibrium with a liquid or solid phase. How will the molecular composition of the gas change with increasing temperature? For simplicity, let us first consider a gaseous phase containing only two species, each

[3]   B. Weinstock, H. H. Claassen, and J. G. Malm, *J. Am. Chem. Soc.*, **79**, 5832 (1957); B. Weinstock, J. G. Malm, and E. E. Weaver, *J. Am. Chem. Soc.*, **83**, 4310 (1961).

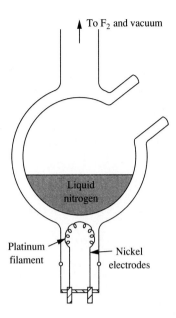

To F$_2$ and vacuum

Liquid
nitrogen

Platinum
filament

Nickel
electrodes

FIGURE 5.1
Glass reactor used for the first synthesis of platinum hexafluoride.

in equilibrium with the same liquid phase. The logarithms of the partial pressures
may be expressed as follows:

$$\frac{d \ln P_1}{dT} = \frac{\Delta H_1^\circ}{RT^2}$$

$$\frac{d \ln P_2}{dT} = \frac{\Delta H_2^\circ}{RT^2}$$

Using these equations, we may write

$$\frac{d \ln(P_1/P_2)}{dT} = \frac{\Delta H_1^\circ - \Delta H_2^\circ}{RT^2}$$

Hence, we see that the gaseous species corresponding to the higher heat of reaction
(more positive ΔH°) will increase in relative importance as the temperature is
raised. Now, as a fair approximation, the values of ΔS° for the reaction forming
the two species are equal.[4] Consequently, the partial pressures of the species will
approach each other as the temperature becomes very large. In Fig. 5.2 we have
plotted the logarithms of the pressures against $1/T$. Obviously, we may state,
as a general rule for a mixture of vapor species in equilibrium with a common
condensed phase, that *whatever species is in relatively small abundance at low
temperatures will gain in relative abundance as the temperature is raised.*

---

[4]   This condition follows from the facts that the entropy of the condensed phase is small relative to
that of the gas and that the major contribution to the entropy of the gas molecules is the translational
entropy, which is the same for each molecule.

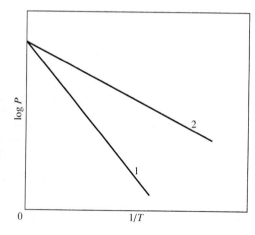

FIGURE 5.2
The partial pressures of two species in equilibrium with the same condensed phase. In this example, $\Delta H_1^\circ > \Delta H_2^\circ$.

In accordance with this rule, it is commonly observed that low-molecular-weight species predominate in saturated vapor at low temperatures, and polymeric species increase in importance with increasing temperature. For example, at low temperatures the main gaseous species in equilibrium with solid molybdenum trioxide is Mo_3O_9, but as the temperature is increased, the proportions of Mo_4O_{12} and Mo_5O_{15} steadily increase. Thus, if one seeks a gaseous system containing a large variety of molecular species of complex structure, it is usually advisable to go to the highest possible temperature at which the saturated system can still exist.[5]

When calculating equilibrium concentrations from free energy data, it is important to be aware of all the reactions that can take place. For example, consider the synthesis of silicon by the reduction of silicon dioxide by carbon:

$$SiO_2 + 2C \quad \rightarrow \quad Si + 2CO$$

At room temperature, this reaction is highly endothermic and is not permitted thermodynamically. However, because ΔS° is positive, ΔG° changes sign and the reaction becomes thermodynamically feasible at high temperatures. The reaction is generally carried out in an electric arc furnace at temperatures around 3000°. Under these conditions, $SiO(g)$, $Si_2C(g)$, $SiC_2(g)$, and $SiC(s)$ are formed in appreciable amounts in addition to $Si(l)$, $Si(g)$, and $Si_2(g)$. Calculation of the equilibrium concentrations using the thermodynamic data for these species (including heat capacities to account for changes in ΔH° and ΔS° with temperature) is such a formidable task that a computer is required.[6] Figure 5.3 shows the computer-calculated mole fractions of these species in a system with an initial C/SiO_2 ratio

[5] L. Brewer, *J. Chem. Educ.*, **35**, 153 (1958).
[6] C. M. Wai and S. G. Hutchison, *J. Chem. Educ.*, **66**, 546 (1989).

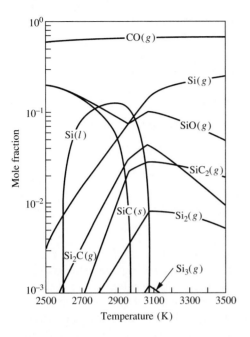

FIGURE 5.3

Distribution of gaseous and condensed species in the Si-O-C system with an initial C/SiO_2 ratio of 2, at 1 atm total pressure. [*Reproduced with permission of C. M. Wai and S. G. Hutchison, J. Chem. Educ., 66, 546 (1989).*]

of 2 at 1 atm total pressure, in the temperature range 2500–3500 K. It can be seen that silicon carbide, SiC, could be an important product if such a system were quenched to room temperature. To avoid carbide formation, the synthesis of silicon is generally carried out in the presence of excess SiO_2.

REDUCTION POTENTIALS [7]

Any oxidation-reduction (redox) reaction can be readily divided into two half-reactions—one in which an element undergoes oxidation and one in which an element undergoes reduction. We shall follow the convention of writing such half-reactions as *reductions* and shall measure their driving forces with *reduction potentials*, thus:

$$Na^+ + e^- \quad \rightarrow \quad Na \quad E° = -2.71 \text{ V}$$

Of course there is no difficulty in applying the same methods, after taking account of the sign change, to *oxidation potentials*:

$$Na \quad \rightarrow \quad e^- + Na^+ \quad E° = +2.71 \text{ V}$$

Obviously, in order to use any potential, it is necessary to know whether it is an oxidation or reduction potential, and whether or not the above sign convention is

---

[7] W. L. Jolly, *J. Chem. Educ.*, **43**, 198 (1966).

used. Particular care must be taken when using potentials from different sources. All the potentials cited in this chapter refer to 25°C.

It should be remembered that the potentials are expressed in volts (a difference in electrical potential). Therefore a potential is a constant which does not change when the coefficients of the half-reaction are changed. Thus both of the following half-reactions have the same potential:

$$\tfrac{1}{2}I_2 + e^- \quad \rightarrow \quad I^- \qquad E^\circ = 0.536 \text{ V}$$

$$I_2 + 2e^- \quad \rightarrow \quad 2I^- \qquad E^\circ = 0.536 \text{ V}$$

For simplicity, we shall hereafter omit the symbol V (for volt) in such expressions involving potentials.

The Use of Tabulated Potentials

Table 5.1 is a table of half-reactions and their standard potentials.[8] The reducing agents are listed in order of decreasing strength (and consequently the oxidizing agents are listed in order of increasing strength) from the top to the bottom of the table. An oxidizing agent should be able to oxidize all reducing agents lying above it in the table, and a reducing agent should be able to reduce all oxidizing agents lying below it in the table. However, because the potentials are "standard" potentials, this statement is strictly correct only when all the species are at unit activity. As will be shown later, exceptions to this rule can be effected by appropriate adjustment of product and reactant concentrations. Because from any pair of reduction potentials one can calculate the driving force for a complete, unique reaction, it is clear that a table of this sort summarizes an enormous amount of information in a small space. One calculates that a table of 100 reduction potentials would yield equilibrium constants for 4950 different reactions.

A tabulation of potentials like Table 5.1 is very handy when looking for oxidizing agents or reducing agents having a certain specified strength. However, a chemist often wishes to know whether a particular reaction can go, or what its equilibrium constant is. For such purposes it is more useful to have the half-reactions listed according to the elements whose oxidation states change.[9] Thus, if one wishes to know the driving force for the reaction

$$H_3AsO_3 + I_3^- + H_2O \quad \rightarrow \quad H_3AsO_4 + 3I^- + 2H^+$$

one looks under the arsenic potentials for

$$H_3AsO_4 + 2H^+ + 2e^- \quad \rightarrow \quad H_3AsO_3 + H_2O \qquad E^\circ = 0.559$$

---

[8]   Most of the potentials given in this chapter are from W. M. Latimer, "Oxidation Potentials," 2d ed., Prentice-Hall, Englewood Cliffs, N.J., 1952.

[9]   Generally there is no ambiguity as to which elements change oxidation state. However, in the case of a half-reaction such as $S_2O_8^{2-} + 2e^- \rightarrow 2SO_4^{2-}$ one might be justified in a double listing.

TABLE 5.1
The reduction potentials of some aqueous half-reactions

Acid solutions		Basic solutions	
Couple	E°	Couple	E°
$Na^+ + e^- \rightarrow Na$	-2.71	$Mg(OH)_2 + 2e^- \rightarrow Mg + 2OH^-$	-2.69
$Mg^{2+} + 2e^- \rightarrow Mg$	-2.37	$H_2AlO_3^- + H_2O + 3e^- \rightarrow Al + 4OH^-$	-2.35
$Al^{3+} + 3e^- \rightarrow Al$	-1.66	$B(OH)_4^- + 2H_2 + 4e^- \rightarrow BH_4^- + 4OH^-$	-1.60
$Zn^{2+} + 2e^- \rightarrow Zn$	-0.76	$HPO_3^{2-} + 2H_2O + 2e^- \rightarrow H_2PO_2^- + 3OH^-$	-1.57
$H_3PO_3 + 2H^+ + 2e^- \rightarrow H_3PO_2 + H_2O$	-0.50	$ZnO_2^{2-} + 2H_2O + 2e^- \rightarrow Zn + 4OH^-$	-1.22
$H_3BO_3 + 7H^+ + 8e^- \rightarrow BH_4^- + 3H_2O$	-0.47	$2SO_3^{2-} + 2H_2O + 2e^- \rightarrow S_2O_4^{2-} + 4OH^-$	-1.12
$Sn^{2+} + 2e^- \rightarrow Sn$	-0.14	$SO_4^{2-} + H_2O + 2e^- \rightarrow SO_3^{2-} + 2OH^-$	-0.93
$2H_2SO_3 + H^+ + 2e^- \rightarrow HS_2O_4^- + 2H_2O$	-0.08	$HSnO_2^- + H_2O + 2e^- \rightarrow Sn + 3OH^-$	-0.91
$2H^+ + 2e^- \rightarrow H_2(g)$	0.00	$2H_2O + 2e^- \rightarrow H_2(g) + 2OH^-$	-0.83
$HCOOH + 2H^+ + 2e^- \rightarrow HCHO + H_2O$	0.06	$AsO_4^{3-} + 3H_2O + 2e^- \rightarrow H_2AsO_3^- + 4OH^-$	-0.67
$S + 2H^+ + 2e^- \rightarrow H_2S$	0.14	$S + 2e^- \rightarrow S^{2-}$	-0.48
$Sn^{4+} + 2e^- \rightarrow Sn^{2+}$	0.15	$CrO_4^{2-} + 4H_2O + 3e^- \rightarrow Cr(OH)_3 + 5OH^-$	-0.13
$SO_4^{2-} + 4H^+ + 2e^- \rightarrow H_2SO_3 + H_2O$	0.17	$O_2(g) + H_2O + 2e^- \rightarrow HO_2^- + OH^-$	-0.08
$Fe(CN)_6^{3-} + e^- \rightarrow Fe(CN)_6^{4-}$	0.36	$PbO_2 + H_2O + 2e^- \rightarrow PbO + 2OH^-$	0.25
$I_3^- + 2e^- \rightarrow 3I^-$	0.54	$O_2(g) + 2H_2O + 4e^- \rightarrow 4OH^-$	0.40
$H_3AsO_4 + 2H^+ + 2e^- \rightarrow H_3AsO_3 + H_2O$	0.56	$IO^- + H_2O + 2e^- \rightarrow I^- + 2OH^-$	0.49
$O_2(g) + 2H^+ + 2e^- \rightarrow H_2O_2$	0.68	$MnO_4^- + 2H_2O + 3e^- \rightarrow MnO_2 + 4OH^-$	0.59
$Fe^{3+} + e^- \rightarrow Fe^{2+}$	0.77	$H_3IO_6^{2-} + 2e^- \rightarrow IO_3^- + 3OH^-$	0.70
$NO_3^- + 4H^+ + 3e^- \rightarrow NO(g) + 2H_2O$	0.96	$BrO^- + H_2O + 2e^- \rightarrow Br^- + 2OH^-$	0.76
$Br_2 + 2e^- \rightarrow 2Br^-$	1.06	$HO_2^- + H_2O + 2e^- \rightarrow 3OH^-$	0.88
$O_2(g) + 4H^+ + 4e^- \rightarrow 2H_2O$	1.23	$ClO^- + H_2O + 2e^- \rightarrow Cl^- + 2OH^-$	0.89
$Cr_2O_7^{2-} + 14H^+ + 6e^- \rightarrow 2Cr^{3+} + 7H_2O$	1.33	$O_3(g) + H_2O + 2e^- \rightarrow O_2(g) + 2OH^-$	1.24
$Cl_2(g) + 2e^- \rightarrow 2Cl^-$	1.36	$S_2O_8^{2-} + 2e^- \rightarrow 2SO_4^{2-}$	2.01
$PbO_2 + 4H^+ + 2e^- \rightarrow Pb^{2+} + 2H_2O$	1.46		
$MnO_4^- + 8H^+ + 5e^- \rightarrow Mn^{2+} + 4H_2O$	1.51		
$H_5IO_6 + H^+ + 2e^- \rightarrow IO_3^- + 3H_2O$	1.60		
$MnO_4^- + 4H^+ + 3e^- \rightarrow MnO_2 + 2H_2O$	1.69		
$H_2O_2 + 2H^+ + 2e^- \rightarrow 2H_2O$	1.77		
$S_2O_8^{2-} + 2e^- \rightarrow 2SO_4^{2-}$	2.01		
$O_3(g) + 2H^+ + 2e^- \rightarrow O_2(g) + H_2O$	2.07		
$F_2(g) + 2e^- \rightarrow 2F^-$	2.87		

and under iodine for

$$I_3^- + 2e^- + \quad \rightarrow \quad 3I^- \qquad E° = 0.536$$

By subtracting the first half-reaction from the second, one gets the desired reaction. Whenever half-reactions are added or subtracted, one should not add or subtract the corresponding $E°$ values, but rather the appropriate number of volt-equivalents, or $nE°$ values (where n is the number of electrons appearing in a half-reaction). For the reaction under question, $nE° = 2(0.536 - 0.559) = -0.046$. From this one can calculate $\Delta G°$ (in calories per mole) or K, using the relations

$$-\Delta G° = 23{,}060nE° \qquad \text{and} \qquad nE° = 0.05916 \log K$$

$$\Delta G° = 1060 \text{ cal mol}^{-1}$$

$$K = 0.17$$

From these results one must not draw the conclusion that the reaction cannot proceed quantitatively in either direction. Indeed, there are volumetric methods of analysis based both on the quantitative oxidation of arsenious acid by triiodide and on the quantitative reduction of arsenic acid by iodide. These are accomplished by appropriately adjusting the hydrogen-ion concentration of the solution. At pH 7,

$$\frac{[I^-]^3[H_3AsO_4]}{[I_3^-][H_3AsO_3]} = \frac{0.17}{(10^{-7})^2} = 1.7 \times 10^{13}$$

and arsenious acid can be titrated with triiodide or vice versa. In 6 M HCl,

$$\frac{[I^-]^3[H_3AsO_4]}{[I_3^-][H_3AsO_3]} \approx \frac{0.17}{6^2} \approx 5 \times 10^{-3} \qquad \text{(neglecting activity coefficients)}$$

and consequently arsenic acid can be quantitatively reduced by an excess of iodide. (The liberated triiodide can be titrated with thiosulfate.)

Dependence of Potentials on pH

The general problem of the pH dependence of reduction potentials is of considerable importance. The potential of any half-reaction changes with the concentration of the species involved according to the Nernst equation,

$$E = E° - \frac{0.05916}{n} \log Q$$

where Q has the same form as the equilibrium constant but is a function of the activities of the actual reactants and products and not those of the equilibrium state. If hydrogen ion or hydroxide ion appears in the half-reaction (and the corresponding ionic activity appears in Q), the potential will change with pH. Thus for the hydrogen-ion–hydrogen half-reaction,

$$H^+ + e^- \quad \rightarrow \quad \tfrac{1}{2}H_2 \qquad E° = 0$$

we write

$$E = -0.05916 \log \frac{P_{H_2}^{1/2}}{[H^+]}$$

and for the oxygen-water half-reaction,

$$\tfrac{1}{2}O_2 + 2H^+ + 2e^- \quad \rightarrow \quad H_2O \qquad E^\circ = 1.23$$

we write

$$E = 1.23 + 0.05916 \log ([H^+]P_{O_2}^{1/4})$$

The potentials for these half-reactions are plotted versus pH in Fig. 5.4. Theoretically no oxidizing agent whose reduction potential lies above the O_2–H_2O line, and no reducing agent whose reduction potential falls below the H^+–H_2 line, can exist in aqueous solutions. Actually, for kinetic reasons, these lines can be extended about 0.5 V, and the dashed lines in Fig. 5.4 are more realistic boundaries for the region of stability of oxidizing and reducing agents in aqueous solutions.

One must take care to use reduction potentials only in the pH ranges for which they are valid. For example, the ferrous ion–iron reduction potential

$$Fe^{2+} + 2e^- \quad \rightarrow \quad Fe \qquad E^\circ = -0.41$$

has little significance in alkaline solutions because the ferrous ion forms an essentially insoluble hydroxide. In order to calculate the appropriate potential for alkaline solutions one must know the solubility product for ferrous hydroxide (8×10^{-16}). Note that by adding the Fe^{2+}–Fe half-reaction to the reaction for the dissolution of ferrous hydroxide, we get the $Fe(OH)_2$–Fe half-reaction:

$$
\begin{array}{lll}
Fe^{2+} + 2e^- \rightarrow Fe & nE^\circ = -0.82 \\
Fe(OH)_2 \rightarrow Fe^{2+} + 2OH^- & nE^\circ = -0.89 \\
\hline
Fe(OH)_2 + 2e^- \rightarrow Fe + 2OH^- & nE^\circ = -1.71
\end{array}
$$

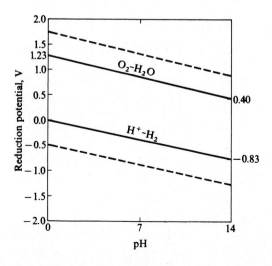

FIGURE 5.4
The H^+–H_2 and O_2–H_2O reduction potentials as a function of pH.

In the case of the $Fe^{2+}-Fe$ half-reaction, $nE°$ is simply obtained by multiplying -0.41 by 2. In the case of the reaction for the dissolution of $Fe(OH)_2$, $nE°$ is obtained from the relation $nE° = 0.05916 \log K$. One adds the volt-equivalents (not the potentials!) for each step to obtain the volt-equivalents corresponding to the new half-reaction. The potential for the new half-reaction is then obtained by dividing by n, the number of electrons. Thus, $E° = -1.71/2 = -0.86$.

Reduction-Potential Diagrams

If an element can exist in several oxidation states, it is convenient to display the reduction potentials corresponding to the various half-reactions in diagrammatic form, as shown in the following diagrams for iron and oxygen.

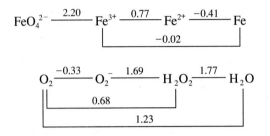

One of the most important facts which can be learned from a reduction-potential diagram is which oxidation states, if any, are unstable with respect to disproportionation. If a given oxidation state is a stronger oxidizing agent than the next higher state, disproportionation can occur. (This is the situation when a reduction potential on the left is more negative than one on the right.) It will be noted that both Fe^{2+} and Fe^{3+} are stable with respect to disproportionation and that H_2O_2 is unstable with respect to disproportionation. Thus the following reactions proceed spontaneously:

$$Fe + 2Fe^{3+} \rightarrow 3Fe^{2+}$$

$$2H_2O_2 \rightarrow 2H_2O + O_2$$

The latter reaction is very slow under ordinary conditions but rapid in the presence of certain catalysts.

Often species which are thermodynamically unstable with respect to disproportionation are intermediates in oxidation-reduction reactions and are the cause of slow reactions. Hydrogen peroxide is such a species. Most oxidations by molecular oxygen proceed with intermediate formation of hydrogen peroxide.[10] So although the O_2/H_2O reduction potential is 1.23 V, the "kinetically effective" reduction potential is only 0.68 V. Consequently, reducing agents in half-reactions with reduction potentials in the range 0.68 to 1.23 V react with oxygen slowly. In

---

[10] Indeed, one-electron reducing agents (including many transition-metal ions) reduce oxygen to an even more unstable species, the superoxide ion, O_2^-. For a discussion of the mechanisms of such reactions, see J. Wilshire and D. T. Sawyer, *Acc. Chem. Res.*, **12**, 105 (1979).

agreement with this generalization bromide (Br_3^-/Br^-, $E° = 1.05$) is oxidized very slowly, and iodide (I_3^-/I^-, $E° = 0.54$) is oxidized rapidly.

Sometimes it is not immediately obvious from an abbreviated reduction-potential diagram that a particular species is unstable with respect to disproportionation. Consider the diagram for phosphorus in basic solution (1 M OH^-):

$$PO_4^{3-} \xrightarrow{-1.12} HPO_3^{2-} \xrightarrow{-1.57} H_2PO_2^- \xrightarrow{-2.05} P_4 \xrightarrow{-0.89} PH_3$$

Hypophosphite ($H_2PO_2^-$) is stable with respect to disproportionation into phosphorus and phosphite. But if we consider the phosphine-hypophosphite half-reaction as shown in the more complete diagram,

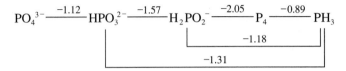

we see that hypophosphite is unstable with respect to the formation of phosphine and phosphite. This reaction takes place when solid hypophosphites are heated, but in hot aqueous solution the principal reaction is the reduction of water to give hydrogen:

$$H_2PO_2^- + OH^- \rightarrow HPO_3^{2-} + H_2$$

It is appropriate at this point to show how one calculates potentials when adding or subtracting two half-reactions to form a third half-reaction. As shown below, the phosphorus-phosphine and hypophosphite-phosphorus half-reactions may be added to give the hypophosphite-phosphine half-reaction

$$
\begin{array}{lll}
\frac{1}{4}P_4 + 3H_2O + 3e^- & \rightarrow \quad PH_3 + 3OH^- & nE° = -2.67 \\
H_2PO_2^- + e^- & \rightarrow \quad \frac{1}{4}P_4 + 2OH^- & nE° = -2.05 \\
\hline
H_2PO_2^- + 3H_2O + 4e^- & \rightarrow \quad PH_3 + 5OH^- & nE° = -4.72
\end{array}
$$

We divide the sum of the volt-equivalents by the number of electrons in the new half-reaction to get the new potential. Thus, $E° = -4.72/4 = -1.18$. It can be seen that the new potential is an equivalents-weighted average of the other two.

One of the nicest applications of reduction-potential diagrams is the prediction of the products of reactions involving elements having several oxidation states. Let us consider the reactions of iodide with permanganate in acid solution. The pertinent diagrams are shown below.

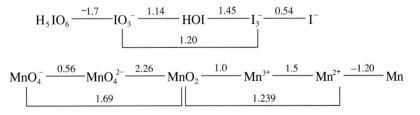

In this case we may correctly assume that the reactions are "thermodynamically controlled," i.e., that equilibria are fairly rapidly achieved.[11] We notice that there are three species in these diagrams which are unstable toward disproportionation: HOI, Mn^{3+}, and MnO_4^{2-}. These are therefore eliminated from consideration. The Mn^{2+}–Mn half-reaction is of no concern to us, because its reduction potential is far too negative for Mn to have any stability when in contact with an acid solution. Therefore we may justifiably consider only the simplified diagrams.

$$H_5IO_6 \xrightarrow{\sim 1.7} IO_3^- \xrightarrow{1.20} I_3^- \xrightarrow{0.54} I^-$$

$$MnO_4^- \xrightarrow{1.69} MnO_2 \xrightarrow{1.239} Mn^{2+}$$

If the reaction between iodide and permanganate is carried out with iodide in excess (as when permanganate is added dropwise to a hydroiodic acid solution), then the products of the reaction must be compatible with the presence of iodide ion. Thus under these conditions iodate cannot be formed, because iodate would react with excess iodide to form triiodide. Similarly, manganese dioxide cannot be formed because it is capable of oxidizing iodide. The observed net reaction is

$$15I^- + 2MnO_4^- + 16H^+ \quad \rightarrow \quad 5I_3^- + 2Mn^{2+} + 8H_2O$$

If the reaction is carried out with permanganate in excess (as when iodide is added dropwise to an acidic permanganate solution), the products of the reaction must be compatible with the presence of permanganate. Thus manganous ion cannot be formed, because it would react with permanganate to form manganese dioxide. The iodide would not be oxidized just to triiodide, because triiodide is capable of reducing permanganate. The fact that the $H_5IO_6 - IO_3^-$ and $MnO_4^- - MnO_2$ half-reactions have potentials of similar magnitude complicates the problem. It turns out that iodide is not cleanly oxidized to either iodate or periodic acid but to a mixture of these products:

$$I^- + 2MnO_4^- + 2H^+ \quad \rightarrow \quad IO_3^- + 2MnO_2 + H_2O$$

$$3I^- + 8MnO_4^- + 11H^+ + 2H_2O \quad \rightarrow \quad 3H_5IO_6 + 8MnO_2$$

Notice that entirely different products are obtained when the reactant in excess is changed.

Reduction-potential diagrams for the more important elements that exist in several oxidation states are given in Table 5.2.

The reduction potential diagram for plutonium in acid solution is worthy of comment. Note that the potentials connecting the +3, +4, +5, and +6 oxidation states are all of similar magnitude: they lie in the narrow range between 0.91 and 1.17 V. Thus these four oxidation states can exist together in solution in equilibrium with each other at appreciable concentrations. This fact makes the study of plutonium chemistry in aqueous solution extraordinarily complicated and fascinating.

---

[11] The oxidation of water by H_5IO_6 and MnO_4^- is slow under ordinary conditions. The potentials for these oxidizing agents barely fall between the dashed lines of Fig. 5.4.

TABLE 5.2
Reduction – potential diagrams*

Potentials in acid solution

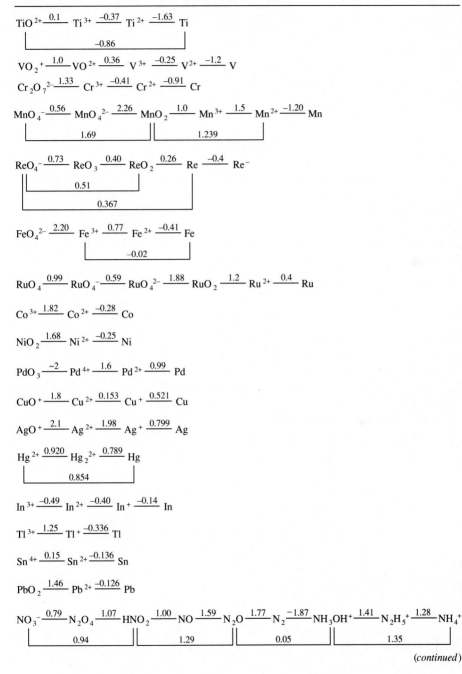

$TiO^{2+} \xrightarrow{0.1} Ti^{3+} \xrightarrow{-0.37} Ti^{2+} \xrightarrow{-1.63} Ti$
$\xrightarrow{-0.86}$

$VO_2^+ \xrightarrow{1.0} VO^{2+} \xrightarrow{0.36} V^{3+} \xrightarrow{-0.25} V^{2+} \xrightarrow{-1.2} V$

$Cr_2O_7^{2-} \xrightarrow{1.33} Cr^{3+} \xrightarrow{-0.41} Cr^{2+} \xrightarrow{-0.91} Cr$

$MnO_4^- \xrightarrow{0.56} MnO_4^{2-} \xrightarrow{2.26} MnO_2 \xrightarrow{1.0} Mn^{3+} \xrightarrow{1.5} Mn^{2+} \xrightarrow{-1.20} Mn$
$\xrightarrow{1.69} \qquad \xrightarrow{1.239}$

$ReO_4^- \xrightarrow{0.73} ReO_3 \xrightarrow{0.40} ReO_2 \xrightarrow{0.26} Re \xrightarrow{-0.4} Re^-$
$\xrightarrow{0.51}$
$\xrightarrow{0.367}$

$FeO_4^{2-} \xrightarrow{2.20} Fe^{3+} \xrightarrow{0.77} Fe^{2+} \xrightarrow{-0.41} Fe$
$\xrightarrow{-0.02}$

$RuO_4 \xrightarrow{0.99} RuO_4^- \xrightarrow{0.59} RuO_4^{2-} \xrightarrow{1.88} RuO_2 \xrightarrow{1.2} Ru^{2+} \xrightarrow{0.4} Ru$

$Co^{3+} \xrightarrow{1.82} Co^{2+} \xrightarrow{-0.28} Co$

$NiO_2 \xrightarrow{1.68} Ni^{2+} \xrightarrow{-0.25} Ni$

$PdO_3 \xrightarrow{\sim 2} Pd^{4+} \xrightarrow{1.6} Pd^{2+} \xrightarrow{0.99} Pd$

$CuO^+ \xrightarrow{1.8} Cu^{2+} \xrightarrow{0.153} Cu^+ \xrightarrow{0.521} Cu$

$AgO^+ \xrightarrow{2.1} Ag^{2+} \xrightarrow{1.98} Ag^+ \xrightarrow{0.799} Ag$

$Hg^{2+} \xrightarrow{0.920} Hg_2^{2+} \xrightarrow{0.789} Hg$
$\xrightarrow{0.854}$

$In^{3+} \xrightarrow{-0.49} In^{2+} \xrightarrow{-0.40} In^+ \xrightarrow{-0.14} In$

$Tl^{3+} \xrightarrow{1.25} Tl^+ \xrightarrow{-0.336} Tl$

$Sn^{4+} \xrightarrow{0.15} Sn^{2+} \xrightarrow{-0.136} Sn$

$PbO_2 \xrightarrow{1.46} Pb^{2+} \xrightarrow{-0.126} Pb$

$NO_3^- \xrightarrow{0.79} N_2O_4 \xrightarrow{1.07} HNO_2 \xrightarrow{1.00} NO \xrightarrow{1.59} N_2O \xrightarrow{1.77} N_2 \xrightarrow{-1.87} NH_3OH^+ \xrightarrow{1.41} N_2H_5^+ \xrightarrow{1.28} NH_4^+$
$\xrightarrow{0.94} \qquad \xrightarrow{1.29} \qquad \xrightarrow{0.05} \qquad \xrightarrow{1.35}$

(continued)

* Most of the potentials given in this chapter are from W. M. Latimer, "Oxidation Potentials," 2d ed., Prentice-Hall, Englewood Cliffs, N.J., 1952.

$$H_3PO_4 \xrightarrow{-0.28} H_3PO_3 \xrightarrow{-0.50} H_3PO_2 \xrightarrow{-0.51} P_4 \xrightarrow{-0.05} PH_3$$

with branches -0.16 and -0.28

$$H_3AsO_4 \xrightarrow{0.559} H_3AsO_3 \xrightarrow{0.247} As \xrightarrow{-0.38} AsH_3$$

$$Sb_2O_5 \xrightarrow{0.58} SbO^+ \xrightarrow{0.212} Sb \xrightarrow{-0.51} SbH_3$$

$$Bi_2O_5 \xrightarrow{\sim 1.6} BiO^+ \xrightarrow{0.32} Bi$$

$$O_2 \xrightarrow{-0.33} O_2^- \xrightarrow{1.69} H_2O_2 \xrightarrow{1.77} H_2O$$

with branches 0.68 and 1.23

$$SO_4^{2-} \xrightarrow{-0.22} S_2O_6^{2-} \xrightarrow{0.57} SO_2(aq) \xrightarrow{0.51} S_4O_6^{2-} \xrightarrow{0.08} S_2O_3^{2-} \xrightarrow{0.50} S \xrightarrow{0.14} H_2S$$

with $HS_2O_4^-$: -0.08 and 0.88

branches 0.17, 0.40, 0.45

$$SeO_4^{2-} \xrightarrow{1.15} H_2SeO_3 \xrightarrow{0.74} Se \xrightarrow{-0.32} H_2Se$$

$$H_6TeO_6(s) \xrightarrow{1.02} TeO_2(s) \xrightarrow{0.53} Te \xrightarrow{-0.44} H_2Te$$

$$ClO_4^- \xrightarrow{1.19} ClO_3^- \xrightarrow{1.21} HClO_2 \xrightarrow{1.64} HOCl \xrightarrow{1.63} Cl_2 \xrightarrow{1.36} Cl^-$$

with branch 1.47

$$BrO_4^- \xrightarrow{1.82} BrO_3^- \xrightarrow{1.49} HOBr \xrightarrow{1.59} Br_2 \xrightarrow{1.07} Br^-$$

with branch 1.51

$$H_5IO_6 \xrightarrow{1.7} IO_3^- \xrightarrow{1.14} HOI \xrightarrow{1.45} I_2 \xrightarrow{0.536} I^-$$

with ICl_2^-: 1.23 and 1.06, branch 1.20

$$H_4XeO_6 \xrightarrow{2.3} XeO_3 \xrightarrow{1.8} Xe$$

$$Ce^{4+} \xrightarrow{1.7} Ce^{3+} \xrightarrow{-2.34} Ce$$

$$Sm^{3+} \xrightarrow{-1.40} Sm^{2+} \xrightarrow{-2.75} Sm$$

$$Eu^{3+} \xrightarrow{-0.34} Eu^{2+} \xrightarrow{-2.82} Eu$$

$$Yb^{3+} \xrightarrow{-1.04} Yb^{2+} \xrightarrow{-2.81} Yb$$

$$PaO_2^+ \xrightarrow{-0.1} Pa^{4+} \xrightarrow{-0.9} Pa$$

(continued)

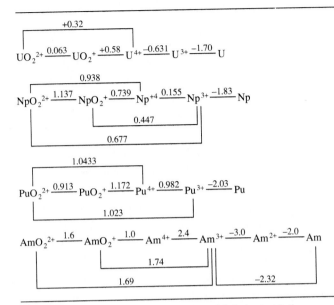

+0.32

$UO_2^{2+} \xrightarrow{0.063} UO_2^{+} \xrightarrow{+0.58} U^{4+} \xrightarrow{-0.631} U^{3+} \xrightarrow{-1.70} U$

0.938

$NpO_2^{2+} \xrightarrow{1.137} NpO_2^{+} \xrightarrow{0.739} Np^{+4} \xrightarrow{0.155} Np^{3+} \xrightarrow{-1.83} Np$

0.447

0.677

1.0433

$PuO_2^{2+} \xrightarrow{0.913} PuO_2^{+} \xrightarrow{1.172} Pu^{4+} \xrightarrow{0.982} Pu^{3+} \xrightarrow{-2.03} Pu$

1.023

$AmO_2^{2+} \xrightarrow{1.6} AmO_2^{+} \xrightarrow{1.0} Am^{4+} \xrightarrow{2.4} Am^{3+} \xrightarrow{-3.0} Am^{2+} \xrightarrow{-2.0} Am$

1.74

1.69   -2.32

Potentials in basic solution

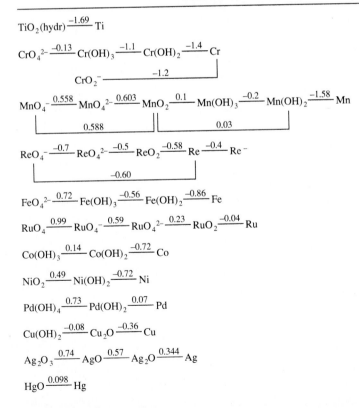

$TiO_2(hydr) \xrightarrow{-1.69} Ti$

$CrO_4^{2-} \xrightarrow{-0.13} Cr(OH)_3 \xrightarrow{-1.1} Cr(OH)_2 \xrightarrow{-1.4} Cr$

$CrO_2^{-} \xrightarrow{\quad -1.2 \quad}$

$MnO_4^{-} \xrightarrow{0.558} MnO_4^{2-} \xrightarrow{0.603} MnO_2 \xrightarrow{0.1} Mn(OH)_3 \xrightarrow{-0.2} Mn(OH)_2 \xrightarrow{-1.58} Mn$

0.588   0.03

$ReO_4^{-} \xrightarrow{-0.7} ReO_4^{2-} \xrightarrow{-0.5} ReO_2 \xrightarrow{-0.58} Re \xrightarrow{-0.4} Re^{-}$

-0.60

$FeO_4^{2-} \xrightarrow{0.72} Fe(OH)_3 \xrightarrow{-0.56} Fe(OH)_2 \xrightarrow{-0.86} Fe$

$RuO_4 \xrightarrow{0.99} RuO_4^{-} \xrightarrow{0.59} RuO_4^{2-} \xrightarrow{0.23} RuO_2 \xrightarrow{-0.04} Ru$

$Co(OH)_3 \xrightarrow{0.14} Co(OH)_2 \xrightarrow{-0.72} Co$

$NiO_2 \xrightarrow{0.49} Ni(OH)_2 \xrightarrow{-0.72} Ni$

$Pd(OH)_4 \xrightarrow{0.73} Pd(OH)_2 \xrightarrow{0.07} Pd$

$Cu(OH)_2 \xrightarrow{-0.08} Cu_2O \xrightarrow{-0.36} Cu$

$Ag_2O_3 \xrightarrow{0.74} AgO \xrightarrow{0.57} Ag_2O \xrightarrow{0.344} Ag$

$HgO \xrightarrow{0.098} Hg$

(continued)

$$In(OH)_3 \xrightarrow{-1.0} In$$

$$Tl(OH)_3 \xrightarrow{-0.05} Tl(OH) \xrightarrow{-0.344} Tl$$

$$Sn(OH)_6^{2-} \xrightarrow{-0.90} HSnO_2^- \xrightarrow{-0.91} Sn$$

$$PbO_2 \xrightarrow{0.28} PbO \xrightarrow{-0.54} Pb$$

$NO_3^- \xrightarrow{-0.86} N_2O_4 \xrightarrow{0.88} NO_2^- \xrightarrow{-0.46} NO \xrightarrow{0.76} N_2O \xrightarrow{0.94} N_2 \xrightarrow{-3.04} NH_2OH \xrightarrow{0.73} N_2H_4 \xrightarrow{0.1} NH_3$

0.01   −0.14   −0.76   0.42

$PO_4^{3-} \xrightarrow{-1.12} HPO_3^{2-} \xrightarrow{-1.57} H_2PO_2^- \xrightarrow{-2.05} P_4 \xrightarrow{-0.89} PH_3$

−1.18

−1.31

$AsO_4^{3-} \xrightarrow{-0.67} H_2AsO_3^- \xrightarrow{-0.68} As \xrightarrow{-1.21} AsH_3$

$Sb(OH)_6^- \xrightarrow{-0.4} SbO_2^- \xrightarrow{-0.66} Sb \xrightarrow{-1.34} SbH_3$

$Bi_2O_5 \xrightarrow{\sim 0.6} Bi_2O_3 \xrightarrow{-0.46} Bi$

$O_2 \xrightarrow{-0.33} O_2^- \xrightarrow{0.17} HO_2^- \xrightarrow{0.87} OH^-$

−0.08

0.401

$SO_4^{2-} \xrightarrow{-0.93} SO_3^{2-} \xrightarrow{-0.80} S_4O_6^{2-} \xrightarrow{0.08} S_2O_3^{2-} \xrightarrow{-0.74} S \xrightarrow{-0.48} S^{2-}$

with $S_2O_4^{2-}$: $\xrightarrow{-1.12} S_2O_4^{2-} \xrightarrow{-0.04}$

−0.58

−0.66

$SeO_4^{2-} \xrightarrow{0.05} SeO_3^{2-} \xrightarrow{-0.37} Se \xrightarrow{-0.84} Se^{2-}$

$TeO_2(OH)_4^{2-} \xrightarrow{0.4} TeO_3^{2-} \xrightarrow{-0.57} Te \xrightarrow{-0.86} Te^{2-}$

$ClO_4^- \xrightarrow{0.36} ClO_3^- \xrightarrow{0.33} ClO_2^- \xrightarrow{0.66} ClO^- \xrightarrow{0.40} Cl_2 \xrightarrow{1.36} Cl^-$

0.50   0.88

$BrO_4^- \xrightarrow{0.99} BrO_3^- \xrightarrow{0.54} BrO^- \xrightarrow{0.45} Br_2 \xrightarrow{1.07} Br^-$

0.76

$H_3IO_6^{2-} \xrightarrow{0.7} IO_3^- \xrightarrow{0.14} IO^- \xrightarrow{0.45} I_2 \xrightarrow{0.54} I^-$

0.49

$HXeO_6^{3-} \xrightarrow{0.9} HXeO_4^- \xrightarrow{0.9} Xe$

PROBLEMS

5.1 Write equations for the principal net reactions which occur in the following cases:

(a) A little potassium iodide is dissolved in a solution of KIO_3 in 6 M HCl.

(b) Aqueous triiodide solution is added to excess cold sodium carbonate solution. (Potentials for half-reactions involving I_3^- are practically the same as those for the corresponding half-reactions involving I_2.)

(c) Cl_2O is passed into a hot solution of sodium hydroxide.

(d) Iodine is added to an excess of aqueous chloric acid.

(e) Hypophosphite solution is added to an excess of an acidic $KMnO_4$ solution.

(f) Arsine is bubbled into an excess of an acidic solution of $Fe(NO_3)_3$.

(g) A suspension of powdered sulfur in an alkaline solution of Na_2SO_3 is boiled.

(h) Arsine is bubbled into a solution of H_3AsO_3.

(i) K_2FeO_4 is added to excess dilute nitric acid.

(j) A solution of Na_2CrO_4 is added to an excess of a solution of V^{3+} in dilute $HClO_4$.

(k) Chlorine is passed into a cold NaOH solution.

(l) Air is bubbled through an Na_2S solution until reaction ceases.

(m) Nitrogen dioxide is bubbled into an NaOH solution.

(n) Anhydrous H_3PO_3 is heated in vacuo.

(o) An excess of powdered arsenic is added to an acidic solution of Ag^+.

(p) Ferrous sulfate solution is added to excess alkaline hypochlorite solution.

(q) Mercurous nitrate solution is added to a solution of arsenious acid.

(r) Na_2CrO_4 is added to an excess of a strongly acidic Mn^{2+} solution.

(s) A solution of V^{3+} is added to excess Fe^{3+} solution.

(t) $Tl(OH)_3$ is added to excess HI solution.

(u) Excess ozone is bubbled through a solution of Ru^{2+}.

5.2 Show that the reduction potentials are consistent with the facts that the ruthenate ion, RuO_4^{2-}, is stable in 1 M OH^-, even if the solution is boiled, whereas if the pH is lowered to about 11, disproportionation to RuO_2 and RuO_4^- occurs.

5.3 Calculate the $MnO_4^- - MnO_2$ reduction potential for pH = 7.

5.4 Calculate the $HPO_3^{2-} - P_4$ reduction potential for $[OH^-] = 1\ M$.

5.5 Calculate the equilibrium constants for the following reactions from reduction potentials.

$$4H^+ + 2Cl^- + MnO_2 \rightarrow Cl_2 + Mn^{2+} + 2H_2O$$
$$O_3(g) + Xe(g) \rightarrow XeO_3(aq)$$

5.6 From data in Appendix E, calculate the equilibrium pressures of CO at 25°C and 420°C for the reaction

$$CaCO_3 + Zn \rightleftharpoons CO + ZnO + CaO$$

What would be the equilibrium pressures of CO_2 at these temperatures?

5.7 From data in Appendix E, calculate the temperature at which the reaction $COCl_2 \rightleftharpoons CO + Cl_2$ is 10 percent complete at equilibrium, at a total pressure of 1 atmosphere.

CHAPTER

6

THE KINETICS
AND MECHANISMS
OF REACTIONS

It is not enough to know the structures and energetics of molecules. It is also important to know the rates at which they react with one another and to have an understanding of the mechanisms of their reactions. In this chapter we will review the principles of chemical kinetics and will study the kinetics and mechanisms of some relatively simple reactions, including some systems in which MO symmetry rules are important. The principles are applicable to all kinds of reactions, and even the mechanisms of the specific reactions discussed can be generalized to many other systems. In later chapters we will discuss the kinetics of more complicated reactions. For example, the kinetics and mechanisms of reactions in solution of transition-metal complexes are covered in Chapter 19, and Chapters 20 and 21 are devoted entirely to catalysis.

Many of the exemplary reactions in this chapter are reactions of molecular hydrogen. These reactions were chosen for the following reasons. (1) The reactions have been thoroughly studied. (2) Hydrogen is a simple molecule, and therefore complicating side reactions are rare. (3) Hydrogen is by far the most abundant element in the universe, and there are probably more known compounds of hydrogen than of any other element. Therefore the reactions of hydrogen and its compounds are important.

RATE LAWS AND MECHANISMS

At any given temperature, the rate of a reaction is some function of the concentrations of the reactants and products. For example, the rate of the reaction of nitric oxide with oxygen,

$$2NO + O_2 \quad \rightarrow \quad 2NO_2$$

is expressed by the following rate law:[1]

$$\frac{-d[O_2]}{dt} = k[NO]^2[O_2]$$

The rate law of a reaction gives us information about the mechanism of the reaction. The formula of the activated complex is the sum of the chemical formulas in the numerator of the rate law minus the sum of the formulas in the denominator (if any), each formula having been multiplied by its corresponding exponent. Thus, in this case the activated complex has the formula N_2O_4. Unfortunately, this restriction does not lead to a unique mechanism. Any mechanism that has a rate-determining step in which the sum of the reactant formulas corresponds to N_2O_4 is consistent with the rate law. Two possible mechanisms are

$$2NO \underset{\text{equil.}}{\overset{\text{fast}}{\rightleftharpoons}} N_2O_2$$

$$N_2O_2 + O_2 \xrightarrow{\text{slow}} 2NO_2$$

and

$$NO + O_2 \underset{\text{equil.}}{\overset{\text{fast}}{\rightleftharpoons}} OONO$$

$$NO + OONO \xrightarrow{\text{slow}} 2NO_2$$

Both of these mechanisms have activated complexes of formula N_2O_4 and correspond to the experimental rate law. A meaningful choice between the two mechanisms cannot be made in the absence of further information, such as would be provided by the identification of the intermediate species. Ambiguities of this type commonly arise in kinetic studies.

The algebraic function of concentrations in a rate law is not always as closely related to the reaction stoichiometry as in the preceding example. The rate of decomposition of aqueous nitrous acid,

$$3HNO_2 \rightarrow H^+ + NO_3^- + 2NO + H_2O$$

is given by the rate law[2]

$$-\frac{d[HNO_2]}{dt} = \frac{k[HNO_2]^4}{[NO]^2}$$

corresponding to an activated complex of formula $N_2O_4 \cdot 2H_2O$ or the kinetic equivalent, N_2O_4 (because of the inability to measure water concentration dependence in ordinary solutions). In this case, the mechanism is believed to proceed as follows:

---

[1] M. Bodenstein, *Z. Elektrochem.*, **24**, 183 (1918); *Z. Physik. Chem.*, **100**, 68 (1922).

[2] D. M. Yost and H. Russell, Jr., "Systematic Inorganic Chemistry," pp. 59–61, Prentice-Hall, Englewood Cliffs, N.J., 1946; W. C. Bray, *Chem. Revs.*, **10**, 161 (1932).

$$2HNO_2 \underset{\text{equil.}}{\overset{\text{fast}}{\rightleftharpoons}} H_2O + NO + NO_2$$

$$2NO_2 \overset{\text{slow}}{\longrightarrow} NO^+ + NO_3{}^-$$

$$NO^+ + H_2O \overset{\text{fast}}{\longrightarrow} HNO_2 + H^+$$

THE STEADY STATE

A rate law with two terms in the denominator, such as

$$\text{Rate} = \frac{k_1[A]}{1 + k_2[B]}$$

corresponds to a reaction mechanism with an intermediate species that reacts by two paths, one of which gives products and the other of which gives nonproducts. Consider, for example, the oxidation of hydrogen peroxide by chlorine,

$$Cl_2 + H_2O_2 \rightarrow 2H^+ + 2Cl^- + O_2$$

for which the rate law is[3]

$$\frac{d[O_2]}{dt} = \frac{k[Cl_2][H_2O_2]}{1 + k'[H^+][Cl^-]}$$

This reaction has two different activated complexes that correspond to different experimental conditions. When the concentrations of H^+ and Cl^- are small, the composition of the activated complex is $Cl_2H_2O_2$; when these concentrations are large, the activated complex has the formula $ClHO_2$. One mechanism that fits these activated-complex criteria is the following, in which the species $HOOCl$ is an intermediate.

$$Cl_2 + H_2O_2 \underset{k_2}{\overset{k_1}{\rightleftharpoons}} H^+ + Cl^- + HOOCl$$

$$HOOCl \overset{k_3}{\longrightarrow} H^+ + Cl^- + O_2$$

Some support for this mechanism is provided by the fact that, when ^{18}O-labeled H_2O_2 is used, all the ^{18}O ends up in the O_2, with none in the H_2O solvent.[4] Calculation of a rate law from a mechanism of this sort, in which a definite activated complex can be specified only for certain limiting conditions (when $k'[H^+][Cl^-] \ll 1$ and when $k'[H^+][Cl^-] \gg 1$), is very difficult unless we make the so-called steady-state approximation. We assume that during most of the re-action the concentration of the intermediate (in this case, $HOOCl$) is essentially

---

[3]   B. Makower and W. C. Bray, *J. Am. Chem. Soc.*, **55**, 4765 (1933); R. E. Connick, ibid., **69**, 1509 (1947).

[4]   See F. A. Cotton and G. Wilkinson, "Advanced Inorganic Chemistry," 5th ed., p. 458, Wiley, New York, 1988.

constant, that is, $d[\text{HOOCl}]/dt \approx 0$. This approximation is obviously valid if the concentration of the intermediate is very small, as is likely the case when the intermediate is a very unstable species. In this case the steady-state approximation gives us

$$\frac{d[\text{HOOCl}]}{dt} = 0 = k_1[\text{Cl}_2][\text{H}_2\text{O}_2] - k_2[\text{H}^+][\text{Cl}^-][\text{HOOCl}] - k_3[\text{HOOCl}]$$

This relation can be solved for [HOOCl]:

$$[\text{HOOCl}] = \frac{k_1[\text{Cl}_2][\text{H}_2\text{O}_2]}{k_3 + k_2[\text{H}^+][\text{Cl}^-]}$$

Inasmuch as the product-producing rate is $k_3[\text{HOOCl}]$, we calculate the overall rate law

$$\frac{d[\text{O}_2]}{dt} = \frac{k_1[\text{Cl}_2][\text{H}_2\text{O}_2]}{1 + (k_2/k_3)[\text{H}^+][\text{Cl}^-]}$$

which has exactly the same form as the experimental rate law if we substitute k for k_1 and k' for k_2/k_3. When the H^+ and Cl^- concentrations are very low, the rate law is essentially $d[\text{O}_2]/dt = k_1[\text{Cl}_2][\text{H}_2\text{O}_2]$, and the rate-determining step is step 1 of the mechanism. When the H^+ and Cl^- concentrations are relatively high, the rate law is essentially as follows:

$$\frac{d[\text{O}_2]}{dt} = \frac{k_1 k_3[\text{Cl}_2][\text{H}_2\text{O}_2]}{k_2[\text{H}^+][\text{Cl}^-]}$$

corresponding to a rapid equilibrium in steps 1 and 2, with step 3 being rate-determining.

CATALYSIS

Catalyzed reactions have rate laws that involve species not appearing in the net reaction. A species that appears in the rate law but not in the equation for the net reaction is called a catalyst. A classic example of catalysis in solution is the decomposition of hydrogen peroxide in the presence of bromide or bromine:[5]

$$2\text{H}_2\text{O}_2 \xrightarrow{\text{Br}^-} 2\text{H}_2\text{O} + \text{O}_2$$

This reaction is so slow as to be essentially undetectable in the absence of catalysts, but in the presence of H^+ and Br^-, the reaction proceeds according to the rate law

$$-\frac{d[\text{H}_2\text{O}_2]}{dt} = 2k_1[\text{H}_2\text{O}_2][\text{H}^+][\text{Br}^-]$$

---

[5] W. C. Bray, *Chem. Revs.*, **10**, 161 (1932).

A plausible mechanism is

$$H_2O_2 + H^+ + Br^- \xrightarrow{k_1} HOBr + H_2O$$

$$H_2O_2 + HOBr \xrightarrow{k_2} O_2 + Br^- + H^+ + H_2O$$

in which a steady-state concentration of HOBr is assumed. At steady state, the rates of the two steps in the mechanism are equal. Thus hydrogen peroxide is reduced in the first reaction as fast as it is oxidized in the second, explaining the factor of 2 in the rate law. Hypobromous acid (HOBr) and bromide are in rapid equilibrium with bromine:

$$HOBr + Br^- + H^+ \rightleftharpoons Br_2 + H_2O$$

Hence steady-state concentrations of all three species are achieved in the reaction, and any one of them may be considered the catalyst for the reaction.

Probably neither of the steps in the mechanism is as simple as it appears. Step 1 probably involves equilibrium formation of the hydroxonium ion, followed by displacement of water by bromide ion:

$$H_2O_2 + H^+ \rightleftharpoons H_3O_2^+$$

$$\begin{matrix} H \\ \diagdown \\ \diagup \\ H \end{matrix} O^+\!\!-\!\!OH + Br^- \longrightarrow H_2O + HOBr$$

And step 2 may involve formation of peroxyhypobromous acid, which then falls apart into oxygen, hydrogen ion, and bromide ion:

$$H_2O_2 + HOBr \rightarrow HOOBr + H_2O$$

$$HOOBr \rightarrow H^+ + O_2 + Br^-$$

The catalysis of organic reactions by transition-metal compounds in homogeneous solution is discussed in Chapter 20. Heterogeneous catalysis, in which the catalyst is a separate phase that provides a surface for facilitating the reaction, is discussed in Chapter 21.

ACTIVATED COMPLEX THEORY[6]

The activated complex must either lead to products or lose its excess energy and return to reactants. For a bimolecular reaction, these steps correspond to

$$A + B \rightleftharpoons (AB)^‡$$

$$(AB)^‡ \xrightarrow{(kT/h)} C + D$$

---

[6] A. A. Frost and R. G. Pearson, "Kinetics and Mechanism: A Study of Homogeneous Chemical Reactions," 2d ed., chap. 5, Wiley, New York, 1961.

where $(AB)^{\ddagger}$ is the activated complex, and the rate constant for the conversion of $(AB)^{\ddagger}$ to products is kT/h (where k is Boltzmann's constant and h is Planck's constant), as calculated from statistical mechanics. It is convenient to assume that A and B are in equilibrium with $(AB)^{\ddagger}$:

$$K^{\ddagger} = \frac{[(AB)^{\ddagger}]}{[A][B]}$$

From simple thermodynamic relations, $K^{\ddagger}$ can be related to the entropy of activation, $\Delta S^{\ddagger}$, and the heat of activation, $\Delta H^{\ddagger}$ (essentially the activation energy, E_a):

$$K^{\ddagger} = e^{\Delta S^{\ddagger}/R} e^{-\Delta H^{\ddagger}/RT}$$

By combining the rate law

$$\frac{d[C]}{dt} = (kT/h)[(AB)^{\ddagger}]$$

with the equilibrium constant expression, we obtain

$$\frac{d[C]}{dt} = k_r[A][B]$$

where k_r is the ordinary rate constant for the reaction. Hence we may write

$$k_r = (kT/h)e^{\Delta S^{\ddagger}/R} e^{-\Delta H^{\ddagger}/RT} \tag{6.1}$$

From a plot of $\log(k_r/T)$ versus $1/T$, it is possible to evaluate $\Delta H^{\ddagger}$ from the slope of the straight line, $-\Delta H^{\ddagger}/(2.303R)$. Of course, the term $(kT/h)e^{\Delta S^{\ddagger}/R}$ changes with temperature relatively slowly compared to the term $e^{-\Delta H^{\ddagger}/RT}$, and the value of E_a obtained from the slope of the more conventional plot of $\log k_r$ versus $1/T$ is practically the same as that of $\Delta H^{\ddagger}$. In either case the entropy of activation may be calculated from Eq. 6.1, using $\Delta H^{\ddagger}$ and values of k_r and T taken from the straight line.

Both $\Delta H^{\ddagger}$ (or E_a) and $\Delta S^{\ddagger}$ are functions of considerable importance in the interpretation of the mechanisms of chemical reactions. From Eq. 6.1 we see that high values of $\Delta H^{\ddagger}$, corresponding to relatively unstable, high-energy activated complexes, are generally associated with slow reactions. We also see that high values of $\Delta S^{\ddagger}$, corresponding to relatively disorderly, high-probability activated complexes, are generally associated with fast reactions (unless very high $\Delta H^{\ddagger}$ values overcompensate).

In reactions between aqueous ions of unlike sign there is usually an entropy increase on going from reactants to the activated complex; for ions of like sign, there is an entropy decrease. These activation entropies can be explained in terms of changes in solvation. When two ions of opposite charge come together to form an activated complex, there is a lowering of net charge and a decrease in solvation. That is, some of the "frozen" solvent molecules are released with an increase in entropy. The opposite occurs for ions of the same charge: there is an increase in net charge and solvation and a corresponding decrease in entropy because of the "freezing" of solvent molecules around the highly charged activated complex. These effects can be seen in the data for the following two reactions. [7]

$$Co(NH_3)_5Br^{2+} + OH^- \rightarrow Co(NH_3)_5OH^{2+} + Br^- \qquad \Delta S^{\ddagger} = 20.1 \text{ eu}$$

$$Co(NH_3)_5Br^{2+} + Hg^{2+} + H_2O \rightarrow Co(NH_3)_5H_2O^{3+} + HgBr^+ \qquad \Delta S^{\ddagger} = -23.6 \text{ eu}$$

REACTIONS OF HYDROGEN WITH HALOGENS

Molecular hydrogen is capable of reacting with many elements and compounds, but at room temperature most of these reactions are extremely slow. This seeming inertness is undoubtedly related to the high dissociation energy of the molecule, 104.2 kcal mol^{-1}. Obviously, when an H_2 molecule reacts, the H—H bond must be broken. However, it is not necessary to supply the entire 104 kcal mol^{-1} to the reacting molecules as activation energy in order to obtain a reaction. Usually, when an H_2 molecule reacts with another molecule, at least one of the hydrogen atoms forms a bond to another atom at the same time that the H—H bond is broken. Thus the bond-breaking contribution to the activation energy is partly compensated by the bond-forming contribution.

For example, consider the reaction of hydrogen with bromine vapor:

$$H_2 + Br_2 \quad \rightarrow \quad 2HBr$$

The reaction rate is negligible in the absence of light (which promotes the reaction) unless the temperature is raised above 200°C. Kinetic studies[8-10] of the reaction have shown that the rate is given by the following expression:

$$\frac{d[HBr]}{dt} = \frac{k[H_2][Br_2]^{1/2}}{1 + k'[HBr]/[Br_2]} \tag{6.2}$$

where k and k' are constants at a given temperature. This rate law is consistent with the following chain mechanism:

$$Br_2 \underset{2}{\overset{1}{\rightleftharpoons}} 2Br$$

$$Br + H_2 \underset{4}{\overset{3}{\rightleftharpoons}} HBr + H$$

$$H + Br_2 \overset{5}{\longrightarrow} HBr + Br$$

Reaction 1, in which radicals are produced, is the chain-initiating step, and reaction 2, in which radicals are consumed, is the chain-terminating step. Reactions

[7] Frost and Pearson, ref. 6, p. 144.

[8] M. Bodenstein and S. C. Lind, *Z. Physik. Chem.*, **57**, 168 (1907).

[9] Frost and Pearson, ref. 6, pp. 236–241.

[10] S. W. Benson, "Foundations of Chemical Kinetics," McGraw-Hill, New York, 1960.

3 to 5 are called chain-propagating steps; the number of radicals is unchanged by these reactions. Note that the sum of reactions 3 and 5 corresponds to the net reaction of H_2 and Br_2 to form two molecules of HBr. If a finite concentration of H and Br atoms is maintained, these steps can proceed as long as the supply of H_2 and Br_2 molecules lasts.

It is not difficult to show that the mechanism is consistent with the experimental rate law. For every molecule of HBr produced by reaction 5, another is produced by reaction 3. Therefore the overall rate of formation of HBr is equal to twice the rate of reaction 5:

$$\frac{d[\text{HBr}]}{dt} = 2k_5[\text{H}][\text{Br}_2] \tag{6.3}$$

If we assume that a low, steady-state concentration of H atoms is achieved, then we may write

$$\frac{d[\text{H}]}{dt} = 0 = k_3[\text{Br}][\text{H}_2] - k_4[\text{HBr}][\text{H}] - k_5[\text{H}][\text{Br}_2]$$

Solving for [H], we obtain

$$[\text{H}] = \frac{k_3[\text{Br}][\text{H}_2]}{k_5[\text{Br}_2] + k_4[\text{HBr}]} \tag{6.4}$$

Substitution in Eq. 6.3 yields

$$\frac{d[\text{HBr}]}{dt} = \frac{2k_3k_5[\text{H}_2][\text{Br}_2][\text{Br}]}{k_5[\text{Br}_2] + k_4[\text{HBr}]} \tag{6.5}$$

In the steady state, the rates of reactions 1 and 2 are equal; i.e., the Br_2 molecules are in equilibrium with the Br atoms. We may write

$$\frac{[\text{Br}]^2}{[\text{Br}_2]} = \frac{k_1}{k_2}$$

and

$$[\text{Br}] = \left(\frac{k_1}{k_2}\right)^{1/2} [\text{Br}_2]^{1/2}$$

Substitution in Eq. 6.5, followed by rearrangement, gives

$$\frac{d[\text{HBr}]}{dt} = \frac{2k_3(k_1/k_2)^{1/2}[\text{H}_2][\text{Br}_2]^{1/2}}{1 + (k_4/k_5)[\text{HBr}]/[\text{Br}_2]}$$

This equation has the same form as Eq. 6.2; it can be seen that $k' = k_4/k_5$ and $k = 2k_3(k_1/k_2)^{1/2}$.

A rate constant may be expressed as $Ae^{-E_a/RT}$, and an equilibrium constant may be expressed as $e^{\Delta S°/R}e^{-\Delta H°/RT}$. Hence the last equation, in which k is represented as a function of the rate constant k_3 and the equilibrium constant k_1/k_2, may be rewritten as follows:

$$Ae^{-E_a/RT} = 2A_3e^{-E_a(3)/RT}e^{\Delta S_1^\circ/2R}e^{-\Delta H_1^\circ/2RT}$$

where E_a is the activation energy of the overall reaction, $E_a(3)$ is the activation energy of reaction 3, and ΔH_1° is the heat of reaction 1 (i.e., the dissociation energy of bromine). Hence

$$E_a = E_a(3) + \tfrac{1}{2}\Delta H_1^\circ \tag{6.6}$$

From studies of the $H_2 + Br_2$ reaction as a function of temperature it is known that k' is temperature-independent and that the overall activation energy E_a is 40.2 kcal mol^{-1}. From Table 3.9, we find that ΔH_1° is 46.1 kcal mol^{-1}. Thus we calculate

$$E_a(3) = 40.2 - 23.0 = 17.2 \text{ kcal mol}^{-1}$$

From data in Table 3.9 we calculate that the energy of reaction 3 is 16.7 kcal mol^{-1}, a value just slightly less than E_a. As a matter of fact, this result is expected. Most exothermic reactions involving radicals are very fast and have low activation energies—typically a few kilocalories per mole. A schematic plot of the energies of the species involved in such a reaction is shown in Fig. 6.1. From the plot we see that the activation energy of an *endothermic* reaction of this type is just slightly greater than the energy of the reaction.

Although reactions 1 to 5 satisfactorily account for the rate law of the reaction, it is of interest to consider why the following reactions are not involved in the reaction mechanism.

$$Br + HBr \xrightarrow{\quad 6 \quad} H + Br_2$$

$$HBr \underset{8}{\overset{7}{\rightleftharpoons}} H + Br$$

$$H_2 \underset{10}{\overset{9}{\rightleftharpoons}} 2H$$

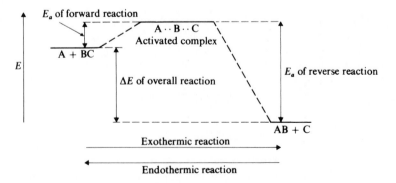

FIGURE 6.1
Energy level diagram for the reaction A + BC $\longrightarrow$ AB + C, where A and C are atoms or radicals.

From data in Table 3.9, we calculate that the activation energies of reactions 6, 7, and 9 are greater than 41.4, 87.5, and 104.2 kcal mol^{-1}, respectively. The activation energy of reaction 6 is much greater than that of any one of the analogous chain-propagating steps, reaction 3, 4, or 5. Similarly, the activation energies of reactions 7 and 9 are much greater than that of the analogous chain-initiating step, reaction 1. Hence the rate constants of reactions 6, 7, and 9 are negligible compared with those of the analogous reactions in the actual mechanism. Rearrangement of Eq. 6.4 gives

$$\frac{[H]}{[Br]} = \frac{k_3[H_2]}{k_5[Br_2] + k_4[HBr]}$$

In the early stages of the reaction, when [HBr] is low, and when H_2 and Br_2 have comparable concentrations, $[H]/[Br] \approx k_3/k_5$. At a typical reaction temperature, say 300°C, $k_3/k_5 \approx e^{-17.2/RT} \approx 10^{-7}$. Therefore reaction of H and Br atoms (reaction 8) would be $\sim 10^{-7}$ as likely as recombination of two Br atoms (reaction 2), and recombination of two H atoms would be $\sim 10^{-14}$ as likely.

   The chain reaction of hydrogen and chlorine, which proceeds by a mechanism analogous to reactions 1 to 5, is faster than that of hydrogen and bromine, whereas the chain reaction of hydrogen and iodine is slower. These qualitative results are readily predictable from the energetics of the steps in the general mechanism. If, as a rough approximation, we equate the energy of the $X + H_2 \rightarrow HX + H$ reaction to the activation energy of that reaction, then from Eq. 6.6 we obtain, for the *overall* activation energy,

$$E_a \approx D(H_2) - D(HX) + \tfrac{1}{2}D(X_2)$$

For chlorine, we calculate

$$E_a \approx 104.2 - 103.2 + 29.1 = 30.1 \text{ kcal mol}^{-1}$$

and, for iodine,

$$E_a \approx 104.2 - 71.3 + 18.0 = 50.9 \text{ kcal mol}^{-1}$$

Notice that the main factor in establishing the trend in the reactivity of the halogens with hydrogen is the trend in the H—X dissociation energy. In the case of chlorine, most of the energy required to break the H—H bond is compensated by the energy released upon formation of the H—Cl bond. In the case of iodine, a relatively small amount of energy is provided by formation of the H—I bond.

   The reaction of hydrogen with iodine by the chain mechanism is so slow that, under ordinary conditions, an entirely different mechanism predominates. The rate law has long been known to be

$$\frac{d[HI]}{dt} = k[H_2][I_2]$$

For many years this rate law was interpreted in terms of a simple bimolecular process with a four-center activated complex:

$$H_2 + I_2 \quad \rightarrow \quad 2HI$$

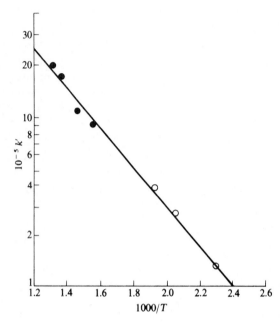

FIGURE 6.2
Plot of log $(10^{-5}k')$ versus $1000/T$ for the $H_2 + I_2$ reaction. Solid circles correspond to k/K values; open circles correspond to directly measured k' values. [*Reproduced with permission from J. H. Sullivan, J. Chem. Phys., 46, 73 (1967).*]

However, relatively recent experimental data suggest that the following, entirely different, mechanism is responsible for the reaction[11]:

$$I_2 \; \rightleftharpoons \; 2I \qquad K = \frac{[I]^2}{[I_2]}$$

$$2I + H_2 \; \xrightarrow{k'} \; 2HI$$

In the latter mechanism, I_2 molecules are in equilibrium with I atoms, and the latter species react with H_2 by a termolecular reaction (or perhaps two bimolecular reactions) to form the product. The corresponding rate law is indistinguishable from that for the simple bimolecular process:

$$\frac{d\,[HI]}{dt} = k'[H_2][I]^2 = k'K[H_2][I_2] = k[H_2][I_2]$$

The termolecular rate constant k' has been measured directly at low temperatures, using photochemically generated I atoms. The same rate constant can also be calculated from the relation $k' = k/K$, using values of k obtained from kinetic data on the $H_2 + I_2$ reaction at high temperatures and values of K obtained from equilibrium thermodynamic data. Both types of k' values have been plotted in the log k' versus $1/T$ plot of Fig. 6.2. Both sets of points fall on the same straight line, indicating that both sets of data yield the same calculated rate constant for any given temperature. This result proves that the simple bimolecular process

---

[11] J. H. Sullivan, *J. Chem. Phys.*, **46**, 73 (1967). However, see G. G. Hammes and B. Widom, *J. Am. Chem. Soc.*, **96**, 7621 (1974); and J. B. Anderson, *J. Chem. Phys.*, **61**, 3390 (1974).

must be relatively unimportant; otherwise k' values calculated from kinetic data on the $H_2 + I_2$ reaction would be higher than k' values obtained from direct measurement of the $2I + H_2$ reaction rate.

ORBITAL SYMMETRY EFFECTS

In recent years chemists have come to realize the importance of molecular orbital symmetry in determining the rates of reactions.[12,13] We now know, for example, that the simple bimolecular $H_2 + I_2$ reaction mechanism involving a trapezoidal activated complex

is symmetry-forbidden, whereas a mechanism involving the preliminary dissociation of iodine is symmetry-allowed.

For a bimolecular reaction, the important orbitals are the highest occupied molecular orbital (HOMO) of one molecule and the lowest unoccupied molecular orbital (LUMO) of the other molecule.[12] In the course of the reaction, electrons flow from the HOMO to LUMO. This process will occur (i.e., the reaction will be allowed) if the following conditions are fulfilled. (1) As the reactants approach one another, the HOMO and LUMO must have a finite net overlap. (2) The energy of the LUMO must be lower than, or no more than about 6 eV higher than, the energy of the HOMO. (3) The HOMO must be either a bonding MO of a bond to be broken or an antibonding MO of a bond to be formed; and vice versa for the LUMO.

Let us apply these rules to the bimolecular $H_2 + I_2$ reaction. The HOMO of H_2 is the $s\sigma_B$ MO, and the LUMO of I_2 is the $p\sigma^*$ MO. These orbitals are illustrated below; the shading indicates electron occupancy.

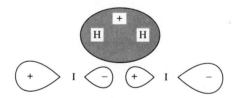

It is obvious that, if the H_2 and I_2 approach each other broadside, the net overlap of these orbitals is zero and the reaction is forbidden. An alternative approach is to consider the HOMO of I_2 (the $p\pi^*$ MO) and the LUMO of H_2 (the $s\sigma^*$ MO):

[12] R. G. Pearson, *Chem. Eng. News*, 66 (Sept. 28, 1970); *Acc. Chem. Res.*, **4**, 152 (1971); *J. Am. Chem. Soc.*, **94**, 8287 (1972); "Symmetry Rules for Chemical Reactions," Wiley-Interscience, New York, 1976.

[13] R. G. Woodward and R. Hoffman, "The Conservation of Orbital Symmetry," Academic, New York, 1969.

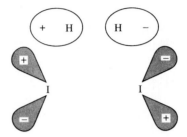

Although these orbitals have appropriate symmetries, i.e., the net overlap is positive, electron flow cannot occur for two reasons: The electron flow would correspond to the *strengthening* rather than the weakening of the I—I bond, and the electron flow from electronegative atoms to relatively electropositive atoms is energetically unfeasible. On the other hand, the reactions of an H_2 molecule with one or two iodine atoms, as in alternative mechanisms of the reaction, are symmetry- and energy-allowed processes:

For another example, let us consider the hydrogenation of ethylene to give ethane:

$$H_2 + C_2H_4 \quad \rightarrow \quad C_2H_6$$

If we assume a symmetric approach of H_2 on one side of the C_2H_4 plane, neither the flow of electrons from the HOMO of H_2 to the LUMO of C_2H_4 nor the flow of electrons from the HOMO[14] of C_2H_4 to the LUMO of H_2 is allowed by symmetry. In each case the net orbital overlap is zero.

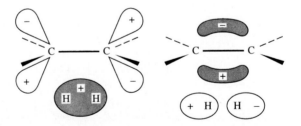

Thus the reaction is symmetry-forbidden. The nonreactivity of hydrogen with ethylene is in striking contrast to the reaction of hydrogen with B_2Cl_4, which proceeds readily even below room temperature.

$$3H_2 + 3B_2Cl_4 \quad \rightarrow \quad 4BCl_3 + B_2H_6$$

---

[14] The HOMO of C_2H_4 is a π bonding MO. The σ MOs associated with the C—H bonds have quite low energies and do not strongly interact with the σ MO of the C—C bond. Thus one does not observe the type of σ- and π-orbital ordering found in the first-row diatomic molecules (see Fig. 4.7).

The first step in this complex reaction probably involves the broadside attack of an H_2 molecule on a B_2Cl_4 molecule to form intermediate $BHCl_2$ molecules:

$$H_2 + B_2Cl_4 \quad \rightarrow \quad 2BHCl_2$$

A series of exchange processes could then lead to the observed products:

$$2BHCl_2 \quad \rightarrow \quad BH_2Cl + BCl_3$$

$$BH_2Cl + BHCl_2 \quad \rightarrow \quad BH_3 + BCl_3$$

$$2BH_3 \quad \rightarrow \quad B_2H_6$$

It is easy to show that the first step of the proposed mechanism is both symmetry- and energy-allowed.[15] Electrons can flow from the $s\sigma_B$ MO of H_2 to the empty $p\pi_B$ MO of B_2Cl_4:

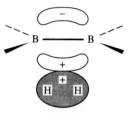

EXPLOSIONS[10]

The temperature of a reaction mixture is stable when the rate of loss of heat by conduction and convection equals the rate of heat production by the reaction. If the rate of heat loss cannot compensate for the rate of heat production, the reaction temperature and rate may increase rapidly, causing a "thermal explosion." Such an explosion can occur for highly exothermic reactions when the reactant concentrations are high. For example, an equimolar mixture of hydrogen and chlorine at a pressure of 50 torr in a glass vessel will explode at temperatures above 400°C. At lower temperatures the reaction rate is so slow that the heat of reaction can be carried off by conduction and convection. Explosion can occur at a lower temperature if the pressure of the reaction mixture is increased, and a higher temperature is required for explosion if the pressure is reduced.

It is interesting that the addition of only 0.5 torr of nitric oxide to 50 torr of $H_2 + Cl_2$ mixture lowers the critical explosion temperature from 400 to 270°C. This sensitization of the explosion is probably caused by an increase in the Cl atom concentration by the reaction

$$NO + Cl_2 \quad \rightarrow \quad NOCl + Cl$$

The latter reaction has an activation energy of only 22 kcal mol^{-1}, whereas the $Cl_2 \rightarrow Cl$ reaction has an activation energy of at least 57 kcal mol^{-1}.

[15] See R. A. Geanangel, *J. Nucl. Inorg. Chem.*, **34**, 1083 (1972).

The reaction of hydrogen with oxygen to produce water is the classical example of a "branching-chain" reaction:

$$2H_2 + O_2 \quad \rightarrow \quad 2H_2O_{(g)} \qquad \Delta H^\circ = -115.6 \text{ kcal mol}^{-1}$$

Many kinetic studies have shown that the following three basic steps are important in the mechanism of this chain reaction.[16]

$$OH + H_2 \xrightarrow{\ 1\ } H_2O + H \qquad k_1 = 2.2 \times 10^{10} e^{-5.15/RT} \text{ L mol}^{-1} \text{ s}^{-1}$$

$$H + O_2 \xrightarrow{\ 2\ } OH + O \qquad k_2 = 2.2 \times 10^{11} e^{-16.8/RT} \text{ L mol}^{-1} \text{ s}^{-1}$$

$$O + H_2 \xrightarrow{\ 3\ } OH + H \qquad k_3 = 3.5 \times 10^{10} e^{-10.0/RT} \text{ L mol}^{-1} \text{ s}^{-1}$$

The first step is chain-propagating and is the major process by which H_2O is formed. The second and third steps differ from the usual chain-propagating steps of chain reactions in that they produce more chain carriers (OH, O, or H) than they consume; thus they are called chain-branching reactions. If these reactions are allowed to predominate, the overall reaction rate will increase essentially without bound; i.e., an explosion will occur. Whether or not an explosion takes place depends on whether or not the chain-terminating steps of the mechanism can compensate for these chain-branching steps.

Reaction steps 1 to 3 have been separately studied over a wide range of temperature by flow-discharge methods, conventional methods, shock tubes, and flames. The kinetic parameters (with the activation energies in kilocalories per mole) are given after the chemical equations, and the log k versus $1/T$ plots are shown in Fig. 6.3. Because step 2 has by far the highest activation energy of the three processes, it is the slowest step and controls the production of chain carriers. Consequently the main radical species in a preexplosion H_2-O_2 mixture is always atomic hydrogen, and the most important chain-terminating steps are those in which atomic hydrogen is consumed.

At low temperatures, atomic hydrogen is lost principally by diffusion to the vessel walls:

$$H + \text{wall} \xrightarrow{\ 4\ } \text{stable species}$$

At low temperatures and pressure, reaction 4 prevents reactions 1 to 3 from proceeding uncontrolledly, and no explosion occurs. However, because reactions 1 to 3 are of higher kinetic order than reaction 4, an increase in pressure causes the combination of reactions 1 to 3 to increase in rate more than reaction 4. If the temperature of the reaction mixture is high enough, increasing the pressure beyond a particular value (called the "first explosion limit") causes explosion. If the pressure is increased considerably beyond the first limit, a second limit is reached, above which no explosion occurs. This quenching of the explosion is caused by the following reaction, which forms the relatively unreactive HO_2 species:

$$H + O_2 + M \xrightarrow{\ 5\ } HO_2 + M$$

---

[16] R. R. Baldwin and R. W. Walker, in "Essays in Chemistry," J. N. Bradley, R. D. Gillard, and R. F. Hudson, eds., vol. 3, pp. 1–37, Academic, London.

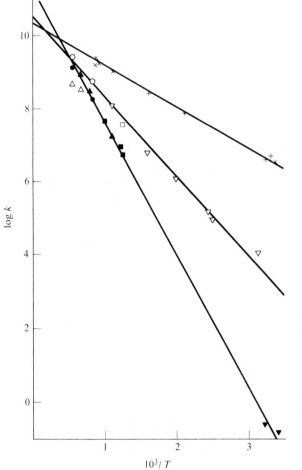

FIGURE 6.3
Plot of log k versus $10^3/T$ for steps 1 to 3 of the $H_2 + O_2$ branching-chain reaction. [*Reproduced with permission from R. R. Baldwin and R. W. Walker, in "Essays in Chemistry," vol. 3. pp. 1–37, Academic, London. Copyright Academic Press Inc. (London) Ltd.*]

The third body (M), which can be any other molecule such as O_2, H_2O, or a spectator gas molecule such as Ar, serves to carry off some of the heat of reaction and prevents the HO_2 from falling apart to OH and O. Note that reaction 5 is of higher kinetic order than reaction 2 and therefore would be expected to predominate at higher pressures. In fact, the concentration of M necessary to prevent explosion (the second explosion limit) can be calculated by equating the rate of chain termination and the rate of chain branching.[17]

$$k_5(H)(O_2)(M) = 2k_2(H)(O_2)$$

---

[17] The factor of 2 arises from the fact that reactions 1, 2, and 3 are stoichiometrically constrained to occur with relative rates of 2:1:1, respectively. Thus two H atoms are produced for every O_2 molecule consumed by step 2, whereas only one H atom is consumed for every O_2 molecule consumed by step 5.

or

$$(M)_{2d\ limit} = \frac{2k_2}{k_5}$$

Even though HO_2 is a radical, process 5 is chain-terminating because HO_2 is too slowly reactive to propagate a chain. It serves to break the chain reaction by decomposition at the wall:

$$HO_2 + HO_2 + wall \xrightarrow{6} H_2O_2 + O_2$$

Therefore near the second explosion limit, the main combustion product is hydrogen peroxide produced by reaction 6, rather than water.

Above the second limit, the reaction rate is relatively slow and proceeds by a complicated mechanism involving steps 1 to 3 and 5, and reactions involving H_2O_2. Increasing the pressure causes a significant amount of HO_2 to react with H_2 in the gas phase:

$$HO_2 + H_2 \xrightarrow{7} H_2O_2 + H$$

Thus the radical production by this reaction competes with the quenching effect of reaction 5, and at a sufficiently high pressure a third limit is reached above which explosion again occurs. The regions of explosion and the various limits are delineated in the pressure-temperature plot of Fig. 6.4.

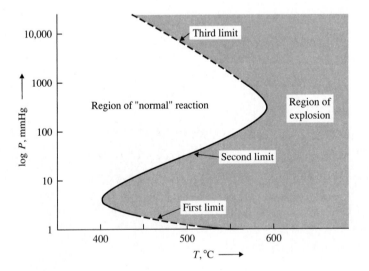

FIGURE 6.4
Plot showing the explosion limits of a stoichiometric mixture of $H_2 + O_2$ in a KCl-coated spherical vessel (7.4-cm diameter). Dashed parts of the curve are extrapolations. *(From S. W. Benson, "Foundations of Chemical Kinetics," McGraw-Hill, New York, 1960.)*

STRATOSPHERIC OZONE REACTIONS

In the troposphere (the layer of atmosphere extending from the earth's surface up to about 10 km altitude) the concentration of ozone in areas far removed from population centers is rather low: 10^{-8} to 10^{-7} mole fraction. However, in the stratosphere (which extends from about 10 to 50 km altitude) ozone is a more abundant natural constituent, with concentrations as high as ≈ 27 percent by weight. This ozone is formed by the action of far-ultraviolet radiation on molecular oxygen:

$$O_2 + h\nu \quad \rightarrow \quad 2O$$

$$O + O_2 + M \quad \rightarrow \quad O_3 + M$$

where M is a molecule of oxygen or nitrogen that serves to absorb some of the energy of the reaction. Ozone is destroyed by near-ultraviolet light:

$$O_3 + h\nu \quad \rightarrow \quad O_2 + O$$

and thus a steady-state concentration is established. The broad optical absorption band about 250 nm which is responsible for this decomposition corresponds to the $\pi \rightarrow \pi*$ transition discussed on p. 120. Because of this absorption of ultraviolet light, stratospheric ozone serves as a protective shield for the earth's surface.

However, the ozone can also be destroyed by catalytic processes involving trace constituents of the atmosphere that reduce the steady-state concentration.[18] For example, chlorofluorocarbons (used as foam-blowing agents, aerosol propellants, and refrigerants) are photochemically dissociated in the stratosphere, giving atomic chlorine:

$$CCl_2F_2 + h\nu \quad \rightarrow \quad CClF_2 + Cl$$

$$CCl_3F + h\nu \quad \rightarrow \quad CCl_2F + Cl$$

The chlorine atoms can destroy ozone in various chain reactions such as the following:

$$Cl + O_3 \quad \rightarrow \quad ClO + O_2$$

$$O + ClO \quad \rightarrow \quad Cl + O_2$$

Note that the net reaction of this cycle corresponds to the destruction of an ozone molecule and an oxygen atom:

$$O + O_3 \quad \rightarrow \quad 2O_2$$

Because of the harmful effects of an increase in the ultraviolet irradiation of the earth's surface, there are continuing attempts to achieve an international ban on the use of chlorofluorocarbons.

---

[18] S. Elliot and F. S. Rowland, *J. Chem. Educ.*, **64**, 387 (1987).

The quantitative prediction of the stratospheric concentration of ozone from kinetic data on the pertinent reactions is exceedingly complicated because of (1) the diversity of catalytic cycles (based on HO_x, NO_x, ClO_x, etc.); (2) the interaction of the catalytic species with one another in reactions such as

$$ClO + NO \quad \rightarrow \quad Cl + NO_2$$

and

$$HO_2 + NO \quad \rightarrow \quad OH + NO_2$$

(3) the conversion of catalytic species into "reservoir" species that are unreactive toward O and O_3, e.g.,

$$Cl + CH_4 \quad \rightarrow \quad HCl + CH_3$$

(4) the conversion of reservoir species back into active catalysts,

$$HCl + OH \quad \rightarrow \quad Cl + H_2O$$

and (5) the influence of heterogeneous catalysts, such as ice crystals in antarctic and arctic clouds, on these processes. When one adds to this chemical complexity the effects of vertical and horizontal transport, the variation of solar radiation as a function of altitude because of absorbing species, and the effects of pressure, it is understandable that the problem is difficult even with the help of powerful computers.[19]

MECHANISMS OF OXYANION REACTIONS[20]

We have already pointed out that, even if thermodynamic data such as reduction potentials indicate that a reaction has a large driving force, without kinetic information one cannot know whether or not the reaction proceeds at a reasonable rate. Let us consider oxidation-reduction reactions involving oxyanions such as BrO_3^-, SO_4^{2-} and OCl^-. Most of these reactions involve the transfer of an oxygen atom (or oxygen atoms) from one atom to another. In the oxyanions, the oxygen atoms are held very tightly, and therefore direct transfer processes such as the following are very slow:

$$NO_2^- + OCl^- \quad \rightarrow \quad NO_3^- + Cl^-$$
$$ClO_3^- + I^- \quad \rightarrow \quad ClO_2^- + OI^-$$

Such reactions are usually acid-catalyzed. When one or two protons are attached to a coordinated oxide ion, the negative charge of the oxide ion is at least partly canceled and an OH^- ion or H_2O molecule can then leave the central atom of the oxyanion relatively easily. Consider the reaction of chlorate with a halide ion:

[19] F. Kaufman, *Science*, **230**, 393 (1985); T. E. Graedel and P. J. Crutzen, *Sci. Am.*, **261**(3), 58 (1989); P. S. Zurer, *Chem. & Eng. News*, Aug. 17, 1987, p. 7.

[20] J. O. Edwards, "Inorganic Reaction Mechanisms," Benjamin, New York, 1964; A. G. Sykes, "Kinetics of Inorganic Reactions," Pergamon, New York, 1966; D. Benson, "Mechanisms of Inorganic Reactions in Solution," McGraw-Hill, London, 1968.

$$ClO_3^- + 6X^- + 6H^+ \rightarrow Cl^- + 3X_2 + 3H_2O$$

The rate law for this reaction is

$$Rate = k[ClO_3^-][X^-][H^+]^2$$

and the initial steps of the mechanism are believed to be as follows:

$$2H^+ + ClO_3^- \rightleftharpoons H_2OClO_2^+ \qquad \text{(fast)}$$

$$X^- + H_2OClO_2^+ \rightarrow XClO_2 + H_2O \qquad \text{(slow)}$$

$$XClO_2 + X^- \rightarrow X_2 + ClO_2^- \qquad \text{(fast)}$$

The ClO_2^- ion is then reduced to Cl^- by a sequence of rapid reactions.

In the $ClO_3^- - Cl^-$ reaction, if the intermediate $ClClO_2$ is allowed to reach a relatively high concentration, as by treating concentrated ClO_3^- solution with concentrated HCl solution, a different reaction occurs, in which two $ClClO_2$ molecules interact to yield chlorine and chlorine dioxide:

$$O_2ClCl + ClClO_2 \rightarrow 2ClO_2 + Cl_2$$

This mechanism has been shown to be consistent with experiments involving isotopically labeled chlorine. When labeled $^*ClO_3^-$ and ordinary HCl are used, all the labeling ends up in the $^*ClO_2$, and none in the Cl_2.[21]

Although the evidence for the species $ClXO_2$, $BrXO_2$, and IXO_2 in halate-halide reactions is purely circumstantial, the compounds $FClO_2$, $FBrO_2$, and FIO_2 have actually been prepared and isolated using the following indicated processes.[22]

$$12KClO_3 + 20BrF_3 \longrightarrow 12KBrF_4 + 4Br_2 + 6O_2 + 12FClO_2$$

$$BrF_5 + 2Br_2 + 10O_3 \longrightarrow 5FBrO_2 + 10O_2$$

$$I_2O_5 + F_2 \xrightarrow{\text{HF}} 2FIO_2 + \tfrac{1}{2}O_2$$

It is conceivable that the first intermediate in the $ClO_3^- - X^-$ reaction, $H_2OClO_2^+$, loses a water molecule before reaction with X^-, but the kinetic data do not give information on this point. The rate law would be unchanged if the rate-determining step were

$$X^- + ClO_2^+ \rightarrow XClO_2$$

It is known that dehydrated intermediates of this type are involved in some reactions. For example, the nitration of an aromatic hydrocarbon by a solution of nitric acid in concentrated sulfuric acid probably proceeds as follows:

$$HNO_3 + H^+ \rightleftharpoons NO_2^+ + H_2O \qquad \text{(fast)}$$

$$NO_2^+ + ArH \rightarrow ArNO_2 + H^+ \qquad \text{(slow)}$$

The electrophilic NO_2^+ ion attacks the π electron system of the hydrocarbon, with

---

[21] H. Taube and H. Dodgen, *J. Am. Chem. Soc.*, **71**, 3330 (1949).

[22] M. Schmeisser and K. Brändle, *Adv. Inorg. Chem. Radiochem.*, **5**, 41 (1963).

displacement of a proton to the solvent. The nitrosation of a secondary amine by nitrous acid is acid-catalyzed and is believed to proceed as follows:

$$HNO_2 + H^+ \ \rightleftharpoons \ NO^+ + H_2O \qquad K = 2 \times 10^{-7}$$

$$NO^+ + R_2NH \ \rightarrow \ R_2NNO + H^+ \qquad (slow)$$

The NO^+ can be detected in HNO_2–H_2SO_4–H_2O mixtures by its ultraviolet absorption spectrum. The exchange of ^{18}O between $H_2{}^{18}O$ and $SO_4{}^{2-}$ is acid-catalyzed and may involve the following mechanism:

$$2H^+ + SO_4{}^{2-} \ \rightleftharpoons \ H_2SO_4$$

$$H_2SO_4 \ \rightarrow \ SO_3 + H_2O \qquad (slow)$$

$$H_2{}^{18}O + SO_3 \ \rightarrow \ 2H^+ + SO_3{}^{18}O^{2-} \qquad (fast)$$

The charge of the central atom of an oxyanion seems to be very important in determining reactivity: the lower the charge, the higher the reactivity.[23] Thus, rates for reactions of the chlorine oxyanions increase in the order $ClO_4{}^- < ClO_3{}^- < ClO_2{}^- < ClO^-$. The rate of oxygen exchange with water increases in the order $ClO_4{}^- < SO_4{}^{2-} < HPO_4{}^{2-} < H_2SiO_4{}^{2-}$. The size of the central atom is also important; the larger it is, the higher the reactivity. For example, iodate reactions are fast, chlorate reactions are slow, and bromate reactions are of intermediate rate.

Proof of the transfer of an oxygen atom from an oxidizing agent to a reducing agent has been obtained in the case of the reaction of nitrite with hypochlorous acid.[24] When ^{18}O-labeled $HOCl$ was used, the ^{18}O ended up in the nitrate product:

$$NO_2{}^- + H^{18}OCl \ \rightarrow \ H^+ + NO_2{}^{18}O^- + Cl^-$$

The reaction may be looked upon as an S_N2 attack of nitrite on the oxygen of HOCl, with displacement of chloride.

The oxidation of sulfite by OCl^- also proceeds by oxygen atom transfer:

$$OCl^- + SO_3{}^{2-} \ \rightarrow \ Cl^- + SO_4{}^{2-}$$

However the oxidation of sulfite by HOCl proceeds by a much faster, entirely different, mechanism that apparently involves Cl^+ ion transfer:[25]

$$HOCl + SO_3{}^{2-} \ \rightarrow \ OH^- + ClSO_3{}^-$$

$$ClSO_3{}^- + H_2O \ \rightarrow \ SO_4{}^{2-} + Cl^- + 2H^+$$

The chlorosulfate ion has been identified as an intermediate in this reaction.

---

[23] It should be remembered that, for a series of closely related species, atomic charge is a function of formal charge and of oxidation state.

[24] H. Taube, *Rec. Chem. Prog. Kresge-Hooker Sci. Libr.*, **17**, 25 (1956).

[25] K. D. Fogelman, D. M. Walker, and D. W. Margerum, *Inorg. Chem.*, **28**, 986 (1989).

TABLE 6.1
Stoichiometries of the oxidation of sulfur dioxide†

Oxidizing agent	Equivalents per SO_2 molecule
1-equivalent oxidants	
Ce^{4+}	1.27–1.44
Co^{3+}	1.04–1.37
Fe^{3+}	~1.2
$IrCl_6{}^{2-}$	1.2–1.87
$Mo(CN)_8{}^{3-}$	1.94–2.00
$Mn(H_2P_2O_7)_3{}^{3-}$	~1.24
1- and 2-equivalent oxidants	
$Cr_2O_7{}^{2-}$	1.84–1.95
$MnO_4{}^-$	1.55–1.80
$PtCl_6{}^{2-}$	~2.00
$VO_2{}^+$	~1.57
2-equivalent oxidants	
I_2	~2.00
Br_2	1.98–2.00
Cl_2	1.99–2.00
$IO_3{}^-$	~2.00
$BrO_3{}^-$	1.77–1.95
H_2O_2	~2.00
Tl^{3+}	~2.00

† W. C. E. Higginson and J. W. Marshall, *J. Chem. Soc.*, 447 (1957);
E. L. Stapp and D. W. Carlyle, *Inorg. Chem.*, **13**, 834 (1974).

Differences Between 1- and 2-Equivalent Oxidizing Agents

Sometimes the product formed when a species is oxidized depends on whether or not one uses a 1-equivalent oxidizing agent or a 2-equivalent oxidizing agent. Examples of 1-equivalent oxidizing agents are Fe^{3+} and Ce^{4+}, which undergo reduction by accepting one electron each. Examples of 2-equivalent oxidizing agents are Tl^{3+} (which accepts two electrons to form Tl^+) and $IO_3{}^-$ (which effectively accepts two electrons by loss of an oxygen atom). The oxidation of aqueous sulfur dioxide can proceed by two main paths:[26]

$$SO_2 + H_2O \;\rightleftharpoons\; e^- + \tfrac{1}{2}S_2O_6{}^{2-} + H^+$$
$$SO_2 + 2H_2O \;\rightleftharpoons\; 2e^- + SO_4{}^{2-} + 4H^+$$

The observed stoichiometries for various oxidizing agents are shown in Table 6.1. It can be seen that the 2-equivalent oxidizing agents almost invariably give sulfate as the only product and that the 1-equivalent oxidizing agents and the oxidizing

---

[26]   W. C. E. Higginson and J. W. Marshall, *J. Chem. Soc.*, 447 (1957).

agents which can act either as 1-equivalent or 2-equivalent generally give both sulfate and dithionate as products. It is interesting to note that the stoichiometry is completely independent of the reduction potential of the oxidant. These results can be explained by mechanisms in which the 2-equivalent oxidants take the S(IV) directly to S(VI),

$$HSO_3^- \xrightarrow{[O]} HSO_4^-$$

and in which the 1-equivalent oxidants give an intermediate radical species which can either dimerize to give dithionate or lose another electron to give sulfate:

$$HSO_3^- \xrightarrow[-H^+]{-e^-} \cdot SO_3^- \longrightarrow \tfrac{1}{2}S_2O_6^{2-}$$
$$\xrightarrow{-e^-} [SO_3] \longrightarrow SO_4^{2-}$$

The latter mechanism is undoubtedly oversimplified; oxidations by some transition-metal complexes probably involve intermediates in which the $\cdot SO_3^-$ groups are stabilized by complexing with the metal atoms or with ligands.[27]

The Landolt Clock Reaction

The reduction of iodate by bisulfite is a fascinating reaction that is often used as an entertaining lecture demonstration.[28] In acid solution iodate is reduced slowly by bisulfite to iodide.

$$IO_3^- + 3HSO_3^- \rightarrow I^- + 3HSO_4^-$$

Iodate also reacts fairly rapidly with iodide to form iodine,

$$IO_3^- + 5I^- + 6H^+ \rightarrow 3I_2 + 3H_2O$$

but the liberated iodine is reduced very rapidly by bisulfite,

$$I_2 + HSO_3^- + H_2O \rightarrow 2I^- + HSO_4^- + 2H^+$$

The last reaction is so rapid that no iodine appears until all the bisulfite has been oxidized. At that point, especailly if a little starch is present, the solution suddenly becomes nearly opaque. The time required for the iodine to appear is a function of concentrations and temperature, and the reaction may be used as a clock. The time T, in seconds at 23°C, is approximately given by the expression

$$T = \frac{3.7 \times 10^{-3}}{[IO_3^-][HSO_3^-]}$$

where the concentrations are expressed in molarities.

---

[27] J. Veprek-Siska et al., *Collect. Czech. Chem. Commun.*, **31**, 1248, 3287 (1966); A. Brown and W. C. E. Higginson, *J. Chem. Soc. Chem. Commun.*, 725 (1967); *J. Chem. Soc. Dalton Trans.*, 166 (1972).

[28] H. Landolt, *Chem. Ber.*, **18**, 249 (1885); **19**, 1317 (1886); **20**, 745 (1887); J. A. Church and S. A. Dreskin, *J. Phys. Chem.*, **72**, 1387 (1968), and references therein.

PROBLEMS

6.1 Would you expect the activated complex of the nitrous acid decomposition reaction to have the same structure as stable N_2O_4? What about the activated complex of the NO oxidation reaction? Propose plausible structures for the activated complexes.

6.2 What is the rate law for the following reaction?

$$2NO + NO_3^- + H^+ + H_2O \longrightarrow 3HNO_2$$

6.3 For the platinum-catalyzed reaction $2SO_2 + O_2 \rightarrow 2SO_3$, $\Delta S^{\ddagger} = 17$ cal deg^{-1} mol^{-1}, and for the reverse reaction, $\Delta S^{\ddagger} = 40$ cal deg^{-1} mol^{-1}. For the reaction as written, $\Delta S° = -44.9$ cal deg^{-1} mol^{-1}. How many molecules of SO_2 are oxidized when one activated complex decomposes?

6.4 The conversion of p-H_2 (nuclear spins antiparallel) to o-H_2 (nuclear spins parallel) follows the rate law: Rate $= k[p\text{-}H_2][H_2]^{1/2}$. Propose a reaction mechanism.

*6.5 Why would you expect the kinetics of the $H_2 + F_2$ reaction to be more complicated than that of the $H_2 + Br_2$ reaction?

*6.6 Under what conditions would you expect the chain mechanism to compete with the nonchain mechanism in the $H_2 + I_2$ reaction?

6.7 Which of the following concerted reactions are "forbidden" for symmetry reasons?

$$2NO \rightarrow N_2 + O_2$$
$$N_2H_2 + C_2H_4 \rightarrow N_2 + C_2H_6$$
$$H_2 + N_2 \rightarrow N_2H_2$$
$$SO_2 + F_2 \rightarrow SO_2F_2$$
$$SF_2 + F_2 \rightarrow SF_4$$
$$O + F_2 \rightarrow OF_2$$
$$O + O_2 \rightarrow O_3 \text{ (terminal addition of O)}$$
$$O + O_2 \rightarrow O_3 \text{ (middle insertion of O)}$$

*6.8 Can singlet methylene, CH_2, react with H_2 by a least-motion concerted reaction to give CH_4? How about a non-least-motion pathway of the following type?

6.9 Explain why a concerted four-center mechanism (with a rectangular activated complex) is unlikely for the reaction $H_2 + D_2 \rightarrow 2HD$.

*6.10 A spark between two wires can set off an explosion in an $H_2 + O_2$ mixture which lies outside the explosion region of Fig. 6.4. Explain.

6.11 Discuss the nuclear fission of ^{235}U in terms of a branching-chain mechanism. Note the analogies between this reaction and the $H_2 + O_2$ reaction.

6.12 The decomposition of Caro's acid (H_2SO_5) to oxygen and bisulfate follows the rate law: Rate $= k[HSO_5^-][SO_5^{2-}]$. Two conceivable "mechanisms" for the reaction are indicated below. Explain how one of these could be eliminated by the use of ^{18}O labeling.

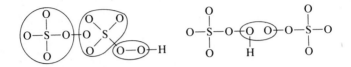

6.13 The rate of exchange of oxygen atoms between water and phosphate passes through a maximum at pH 5. Propose a plausible mechanism involving a reactive intermediate.

6.14 Number the following aqueous reactions in order of increasing rate in 1 M H^+ at room temperature.

(a) $IO_3^- + 8I^- + 6H^+ \rightarrow 3I_3^- + 3H_2O$

(b) $BrO_3^- + 5Br^- + 6H^+ \rightarrow 3Br_2 + 3H_2O$

(c) $ClO_4^- + 12I^- + 8H^+ \rightarrow 4I_3^- + Cl^- + 4H_2O$

(d) $HOI + 2I^- + H^+ \rightarrow I_3^- + H_2O$

CHAPTER
7

COMPOUNDS OF HYDROGEN

In Chap. 6 we pointed out that there are probably more known compounds of hydrogen than of any other element in the periodic table. There is no question that hydrogen-containing compounds (hydrides) constitute the majority of the compounds that we work with and that we should be familiar with their physical and chemical properties. In this chapter we describe the various kinds of hydrides and discuss periodic-table trends in their reactivities.

CLASSIFICATION OF THE HYDRIDES

Hydrides represent a wide variety of compounds having many different types of structure and bonding. This variety in structure and bonding is attributable to the unique electronic structure of the hydrogen atom: The atom possesses no core electrons, it has only one valence electron (and thus is *formally* analogous to the alkali-metal atoms), and it is only one electron short of a stable rare-gas electron configuration (and thus is analogous to halogen atoms). The known hydrides may be roughly categorized as saline hydrides, metallic hydrides, transition-metal hydride complexes, and nonmetal hydrides.

Saline Hydrides

At high temperatures hydrogen reacts with the alkali metals and the alkaline-earth metals heavier than beryllium to form salts, such as NaH and CaH_2, containing the hydride ion H^-. Inasmuch as the ionic radii of H^- and F^- are very similar[1] (1.40

---

[1] T. R. P. Gibb, *Prog. Inorg. Chem.*, **3**, 315 (1962).

and 1.36 Å, respectively, on the Pauling scale), the physical properties of the saline hydrides are similar to those of the corresponding fluorides. Even the periodic-table trend in the heats of formation of the alkali-metal hydrides resembles that for the alkali-metal fluorides. In each case, the lithium compound is more stable than the cesium compound. (See Table 7.1.)

Striking evidence for the negative charge of the hydrogen in lithium hydride is obtained when molten lithium hydride (mp 691°C) is electrolyzed: Lithium metal forms at the cathode, and hydrogen is evolved at the anode. The saline hydrides react vigorously with protonic solvents such as water, alcohol, and liquid ammonia to form molecular hydrogen and the basic anions of the solvents:

$$CaH_2 + 2H_2O \rightarrow Ca^{2+} + 2OH^- + 2H_2$$

$$LiH + C_2H_5OH \rightarrow Li^+ + C_2H_5O^- + H_2$$

$$KH + NH_3 \rightarrow K^+ + NH_2^- + H_2$$

Sodium hydride finds considerable use as a reducing agent in synthetic chemistry. Sodium hydroborate, $NaBH_4$, is prepared by the reaction of sodium hydride with methyl borate at elevated temperatures:

$$4NaH + B(OCH_3)_3 \rightarrow NaBH_4 + 3NaOCH_3$$

Aqueous solutions of sodium hydroborate are kinetically stable at high pH; however, acidic solutions decompose rapidly to boric acid and hydrogen:

$$BH_4^- + H^+ + 3H_2O \rightarrow 4H_2 + H_3BO_3$$

Sodium hydroborate is often used as a reducing agent in both inorganic and organic syntheses. When ether solutions of sodium hydroborate are heated at 100° with diborane, the octahydrotriborate ion, $B_3H_8^-$, is formed:

$$B_2H_6 + NaBH_4 \rightarrow NaB_3H_8 + H_2$$

TABLE 7.1
Properties of the saline hydrides†

Formula	Density, g ml^{-1}	$\Delta H°$ of formation kcal mol^{-1}	Crystal structure analog
LiH	0.778	−21.67	NaCl
NaH	1.36	−13.49	NaCl
KH	1.43	−13.82	NaCl
RbH	2.60	−11.3	NaCl
CsH	3.41	−11.92	NaCl
MgH$_2$	1.42	−17.79	TiO$_2$
CaH$_2$	1.9	−41.65	Distorted PbCl$_2$
SrH$_2$	3.27	−42.3	Distorted PbCl$_2$
BaH$_2$	4.15	−40.9	Distorted PbCl$_2$

† K. M. Mackay, "Hydrogen Compounds of the Metallic Elements," Spon, London, 1966.

By suitable changes in reaction time, temperature, diborane pressure, and solvent, these same reagents will yield various anions of the general formula $B_nH_n^{2-}$ (n ranging from 6 to 12). These ions are remarkably resistant toward hydrolysis, and their boron skeletons remain intact throughout a wide variety of reactions in which the hydrogen atoms are replaced by other groups. (See Chap. 10 for further discussion of the boron hydrides and their derivatives.)

Lithium hydride is used in the preparation of lithium aluminum hydride (lithium hydroaluminate), $LiAlH_4$:

$$4LiH + AlCl_3 \longrightarrow LiAlH_4 + 3LiCl$$

Both $LiAlH_4$ and $NaAlH_4$ can be prepared directly from the elements:

$$M + Al + 2H_2 \xrightarrow[\substack{1000-5000 \text{ lb in}^{-2} \\ 120-150°C}]{\text{ether solvent}} MAlH_4$$

Lithium aluminum hydride is an important reagent in synthetic chemistry; it is a much more powerful reducing agent than an alkali-metal hydroborate, and it reacts vigorously with protonic solvents such as alcohols.

Metallic Hydrides[2]

The binary hydrides of the transition metals have metallic properties and are usually obtained as powders or brittle solids which are dark and metallic in appearance and which have electric conductivities and magnetic properties characteristic of metals. Most of these hydrides have nonstoichiometric compositions; the stoichiometries and properties depend on the purity of the metals used in the preparations. Some typical limiting formulas are ScH_2, LaH_2, SmH_2, ThH_2, PaH_3, UH_3, VH, and CrH.

Uranium metal reacts exothermically and reversibly with hydrogen at 250–300°C to form a pyrophoric black powder with a stoichiometry approaching UH_3. The hydride can be decomposed at somewhat higher temperatures and low pressures to give extremely reactive, finely divided metal. The isostructural deuteride has a structure in which each D atom is surrounded by four U atoms at a distance of 2.32 Å and in which each U atom is surrounded by 12 D atoms in an irregular icosahedral arrangement. No U—U bonds appear to be present.

Palladium is unique among the metals in being able to dissolve large quantities of hydrogen while retaining considerable ductility. This property and the high rate of diffusion of hydrogen in the lattice have been exploited in the use of palladium and some of its alloys as hydrogen diffusion membranes.[3] The permeability of the metal to all other gases (except deuterium) is so low as to be

---

[2] W. M. Mueller, J. P. Blackledge, and G. G. Libowitz, "Metal Hydrides," Academic Press, New York, 1968.

[3] A. G. Knapton, *Plat. Metals Rev.*, **21**, 44 (1977).

negligible, and the metal therefore functions as a specific filter for the production of ultra-pure hydrogen or for removing hydrogen from gaseous mixtures. The palladium-hydrogen phase diagram is shown in Fig. 7.1. The metal atoms in the beta phase have the same face-centered cubic structure as in the alpha (metal) phase, but with an expanded lattice parameter. At temperatures below 300°C, increasing the hydrogen concentration leads to the formation of the beta phase, which can coexist with the alpha phase. One way to avoid the phase change is to ensure that the palladium membrane is always maintained above 300°C. Another way is to use an alloy containing 20–25 weight percent silver. In such alloys the α-β miscibility gap is depressed to well below room temperature, and the diffusion coefficient for hydrogen is nearly twice as great as in pure palladium.

It should be noted that when palladium is made the cathode in the electrolysis of an aqueous solution, much of the hydrogen formed dissolves in the metal; the method thus serves for preparing palladium hydride.

The nature of the bonding in metallic hydrides is poorly understood. In one theory, perhaps applicable to hydrides such as YH_2 and UH_3, the hydrogen atoms are assumed to have acquired electrons from the metal conduction band and to be present as hydride ions, much as in the saline hydrides. The partially depleted conduction band gives residual metallic bonding in the compound. Thus

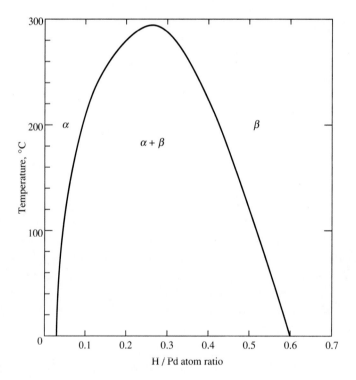

FIGURE 7.1
Temperature-composition phase diagram for the hydrogen-palladium system, showing the two-phase region, or α-β miscibility gap.

the compound can be described as a saline hydride with metallic properties. In another theory, perhaps applicable to materials such as palladium hydride, the hydrogen atoms are assumed to have lost their electrons to the d orbitals of the metal atoms and to be present in the lattice as mobile protons. Thus one can explain (a) the mobility of hydrogen in PdH_x, (b) the fact that the magnetic susceptibility of palladium falls as hydrogen is added, and (c) the fact that if an electric potential is applied across a filament of PdH_x, the hydrogen migrates toward the negative electrode.

Hydrogen Storage with Metallic Hydrides

When the time comes that the combustion of petroleum is no longer an economical source of energy, hydrogen may be an important fuel. Hydrogen is an attractive fuel because it has an extremely high density of energy per unit weight, it is nonpolluting (the main combustion product is water), and it can be oxidized in a variety of devices ranging from internal combustion engines to fuel cells. In the past, the problems of storing hydrogen safely and compactly have retarded the development of hydrogen as a fuel. However, solutions to those problems may be found in the form of certain metallic hydrides which can be reversibly decomposed to give hydrogen gas and the metals.[4] It is possible to store more hydrogen in the form of these hydrides than in the same volume of liquid hydrogen. The hydrogen storage capacities of some representative hydrogen storage systems can be compared in Table 7.2.

Consider the various steps in the absorption of hydrogen by a metal. A small amount of hydrogen first simply dissolves in the metal. In this state the hydrogen exists as atoms in interstitial lattice sites. As the pressure of hydrogen gas in contact with a metal is increased, more and more hydrogen atoms are forced into

---

[4]   J. J. Reilly and G. D. Sandrock, *Sci. Am.*, **242**, 118 (February 1980); R. L. Cohen and J. H. Wernick, *Science*, **214**, 1081 (1981).

TABLE 7.2
Hydrogen storage capacities of some hydrogen storage systems

Storage medium	Weight percent hydrogen	Hydrogen density, g ml^{-1}
MgH_2	7	0.101
Mg_2NiH_4	3.16	0.081
VH_2	2.07	0.095
$FeTiH_{1.95}$	1.75	0.096
$LaNi_5H_7$	1.37	0.089
Liquid H_2	100	0.07
Gaseous H_2 (100 atm)	100	0.008

the metal. At some critical concentration and pressure the metal becomes saturated, and a new phase, the metal hydride, begins to form. Then further hydrogen is absorbed with essentially no increase in pressure. Ultimately all the hydrogen-saturated metal phase is converted into the hydride, and further hydrogen can be dissolved in the hydride phase only by increasing the pressure.

Pressure-composition isotherms for the hydrogen-iron-titanium system are shown in Fig. 7.2. This system is an example of the formation of a ternary hydride from an intermetallic compound. Most of the hydrides that are of interest for storing hydrogen are of this type. Plots of the midplateau hydrogen dissociation pressure versus $1000/T$ for several metals and intermetallics are shown in Fig. 7.3.

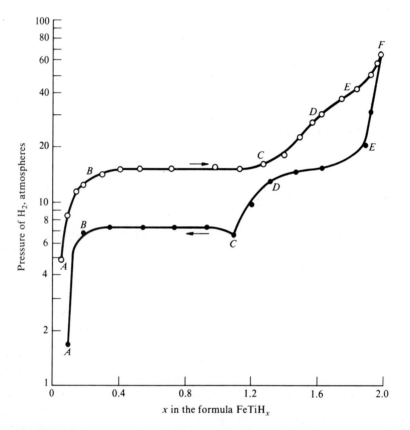

FIGURE 7.2
Pressure-composition isotherms for the hydrogen-iron-titanium system. The upper curve corresponds to the equilibrium pressure as hydrogen was added stepwise to the alloy; the lower curve corresponds to the equilibrium pressure as hydrogen was removed stepwise from the hydride. The cause of this hysteresis is unknown. Regions *A-B* correspond to solid solutions of hydrogen in metal; regions *B-C* to mixtures of the saturated metal and the monohydride; regions *C-D* to solid solutions of hydrogen in the monohydride; regions *D-E* to mixtures of the monohydride and dihydride phases, and regions *E-F* to solutions of hydrogen in the dihydride. [*Adapted from J. J. Reilly and G. D. Sandrock, Sci. Am.,* **242***, 118 (Feb. 1980).*]

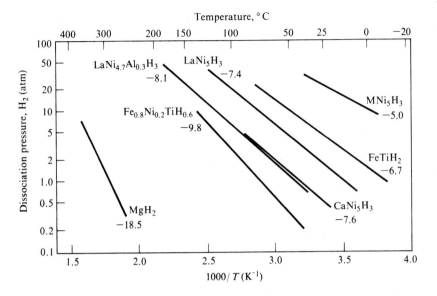

FIGURE 7.3
Dissociation pressure versus $1000/T$ for several hydrides. Calculated heats of reaction (kcal per mole of H_2) are indicated. [*Reproduced with permission from E. L. Huston and G. D. Sandrock, J. Less Common Met.,* **74**, *435 (1980)*.]

Of the various hydrides which have been considered for hydrogen storage, the leading contender is $FeTiH_{1.95}$, because of its cheapness and its relatively high hydrogen density. Magnesium hydride is only a borderline possibility for such use because a rather high temperature (289°C) is required to liberate its hydrogen at 1 atm pressure.

Transition-Metal Hydride Complexes

Until the late 1950s, very few hydrides were known in which hydrogen atoms were covalently bonded directly to transition-metal atoms. Since then, numerous compounds of this type have been characterized. Most of them have the general formula MH_xL_y, where M is a transition metal and L is a ligand capable of acting both as a σ donor and a π acceptor. Examples are $K_3[CoH(CN)_5]$ (a homogeneous hydrogenation catalyst for the conversion of alkynes to alkenes), *trans*-$PtHCl[P(C_2H_5)_3]_2$ (a compound so stable that it can be sublimed in a vacuum), $OsHCl_2[P(C_4H_9)_2C_6H_5]_3$ (a complex with an odd number of electrons and therefore paramagnetic), and K_2ReH_9 (remarkable because the transition metal is bonded to an unusually large number of hydrogen atoms, and no other ligands[5]).

---

[5] Similar complex ions, FeH_6^{4-} and RuH_6^{4-}, have been identified. See R. Bau, D. M. Ho, and S. G. Gibbons, *J. Am. Chem. Soc.*, **103**, 4960 (1981); and R. O. Moyer, Jr., et al., *Inorg. Chem.*, **24**, 3890 (1985).

Some transition-metal ions and complexes react directly with molecular hydrogen to form complexes containing metal-hydrogen bonds. These reactions will be discussed in Chap. 20. In general, transition-metal hydride complexes are prepared by the reduction of transition-metal compounds in the presence of π-bonding ligands. Besides the obvious reducing agents BH_4^- and $LiAlH_4$, reagents such as alcoholic KOH, hydrazine, and hypophosphorous acid have been used. A typical procedure is to treat the anhydrous halide with $NaBH_4$ or $LiAlH_4$ in an ether solvent. For example,

$$C_5H_5Fe(CO)_2Cl \xrightarrow[\text{tetrahydrofuran}]{NaBH_4} C_5H_5Fe(CO)_2H$$

$$trans\text{-}Pt(PR_3)_2Cl_2 \xrightarrow[\text{tetrahydrofuran}]{LiAlH_4} trans\text{-}Pt(PR_3)_2(H)Cl$$

Many hydrides and carbonyl hydrides can be prepared by the treatment of anhydrous halides with triphenylphosphine or triphenylarsine in an alcohol. For example,

$$(NH_4)_2OsCl_6 \xrightarrow[\substack{\text{diethylene glycol}\\\text{monomethyl ether}}]{PPh_3,\ 165°C} HOsCl(CO)(PPh_3)_3$$

Some transition-metal hydride complexes have been prepared by simply heating the metal together with a saline hydride and hydrogen gas:

$$Ru + 2BaH_2 + H_2 \xrightarrow{700°C} Ba_2RuH_6$$

In early research, many of the transition-metal hydride complexes were believed to involve low oxidation states of the metals, and the presence of hydrogen was not suspected. It is now known that the hydrogen in such compounds can be identified by an infrared absorption, due to the M—H stretch, in the range 1600 to 2250 cm^{-1} and a proton magnetic resonance signal at very high field, corresponding to τ values in the range 15 to 50.

Nonmetal Hydrides

Nonmetal hydrides contain hydrogen atoms covalently bonded to the elements of main groups III to VII of the periodic table. The compounds generally consist of discrete molecules having considerable volatility. Properties of some of the volatile nonmetal hydrides are listed in Table 7.3.

A wide variety of synthetic methods have been used in the synthesis of nonmetal hydrides. Some of these methods and typical examples are

1. The reaction of lithium aluminum hydride with a nonmetal halide,

$$LiAlH_4 + SiCl_4 \xrightarrow{\text{ether}} SiH_4 + LiCl + AlCl_3$$

2. The reduction of a nonmetal oxyacid with borohydride in aqueous solution,

$$3BH_4^- + 4H_3AsO_3 + 3H^+ \xrightarrow{H_2O} 4AsH_3 + 3H_3BO_3 + 3H_2O$$

TABLE 7.3
Properties of some volatile hydrides†‡

Formula	Melting point, °C	Boiling point, °C	$\Delta H°$ of formation of gas, kcal mol^{-1}[*]
B_2H_6	-165	-90	7.5
B_4H_{10}	-120	16.1	13.8
B_5H_9	-46.1	58.4	15.0
B_5H_{11}	-123.3	63	22.2
B_6H_{10}	-63.2	108	19.6
$B_{10}H_{14}$	100	....	4.4
SiH_4	-185	-111.2	7.3
Si_2H_6	-129.3	-14.2	17.1
Si_3H_8	-114.8	53.0	25.9
GeH_4	-165.9	-88.51	21.6
Ge_2H_6	-109	29	38.7
Ge_3H_8	-101.8	110.5	53.6
SiH_3GeH_3	-119.7	7.0	27.8
SnH_4	-150	-51.8	38.9
PbH_4	......	~-13	60
PH_3	-133.8	-87.74	1.3
P_2H_4	-99	63.5	5.0
SiH_3PH_2	<-135	12.7	2
AsH_3	-116.9	-62.5	15.9
SbH_3	-88	-18.4	34.7
BiH_3	......	~17	66

† M. F. Hawthorne, in "The Chemistry of Boron and Its Compounds," E. L. Muetterties, ed., pp. 223–323, Wiley, New York, 1967.

‡ W. L. Jolly and A. D. Norman, *Prep. Inorg. React.*, **4**, 1 (1968).

* 1 kcal = 4.1840 kJ.

3. The solvolysis of a binary compound of a nonmetal (e.g., calcium phosphide) in acid solution,

$$Ca_3P_2 + 6H^+ \xrightarrow{H_2O} 3Ca^{2+} + 2PH_3$$

4. The pyrolysis of a nonmetal hydride,

$$B_2H_6 \xrightarrow{\Delta} H_2 + B_4H_{10} + B_5H_{11} + \cdots$$

5. The treatment of a nonmetal hydride, or a mixture of such hydrides, with an electric discharge,

$$GeH_4 + PH_3 \xrightarrow[\text{discharge}]{\text{electric}} GeH_3PH_2 + Ge_2H_6 + P_2H_4 + \cdots$$

6. The coupling of an alkali-metal salt of a nonmetal hydride with a halogen derivative of a nonmetal hydride,

$$KSiH_3 + SiH_3Br \rightarrow KBr + Si_2H_6$$

The study of the hydrides of carbon and their derivatives constitutes the enormous discipline of organic chemistry. Carbon atoms are remarkable for their ability to catenate, i.e., to form chains, as in the aliphatic hydrocarbons, or alkanes. Other elements show this ability to a much lesser extent. Silicon and germanium hydrides analogous to the organic alkanes have been characterized up to $Si_{10}H_{22}$ and $Ge_{10}H_{22}$, that is, up to silicon and germanium chain lengths of 10. The longest molecular-chain hydrides which have been definitely characterized for other elements are Sn_2H_6, HN_3, P_3H_5, As_3H_5, Sb_2H_4, H_2O_3, and H_2S_6.

HYDRIDIC AND PROTONIC CHARACTER

The hydrides of boron, a relatively electropositive nonmetal, react as if the hydrogen atoms had "hydridic" character (i.e., as if they were H^- ions). Thus these hydrides react with water to form hydrogen: $B_2H_6 + 6H_2O \rightarrow 6H_2 + 2H_3BO_3$. Hydrides of the more electronegative elements such as those of groups VI and VII react as if the hydrogen atoms had protonic (H^+) character. Thus these react with bases to form salts or adducts: $HCl + C_5H_5N \rightarrow (C_5H_5NH)Cl$. Hydrides of the elements of intermediate electronegativity, such as those of group IV, have no marked hydridic or protonic character. These hydrides may show, depending on the reagent, either hydridic or protonic character. Thus germane, GeH_4, shows hydridic character in the reaction with hydrogen bromide, $GeH_4 + HBr \rightarrow GeH_3Br + H_2$, and shows protonic character in its reaction with sodium amide, $GeH_4 + NaNH_2 \rightarrow NaGeH_3 + NH_3$.

It is instructive to study the trends in the hydridic and protonic characters of the binary molecular hydrides from a thermodynamic viewpoint. The hydridic character of a hydride H_nX can be measured by the energy of the following gas-phase process:

$$H_nX(g) \quad \rightarrow \quad H^-(g) + H_{n-1}X^+(g) \tag{7.1}$$

The question is: How does this energy change as we change element X by moving from one place to another in the periodic table? The problem is somewhat simplified if we consider the energy of reaction 7.1 as a combination of energies. The energy can be broken up into the energies of the following separate processes:

$$H_nX(g) \quad \rightarrow \quad H(g) + H_{n-1}X(g) \qquad D(H\!-\!XH_{n-1})$$

$$H(g) + e^-(g) \quad \rightarrow \quad H^-(g) \qquad -EA(H)$$

$$H_{n-1}X(g) \quad \rightarrow \quad H_{n-1}X^+(g) + e^-(g) \qquad IE(H_{n-1}X)$$

These energies are (1) the energy required to remove a hydrogen atom from the gaseous molecular hydride, (2) the negative electron affinity of hydrogen, and (3) the ionization energy of the $H_{n-1}X$ radical. When we go from one hydride to another, the change in hydridic character is the sum of the changes in $D(H\!-\!XH_{n-1})$ and $IE(H_{n-1}X)$. Qualitative change in these latter energies are fairly easy to predict using concepts learned in Chaps. 1 and 3. Both the bond dissociation energy and the ionization energy would be expected to increase from left to right in a *series*, or from bottom to top in a *family*, of the periodic table. Hence, we would expect the energy of reaction 7.1 to show the same trends. Hydridic character should be at a maximum for the gaseous molecule FrH and at a minimum for

HF. Actual values of the energy of the reaction in Eq. 7.1 for various molecular hydrides are given in Table 7.4; it can be seen that the data are consistent with our expectations.

The protonic character of a hydride H_nX can be measured by the energy of the process

$$H_nX(g) \quad \rightarrow \quad H^+(g) + H_{n-1}X^-(g) \tag{7.2}$$

This energy can be broken up into the energies of the following processes:

$$H_nX(g) \quad \rightarrow \quad H(g) + H_{n-1}X(g) \qquad D(H{—}XH_{n-1})$$

$$H(g) \quad \rightarrow \quad H^+(g) + e^-(g) \qquad IE(H)$$

$$e^-(g) + H_{n-1}X(g) \quad \rightarrow \quad H_{n-1}X^-(g) \qquad -EA(H_{n-1}X)$$

We have already pointed out that $D(H{—}XH_{n-1})$ would be expected to increase toward the upper right-hand corner of the periodic table. By recognizing that the electron affinity of a species is the ionization potential of the negative ion, we predict the same sort of trend for $EA(H_{n-1}X)$. However, a change in protonic character is equal to the corresponding change in the *difference* $D(H{—}XH_{n-1}) - EA(H_{n-1}X)$, and therefore it is not possible to predict trends in protonic character without quantitative data on the trends in $D(H{—}XH_{n-1})$ and $EA(H_{n-1}X)$. Values of $D(H{—}XH_{n-1})$, $EA(H_{n-1}X)$, and the energy of reaction 7.2 for various molecular hydrides are given in Table 7.5. It can be seen that, within a given family (vertical column of the periodic table), the dissociation energy changes more rapidly than the electron affinity. Hence the general tendency is for the gas-phase acidity to increase on descending a given family. However, from left to right in the periodic table, the change in electron affinity is generally slightly greater than the corresponding change in dissociation energy, at least in the case of the nonmetal hydrides. Hence there is a small increase in gas-phase acidity on going from left to right in the table. Tables 7.4 and 7.5 reveal some interesting facts. For example, the gas-phase acidity of NaH is greater than that

TABLE 7.4
Values of $\Delta H°$ of the reaction $H_nX \quad \rightarrow \quad H^- + H_{n-1}X^+$ (in kilocalories per mole)†

LiH	...	CH$_4$	NH$_3$	H$_2$O	HF
165	...	312	349	404	521
NaH	...	SiH$_4$	PH$_3$	H$_2$S	HCl
149	...	~260	~290	315	386
KH	...	...	...	...	HBr
112	...	...	...	...	344
					HI
					295

1 kcal = 4.1840 kJ.

† J. L. Franklin et al., *Ionization Potentials, Appearance Potentials, and Heats of Formation of Gaseous Positive Ions*, *Natl. Bur. Std. (U.S.), Natl. Std. Ref. Data Ser.*, NSRDS-NBS 26, 1969; R. S. Berry, *Chem. Rev.*, **69**, 533 (1969).

TABLE 7.5
Thermodynamic data for molecular binary hydrides, H_nX

Dissociation energies, electron affinities of $H_{n-1}X$, and energies for the reaction $H_nX \rightarrow H^+ + H_{n-1}X^-$, all in kilocalories per mole†,‡

	LiH	CH$_4$	NH$_3$	H$_2$O	HF
D	58	104	103	119	136
EA	14	2	17	42	78
$\Delta H°$	358	416	400	391	372
	NaH	SiH$_4$	PH$_3$	H$_2$S	HCl
D	48	91	84	93	103
EA	13	33	29	54	83
$\Delta H°$	349	372	369	353	334
	KH	GeH$_4$	AsH$_3$	H$_2$Se	HBr
D	44	87	~ 73	~ 75	88
EA	12	40	29	51	78
$\Delta H°$	346	361	358	338	324
					HI
D	....	....	....	....	71
EA	....	....	....	....	71
$\Delta H°$	....	....	....	....	314

1 kcal = 4.1840 kJ.

† J. L. Franklin et al., Ionization Potentials, Appearance Potentials, and Heats of Formation of Gaseous Positive Ions, *Natl. Std. Bur. (U.S.), Natl. Std. Ref. Data Ser.*, NSRDS-NBS 26, 1969; R. S. Berry, *Chem. Rev.*, **69**, 533 (1969).

‡ J. I. Brauman et al., *J. Am. Chem. Soc.*, **93**, 6360 (1971); K. C. Smyth and J. I. Brauman, *J. Chem. Phys.*, **56**, 1132 (1972); J. I. Brauman, private communication.

of HF! Note, however, that the ionization of NaH to give Na^+ and H^- is favored by 200 kcal mol^{-1} over the ionization to give H^+ and Na^-. On the other hand, the ionization of HF to give H^+ and F^- is favored by 149 kcal mol^{-1} over the ionization to give F^+ and H^-.

PROBLEMS

7.1 Write equations for the net reactions which occur in the following cases:

(a) Calcium hydride is added to an aqueous solution of sodium carbonate.

(b) Sodium hydride is heated to 500°C in vacuo.

(c) LiAlH$_4$ and $(C_2H_5)_2O·BF_3$ react in ether to form diborane.

(d) Germane is passed into a stirred suspension of finely divided potassium in a high-molecular-weight ether such as diglyme.

(e) Arsine is passed into a liquid ammonia solution of KGeH$_3$.

(f) Excess of water is added to an ether solution of LiAlH$_4$.

7.2 Suggest a sequence of reactions for preparing GeH$_3$AsH$_2$ from As$_2$O$_3$ and GeCl$_4$.

7.3 Suggest a sequence of reactions for preparing CH$_3$GeH$_2$Br from GeH$_4$ and CH$_3$I.

*7.4 In the synthesis of Ge$_2$H$_6$ from GeH$_3$Br and KGeH$_3$, large amounts of $(GeH_2)_x$ and GeH$_4$ form as by-products. Explain. [Hint: GeH$_2$Br$^-$ would be expected to decompose to give $(GeH_2)_x$ and Br$^-$.]

CHAPTER
8

ACID-BASE
REACTIONS

We have learned that protonic character, or acidity, is an important feature of many hydrogen compounds. In this chapter we will systematically study various aspects of protonic acid-base reactions, including hydrogen bonding, the kinetics of proton-transfer reactions, and reactions in protonic nonaqueous solvents. We will conclude with a study of aprotic acid-base systems (i.e., acid-base systems not involving protons) and the completely general Lewis acid-base theory.

GAS-PHASE PROTON AFFINITIES

Proton Affinities of Anions

In Chap. 7 we considered the factors which influence the heats of ionization of the simple binary hydrides. A more extensive list of heats of ionization for gas-phase protonic acids, i.e., the "proton affinities" of the corresponding anions, is given in Table 8.1. These values were derived from various experimental data, including ionization energies and appearance energies measured by electron impact and photoionization mass spectrometry, and equilibrium data for ion-molecule reactions studied by high-pressure mass spectrometry, flowing afterglow, and ion cyclotron resonance spectrometry. The data in this table give us information regarding substituent effects free of the complications of solvation. Let us consider the effects of the substituents, R, on the acidity of an acid HAR_n.

$$HAR_n \quad \rightarrow \quad H^+ + AR_n^-$$

When considering the relative acid-strengthening or acid-weakening effects of various substituents, it is helpful to break up the ionization process into two hypothetical steps. In the first step, the proton is removed from the acid without

TABLE 8.1
Heats of ionization of protonic acids in the gas phase
(proton affinities of anions†)

Acid	ΔH°_{ion}, kcal mol^{-1}	Acid	ΔH°_{ion}, kcal mol^{-1}
CH_4	416.6	CH_3SH	359.0
H_2	400.4	AsH_3‡	~359
NH_3	399.6	CH_3NO_2	358.7
H_2O	390.8	C_2H_5SH	357.4
CH_3OH	379.2	Cyclopentadiene	356.1
$C_6H_5CH_3$	379.0	H_2S	353.4
C_2H_5OH	376.1	HCN	353.1
HCF_3	375.6	C_6H_5OH	351.4
C_2H_2	375.4	CH_3COOH‡	348.5
$n\text{-}C_3H_7OH$	374.7	HCOOH‡	345.2
CH_3CN	372.2	H_2Se	338.7
HF	371.5	HNO_2*	338.2
SiH_4	371.5	HCl	333.3
PH_3	370.4	$HRe(CO)_5$*	332
CH_3COCH_3	368.8	HNO_3‡	324.6
$C_6H_5NH_2$	367.1	HBr	323.6
$CH_3SO_2CH_3$	366.6	CF_3COOH‡	322.7
CH_3CHO	366.4	$HMn(CO)_5$*	317
$HCCl_3$‡	~362	HI	314.3
GeH_4	360.7	HPO_3*	311

† Except as noted, from J. E. Bartmess, J. A. Scott, and R. T. McIver, *J. Am. Chem. Soc.*, **101**, 6046 (1979).

‡ J. E. Bartmess and R. T. McIver, in "Gas-Phase Ion Chemistry," M. T. Bowers, ed., chap. 11, p. 87, Academic, New York, 1979.

* A. E. S. Miller, A. R. Kawamura, and T. M. Miller, *J. Am. Chem. Soc.*, **112**, 457 (1990).

any rearrangement of the electrons in the remaining anion. That is, the electrons of the anion are "frozen" as they were in the neutral acid.

$$HAR_n \quad \rightarrow \quad H^+ + AR_n^{-*}$$

The asterisk indicates that the anion has this frozen electron configuration. The energy of this first step is merely the coulombic interaction energy between the proton in HAR_n and all the electrons and nuclei. To a first approximation, this energy is determined by the charge on atom A, which, in turn, is determined by the electronegativity of the substituent R. The more electronegative (electron-withdrawing) R is, the more positively charged A is and the lower the energy of this first step is. Thus, other things being equal, more electronegative substituents increase acidity. In the second hypothetical step, electronic relaxation occurs in the anion; i.e., the electrons of AR_n^{-*} redistribute to give the distribution of the ground-state anion AR_n^-:

$$AR_n^{-*} \quad \rightarrow \quad AR_n^-$$

This electronic relaxation step consists principally of a flow of electron density from atom A to the substituents R. The more readily the substituents can delocalize the negative charge of the anion, the more exothermic is the electronic relaxation step and the stronger is the acid HAR_n. In principle, a third step, corresponding to the stereochemical rearrangement of the atoms in $AR_n{}^-$, should be considered, but the energy of this step is usually relatively small and can be ignored as a first approximation.

Table 8.1 includes data for various substituted methanes which can lose a proton from the carbon atom. All the substituted methanes in the table are stronger acids than methane itself. Fluoroform, HCF_3, is a relatively weak acid even though the three fluorine atoms undoubtedly make the carbon atom highly positive in charge. This surprisingly weak acidity of HCF_3 indicates a low electronic relaxation energy; apparently there is relatively little delocalization of the negative formal charge on the carbon atom in $:CF_3{}^-$. On the other hand, nitromethane, CH_3NO_2, is a relatively strong carbon acid. In this case much of the acidity may be due to extensive electronic relaxation in which the carbon lone pair is delocalized into the π^* orbital of the NO_2 group.

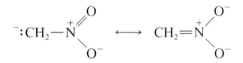

It is interesting to note the effects of methyl substitution on H_2S and H_2O. The methyl group is generally considered to be electron-donating relative to hydrogen. Therefore if we were to consider only the electrostatic potential term, we would predict both CH_3SH and CH_3OH to be weaker acids than their parent hydrides. In fact, CH_3SH is a weaker acid than H_2S, but CH_3OH is a stronger acid than H_2O. The increased acid strength of CH_3OH is probably due to extensive delocalization of negative charge onto the hydrogen atoms of the methyl group, either mainly in the anion,

$$CH_3OH \quad \rightarrow \quad H^+ + H^-H_2C{=}O$$

or in both the neutral molecule and the anion,

$$H^-H_2C{=}O^+{-}H \quad \rightarrow \quad H^+ + H^-H_2C{=}O$$

However, in the case of CH_3SH, the data suggest that $\pi-\sigma^*$ overlap in the C—S bond is so weak (because of the great difference in size between carbon and sulfur) that such acid-strengthening resonance is relatively unimportant.

Although the replacement of a hydrogen by a methyl group is sometimes acid-strengthening and other times acid-weakening, the evidence suggests that increasing the size of the alkyl group is always acid-strengthening because of an increase in polarizability and a consequent increase in the electronic relaxation energy. In Table 8.1 this effect can be seen in the decreasing ΔH°_{ion} values in the series CH_3OH, C_2H_5OH, $n\text{-}C_3H_7OH$.

Proton Affinities of Neutral Molecules

The energy required to dissociate a proton from a gas-phase *cationic* species, BH^+, is called the "proton affinity" of the base B and is a measure of both the acidity of BH^+ and the basicity of B.

$$BH^+ \quad \rightarrow \quad H^+ + B \qquad \Delta E = PA(B)$$

A wide variety of proton affinities are listed in Table 8.2. Most of the bases in Table 8.2 possess nonbonding lone-pair electrons which become bonding electrons in the protonated species. However, some of the bases contain no lone-pair electrons; these are σ or π bond-pair bases which react with protons to form species containing three-center bonds. For example, the protonations of methane and hydrogen may be written as follows:

$$CH_4 + H^+ \longrightarrow H_3C \overset{H^+}{\underset{H}{\diagdown}}$$

$$H_2 + H^+ \longrightarrow H \overset{}{\cdots} H^+$$
$$ H$$

In Chap. 10, the concept of three-center bonding is discussed further.

We have shown that the energy of ionization, ΔH°_{ion}, of a neutral acid HAR_n can be expressed as follows:

$$\Delta H^\circ_{ion} = D(H\!-\!AR_n) + IE(H) - EA(AR_n)$$

On going from one acid to another, IE(H) of course is constant and $D(H\!-\!AR_n)$ might be expected to be fairly closely correlated with ΔH°_{ion}, particularly when the molecules HAR_n have similar electronic structures. Hence for the isoelectronic sets of AH_n molecules CH_4, NH_3, H_2O, HF and SiH_4, PH_3, H_2S, HCl, the heat of ionization would be expected to be linearly related to the electron affinity of the radical. This expectation is confirmed by Fig. 8.1, which is a plot of $\Delta H^\circ_{ion}(AH_n)$ versus $EA(AH_{n-1})$.

Similarly, the proton affinity of a base B can be expressed as follows:

$$PA(B) = D(H\!-\!B^+) + IE(H) - IE(B)$$

where $D(H\!-\!B^+)$ is the energy required to dissociate a hydrogen atom from the protonated base and IE(B) is the ionization energy of B. For reasons analogous to those given above, one would expect the proton affinities of a closely related set of bases to be linearly related to the corresponding ionization energies. Such linear correlations are shown for the isoelectronic bases NH_3, H_2O, HF, Ne and PH_3, H_2S, HCl, Ar in Fig. 8.2. DeKock and Barbachyn[1] have discussed the correlation of proton affinity with ionization energy for a wide variety of bases.

---

[1]   R. L. DeKock and M. R. Barbachyn, *J. Am. Chem. Soc.*, **101**, 6516 (1979).

TABLE 8.2
Proton affinities of some neutral molecules and atoms†

Molecule or atom	Proton affinity		Molecule or atom	Proton affinity	
	eV	kcal mol^{-1}		eV	kcal mol^{-1}
He	1.84	42.5	C_2H_5OH	8.17	188.3
Ne	2.09	48.1	CH_3CN	8.17	188.4
Ar	3.84	88.6	PH_3	8.18	188.6
O_2	4.38	100.9	NCl_3‡	~8.2	~189
H_2	4.39	101.3	C_2H_5SH	8.27	190.8
Kr	4.41	101.6	n-PrOH	8.27	190.8
HF	5.07	117	n-PrSH	8.31	191.6
N_2	5.13	118.2	NH_2OH‡	~8.3	~192
Xe	5.14	118.6	$(CH_3)_2O$	8.33	192.1
NO	~5.5	~127	$\overline{OP(OCH_2)_2}CH$	8.41	194.0
CO_2	5.68	130.9	$B_3N_3H_6$	8.42	194.1
CH_4	5.72	132.0	$(CH_3)_2CO$	8.53	196.7
HCl	5.85	134.8	$(C_2H_5)_2O$	8.68	200.2
HBr	5.9	136	$(CH_3)_2S$	8.70	200.6
N_2O	5.92	136.5	$Fe(CO)_5$	~8.8	~202
CO	6.15	141.9	NH_3	8.85	204.0
C_2H_6	6.23	143.6	CH_3PH_2	8.85	204.1
NF_3	6.24	144	N_2H_4	8.88	204.7
HI	6.50	150	$(C_2H_5)_2S$	8.89	205.0
OCS	6.55	151	$1,6$-$C_2B_4H_6$	8.98	207
C_2H_2	6.65	153.3	$C_6H_5NH_2$	9.08	209.5
AsF_3	6.72	155	$P(OCH_2)_3CCH_3$	9.11	210.0
SiH_4	~6.7	~155	$Fe(C_5H_5)_2$	~9.1	~210
SO_2	7.01	161.6	$(CH_3)_2SO$	9.16	211.3
H_2O_2	7.02	162	$HCON(CH_3)_2$	9.17	211.4
C_2H_4	7.05	162.6	$OP(OCH_3)_3$	9.19	212.0
PF_3	7.22	166.5	$As(CH_3)_3$	9.25	213.4
H_2O	7.22	166.5	CH_3NH_2	9.28	214.1
CS_2	7.25	167.1	$SP(OCH_3)_3$	9.30	214.5
OPF_3	7.28	167.8	$(CH_3)_2PH$	9.38	216.3
$2,4$-$C_2B_5H_7$	7.28	168	$P(OCH_3)_3$	9.57	220.6
H_2S	7.38	170.2	$(CH_3)_2NH$	9.57	220.6
H_2Se	7.43	171.3	C_5H_5N	9.57	220.8
HCN	7.43	171.4	$(CH_3)_3N$	9.76	225.1
H_2CO	7.45	171.7	$(CH_3)_3P$	9.85	227.1
CS	7.59	175	$(C_2H_5)_3P$	10.05	231.7
ClCN	7.62	175.7	$(C_2H_5)_3N$	10.07	232.3
AsH_3	7.77	179.2	LiOH*	10.45	241
C_6H_6	7.86	181.3	NaOH*	10.75	248
CH_3OH	7.89	181.9	KOH*	10.40	263
CH_3SH	8.13	187.4	CsOH*	11.66	269

† Except as noted, data are from S. G. Lias, J. F. Liebman, and R. D. Levin, *J. Phys. Chem. Ref. Data,* **13**(3), 695 (1984).

‡ W. L. Jolly and C. Gin, *Int. J. Mass Spectr. Ion Phys.,* **25**, 27 (1977).

* P. Kebarle, *Ann. Rev. Phys. Chem.,* **28**, 445 (1977).

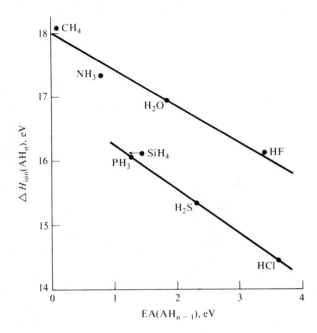

FIGURE 8.1

Plot of heat of ionization for the acids AH_n versus electron affinity for the AH_{n-1} radicals. The lines correspond to isoelectronic series.

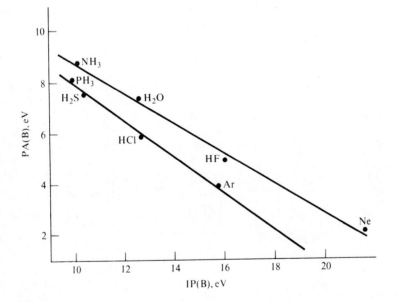

FIGURE 8.2

Plot of proton affinity versus ionization energy for isoelectronic sets of bases. [*Adapted with permission from R. L. DeKock and M. R. Barbachyn, J. Am. Chem. Soc., 101, 6516 (1979). Copyright 1979 American Chemical Society.*]

They found that the data for lone-pair bases are approximately represented by the following equation:

$$PA = 14.7 - 0.597IE$$

and that the data for π and σ bond-pair bases are approximately represented by the equation

$$PA = 13.3 - 0.593IE$$

AQUEOUS ACIDITIES OF PROTONIC ACIDS

Binary Hydrides

The aqueous pK values ($pK = -\log K_a$) of the binary hydrides of the nonmetals are listed in Table 8.3. These values have been obtained by various methods which we shall briefly describe. The pK values for HF, H_2O, H_2S, H_2Se, and H_2Te were obtained by conventional electrochemical methods generally involving emf data from cells with pH-sensitive glass electrodes. The values for HCl, HBr, and HI were estimated from thermodynamic data, making the assumption that the standard free energy of dissolution of gaseous HX to form undissociated aqueous HX is zero.[2] Thus, in the case of HBr,

$$
\begin{array}{llll}
HBr(g) & \rightarrow & H^+(aq) + Br^-(aq) & \Delta G^\circ = -11.85 \text{ kcal mol}^{-1} \\
HBr(aq) & \rightarrow & HBr(g) & \Delta G^\circ \approx 0 \\
\hline
HBr(aq) & \rightarrow & H^+(aq) + Br^-(aq) & \Delta G^\circ \approx -11.85 \text{ kcal mol}^{-1}
\end{array}
$$

Hence

$$pK = -\log K \approx \frac{-11.85}{2.3RT} = \frac{-11.85}{1.364} = -8.7$$

---

[2]   R. T. Myers [*J. Chem. Educ.*, **53**, 17, 802 (1976)] has made the assumption that the molecules HCl, HBr, and HI all form hydrogen bonds with water in the same way that HF does and that therefore the standard free energy of solution has the same value, -5.6 kcal mol^{-1}. Thus he obtains pK values for HCl, HBr, and HI which are 4.1 units greater than those in Table 8.3. However, see the comments of L. Pauling [*J. Chem. Educ.*, **53**, 762 (1976)].

TABLE 8.3
Aqueous pK values of the binary hydrides
of the nonmetals

CH_4	NH_3	H_2O	HF
~ 44	39	15.74	3.15
SiH_4	PH_3	H_2S	HCl
~ 35	27	6.89	-6.3
GeH_4	AsH_3	H_2Se	HBr
25	≤ 23	3.7	-8.7
		H_2Te	HI
		2.6	-9.3

The pK value for PH$_3$ was estimated from kinetic data on the base-catalyzed exchange of hydrogen and deuterium atoms between PH$_3$ and D$_2$O.[3] The rate of reaction was found to be first order in both PH$_3$ and OD$^-$, and the mechanism of the reaction was assumed to be

$$PH_3 + OD^- \xrightarrow{\text{slow}} PH_2^- + HOD$$

$$PH_2^- + D_2O \xrightarrow{\text{fast}} PH_2D + OD^-$$

The rate constant for the exchange, and presumably for the slow step of the proposed mechanism, is 0.4 M^{-1} s^{-1} at 25°C. Now, in ordinary aqueous solution, exchange probably proceeds by the same sort of mechanism, i.e.,

$$PH_3 + OH^- \underset{k_2}{\overset{k_1}{\rightleftharpoons}} PH_2^- + H_2O$$

If we neglect isotope effects, then $k_1 = 0.4$ M^{-1} s^{-1}, and, because the reverse reaction is an extremely exothermic proton transfer, we may assume $k_2 \approx 10^{11} M^{-1}$ s^{-1}. Hence for the latter reaction we can calculate the equilibrium constant, $K' = k_1/k_2 = 4 \times 10^{-12}$. This value is based on the concentrations of all species expressed as molarity. To obtain the value based on the usual convention of unit activity for water, we must divide K' by the molarity of water; hence, $K = (4/55.5) \times 10^{-12} \approx 10^{-13}$. Multiplication by the ionization constant of water yields $K_a = 10^{-27}$ for

$$PH_3 \rightleftharpoons H^+ + PH_2^-$$

The relative pK values in liquid ammonia for PH$_3$, GeH$_4$, and AsH$_3$ were determined from equilibrium measurements.[4] The tabulated values for the aqueous pKs of GeH$_4$ and AsH$_3$ are based on the assumption that the relative pK values of the acids are the same in water and liquid ammonia. The pK value for NH$_3$ is based on the measured self-ionization constant for liquid ammonia and the general observation that pK values of acids are about 10 units lower in ammonia than in water. The value for CH$_4$ is an estimate based on correlations between base-catalyzed hydrogen exchange rates for hydrocarbons and the corresponding pK values. The value for SiH$_4$ is obtained from a simple interpolation.

From the data in Table 8.3 it is obvious that the trend in the aqueous acidity of binary hydrides within a family or group of elements is the same as the corresponding trend in gas-phase acidity; i.e., acidity increases on descending the family. It is also obvious that there is a marked trend toward greater acidity from left to right within a horizontal series of elements. This horizontal trend is in contrast to the absence of a marked trend in the corresponding gas-phase acidities. Undoubtedly the strong horizontal trend in aqueous acidities is due to the increase in anionic hydration energy from left to right (i.e., from larger ions to smaller

[3]   R. E. Weston, Jr., and J. Bigeleisen, *J. Am. Chem. Soc.*, **76**, 3078 (1954).

[4]   T. Birchall and W. L. Jolly, *Inorg. Chem.*, **5**, 2177 (1966).

TABLE 8.4
Aqueous pK values of some derivatives
of water, ammonia, and methane

	pK		
X	HOX	NH$_2$X	CH$_3$X
H	16	39	~44
C$_6$H$_5$	10	27	38
CH$_3$CO	5	15	~20
CN	4	10.5	20
NO$_2$	−2	7	10

ions). The interaction energy of a point charge with a dipole increases with decreasing distance between the charge and the dipole; therefore ionic hydration energy is expected to increase with decreasing ionic radius. For most practical purposes, only the hydrides of groups VI and VII show any detectable acidity in aqueous solution. The acids with negative pK values (HCl, HBr, and HI) are essentially completely ionized at all concentrations.

The acidity of a hydride of group IV, V, or VI can be increased by replacing one (or more, when possible) of the hydrogen atoms in the molecule by a relatively electronegative atom or group. The relative effects of various substituents on the aqueous pK values of water, ammonia, and methane can be seen in the data of Table 8.4. Notice that substitution of ammonia with the acetyl group, CH$_3$CO, yields a compound (acetamide) which has a barely detectable acidity in aqueous solution. Methane must be substituted with the much more electronegative nitro group, NO$_2$, to obtain a compound (nitromethane) with detectable acidity in aqueous solution. It is significant that the order of the acidifying strength of the various groups is the same for the aquo acids, the ammono acids, and the methane acids and that for a given substituent, X, the acidity always increases in the order CH$_3$X < NH$_2$X < HOX.

Hydroxy Acids

By far the most important family of acids in aqueous solutions are those of the general formula HOX, called "aquo," "oxo," or simply "hydroxy" acids. The pK values of some of these acids are given in Table 8.5. In this table the acids are categorized according to the formal charge on the atom bonded to the hydroxyl group. It can be seen that, as that formal charge increases, the pK value decreases (roughly 6 pK units per formal charge unit). It is also obvious from the table that, for polybasic acids, the pK value increases with the degree of ionization of the acid—about 5 pK units for each ionization step. However, the pK values for H$_2$CO$_3$, HNO$_2$, and HNO$_3$ are exceptionally low when compared with the values for the other acids in their respective formal charge categories. Significantly, these exceptionally strong acids are the only ones in the table for which multiple bonding

TABLE 8.5
pK values for some hydroxy acids

Formal charge 0†			Formal charge +1†				Formal charge +2†			Formal charge +3†	
	pK_1	pK_2		pK_1	pK_2	pK_3		pK_1	pK_2		pK
$HOCH_3$	~16		H_2TeO_3	2.7	8.0		H_2SO_4	<0	1.9	$HClO_4$	≪0
HOH	15.7		H_2SeO_3	2.6	8.3		H_2SeO_4	<0	2.0	$HMnO_4$	≪0
HOI	10.7		H_3AsO_4	2.3	7.0	13.0					
H_3AsO_3	9.2		H_3PO_4	2.2	7.1	12.3					
$HOBr$	8.7		$HClO_2$	2.0							
H_4GeO_4	8.6	12.7	H_3PO_2	2.0							
H_6TeO_6	7.8	11.2	H_2SO_3	—	7.2						
$HOCl$	7.5		H_3PO_3	1.8	6.2						
H_2CO_3	3.6*	10.3	H_5IO_6	1.6	7.0						
HNO_2	3.2		HNO_3	-1.3							

† Formal charge of atom bonded to hydroxyl group. Lewis octet structures are assumed for all acids except H_6TeO_6 and H_5IO_6, in which the central atoms are assumed to be single-bonded to the six peripheral oxygen atoms.

* Corrected for unhydrated CO_2.

is involved in the Lewis structures. The nonprotonated oxygen atoms in the anions of these acids have average formal charges which are more positive than -1 and thus attract protons less strongly than do the oxygens with -1 formal charges in the anions of the other acids. The following equation can be used to estimate the pK of an aqueous hydroxy acid:

$$pK = 8.3 - 6F - 5C - 10(1 + f) \tag{8.1}$$

where F is the formal charge on the atom bonded to the hydroxyl group, C is the net charge on the acid, and f is the average formal charge on the nonprotonated oxygen atoms of the conjugate base. It should be noted that the last term in Eq. 8.1 can be ignored for most acids, for which $f = -1$.

Consider the following applications of Eq. 8.1. In the case of $H_2PO_4{}^-$,

$$
\begin{array}{c}
OH \\
| \\
HO-P^+\!\!-H \\
| \\
O^-
\end{array}
$$

$F = 1$, $C = -1$, and $f = -1$. Hence we calculate $pK = 7.3$, in good agreement with the experimental value, 7.1. Next consider HNO_2,

$$HO-N{=}O$$

for which $F = 0$, $C = 0$, and $f = -0.5$, and for which we calculate $pK = 3.3$, in good agreement with the experimental value, 3.2.

However, from Table 8.5 we see that Eq. 8.1 sometimes gives poor results, as in the case of the weak acids $HOCH_3$ and HOH. Clearly, the pK values are influenced by more factors than just F, C, and f. It can be seen that, for acids having analogous structures and the same formal charges, the acidity increases with increasing electronegativity of the atom bonded to the hydroxyl group. When greater predictive accuracy is required, the following equation of Branch and Calvin[5] can be used for estimating the pK of a hydroxy acid with $f = -1$.

$$pK = 16 - \sum I_\alpha \left(\frac{1}{2.8}\right)^i - \sum I_c \left(\frac{1}{2.8}\right)^i - \log \frac{n}{m}$$

The summations are carried out over all the atoms in the group attached to the OH group. The value of I_α is a measure of the electron-attracting ability of the atom; values of I_α for several important elements are given in Table 8.6. The term I_c is 12.3 times the formal charge of the atom, and i is the number of atoms separating the OH group from the atom in question. The last term in the equation accounts for the increase in acidity due to multiple —OH groups in the acid and

---

[5] G. E. K. Branch and M. Calvin, "The Theory of Organic Chemistry," Prentice-Hall, Englewood Cliffs, N.J., 1941.

TABLE 8.6
Values of I_α for use in Branch and Calvin's equation†

H	0	O	4.0	P		1.1
Cl	8.5	S	3.4	As		1.0
Br	7.5	Se	2.7	C		−0.4
I	6.0	N	1.3	C_6H_5		2.0

† G. E. K. Branch and M. Calvin, "The Theory of Organic Chemistry," Prentice-Hall, Englewood Cliffs, N.J., 1941.

the decrease in acidity due to multiple —O⁻ groups in the anion; n is the number of —OH groups in the acid, and m is the number of —O⁻ groups in the anion. We shall illustrate the use of this equation in the estimation of the pK of the dihydrogen phosphate ion,

$$\sum I_\alpha \left(\frac{1}{2.8}\right)^i = 1.1\left(\frac{1}{2.8}\right)^0 + 3(4)\left(\frac{1}{2.8}\right)^1 = 5.3_8$$

$$\sum I_c \left(\frac{1}{2.8}\right)^i = 12.3\left(\frac{1}{2.8}\right)^0 - 2(12.3)\left(\frac{1}{2.8}\right)^1 = 3.5_2$$

$$\log \frac{n}{m} = \log \frac{2}{3} = -0.1_8$$

$$pK = 16 - 5.3_8 - 3.5_2 + 0.1_8 = 7.3$$

The calculated value, 7.3, is in good agreement with the experimental value, 7.1.

Qualitative trends in acidity are not always the same in aqueous solution and in the gas phase. For example, in aqueous solution, the acid strengths of the haloacetic acids decrease in the order[6] $FCH_2COOH > ClCH_2COOH > BrCH_2COOH$, whereas in the gas phase the acid strengths decrease in the opposite order[7] $BrCH_2COOH > ClCH_2COOH > FCH_2COOH$. In the gas phase, these acidities do not show the trend that one would predict from the relative electronegativities of the halogens. Obviously the gas-phase trend must be established by the relative electronic relaxation energies accompanying the ionizations: The relaxation energy and the stability of the anion increase as the polarizability of

[6] R. T. Morrison and R. N. Boyd, "Organic Chemistry," 3d ed., p. 600, Allyn and Bacon, Boston, 1973.
[7] H. Hiraoka, R. Yamdagni, and P. Kebarle, *J. Am. Chem. Soc.*, **95**, 6833 (1973).

the halogen atom increases. However, in aqueous solution, the relaxation energy is affected by the polarization of the solvent in addition to the polarization of the halogen atom. The solvent polarization is greatest for the smallest halogen because the solvent molecules are closest to the center of the negatively charged halogen atom in this case. In aqueous solution, the polarization of the solvent dominates over the polarization of the halogen atom, and the order of acidity is the opposite of that in the gas phase.

Transition-Metal Carbonyl Hydrides

Hydrogen atoms bonded to transition-metal atoms generally do not exhibit significant acidic character, but when the other ligands of the metal atom are strongly electron-withdrawing, acidic behavior is observed. Thus the carbonyl hydrides, in which the metal atoms have rather high positive charges because of back bonding to the CO groups (see Chap. 18), are acids in aqueous and alcohol solutions. The pK values in methanol of some carbonyl hydrides are given in Table 8.7. It can be seen that replacement of a CO in $HCo(CO)_4$ or $HV(CO)_6$ by a much better σ donor and poorer π acceptor, $P(C_6H_5)_3$, causes a great reduction in acidity.

The acidity of these compounds should not be taken as evidence that the hydrogen atoms in the compounds are positively charged. In fact, x-ray photoelectron spectroscopic studies of such compounds indicate that the hydrogen atoms are negatively charged.[8] It is interesting that $HMn(CO)_5$, which is a fairly strong acid ($pK = 7.1$), shows hydridic behavior when treated with a sufficiently strong acid. Neat trifluoromethanesulfonic acid reacts with $HMn(CO)_5$ to give hydrogen and a solution containing $Mn(CO)_5^+$ ions:[9]

---

[8] H. W. Chen, W. L. Jolly, J. Kopf, and T. H. Lee, *J. Am. Chem. Soc.*, **101**, 2607 (1979).

[9] W. C. Trogler, *J. Am. Chem. Soc.*, **101**, 6459 (1979).

TABLE 8.7
pK values of some transition-metal carbonyl hydrides in water at 25°C†

Compound	pK
$HCo(CO)_4$	<0
$HV(CO)_6$	<0
$H_2Fe(CO)_4$	4.4
$HCo(CO)_3P(OPh)_3$	5.0
$HFe(NO)(CO)_3$	~5.1
$HV(CO)_5PPh_3$	6.8
$HCo(CO)_3PPh_3$	7.0
$HMn(CO)_5$	7.1
$HFe(CO)_4^-$	~14
$HRe(CO)_5$	Very weak

† Data from R. G. Pearson, *Chem. Rev.*, **85**, 41 (1985).

$$HMn(CO)_5 + HO_3SCF_3 \xrightarrow{\text{HO}_3\text{SCF}_3} H_2 + [Mn(CO)_5](O_3SCF_3)$$

Of course this result does not imply a mechanism with the improbable step $H^+ + H^- \rightarrow H_2$. However, the observed products are those expected from the reaction of a proton from HO_3SCF_3 with a hydride ion from $HMn(CO)_5$.

HYDROGEN BONDING[10]

In some compounds, individual hydrogen atoms are bound simultaneously to two (or sometimes more) electronegative atoms. The electronegative atoms are said to be "hydrogen-bonded." The first proposal of such bonding was made by Moore and Winmill,[11] who explained the weakness of trimethylammonium hydroxide relative to tetramethylammonium hydroxide by postulating the structure

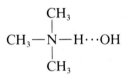

Of course it is now recognized that the hydrogen atom, with only one valence atomic orbital, cannot exceed a total covalency of 1. Hydrogen bonds are generally indicated by dotted lines, as follows:

$$\begin{array}{c} CH_3 \\ | \\ CH_3-N-H\cdots OH \\ | \\ CH_3 \end{array}$$

Evidence for hydrogen bonding is found in a wide variety of experimental data. In the following paragraphs we shall describe some of this evidence.

The data in Table 8.1 show that the gas-phase acidity of toluene is greater than that of water. That is, the proton affinity of OH^- is greater than that of $C_6H_5CH_2^-$. Yet every chemist knows that toluene shows no significant ionization in aqueous solution, whereas water itself self-ionizes to an important extent. It has been estimated that the pK_a of toluene in water is about 38; this should be compared with the pK_a of water which, when the concentration of water is expressed as molarity, is about 16. The enormous change in the relative acidities upon going to aqueous solution is due to the relatively strong stabilization of the aqueous hydroxide ion by hydrogen bonding. It is believed that the aqueous

[10]   L. Pauling, "The Nature of the Chemical Bond," 3d ed., pp. 449–504, Cornell University Press, Ithaca, N.Y., 1960; G. C. Pimentel and A. L. McClellan, "The Hydrogen Bond," Freeman, San Francisco, 1959.

[11]   T. S. Moore and T. F. Winmill, *J. Chem. Soc.*, **101**, 1635 (1912). Also see W. M. Latimer and W. H. Rodebush, *J. Am. Chem. Soc.*, **42**, 1419 (1920); and M. L. Huggins, *Phys. Rev.*, **18**, 333 (1921); **19**, 346 (1922).

hydroxide ion is hydrated to form species of the type

$$\left[\begin{array}{l} \text{H} \\ \text{OH}\cdots\text{OH} \end{array}\right]^{-} \qquad \left[\begin{array}{l} \text{OH} \\ \text{H}\quad\;\;\cdots \\ \quad\quad\text{OH} \\ \text{OH}\;\cdots \\ \text{H} \end{array}\right]^{-} \qquad \text{etc.}$$

To some extent, this hydrogen bonding may be looked upon as an interaction of the dipole moment of the O—H bonds in the water molecules with the negative charge of the OH$^-$ ion. Because the OH$^-$ ion is very small, the water dipoles can approach the ion very closely and interact strongly. On the other hand, the benzyl anion is relatively large, and the energy of interaction with water molecules is relatively small.

The effect of anion size on gas-phase hydration energy is shown by the data in Table 8.8. Notice that the fluoride ion, which has a size comparable to that of the hydroxide ion, also has very high hydration energies. One would expect that the $-\Delta H^\circ$ for the reaction $X^-(H_2O)_{n-1}(g) + H_2O(g) \rightarrow X^-(H_2O)_n(g)$ would approach the heat of vaporization of water, 10.5 kcal mol^{-1}, as n approached ∞. The data of Table 8.8 show that the $-\Delta H^\circ$ values for iodide have essentially this value even for low values of n. In other words, a water molecule interacts with an iodide ion with about the same energy that it does with other water molecules in liquid water.

Because of the strong hydrogen bonding of the hydroxide ion in water and other hydroxylic solvents such as alcohols, much of the intrinsic basicity of the ion is lost when it is dissolved in such solvents. However, by using a solid alkali-metal hydroxide it is possible to take advantage of the strong basicity of the hydroxide ion. Thus a suspension of potassium hydroxide in a nonhydroxylic solvent such as dimethyl sulfoxide can be used to deprotonate extremely weak acids, such as triphenylmethane (aqueous p$K \approx 30$), phosphine (aqueous p$K \approx 27$), and germane (aqueous p$K \approx 25$).[12]

---

[12] W. L. Jolly, *J. Chem. Educ.*, **44**, 304 (1967); *Inorg. Synth.*, **11**, 113 (1968).

TABLE 8.8
Gas-phase hydration energies of anions†
$-\Delta H^\circ$ in kilocalories per mole for $X^-(H_2O)_{n-1}(g) + H_2O(g) \rightarrow X^-(H_2O)_n(g)$

n	OH$^-$	F$^-$	Cl$^-$	Br$^-$	I$^-$
1	22.5	23.3	13.1	12.6	10.2
2	16.4	16.6	12.7	12.3	9.8
3	15.1	13.7	11.7	11.5	9.4
4	14.2	13.5	11.1	10.9	

† M. Arshadi, R. Yamdagni, and P. Kebarle, *J. Phys. Chem.*, **74**, 1475 (1970); M. Arshadi and P. Kebarle, *J. Phys. Chem.*, **74**, 1483 (1970).

$$2KOH(s) + HA \quad \rightarrow \quad K^+ + A^- + KOH \cdot H_2O(s)$$

This result is particularly impressive when one remembers that, in aqueous solution, the hydroxide ion is incapable of quantitatively deprotonating any acid with a pK higher than about 14.

The compounds HF, H_2O, and NH_3 are hydrogen-bonded in both the solid and liquid states. However, only in the case of the H_2O molecule, which contains two hydrogen atoms and one oxygen atom in a nonlinear arrangement, is the bonding ideally suited for forming a three-dimensional network of hydrogen bonds.[13] Each water molecule can act both as a donor and as an acceptor of two hydrogen atoms. This dual ability is clearly illustrated by the crystal structure of ordinary hexagonal ice, shown in Fig. 8.3. In ice each oxygen atom is surrounded

---

[13] F. H. Stillinger, *Science,* **209,** 451 (1980).

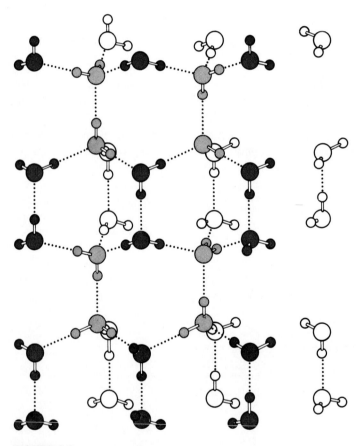

FIGURE 8.3

Structure of ice, showing the hydrogen bonds. Notice the irregular arrangement of the O—H $\cdots$ O bonds. *(From L. Pauling, "The Nature of the Chemical Bond," 3d ed., pp. 449–504, Cornell University Press, Ithaca, N.Y., 1960. Used by permission of Cornell University Press.)*

by four other oxygen atoms, and the hydrogen atoms are located between the oxygen atoms, but not midway. Each oxygen atom is directly bonded to two hydrogen atoms at a distance of 1.01 Å and is hydrogen-bonded to the other two hydrogen atoms at a distance of 1.75 Å. (This asymmetry is typical of most hydrogen bonds.) Except for the restriction that there are always two hydrogen bonds and two normal O—H bonds to each oxygen atom, the distribution of these bonds throughout the lattice is random, even in ice which has been slowly cooled to the neighborhood of 0 K. This randomness is responsible for a residual entropy of $R \ln \frac{3}{2} = 0.81$ cal deg^{-1} mol^{-1} in ice at 0 K. From Fig. 8.3 it can be seen that ice has a rather open structure, with many holes in the lattice. When ice melts, many of the hydrogen bonds are broken, permitting the structure to collapse and causing an increase in density. Because the processes of melting and vaporization for HF, H_2O, and NH_3 require the breaking of hydrogen bonds, these hydrides have abnormally high heats of melting and vaporization as well as high melting points and boiling points. In Fig. 8.4 the boiling points for various nonmetal hydrides are plotted against the periodic-table positions of the nonmetals. Clearly the points for HF, H_2O, and NH_3 deviate markedly from the positions expected if these compounds were unassociated in the liquid state. The point for HCl also shows a significant deviation, probably attributable to a small amount of hydrogen bonding.

The hydrogen bonding in HF is so strong that it persists even in the gaseous state. At 20°C and 745 torr, 80 percent of the HF molecules are in the form of $(HF)_6$ polymers. In aqueous solution, HF is a weak acid which interacts with the

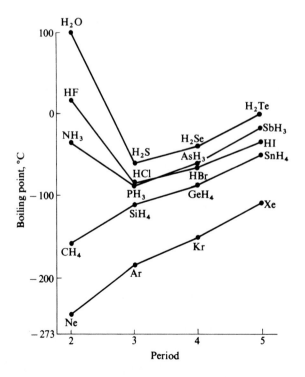

FIGURE 8.4
Boiling points of some molecular hydrides and, for comparison, the noble gases. *(Reproduced with permission from W. L. Jolly, "The Chemistry of the Non-Metals," Prentice-Hall, Englewood Cliffs, N.J., 1966.)*

fluoride ion to form the bifluoride ion HF_2^-:

$$F^- + HF \rightleftharpoons HF_2^- \qquad K = 3.9$$

In salts containing this anion, the proton is located midway between the two fluoride ions; hence this hydrogen-bonded system has no residual entropy at 0 K.

The open structure of ice is a consequence of the hydrogen bonding. Many other compounds have structures based on hydrogen bonding. For example, the helical configurations of polypeptide chains would not be possible if the chains were not "knit" together by N—H···O hydrogen bonds. (See Fig. 8.5.) Crystalline boric acid contains layers of $B(OH)_3$ molecules held together by hydrogen bonds, as shown in Fig. 8.6. Crystalline potassium dihydrogen phosphate, KH_2PO_4, contains PO_4 tetrahedra, each of which is surrounded tetrahedrally by four other PO_4 groups. A drawing of the structure is shown in Fig. 8.7. A feature not shown by the drawing is that the PO_4 groups are connected by hydrogen bonds, with two

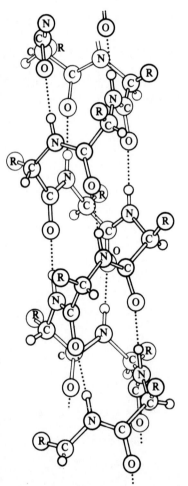

FIGURE 8.5
The α helix of polypeptide chains, showing the N—H···O hydrogen bonds. (*From L. Pauling, "The Nature of the Chemical Bond," 3d ed., Cornell University Press, Ithaca, N.Y., 1960. Used by permission of Cornell University Press.*)

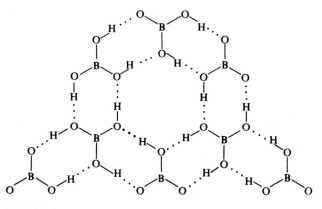

FIGURE 8.6
Arrangement of atoms in a layer of crystalline boric acid.

short O—H bonds and two long O · · · H bonds to each PO_4 group. Below 121 K, all the short O—H bonds are on the same side of the PO_4 groups, as indicated schematically in the following:

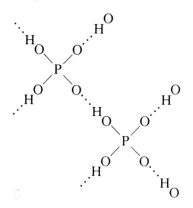

By appropriate application of an electric field to the crystal, the polarization of the hydrogen bonds can be reversed, as indicated in the following:

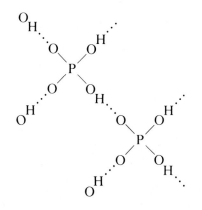

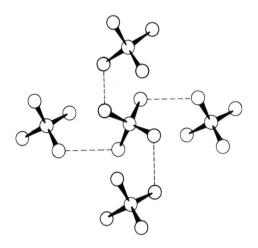

FIGURE 8.7
The environment of an $H_2PO_4^-$ ion in KH_2PO_4.

Because of this spontaneous and reversible electric polarization of KH_2PO_4, it is a member of the class of materials known as "ferroelectrics." Above 121 K, thermal agitation breaks down the ordering of the hydrogen bonds to produce a random structure with no net spontaneous polarization. Above this temperature, the crystal is said to be "paraelectric." Evidence for the ferroelectric-paraelectric transition can be seen in the plot of heat capacity versus temperature shown in Fig. 8.8. The shaded area under the peak around 121 K corresponds to the energy required to randomize the hydrogen bonds.

The oxonium ion, H_3O^+, is a species found in many solid acid hydrates.[14] Thus $HClO_4 \cdot H_2O$ is really a salt, $H_3O^+ClO_4^-$. Similarly, $HCl \cdot H_2O$ is $H_3O^+Cl^-$ and $HNO_3 \cdot 3H_2O$ is $H_3O^+NO_3^- \cdot 2H_2O$. In these and similar compounds, the

---

[14]   W. C. Hamilton and J. A. Ibers, "Hydrogen Bonding in Solids," Benjamin, New York, 1968.

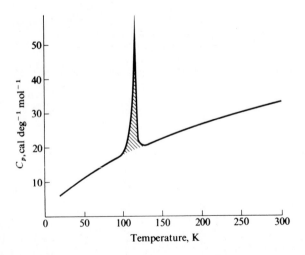

FIGURE 8.8
Plot of heat capacity versus temperature for KH_2PO_4. The peak corresponds to the ferroelectric-paraelectric transition. *(Reproduced with permission from W. C. Hamilton and J. A. Ibers, "Hydrogen Bonding in Solids," Benjamin, New York, 1968.)*

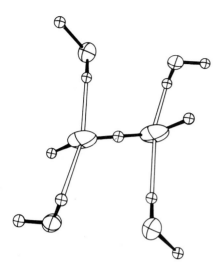

FIGURE 8.9
A perspective view of the $H_3O_2^-$ anion, showing the weak hydrogen bonds between it and four adjacent waters of crystallization. [*Adapted with permission from K. Abu-Dari, K. N. Raymond, and D. P. Freyberg, J. Am. Chem. Soc.,* **101,** *3688 (1979). Copyright 1979 American Chemical Society.*]

H_3O^+ ion is pyramidal, with an H—O—H bond angle of approximately 115°. The ^{17}O nmr spectrum of a solution, in liquid SO_2 at $-15°C$, of approximately equimolar amounts of ^{17}O-enriched H_2O, HF, and SbF_5, consists of a 1:3:3:1 quartet indicative of the three equivalent protons in the H_3O^+. The magnitude of the ^{17}O—H coupling constant (i.e., the separation of the lines in the quartet) corresponds to essentially sp^2 hybridization of the oxygen and is suggestive of a planar or near-planar oxonium ion.[15] In aqueous solutions of strong acids, the oxonium ion undoubtedly exists in a highly hydrated, or hydrogen-bonded, form. Cations of the general formula $H(H_2O)_n^+$ have been identified in the gas phase by mass spectroscopy and in various solid compounds by x-ray and neutron diffraction. For example, the ions $H_5O_2^+, H_7O_3^+, H_9O_4^+$, and $H_{13}O_6^+$ have been structurally characterized in crystalline hydrates.

Relatively few hydrates of the hydroxide ion have been characterized. However, a complex mixed salt containing sodium and triethylmethylammonium cations and tris(thiohydroximato)chromate(III) and hydroxide anions has been shown by an x-ray study to contain the hydroxide analog of the bifluoride ion, $H(OH)_2^-$ or $H_3O_2^-$, shown in Fig. 8.9.[16] The O—O distance of 2.29 Å is remarkably short; the proton of the hydrogen bond probably lies midway between the oxygen atoms, analogously to the proton in HF_2^-. It has been suggested[17] on the basis of spectral data that the anion in the monohydrate of tetramethylammonium hydroxide is $H_6O_4^{2-}$, with a tetrahedral arrangement of oxygen atoms, each of which is hydrogen-bonded to the other three oxygen atoms, as shown in the following:

[15]   G. D. Mateescu and G. M. Benedikt, *J. Am. Chem. Soc.,* **101,** 3959 (1979).

[16]   K. Abu-Dari, K. N. Raymond, and D. P. Freyberg, *J. Am. Chem. Soc.,* **101,** 3688 (1979); A. Bino and D. Gibson, *J. Am. Chem. Soc.,* **103,** 6741 (1981).

[17]   I. Gennick, K. M. Harmon, and J. Hartwig, *Inorg. Chem.,* **16,** 2241 (1977).

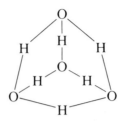

Infrared spectroscopy is commonly used to detect, and to measure quantitatively, hydrogen bonding. A change in the frequency, width, and intensity of an X—H stretching band is particularly diagnostic. For example, the stretching frequency band of an isolated O—H bond might occur around 3500 cm^{-1} and have a width of less than 10 cm^{-1}. After strong hydrogen bonding, the frequency might occur around 1700 cm^{-1}, the bandwidth would be several hundred wave numbers, and the band intensity would probably have increased markedly.

The data in Table 8.9 show that the aqueous ionic conductances of the hydrogen ion and the hydroxide ion are extremely high in comparison with those of other ions.

It is believed that these abnormally high conductances are due to the hydrogen bonding in water and to the fact that the H$^+$ and OH$^-$ ions are strongly hydrogen-bonded. Consider the following picture of an oxonium ion and some of its associated water molecules:

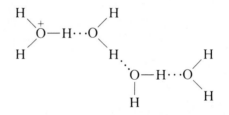

By the slight shift of the protons in three hydrogen bonds, the oxonium ion can in effect move to the other end of the chain of water molecules:

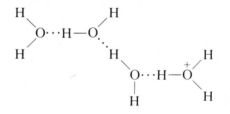

A similar conduction mechanism is possible for the hydroxide ion:

TABLE 8.9
Ionic conductances in water at 25°C

H^+	350		OH^-	192
Na^+	51		Cl^-	76
K^+	74		NO_3^-	71
Ag^+	64		$C_2H_3O_2^-$	41

Thus these ions can migrate through an aqueous solution without having to shove aside water molecules and without having to carry along shells of coordinated water molecules, as is necessary in the migration of other aqueous ions.

KINETICS OF PROTON-TRANSFER REACTIONS

The migration of a proton through water involves steps corresponding to the transfer of a proton from one water molecule to another. This process may be written as a chemical reaction:

$$H_2O + H_3O^+ \quad \rightarrow \quad H_3O^+ + H_2O$$

The rate constant for this bimolecular aqueous reaction has been determined by nmr line width measurements[18] to be $k = 6.0 \times 10^{11} e^{-2.4/RT}$ M^{-1} s^{-1} or $k = 1 \times 10^{10}$ M^{-1} s^{-1} at 25°C. The migration of a hydroxide ion through water likewise corresponds to successive transfers of protons, but between *hydroxide* ions, $H_2O + OH^- \rightarrow OH^- + H_2O$. The nmr-determined[18] rate constant for this reaction is $k = 1.0 \times 10^{11} e^{-2.1/RT}$ M^{-1} s^{-1}, or $k = 3 \times 10^9$ M^{-1} s^{-1} at 25°C. Both of these reactions are among the fastest known reactions in aqueous solution. The fastest aqueous reaction is the transfer of a proton from a water molecule to a hydroxide ion: $H_3O^+ + OH^- \rightarrow H_2O + H_2O$. This reaction, for which $k = 1.4 \times 10^{11}$ M^{-1} s^{-1} at 25°C, is the net reaction which occurs whenever a strong acid is neutralized by a strong base.[19]

We may state as a general rule that any hydrogen-bond proton-transfer reaction for which $\Delta H° < 0$ has a very low activation energy and is very fast, with $k \sim 10^{10}$ to 10^{11} M^{-1} s^{-1}. Consequently endothermic proton-transfer reactions have activation energies slightly greater than $\Delta H°$ and rate constants approximately equal to $K_{eq} \times 10^{10}$, where K_{eq} is the equilibrium constant of the reaction. The forward and reverse rate constants for various reactions of the type

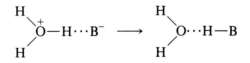

[18] Z. Luz and S. Meiboom, *J. Am. Chem. Soc.*, **86**, 4768 (1964).
[19] M. Eigen, *Angew. Chem. Int. Ed.*, **3**, 1 (1964).

and

$$\begin{array}{ccc} \overset{\displaystyle H}{\underset{\displaystyle |}{}} & & \overset{\displaystyle H}{\underset{\displaystyle |}{}} \\ {}^{-}O{\cdots}H{-}B & \longrightarrow & O{-}H{\cdots}B^{-} \end{array}$$

are given in Table 8.10.

A technique based on the measurement of nmr line widths has been used to determine the rates of symmetric proton exchange for ammonia and various amines.[20] Two types of processes were detected: direct exchange,

$$H_3\overset{+}{N}{-}H{\cdots}NH_3 \xrightarrow{\ k_1\ } H_3N{\cdots}H{-}\overset{+}{N}H_3$$

and exchange via an intervening water molecule,

$$H_3\overset{+}{N}{-}H{\cdots}\underset{\underset{\displaystyle H}{|}}{O}{-}H{\cdots}NH_3 \xrightarrow{\ k_2\ } H_3N{\cdots}H{-}\underset{\underset{\displaystyle H}{|}}{O}{\cdots}H{-}\overset{+}{N}H_3$$

The rate constants k_1 and k_2 are listed in Table 8.11. It can be seen that, as the hydrogen atoms of ammonia are replaced by methyl groups, the rate of the direct process decreases relative to that of the indirect process.

---

[20]   A. Loewenstein and S. Meiboom, *J. Chem. Phys.,* **27,** 1067 (1957); S. Meiboom, A. Loewenstein, and S. Alexander, *J. Chem. Phys.,* **29, 969** (1958).

TABLE 8.10
Rate constants for aqueous proton-transfer reactions†

Reaction	$k_f, M^{-1}\,s^{-1}$	k_r, s^{-1} or $M^{-1}\,s^{-1}$
$H^+ + OH^- \rightarrow H_2O$	1.4×10^{11}	2.5×10^{-5}
$H^+ + NH_3 \rightarrow NH_4^+$	4.3×10^{10}	24
$H^+ + HS^- \rightarrow H_2S$	7.5×10^{10}	8.3×10^3
$H^+ + OAc^- \rightarrow HOAc$	4.5×10^{10}	8.4×10^5
$H^+ + F^- \rightarrow HF$	1.0×10^{11}	7×10^7
$H^+ + SO_4^{2-} \rightarrow HSO_4^-$	1×10^{11}	1×10^9
$H^+ + H_2O \rightarrow H_3O^+$	1×10^{10}	1×10^{10}
$H^+ + SO_2(NH_2)_2 \rightarrow NH_2SO_2NH_3^+$	3×10^7	$> 10^8$
$OH^- + H^+ \rightarrow H_2O$	1.4×10^{11}	2.5×10^{-5}
$OH^- + HATP^{3-} \rightarrow H_2O + ATP^{4-}$‡	1.2×10^9	38
$OH^- + NH_4^+ \rightarrow H_2O + NH_3$	3.4×10^{10}	6×10^5
$OH^- + HCO_3^- \rightarrow H_2O + CO_3^{2-}$	6×10^9	1.2×10^6
$OH^- + HPO_4^{2-} \rightarrow H_2O + PO_4^{3-}$	2×10^9	4×10^7
$OH^- + SO_2(NH_2)_2 \rightarrow H_2O + NH_2SO_2NH^-$	10^{11}	
$OH^- + H_2O \rightarrow H_2O + OH^-$	3×10^9	3×10^9
$OH^- + SO_3NH_2^- \rightarrow H_2O + SO_3NH^{2-}$	$\sim 10^8$	
$OH^- + CH_3OH \rightarrow H_2O + CH_3O^-$	3×10^6	

† M. Eigen, *Angew. Chem. Int. Ed.,* **3,** 1 (1964).

‡ ATP = adenosine triphosphate.

TABLE 8.11
Rate constants for proton-exchange reactions†

Reaction	$k_1, M^{-1}s^{-1}$	$k_2, M^{-1}s^{-1}$	$k_2 k_1$
$NH_4^+ + NH_3$	10.6×10^8	0.9×10^8	0.09
$CH_3NH_3^+ + CH_3NH_2$	2.5×10^8	3.4×10^8	1.4
$(CH_3)_2NH_2^+ + (CH_3)_2NH$	0.4×10^8	5.6×10^8	14
$(CH_3)_3NH^+ + (CH_3)_3N$	$<0.3 \times 10^8$	3.1×10^8	>10

† A. Loewenstein and S. Meiboom, *J. Chem. Phys.*, **27**, 1067 (1957); S. Meiboom, A. Loewenstein, and S. Alexander, *J. Chem. Phys.*, **29**, 969 (1958).

THE ENTROPY OF IONIZATION OF AQUEOUS ACIDS

Protonic acids differ in acidity principally because of corresponding differences in the enthalpy of ionization. This statement of course implies that the entropy of ionization is essentially a constant for aqueous acids. Indeed, this generalization was stated many years ago by Pitzer,[21] who pointed out that the entropies of ionization of weak acids generally have values around -21 cal deg^{-1} mol^{-1}. In Table 8.12 the enthalpies, free energies, and entropies of ionization of some common acids are listed. It can be seen that Pitzer's rule is a fairly good approximation. It should be noted that, at 25°C, a change in $\Delta S°$ of 3 cal deg^{-1} mol^{-1} corresponds to a change in $T\Delta S°$ of almost 1 kcal mol^{-1}; obviously the total range of $\Delta H°$ values in Table 8.12 is much greater than the total range of $T\Delta S°$ values. The approximate constancy of $\Delta S°$ is due to the fact that the difference in entropy

[21]   K. S. Pitzer, *J. Am. Chem. Soc.*, **59**, 2365 (1937).

TABLE 8.12
Thermodynamic constants for the ionization of aqueous acids†

Acid	$\Delta H°$, kcal mol^{-1}	$\Delta G°$, kcal mol^{-1}	$\Delta S°$, cal deg^{-1} mol^{-1}
$HClO_2$	-4.10	2.67	-22.7
Sulfurous ($SO_2 + H_2O$)	-4.00	2.72	-22.6
H_3PO_4	-1.88	2.90	-16.0
$ClCH_2COOH$	-1.17	3.90	-17.0
Butyric	-0.69	6.57	-24.4
Propionic	-0.17	6.65	-22.9
CH_3COOH	-0.11	6.49	-22.1
HCOOH	-0.01	5.12	-17.2
Carbonic ($CO_2 + H_2O$)	1.84	8.68	-22.9
HOCl	3.32	10.10	-22.8
H_2O	13.36	19.09	$-19.2 \ (-27\ddagger)$

† K. S. Pitzer, *J. Am. Chem. Soc.*, **59**, 2365 (1937).

‡ For the 1 *M* standard state of H_2O.

between a weak acid HA and its anion A^- is approximately independent of the nature of A. The total entropy difference is negative because of the considerable structure introduced into the solvent by hydrogen-bonding to the proton and the anion A^-. It should be noted that the entropy of ionization of water itself, when the hypothetical 1 m solution is taken as the standard state (as in the case of all the other acids), is considerably more negative than the average value. This very negative value for water is probably attributable to the fact that the charge of the hydroxide ion is highly localized on the oxygen atom, causing extraordinarily strong hydrogen bonding.

Pitzer's idea can be extended to cationic and anionic protonic acids. An entropy of dissociation of zero is a good approximation for acids with a $+1$ charge, as shown by the values for H_3O^+ and NH_4^+:

$$H_3O^+ \rightleftharpoons H^+ + H_2O \qquad \Delta S° = 0$$

$$NH_4^+ \rightleftharpoons H^+ + NH_3 \qquad \Delta S° = -0.7 \text{ cal deg}^{-1} \text{ mol}^{-1}$$

In the case of acids with a -1 charge (second ionizations) the entropy of ionization appears to be around -30 cal deg^{-1} mol^{-1}, as seen from the following data for HCO_3^-, $H_2PO_4^-$, and HSO_4^-.

$$HCO_3^- \rightleftharpoons H^+ + CO_3^{2-} \qquad \Delta S° = -35 \text{ cal deg}^{-1} \text{ mol}^{-1}$$

$$H_2PO_4^- \rightleftharpoons H^+ + HPO_4^{2-} \qquad \Delta S° = -30 \text{ cal deg}^{-1} \text{ mol}^{-1}$$

$$HSO_4^- \rightleftharpoons H^+ + SO_4^{2-} \qquad \Delta S° = -26 \text{ cal deg}^{-1} \text{ mol}^{-1}$$

SOME SPECIAL AQUEOUS ACIDS

It is often assumed, for convenience, that in an aqueous solution of carbon dioxide the dissolved gas exists entirely as carbonic acid, H_2CO_3. The ionization constant for carbonic acid that is usually quoted is based on that assumption.

$$H_2CO_3 \rightleftharpoons H^+ + HCO_3^- \qquad K = 4.16 \times 10^{-7}$$

Actually this equilibrium constant is incorrect because most of the dissolved carbon dioxide exists as loosely hydrated CO_2, and the true ionization constant of the small amount of H_2CO_3 present is much greater, 2.5×10^{-4}. The latter value is more in agreement with that expected for an acid with the structure $(HO)_2CO$. (Compare CH_3COOH, $K = 1.8 \times 10^{-5}$.) Dissolved carbon dioxide comes to equilibrium with H_2CO_3 and its dissociation products at a measurably slow rate. This slow reaction can be demonstrated by separately adding a sodium hydroxide solution containing phenolphthalein to a CO_2 solution and to an acetic acid solution. The decolorization of the phenolphthalein takes several seconds with the CO_2 solution, whereas it occurs instantaneously with the acetic acid. The slowness of the reaction of CO_2 is of considerable importance in biological processes, and in Chap. 22 we shall discuss the function of the enzyme carbonic anhydrase in catalyzing the reaction.

The rate constants for the following processes have been determined.[22]

$$CO_2 + H_2O \underset{k_{-1}}{\overset{k_1}{\rightleftharpoons}} H_2CO_3$$

$$OH^- + CO_2 \underset{k_{-2}}{\overset{k_2}{\rightleftharpoons}} HCO_3^-$$

$$k_1 = 3.9 \times 10^{-2} \text{ s}^{-1}$$

$$k_{-1} = 23 \text{ s}^{-1}$$

$$k_2 = 9.4 \times 10^3 \ M^{-1} \text{ s}^{-1}$$

$$k_{-2} = 2.2 \times 10^{-4} \text{ s}^{-1}$$

The second ionization constant of carbonic acid has a rather low value,

$$HCO_3^- \rightleftharpoons H^+ + CO_3^{-2} \qquad K = 4.84 \times 10^{-11}$$

corresponding to a difference in the true pK values of H_2CO_3 and HCO_3^- of 6.7, a value considerably greater than the usual difference of about 5 for consecutive ionizations. Probably this big difference is due to the π bonding between the nonprotonated oxygens and the carbon atom in HCO_3^- and CO_3^{2-}, which causes the formal charge on the nonprotonated oxygens in CO_3^{2-} to be more negative than in HCO_3^- ($-\frac{2}{3}$ as opposed to $-\frac{1}{2}$).

Solutions of sulfur dioxide in water have long been referred to as solutions of "sulfurous acid," H_2SO_3. However, Raman spectra of the solutions show only the presence of dissolved SO_2, with no evidence for the hydroxy acid $(HO)_2SO$. Hence the first ionization of sulfurous acid should be represented as follows:

$$H_2O + SO_2 \rightleftharpoons H^+ + HSO_3^- \qquad K_1 = 1.3 \times 10^{-2}$$

In crystalline bisulfite salts for which structures have been determined, the HSO_3^- ion has a threefold axis of symmetry, with the hydrogen atom attached to the sulfur atom, not an oxygen atom. The presence of a strong band in the S—H stretching region around 2500 cm^{-1} in Raman spectra of bisulfite solutions indicates that the same structure exists in aqueous bisulfite solutions.[23] In solution, bisulfite ions are in equilibrium with $disulfite$ ions[24];

$$2HSO_3^- \rightleftharpoons O_2S{-}SO_3^{2-} + H_2O \qquad K = 0.088 \text{ at } \mu = 1.0$$

The disulfite ion has a sulfur-sulfur bond with the indicated asymmetric structure. Indeed, commercial solid sodium and potassium "bisulfites" are actually the corresponding disulfites. Of course, in strongly alkaline solutions both bisulfite and disulfite are converted to the sulfite ion:

$$HSO_3^- \rightleftharpoons H^+ + SO_3^{2-} \qquad K_2 = 6.2 \times 10^{-8}$$

[22]   M. J. Welch, J. F. Lifton, and J. A. Seck, *J. Phys. Chem.*, **73**, 3351 (1969).

[23]   A. Simon and H. Kriegsmann, *Chem. Ber.*, **89**, 2442 (1956). Also see earlier papers cited therein.

[24]   R. E. Connick, T. M. Tam, and E. von Deuster, *Inorg. Chem.*, **21**, 103 (1982).

The structure of solid boric acid, $B(OH)_3$, is well established from x-ray diffraction studies. The individual molecules are trigonal planar, and the crystal is sheetlike, with the molecules being held together by hydrogen bonds in the common molecular plane. (See Fig. 8.6.) Raman spectra of aqueous boric acid solutions are very similar to those of the solid acid, and there is little doubt that the species in solution is like that in the solid, i.e., that aqueous solutions contain planar $B(OH)_3$ units, with no significant formation of tetrahedral $\overset{-}{B}(OH)_3\overset{+}{OH}_2$ species.[25] In strongly alkaline solutions, Raman spectra indicate that the borate ion exists as the mononuclear tetrahedral $B(OH)_4{}^-$. Thus we must conclude that $B(OH)_3$ does not lose a proton upon ionization; it abstracts a hydroxide ion from water:

$$B(OH)_3 + H_2O \;\rightleftharpoons\; H^+ + B(OH)_4{}^- \qquad K = 5.8 \times 10^{-10}$$

At boron concentrations greater than about $0.025\ M$ and pH values between 7 and 11, polyborate species are formed.

The important equilibria and the probable polyborate structures follow.[26]

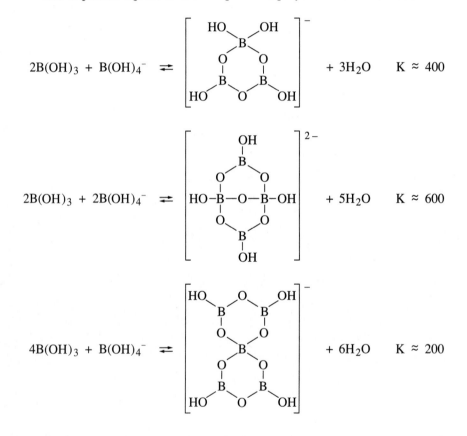

[25]   V. F. Ross, J. O. Edwards, R. P. Bell, and R. B. Jones, in "The Chemistry of Boron and Its Compounds," E. L. Muetterties, ed., chaps. 3 and 4, Wiley, New York, 1967.
[26]   L. Maya, *Inorg. Chem.*, **15**, 2179 (1976).

One of the acids with which most beginning students of chemistry become familiar is hydrogen sulfide. The aqueous equilibria between this dibasic acid and various slightly soluble sulfides are commonly used as examples in exercises in which students calculate, for example, the solubility of a sulfide in an H_2S solution at a particular pH. The first ionization constant of H_2S at 25°C is fairly well established as 1.3×10^{-7}, and the second ionization constant, although less certain, is reported fairly consistently in the older literature to have a value in the vicinity of 10^{-14}. However, most of the early studies of the $HS^- + OH^- \rightleftharpoons H_2O + S^{2-}$ equilibrium are probably invalid because of air oxidation of the sulfide ion to polysulfide ions.

$$\left(\frac{x-1}{2}\right) O_2 + x S^{2-} + (x-1)H_2O \rightarrow S_x^{2-} + (2x-2)OH^-$$

Data obtained using improved techniques to exclude oxygen from strongly alkaline sulfide solutions are consistent with a pK_2 value of 19 ± 2.[27] It is clear that most of the tabulated solubility products of metal sulfides, which have been calculated using pK_2 values around 14, are grossly in error. In fact, this predicament shows that instead of tabulating equilibrium constants for reactions such as

$$CuS \rightleftharpoons Cu^{2+} + S^{2-}$$

it would be better to tabulate equilibrium constants for reactions such as

$$CuS + H_2O \rightleftharpoons Cu^{2+} + OH^- + HS^-$$

or

$$CuS + 2H^+ \rightleftharpoons Cu^{2+} + H_2S$$

for which the equilibrium constants are independent of pK_2 for H_2S.

ACIDS AND BASES IN PROTONIC SOLVENTS

Under ordinary conditions, the ionization of a protonic acid does not involve the liberation of free protons. Earlier in this chapter we have seen that such a process occurring in aqueous solution is best looked upon as the transfer of a proton to a water molecule (or, better, a group of water molecules). In fact, almost all reactions of protonic acids consist of transfers of protons from one base to another. Long ago Brönsted[28] pointed out that acids and bases may be related by the following half reaction:

$$\text{Acid} \rightleftharpoons H^+ + \text{base}$$

[27]   W. Giggenbach, *Inorg. Chem.,* **10,** 1333 (1971); B. Meyer, K. Ward, K. Koshlap, and L. Peter, *Inorg. Chem.,* **22,** 2345 (1983); R. J. Myers, *J. Chem. Educ.,* **63,** 687 (1986).
[28]   J. N. Brönsted, *Rec. Trav. Chim.,* **42,** 718 (1923).

Thus an acid is a species which can act as a proton donor, and a base is a species which can act as a proton acceptor. A typical acid-base reaction may then be written

$$\text{Acid 1} + \text{base 2} \quad \rightleftharpoons \quad \text{Acid 2} + \text{base 1}$$

By removing a proton from acid 1, we obtain base 1. Acid 1 is said to be the conjugate acid of base 1 and base 1 the conjugate base of acid 1. Similar relationships exist between acid 2 and base 2.

Certain solvents resemble water in that they self-ionize to give solvated protons and anions and dissolve salts to give conducting solutions. A list of such solvents, with their physical properties, is given in Table 8.13. In these solvents the Brönsted concept is quite applicable. For example, in liquid ammonia[29] the following self-ionization, or "autoprotolysis," occurs:

$$2NH_3 \quad \rightleftharpoons \quad NH_4{}^+ + NH_2{}^- \qquad K_{-33°C} \approx 10^{-30}$$

In liquid ammonia any species which gives (or reacts to give) ammonium ions is an acid, and any species which gives (or reacts to give) amide ions is a base. Thus the following reactions are acid-base reactions in ammonia.

$$H_2NCONH_2 + NH_2{}^- \quad \rightarrow \quad H_2NCONH^- + NH_3$$

$$NH_4{}^+ + KOH(s) \quad \rightarrow \quad K^+ + NH_3 + H_2O$$

$$H_2O + K^+ + NH_2{}^- \quad \rightarrow \quad KOH(s) + NH_3$$

In anhydrous sulfuric acid,[30] the following autoprotolysis occurs:

$$H_2SO_4 \quad \rightleftharpoons \quad H^+ + HSO_4{}^-$$

For simplicity, we write H^+ for the solvated proton in H_2SO_4 rather than $H_3SO_4{}^+$ or $H(H_2SO_4)_n{}^+$. When water is dissolved in anhydrous sulfuric acid, the following solvolysis takes place:

$$H_2O + H_2SO_4 \quad \rightleftharpoons \quad H_3O^+ + HSO_4{}^-$$

Hence water is a base in sulfuric acid. Similar reactions occur with many other oxygen-containing compounds. Often dehydration accompanies the basic hydrolysis:

$$HONO_2 + 2H_2SO_4 \quad \rightarrow \quad NO_2{}^+ + H_3O^+ + 2HSO_4{}^-$$

A major, reactive constituent of solutions of nitric acid in concentrated sulfuric acid, which are often used in the nitration of aromatic compounds, is the

---

[29]   W. L. Jolly and C. J. Hallada, in "Non-aqueous Solvent Systems," T. C. Waddington, ed., pp. 1–45, Academic, London, 1965; J. J. Lagowski and G. A. Moczygemba, in "The Chemistry of Non-aqueous Solvents," J. J. Lagowski, ed., vol. 2, pp. 320–371, Academic, New York, 1967.

[30]   R. J. Gillespie and E. A. Robinson, in "Non-aqueous Solvent Systems," T. C. Waddington, ed., pp. 117–210, Academic, London, 1965.

TABLE 8.13
Physical properties of some protonic solvents†

Solvent	Melting point, °C	Boiling point, °C	Dielectric constant at 25°C	Electric conductivity, mho cm	$-\log K_{ion}$ at 25°C	ΔS_{vap}	Viscosity 1000η
HSO_3F	−89.0	162.7	~120	$1.1 \times 10^{-4}(25°C)$	7.4	····	15.6(25°C)
H_2SO_4	10.37	328.6	100	$1.04 \times 10^{-2}(20°C)$	3.4(10°C)	24.6	245(25°C)
HNO_3	−41.6	86	·········	$8.9 \times 10^{-3}(0°C)$	~−1.7(−40°C)	····	8.9(20°C)
H_3PO_4	42.35	····	~61	$4.6 \times 10^{-2}(25°C)$	0.9	····	1780(25°C)
HF	−89.4	19.51	60(19°C)	$1.4 \times 10^{-5}(-15°C)$	≥11.7(0°C)	6.1	2.4(6°C)
HCl	−114.6	−84.1	9.28(−95°C)	$3.5 \times 10^{-9}(-85°C)$	····	20.4	5.1(−95°C)
HOOCH	8.3	100.5	57.9(20°C)	6×10^{-5}	6.2	14.2	18.0(20°C)
HOAc	16.6	118.2	6.19	$5 \times 10^{-9}(25°C)$	14.45	14.9	11.6(25°C)
$HOOCCF_3$	−15.4	71.8	·········	·········	····	····	····
HCN	−13.2	25.7	106.8	$5 \times 10^{-7}(0°C)$	~−18.7(12°C)	20.2	2.0(20°C)
H_2S	−85.5	−60.3	10.2(−60°C)	$3.7 \times 10^{-11}(-78°C)$	·········	21.0	4.3(−60°C)
H_2O_2	−0.9	151.4	93.7	$2 \times 10^{-6}(25°C)$	13	27.3	····
H_2O	0.0	100.0	78.5	$4 \times 10^{-8}(18°C)$	14.0	26.0	10.1(20°C)
CH_3OH	−97.9	64.7	32.6	$1.5 \times 10^{-9}(25°C)$	16.6	25.0	5.45(25°C)
C_2H_5OH	−114.6	78.5	24.3	$1.35 \times 10^{-9}(25°C)$	18.9	26.2	10.8(25°C)
$n\text{-}C_3H_7OH$	−127	97.8	19.7	·········	····	····	22.5(20°C)
$iso\text{-}C_3H_7OH$	−85.8	82.5	18.3	·········	····	····	17.7(30°C)
C_6H_5OH	41	182	9.8(60°C)	·········	····	····	34.9(50°C)
N_2H_4	1.5	113.5	51.7	$2.3 \times 10^{-6}(25°C)$	24.7	25.9	9.0(25°C)
NH_3	−77.7	−33.38	16.9	$1 \times 10^{-11}(-33°C)$	27	23.3	2.5(−33°C)
$iso\text{-}C_3H_7NH_2$	−101.2	34	5.5(20°C)	·········	····	····	·········
$C_6H_5NH_2$	−6.2	184.4	6.89(20°C)	·········	····	····	37.1(25°C)

† Reproduced with permission from W. L. Jolly, "The Synthesis and Characterization of Inorganic Compounds," p. 99, Prentice-Hall, Englewood Cliffs, N.J., 1970.

nitryl ion, NO_2^+. The NO_2^+ ion can be detected by Raman spectroscopy in HNO_3–H_2SO_4–H_2O mixtures. Because of the high acidity of sulfuric acid itself, relatively few substances are acids in this solvent. However, SO_3 does dissolve to form disulfuric acid, which is a slightly stronger acid than sulfuric acid.

$$SO_3 + H_2SO_4 \rightarrow H_2S_2O_7$$

$$H_2S_2O_7 \rightleftharpoons H^+ + HS_2O_7^-$$

Anhydrous hydrogen fluoride self-ionizes to give the solvated proton and the bifluoride ion.

$$2HF \rightleftharpoons H^+ + HF_2^-$$

Hence strong fluoride acceptors such as antimony pentafluoride are acids in hydrogen fluoride,

$$SbF_5 + HF \rightarrow H^+ + SbF_6^-$$

and proton acceptors such as water and nitric acid are bases in hydrogen fluoride,

$$H_2O + 2HF \rightleftharpoons H_3O^+ + HF_2^-$$

$$HNO_3 + 2HF \rightleftharpoons H_2NO_3^+ + HF_2^-$$

If one wishes to have a protonic acid and its anion present in comparable amounts at equilibrium in an aqueous solution, the pK value of the acid must lie within, or near, the range 0 to 14. If the pK is much less than 0, the acid will be completely ionized even in a strongly acidic solution, and if the pK is much greater than 14, the acid will be completely nonionized even in a strongly basic solution. To have a measurable acid-anion equilibrium for $pK \ll 0$, one must use a solvent that is less basic than water, such as sulfuric acid or acetic acid. To have a measurable acid-anion equilibrium for $pK \gg 14$, one must use a solvent that is more basic than water, such as liquid ammonia. In Fig. 8.10, the effective pH ranges corresponding to aqueous solutions that can be achieved in several protonic solvents are shown graphically. By the "effective pH" of a nonaqueous solution we mean the pH of a hypothetical aqueous solution in which a dissolved test acid would have the same equilibrium acid-anion ratio as found in the nonaqueous solution. It can be seen that in a solvent such as formic acid one can differentiate the acidities of HCl ($pK \approx -6$) and HBr ($pK \approx -9$). On the other hand, in a solvent such as liquid ammonia, one can differentiate the acidities of PH_3 ($pK \approx 27$) and GeH_4 ($pK \approx 25$).

In aqueous solutions, the hydrogen-ion activity is closely proportional to the concentration of strong acid down to about pH 0 and closely proportional to the reciprocal of the concentration of strong base up to about pH 14. Beyond these pH limits, the hydrogen-ion activity and the effective pH change much more rapidly than one would calculate from changes in the concentration of strong acid or strong base. In such solutions, and even in completely nonaqueous solutions, it is possible to define an "acidity function"[31] which is the effective aqueous pH.

---

[31] L. P. Hammett, "Physical Organic Chemistry," McGraw-Hill, New York, 1940.

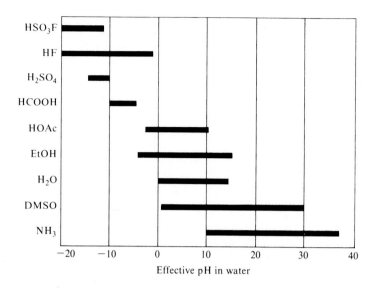

FIGURE 8.10
Effective pH ranges of various protonic solvents.

The acidity function H_0 is defined in terms of a system buffered with a base B and its cationic conjugate acid BH^+:

$$H_0 = pK_a + \log \frac{C_B}{C_{BH^+}}$$

where pK_a is the ionization constant of BH^+, and C_{BH^+} and C_B are the concentrations of BH^+ and B, respectively. The function H_- is defined in terms of a system buffered with a base B^- and its neutral conjugate acid HB:

$$H_- = pK_a + \log \frac{C_{B^-}}{C_{HB}}$$

By appropriate choices of solvent and acid or base concentration, it is possible to prepare solutions having overlapping H_0 or H_- ranges that collectively span an effective pH range from -21 to $+21$. The compositions and H_0 or H_- values for some representative solutions are listed in Table 8.14.

Strongly acidic media such as solutions of SbF_5 in either HSO_3F or HF have been used to prepare stable solutions of a variety of organic carbonium ions, oxonium ions, acyl cations, etc., which are too reactive to exist in less acidic solvents.[32] Some typical reactions follow.

[32]   G. A. Olah, G. K. S. Prakash, and J. Sommer, *Science,* **206,** 13 (1979); G. A. Olah, D. G. Parker, and N. Yoneda, *Angew. Chem. Int. Ed.,* **17,** 909 (1978).

TABLE 8.14
Compositions of some solutions with high $-H_0$ and H_- values

Solution	$-H_0$†	Solution	H_-‡
0.6% SbF₅ in HF	21.13	95% DMSO*,5%	
25% SbF₅ in HSO₃F	21.0	EtOH; 10^{-2} M KOEt	20.68
7% SbF₅ · 3SO₃ in HSO₃F	19.35	90% DMSO*,10%	
10% SbF₅ in HSO₃F	18.94	EtOH; 10^{-2} M KOEt	19.68
HSO₃F	15.07	80% DMSO*,20%	
H₂S₂O₇	14.14	EtOH; 10^{-2} M KOEt	18.97
100% H₂SO₄	11.93	15 M KOH (aqueous)	18.23
100% HF	15.1	10 M KOH (aqueous)	16.90
98% H₂SO₄	10.44	5 M KOH (aqueous)	15.44
90% H₂SO₄	8.92		
60% H₂SO₄	4.46		

† R. J. Gillespie, T. E. Peel, and E. A. Robinson, *J. Am. Chem. Soc.*, **93**, 5083 (1971); R. J. Gillespie and T. E. Peel, *J. Am. Chem. Soc.*, **95**, 5173 (1973); M. J. Jorgenson and D. R. Harrter, *J. Am. Chem. Soc.*, **85**, 878 (1963); G. A. Olah, G. K. S. Prakash and J. Sommer, *Science*, **206**, 13 (1979); R. J. Gillespie and J. Liang, *J. Am. Chem. Soc.*, **110**, 6053 (1988).

‡ K. Bowden and R. Stewart, *Tetrahedron*, **21**, 261 (1965); G. Yagil, *J. Phys. Chem.*, **71**, 1034 (1967).

* DMSO = dimethyl sulfoxide.

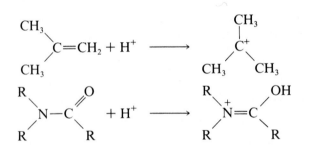

Unusual cations of the nonmetals can also be prepared in such solvents.[33] A blue solution of I_2^+ may be generated by the oxidation of I_2 with $S_2O_6F_6$ in HSO_3F:

$$2I_2 + S_2O_6F_2 \quad \rightarrow \quad 2I_2^+ + 2SO_3F^-$$

---

[33] R. J. Gillespie and M. J. Morton, *Quart. Rev., 1971*, 553; R. J. Gillespie and J. Passmore, *Acc. Chem. Res.*, **4**, 413 (1971).

Elemental sulfur, selenium, and tellurium give colored solutions when dissolved in a number of strongly acidic media. The species S_{16}^{2+}, S_8^{2+}, S_4^{2+}, Se_8^{2+}, Te_4^{2+}, and Te_6^{2+} are present in such solutions. Typical reactions are

$$2S_8 + S_2O_6F_2 \xrightarrow{\text{HSO}_3\text{F}} S_{16}^{2+} + 2SO_3F^-$$

$$8Se + 6H_2S_2O_7 \longrightarrow Se_8^{2+} + HS_3O_{10}^- + 5H_2SO_4 + SO_2$$

ACIDS AND BASES IN APROTIC SOLVENTS

Some solvents do not self-ionize to any significant extent to give solvated protons. This category of "aprotic" solvents includes most hydrocarbons, ethers, and of course solvents containing no hydrogen such as bromine trifluoride and sulfur dioxide. A list of such solvents, and their physical properties, is given in Table 8.15. Obviously the Brönsted concept of acids and bases is inapplicable in these solvents. If an aprotic solvent undergoes self-ionization, the cation formed in the self-ionization is considered the acidic species, and the anion formed in the self-ionization is considered the basic species. For example, BrF_2^+ and BrF_4^- are the acidic and basic species, respectively, in bromine trifluoride.

$$2BrF_3 \rightleftharpoons BrF_2^+ + BrF_4^-$$

Thus antimony pentafluoride, which reacts with BrF_3 as follows,

$$SbF_5 + BrF_3 \rightarrow BrF_2^+ + SbF_6^-$$

is an acid in BrF_3. Potassium fluoride, which reacts as follows,

$$KF + BrF_3 \rightarrow K^+ + BrF_4^-$$

is a base in BrF_3. When solutions of these reagents are mixed, a neutralization reaction occurs:

$$BrF_2^+ + BrF_4^- \rightleftharpoons 2BrF_3$$

In the case of aprotic solvents which do not self-ionize, acid-base reactions are best treated in terms of the very general Lewis acid-base theory, which we shall now discuss.

LEWIS ACID-BASE THEORY

The Lewis acid-base theory is completely independent of solvent considerations. Indeed, many Lewis acid-base reactions proceed in the gas phase. In a Lewis acid-base reaction, a pair of electrons from one species is used to form a covalent bond to another species. The species that "donates" the electron pair is the base, and the species that "accepts" the electron pair is the acid. The reaction may be written

$$A + :B \rightarrow A—B$$

Sometimes the acid is referred to as an "electron acceptor" or an "electrophile" and the base is referred to as an "electron donor" or a "nucleophile." Many reactions

TABLE 8.15
Physical properties of some aprotic solvents†

Solvent	Formula	Melting point, °C	Boiling point, °C	Dielectric constant at 25°C	Viscosity 1000η
Sulfur dioxide	SO_2	−72.7	−10.2	12.3 (22°C)	4.28 (−10°C)
Dinitrogen tetraoxide	N_2O_4	−11.2	21.15	2.4 (18°C)	⋯
Arsenic trifluoride	AsF_3	−8.5	63	5.7 (< −6°C)	⋯
Iodine pentafluoride	IF_5	9.6	98	36.2 (35°)	21.9 (25°C)
Bromine trifluoride	BrF_3	8.8	127.6	⋯	22.2 (25°C)
Phosphorus trichloride	PCl_3	−91	76.0	3.43 (25°C)	⋯
Disulfur dichloride	S_2Cl_2	−80	135.6	4.79 (15°C)	⋯
Arsenic trichloride	$AsCl_3$	−18	130.2	12.8 (20°C)	12.25 (20°C)
Antimony trichloride	$SbCl_3$	73	221	33.0 (75°C)	33 (95°C)
Tin tetrachloride	$SnCl_4$	−30.2	114.1	2.87 (20°C)	⋯
Iodine monochloride	ICl	27.2	100	⋯	41.9 (28°C)
Nitrosyl chloride	$NOCl$	−61.5	−5.4	19.7 (−10°C)	5.47 (−20°C)
Carbonyl chloride	$COCl_2$	−104	8.3	4.3 (22°C)	⋯
Thionyl chloride	$SOCl_2$	−104.5	79	9.0 (22°C)	⋯
Phosphorus oxychloride	$POCl_3$	1.25	105.3	13.9 (22°C)	11.5 (25°C)
Selenium oxychloride	$SeOCl_2$	10.9	176.4	46 (20°C)	⋯
Antimony tribromide	$SbBr_3$	97	280	20.9 (100°C)	68.1 (100°C)
Arsenic tribromide	$AsBr_3$	35	220	8.8 (35°C)	54.1 (35°C)
Iodine monobromide	IBr	41	∼116	⋯	⋯
Mercury(II) bromide	$HgBr_2$	238	320	9.8	⋯
Tin tetraiodide	SnI_4	143.5	340	⋯	⋯
Iodine	I_2	113.7	184.3	11.1 (118°C)	19.8 (116°C)
n-Hexane	$n\text{-}C_6H_{14}$	−94.3	69.0	1.90	2.94 (25°C)
Isooctane	$iso\text{-}C_8H_{18}$	−111.3	117.2	∼1.94 (20°C)	⋯
Kerosene	$\sim C_{12}H_{26}$	⋯	210	∼2.0	⋯
Benzene	C_6H_6	5.5	80.1	2.27	6.52 (20°C)
Toluene	$C_6H_5CH_3$	−95	110.6	2.38	5.90 (20°C)
o-Xylene	$o\text{-}C_6H_4(CH_3)_2$	−25	144.4	2.57 (20°C)	8.10 (20°C)

Carbon tetrachloride	CCl_4	-22.8	76.8	2.23	9.69 (20°C)
Chloroform	$CHCl_3$	-63.5	61.3	4.70	5.42 (25°C)
Dichloromethane	CH_2Cl_2	-96.7	40.1	8.9	3.9 (30°C)
Ethylene chloride	$ClCH_2CH_2Cl$	-35.3	83.7	10.36	8.0 (20°C)
Tetrachloroethane	$Cl_2CHCHCl_2$	-43.8	146.3	...	18.4 (15°C)
Chlorobenzene	C_6H_5Cl	-45.2	132	5.62	7.99 (20°C)
Perfluoroheptane	$n\text{-}C_7F_{16}$	-51	82.5	...	...
Diethyl ether	$(C_2H_5)_2O$	-116.3	34.6	4.22	2.22 (25°C)
Tetrahydrofuran	$OCH_2CH_2CH_2CH_2$	...	65.4	7.39	...
1,2-Dimethoxyethane	$CH_3OCH_2CH_2OCH_3$	-69	85.2	3.5–6.8	11 (20°C)
Diglyme	$CH_3(OCH_2CH_2)_2OCH_3$	-64	162.0	...	20 (20°C)
Dioxane	$OCH_2CH_2OCH_2CH_2$	11.7	101.5	2.21	...
Ethyl acetate	$C_2H_5OCOCH_3$	-83.6	77.1	6.02	4.41 (25°C)
Acetone	CH_3COCH_3	-95.4	56.2	20.7	3.16 (25°C)
Acetonitrile	CH_3CN	-45.7	81.6	36.2	3.45 (25°C)
Pyridine	C_5H_5N	-41.8	115.5	12.3	9.45 (20°C)
Nitromethane	CH_3NO_2	-28.5	101.2	38.6 (20°C)	6.08 (25°C)
Nitrobenzene	$C_6H_5NO_2$	5.7	210.9	34.6	20.3 (20°C)
Formamide	$HCONH_2$	2.55	193	109.5	37.6 (20°C)
N-Methylformamide	$HCONHCH_3$	...	111.2	182.4	16.5 (25°C)
Dimethylformamide	$HCON(CH_3)_2$	-61	153.0	36.7	7.96 (25°C)
N-Methylacetamide	$CH_3CONHCH_3$	29.8	206	178.9 (30°C)	38.85 (30°C)
Dimethylacetamide	$CH_3CON(CH_3)_2$	-20	166.1	37.8	9.2 (25°C)
Carbon disulfide	CS_2	-111.6	46.3	2.64	3.76 (20°C)
Dimethyl sulfoxide	$(CH_3)_2SO$	18.55	189.0	47.6 (23°C)	19.8 (25°C)
Sulfolane	$(CH_2)_4SO_2$	28.37	283	44.0 (30°C)	98.7 (30°C)
Dimethyl carbonate	$CH_3OCOOCH_3$	5.0	90.4	...	...
Ethylene carbonate	OCH_2CH_2OCO	...	244	...	...
Propylene carbonate	$OCHCH_3CH_2OCO$	-49.2	241.7	65.1	25.3 (25°C)
Tri-n-butyl phosphate	$(n\text{-}C_4H_9O)_3PO$	< -80	289 d.	...	...
Hexamethylphosphoryltriamide	$[(CH_3)_2N]_3PO$	7.2	235	30 (20°C)	35 (60°C)

† Reproduced with permission from W. L. Jolly, "The Synthesis and Characterization of Inorganic Compounds," pp. 100–101, Prentice-Hall, Englewood Cliffs, N.J., 1970.

involve the transfer of an acid from one base to another, or vice versa, or both transfers at once. In such cases the net reactions are written

$$B + A{—}B' \quad \rightarrow \quad B' + A{—}B$$

$$A + A'{—}B \quad \rightarrow \quad A' + A{—}B$$

$$A{—}B + A'{—}B' \quad \rightarrow \quad A{—}B' + A'{—}B$$

A few examples of some Lewis acids, bases, and the known adducts of these are listed in Table 8.16.

There is nothing inconsistent with a molecule acting both as a Lewis acid and a Lewis base. All that is required is an atom which can accept a pair of electrons and which has a lone pair of electrons. For example, $SnCl_2$ undergoes the following reactions in which it acts in turn as a Lewis acid and as a Lewis base:

$$SnCl_2 \xrightarrow{Cl^-} SnCl_3^- \xrightarrow{PtCl_4^{2-}} Pt(SnCl_3)_5{}^{3-}$$

$$SnCl_2 \xrightarrow{N(CH_3)_3} Cl_2SnN(CH_3)_3 \xrightarrow{BF_3} \underset{Cl \quad\quad Cl}{\overset{F_3B \diagdown \quad \diagup N(CH_3)_3}{Sn}}$$

One of the difficulties of systematizing the relative basicities of Lewis bases is the fact that different trends in K, $\Delta G°$, or $\Delta H°$ are obtained when different reference acids are used. For example, the complexing ability of the halide ions toward Al^{3+} increases in the order $I^- < Br^- < Cl^- < F^-$. On the other hand, the order is $F^- < Cl^- < Br^- < I^-$ for Hg^{2+}. A similar reversal can be seen in the heats of reaction of the acids I_2 and C_6H_5OH with the bases $(C_2H_5)_2O$ and $(C_2H_5)_2S$.[34] The heat of reaction of I_2 with $(C_2H_5)_2S$ is greater than that with $(C_2H_5)_2O$, whereas the heat of reaction of C_6H_5OH with $(C_2H_5)_2S$ is less than that with $(C_2H_5)_2O$. To resolve these problems, Pearson proposed the principle of

---

[34]   R. J. Niedzielski, R. S. Drago, and R. L. Middaugh, *J. Am. Chem. Soc.*, **86**, 1694 (1964).

TABLE 8.16
Some Lewis acids, bases, and their adducts

Lewis acids	Lewis bases				
	H^-	OH^-	NH_3	Cl^-	C_6H_6
H^+	H_2	H_2O	$NH_4{}^+$	HCl	$[C_6H_7{}^+]$
$B(CH_3)_3$	$B(CH_3)_3H^-$	$B(CH_3)_3OH^-$	$B(CH_3)_3NH_3$	...	$B(CH_3)_3 \cdot C_6H_6$
SO_3	HSO_3^-	HSO_4^-	NH_3SO_3	SO_3Cl^-	...
Ag^+	...	$[AgOH]$	$Ag(NH_3)_2{}^+$	$AgCl_2^-$	$AgC_6H_6{}^+$
I_2	...	I_2OH^-	I_2NH_3	I_2Cl^-	$I_2C_6H_6$

TABLE 8.17
The classification of some Lewis acids†

Hard	Borderline	Soft
H^+, Li^+, Na^+, K^+	Fe^{2+}, Co^{2+}, Ni^{2+}	Cu^+, Ag^+, Au^+, Tl^+, Hg^+
Be^{2+}, Mg^{2+}, Ca^{2+}, Sr^{2+}, Mn^{2+}	Cu^{2+}, Zn^{2+}, Pb^{2+}	Pd^{2+}, Cd^{2+}, Pt^{2+}, Hg^{2+}
Al^{3+}, Sc^{3+}, Ga^{3+}, In^{3+}, La^{3+}	Sn^{2+}, Sb^{3+}, Bi^{3+}, Rh^{3+}	CH_3Hg^+, $Co(CN)_5^{2-}$, Pt^{4+}
N^{3+}, Cl^{3+}, Gd^{3+}, Lu^{3+}, Cr^{3+}	Ir^{3+}, $B(CH_3)_3$, SO_2	Te^{4+}
Co^{3+}, Fe^{3+}, As^{3+}, CH_3Sn^{3+}	NO^+, Ru^{2+}, Os^{2+}	Tl^{3+}, $Tl(CH_3)_3$, BH_3, $Ga(CH_3)_3$
Si^{4+}, Ti^{4+}, Zr^{4+}, Th^{4+}, U^{4+}	R_3C^+, $C_6H_5^+$, GaH_3	$GaCl_3$, GaI_3, $InCl_3$
Pu^{4+}, Ce^{3+}, Hf^{4+}, WO^{4+}, Sn^{4+},		RS^+, RSe^+, RTe^+
UO_2^{2+}, $(CH_3)_2Sn^{2+}$		I^+, Br^+, HO^+, RO^+
VO^{2+}, MoO^{3+}, $BeMe_2$, BF_3		I_2, Br_2, ICN, etc.
$B(OR)_3$, $Al(CH_3)_3$, $AlCl_3$		Trinitrobenzene, etc.
AlH_3, RPO_2^+, $ROPO_2^+$		Chloranil, quinones, etc.
RSO_2^+, $ROSO_2^+$, SO_3		Tetracyanoethylene, etc.
I^{7+}, I^{5+}, Cl^{7+}, Cr^{6+}, RCO^+		O, Cl, Br, I, N, RO, RO_2
CO_2, NC^+		M^0 (metal atoms)
HX (hydrogen-bonding		Bulk metals
molecules)		CH_2, carbenes

† R. G. Pearson, *J. Am. Chem. Soc.*, **85**, 3533 (1963); *Science*, **151**, 172, (1966); *Chem. Br.*, **3**, 103 (1967); *J. Chem. Educ.*, **45**, 581, 643 (1966).

hard and soft acids and bases.[35] A soft acid or base is one in which the valence electrons are easily polarized or removed, and a hard acid or base is one which holds its valence electrons tightly and is not easily distorted. The principle is: Hard acids prefer to coordinate to hard bases, and soft acids prefer to coordinate to soft bases. Tables 8.17 and 8.18 list some examples of acids and bases in the hard, soft, and borderline categories. The opposing trends in the reactivity of the halide ions toward Al^{3+} and Hg^{2+} are now easily rationalized. The Al^{3+} ion is a hard acid that prefers to bond to hard bases (such as F^-), and the Hg^{2+} ion is a

[35]   R. G. Pearson, *J. Am. Chem. Soc.*, **85**, 3533 (1963); *Science*, **151**, 172 (1966); *Chem. Br.*, **3**, 103 (1967); *J. Chem. Educ.*, **45**, 581, 643 (1968).

TABLE 8.18
The classification of some Lewis bases†

Hard	Borderline	Soft
H_2O, OH^-, F^-	$C_6H_5NH_2$, C_5H_5N, N_3^-	R_2S, RSH, RS^-
$CH_3CO_2^-$, PO_4^{3-}, SO_4^{2-}	Br^-, NO_2^-, SO_3^{2-}, N_2	I^-, SCN^-, $S_2O_3^{2-}$, R_3P, R_3As
Cl^-, CO_3^{2-}, ClO_4^-, NO_3^-		$(RO)_3P$
ROH, RO^-, R_2O		CN^-, RNC, CO
NH_3, RNH_2, N_2H_4		C_2H_4, C_6H_6
		H^-, R^-

† R. G. Pearson, *J. Am. Chem. Soc.*, **85**, 3533 (1963); *Science*, **151**, 172 (1966); *Chem. Br.*, **3**, 103 (1967); *J. Chem. Educ.*, **45**, 581, 643 (1968).

soft acid that prefers to bond to soft bases (such as I^-). We can similarly explain the thermal data for I_2, C_6H_5OH, $(C_2H_5)_2O$, and $(C_2H_5)_2S$. Iodine and $(C_2H_5)_2S$ interact relatively strongly because they are both soft, and C_6H_5OH and $(C_2H_5)_2O$ interact strongly because they are both hard.

The principle of hard and soft acids and bases has many qualitative applications in inorganic chemistry. Because the hardness of an element usually increases with increasing oxidation state, to stabilize an element in a very high oxidation state, the element should be coordinated to hard bases, such as O^{2-}, OH^-, and F^-. Thus iron(VI), silver(III), and platinum(VI) can be obtained in the compounds K_2FeO_4, AgO, and PtF_6, respectively. To stabilize an element in a low oxidation state, it should be coordinated to soft bases, such as CO and PR_3. Thus cobalt($-$I) and platinum(0) are found in compounds such as $Na[Co(CO)_4]$ and $Pt[P(CH_3)_3]_4$. In the series of compounds $R_3SiI \rightarrow (R_3Si)_2S \rightarrow R_3SiBr \rightarrow R_3SiNC \rightarrow R_3SiCl \rightarrow R_3SiNCS \rightarrow R_3SiNCO \rightarrow R_3SiF$, a given compound can be converted to any other on its right by refluxing it with the appropriate silver salt.[36] However, a compound cannot be converted in good yield to a compound on its left by this method. For example, the following reaction is irreversible:

$$Et_3SiI + AgBr \quad \rightarrow \quad Et_3SiBr + AgI$$

The soft silver ion prefers to bond to a relatively soft group, and the hard silicon atom prefers to bond to a relatively hard group.

The concept of absolute hardness, η, has been defined as follows:

$$\eta = \frac{1}{2}\left[\frac{\partial^2 E}{\partial N^2}\right] \approx \frac{IE - EA}{2}$$

where, for a molecule, atom, or ion, E is the electronic energy, IE the ionization energy, EA the electron affinity, and N the number of electrons.[37] Absolute softness is simply the reciprocal of η. Although it is thus possible to quantify hardness and softness, such quantitative data are not much more valuable than the qualitative information given in Tables 8.17 and 8.18. The reason is that, if one does not take into account the *strengths* of Lewis acids and bases as well as their hardness or softness, wrong conclusions regarding the driving forces of acid-base reactions can be drawn. For example, consider the reaction

$$CHCl_3 \cdot NH_3 + B(CH_3)_3 \cdot C_6H_6 \quad \rightleftharpoons \quad CHCl_3 \cdot C_6H_6 + B(CH_3)_3 \cdot NH_3$$

According to Tables 8.17 and 8.18 $CHCl_3$ and NH_3 are both hard, $B(CH_3)_3$ is borderline, and C_6H_6 is soft. Therefore one might predict that the reaction would go in the reverse direction. However, the reaction actually goes in the forward direction because of the very strong interaction between $B(CH_3)_3$ and

---

[36] C. Eaborn, *J. Chem. Soc.*, 3077 (1950); A. G. MacDiarmid, *Prep. Inorg. React.*, **1**, 165 (1964).

[37] R. G. Parr and R. G. Pearson, *J. Am. Chem. Soc.*, **105**, 7512 (1983); R. G. Pearson, ibid., **107**, 6801 (1985); **110**, 7684 (1988).

NH$_3$. Obviously at least two parameters must be used to adequately characterize a Lewis acid or base.

We have pointed out that the interaction of a proton with a Lewis base can be divided into two hypothetical parts: (1) the electrostatic interaction between the proton and the electronically frozen base molecule, and (2) the electronic relaxation, or flow of electron density toward the proton. This concept can be extended to the general interaction of Lewis acids with Lewis bases. Any such interaction can be broken up into an electrostatic part and a relaxation part. These two parts are referred to in the literature by a variety of terms. Thus the electrostatic interaction has been referred to as "the initial state interaction," "the frozen orbital potential," and "ionic interaction," and the relaxation part of the interaction has been referred to as "covalent bonding," "the final state interaction," "electron flow," and "polarization." Although there are some differences in the meanings of the expressions in each set, the expressions within each set are more or less interchangeable.

One way of quantitatively accounting for both the electrostatic and relaxation parts of acid-base interactions is by assigning two empirical parameters to each acid and each base and by equating the interaction energy of a given acid-base pair to a suitable algebraic function of the four parameters involved. For example, Drago et al.[38] have shown that the heat of reaction of an acid A and a base B to form an adduct AB can be calculated from the equation

$$-\Delta H^\circ = E_A E_B + C_A C_B \qquad \text{kcal mol}^{-1}$$

where E_A, E_B, C_A, and C_B are empirical constants characteristic of the acid A and the base B. Values of these parameters are given in the literature[38,39] for a wide variety of Lewis acids and bases. The method has not achieved much popularity, perhaps because (a) many experimental data must be available to establish a useful set of parameters and (b) there is considerable arbitrariness in assigning values to the parameters.

PROBLEMS

*8.1 The base-catalyzed exchange reaction of PH$_3$ and D$_2$O may possibly proceed by a one-step process such as the following:

$$D_2O + PH_3 + OD^- \;\rightarrow\; \left[DOD\cdots \overset{\displaystyle H}{\underset{\displaystyle H}{P}}{-}H\cdots OD \right]^{\ddagger} \;\rightarrow\; DO^- + PH_2D + HOD$$

Consequently what may one say about the pK of PH$_3$?

[38] R. S. Drago and B. B. Wayland, *J. Am. Chem. Soc.*, **87**, 3571 (1965); R. S. Drago, *Chem. Br.*, **3**, 516 (1967); *Struct. Bonding (Berlin)*, **15**, 73 (1973); *J. Chem. Educ.*, **51**, 300 (1974), and references therein.

[39] W. L. Jolly, J. D. Illige, and M. H. Mendelsohn, *Inorg. Chem.*, **11**, 869 (1972); D. R. McMillin and R. S. Drago, ibid., **11**, 872 (1972).

8.2 How do we know that phosphorous acid does not have the following structure?

8.3 Arrange the following aqueous acids in order of increasing acidity: HSO_4^-, H_3O^+, C_2H_5OH, H_4SiO_4, $HSeO_4^-$, CH_3GeH_3, NH_3, AsH_3, $HClO_4$, HSO_3F.

8.4 According to the Branch and Calvin equation for estimating the pK of hydroxy acids, by what amount should the successive pK values of a polybasic acid differ?

8.5 Which of the following aqueous reactions would you expect to be the slowest? The fastest?

$$NH_4^+ \quad \rightarrow \quad NH_3 + H^+$$

$$NH_3 + OH^- \quad \rightarrow \quad H_2O + NH_2^-$$

$$HS^- \quad \rightarrow \quad H^+ + S^{2-}$$

$$HSO_4^- + OH^- \quad \rightarrow \quad H_2O + SO_4^{2-}$$

8.6 Predict the pK of diamidophosphoric acid, $HOPO(NH_2)_2$.

*8.7 From the sublimation energy of ice (11.9 kcal mol^{-1}) estimate the O—H $\cdots$ O hydrogen bond energy.

*8.8 Show that one might expect a significant fraction of the aqueous bisulfite ion to exist in the form of the $HOSO_2^-$ isomer.

8.9 Explain why a solution of borax, $Na_2B_4O_7 \cdot 10H_2O$, is a good buffer solution.

*8.10 Because fluorine is more electronegative than chlorine, one might expect the boron atom in BF_3 to be more electron-deficient than that in BCl_3. However, BCl_3 is a stronger acid toward $N(CH_3)_3$ than BF_3. Explain.

8.11 Write equations for the net reactions which occur in the following cases:

(*a*) Aniline is added to a solution of potassium amide in liquid ammonia.

(*b*) Lithium nitride is added to a solution formed by the addition of acetic acid to excess liquid ammonia.

(*c*) NO_2Cl is formed by passing HCl into a solution of nitric acid in concentrated sulfuric acid.

(*d*) Boric acid is dissolved in sulfuric acid. (Six moles of dissolved species, including 2 mol of bisulfate ion, are formed per mole of boric acid.)

8.12 Explain how potassium acid phthalate can be a primary standard acid in water, and a primary standard base in anhydrous acetic acid.

8.13 What is the ammonium ion concentration in pure liquid ammonia?

8.14 Explain the fact that KOH is much more soluble in liquid ammonia containing a little water than in pure liquid ammonia.

8.15 Explain the fact that $FeCl_4^-$ is formed when $FeCl_3$ is dissolved in $OP(OEt)_3$.

8.16 Which of the following reactions have equilibrium constants greater than 1 at ordinary temperatures? (All species may be assumed to be gaseous unless otherwise indicated.)

(*a*) $R_3PBBr_3 + R_3NBF_3 \quad \rightleftharpoons \quad R_3PBF_3 + R_3NBBr_3$

(b) $SO_2 + (C_6H_5)_3PHOC(CH_3)_3 \rightleftharpoons (CH_3)_3COH + (C_6H_5)_3PSO_2$

(c) $AgCl_2^-(aq) + 2CN^-(aq) \rightleftharpoons Ag(CN)_2^-(aq) + 2Cl^-(aq)$

(d) $MeOI + Et(CO)OMe \rightleftharpoons MeOOMe + Et(CO)I$

8.17 Explain why the entropies of vaporization of HF, HOOCH, and HOAc are abnormally low (see Table 8.13). Consider the state of aggregation in the vapor phase.

*8.18 The heats of reaction of $B(CH_3)_3$ with NH_3, CH_3NH_2, $(CH_3)_2NH$, and $(CH_3)_3N$ are 13.8, 17.6, 19.3, and 17.6 kcal mol^{-1}, respectively. Can you explain why the value for $(CH_3)_3N$ is out of line?

8.19 Both BH_3 and BF_3 form an adduct with $(CH_3)_2NPF_2$. In one adduct, the boron atom is bonded to the nitrogen atom, and in the other, to the phosphorus atom. Predict and rationalize the structures of the adducts.

*8.20 Why are strongly acidic solvents used to prepare cationic species such as I_2^+ and Se_8^{2+} and strongly basic solvents used to prepare anionic species such as S_4^{2-} and Pb_9^{4-}?

8.21 Cite other examples (besides those in this chapter) of species which can act both as Lewis acids and Lewis bases.

CHAPTER
9

SOLVATED ELECTRONS

The electron is the simplest and most fundamental chemical species. We now know that it can exist in solution much like any other anion and that it can engage in a wide variety of reactions. First we will discuss the production of the aqueous electron and the methods used to study its very fast reactions with other aqueous species. Then we will study the remarkable metastable solutions of alkali metals in liquid ammonia, which allow us to use the ammoniated electron as a valuable reagent in synthetic chemistry. Finally we will consider alkali-metal anions, which are legitimate examples of alkali metals in the -1 oxidation state.

THE HYDRATED ELECTRON[1]

When an aqueous solution is irradiated with x-rays, gamma rays, or accelerated electrons, water molecules are ionized as follows:

$$H_2O \xrightarrow[\text{radiation}]{\text{high energy}} H_2O^+ + e^-$$

An ejected electron generally has a high energy which is rapidly dissipated by collisions in which electrons are knocked out of other water molecules. In about 10^{-11} s the ejected electron reaches thermal equilibrium and becomes hydrated; we represent the hydrated species by e^-_{aq}. In the meantime H_2O^+ is transformed into OH by the reaction

$$H_2O^+ + H_2O \quad \rightarrow \quad H_3O^+ + OH$$

[1] E. J. Hart, *Acc. Chem. Res.*, **2**, 161 (1969); E. J. Hart and M. Anbar, "The Hydrated Electron," Wiley-Interscience, New York, 1970; M. Anbar and P. Neta, *Int. J. Appl. Radiat. Isot.*, **18**, 493 (1967).

Hence the net ionization reaction is

$$2H_2O \quad \rightarrow \quad H_3O^+ + OH + e^-_{aq}$$

The radiation promotes some water molecules into highly excited electronic states. These excited water molecules decompose into H and OH radicals:

$$H_2O* \quad \rightarrow \quad H + OH$$

Thus the irradiation of an aqueous solution gives H, OH, and e^-_{aq} as the major primary reactive species. These species may react with water, with themselves, or with other reagents in the solution to give various products. For example, the hydrated electron is known to react with H_2O_2, O_2, H^+, and NO_2^- as shown in the following reactions:

$$e^-_{aq} + H_2O_2 \quad \rightarrow \quad OH + OH^- \qquad (9.1)$$

$$e^-_{aq} + \quad O_2 \quad \rightarrow \quad O_2^- \qquad (9.2)$$

$$e^-_{aq} + \quad H^+ \quad \rightarrow \quad H \qquad (9.3)$$

$$e^-_{aq} + NO_2^- \quad \rightarrow \quad NO_2^{2-} \qquad (9.4)$$

Before 1960, many reduction reactions of this type were mistakenly ascribed to atomic hydrogen. In 1962 Czapski and Schwarz[2] reported data which proved that, in each of the four reactions above, the primary reducing species has a -1 charge and therefore probably is e^-_{aq}. They had no facilities for measuring the absolute rate constants for these reactions, but they were able to measure the ratios of rate constants by determining the relative yields of final products in systems in which two of the reactions were competing. They determined rate-constant ratios as a function of ionic strength and used the following Brönsted-Bjerrum equation[3] to evaluate the charge of the reducing species:

$$\log \frac{k}{k_0} = \frac{1.02 Z_a Z_b \mu^{1/2}}{1 + \mu^{1/2}}$$

In this equation k is the rate constant at ionic strength μ, k_0 is the rate constant at infinite dilution, and Z_a and Z_b are the charges of the species involved in the second-order reaction. In Fig. 9.1, $\log[(k/k_0)_n/(k/k_0)_1]$ is plotted versus $\mu^{1/2}/(1 + \mu^{1/2})$ for $n = 2, 3$, and 4 (i.e., for the rate constants of reactions 9.2, 9.3, and 9.4 relative to that for reaction 9.1). The points for reaction 9.2 form a line of zero slope corresponding to $Z_a Z_b = 0$. Because the charge of the O_2 molecule is zero, this result is expected. The points for reaction 9.3 form a line of -1 slope, as expected for Z_a and Z_b values of opposite sign. The points for reaction 9.4

[2]   G. Czapski and H. A. Schwarz, *J. Phys. Chem.*, **66**, 471 (1962).

[3]   A. A. Frost and R. G. Pearson, "Kinetics and Mechanism," 2d ed., pp. 150–153, Wiley, New York, 1961.

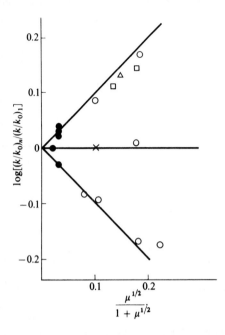

FIGURE 9.1

Plot of $\log[(k/k_0)_n/(k/k_0)_1]$ versus $\mu^{1/2}/(1 + \mu^{1/2})$. Upper curve, $n = 4$; middle curve, $n = 2$; lower curve, $n = 3$. The ionic strength was varied with LiClO$_4$ (○), KClO$_4$ (□), NaClO$_4$ (△), and MgSO$_4$ (×). The closed circles represent no added salt other than reactants. [*Reproduced with permission from G. Czapski and H. A. Schwarz, J. Phys. Chem.,* **66,** *471 (1962). Copyright 1962 American Chemical Society.*]

form a line of $+1$ slope, as expected for $Z_a = Z_b = -1$. Thus the variation of the rate constants with ionic strength is consistent with a -1 charge for the reducing species in these reactions.

Although the hydrated electron decays very rapidly in aqueous solutions, it is possible to obtain its absorption spectrum by fast-scan spectrophotometry of solutions which have been exposed to an intense pulse of radiation. The absorption spectrum is shown in Fig. 9.2. Notice that e^-_{aq} has a broad absorption band at

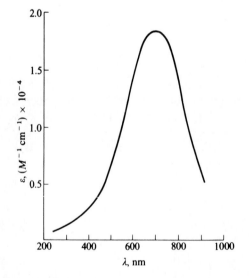

FIGURE 9.2

Absorption spectrum of e^-_{aq} *(Adapted with permission from E. J. Hart and M. Anbar, "The Hydrated Electron," Wiley-Interscience, New York, 1970.)*

TABLE 9.1
Rate constants of some e^- reactions†

Reaction	Rate constant, $M^{-1}\,s^{-1}$
$e^-_{aq} + H_2O \rightarrow H + OH^-$	16
$e^-_{aq} + e^-_{aq} \xrightarrow{2H_2O} H_2 + 2OH^-$	6.0×10^9
$e^-_{aq} + H \xrightarrow{H_2O} H_2 + OH^-$	2.5×10^{10}
$e^-_{aq} + OH \rightarrow OH^-$	3.0×10^{10}
$e^-_{aq} + H^+ \rightarrow H$	2.1×10^{10}
$e^-_{aq} + H_2O_2 \rightarrow OH + OH^-$	1.2×10^{10}

† J. Hart, *Acc. Chem. Res.*, **2**, 161 (1969); E. J. Hart and M. Anbar, "The Hydrated Electron," Wiley-Interscience, New York, 1970; M. Anbar and P. Neta, *Int. J. Appl. Radiat, Isot.*, **18**, 494 (1967).

715 nm, corresponding to a blue solution. By following the absorbance of e^-_{aq} as a function of time (during an interval of approximately 1 millisecond), it is possible to study directly the kinetics of the reaction of e^-_{aq} with various species in the solution. The second-order rate constants for the reaction of e^-_{aq} with some representative species are given in Tables 9.1 and 9.2.

In alkaline solutions, hydrogen atoms are converted into hydrated electrons:

$$H + OH^- \rightarrow e^-_{aq} + H_2O$$

TABLE 9.2
Rate constants of e^-_{aq} and H-atom reactions†

Reduced species	$k, M^{-1}\,s^{-1}$	
	e^-_{aq}	H atom
Ag^+	3.2×10^{10}	1.1×10^{10}
Zn^{2+}	1.4×10^9	$<1.0 \times 10^5$
Cd^{2+}	5.2×10^{10}	$<1.0 \times 10^5$
O_2	1.9×10^{10}	1.9×10^{10}
N_2O	5.6×10^9	$\sim 10^5$
MnO_4^-	3.0×10^{10}	2.6×10^{10}
Gd^{3+}	5.5×10^8	
Eu^{3+}	6.1×10^{10}	
Cl^-	$< 10^5$	
ClO^-	7.2×10^9	
C_6H_6	1.4×10^7	1.1×10^9
C_6H_{10}	$<1.0 \times 10^6$	3×10^9

† E. J. Hart, *Acc. Chem. Res.*, **2**, 161 (1969); E. J. Hart and M. Anbar, "The Hydrated Electron," Wiley-Interscience, New York, 1970; M. Anbar and P. Neta, *Int. J. Appl. Radiat. Isot.*, **18**, 494 (1967).

From the measured rate constants for this reaction ($2.0 \times 10^7\ M^{-1}\ s^{-1}$) and the reverse reaction ($16\ M^{-1}\ s^{-1}$) we can calculate the equilibrium constant:

$$\frac{[e^-_{aq}]}{[H][OH^-]} = K = \frac{k_f}{k_r} = \frac{2.0 \times 10^7}{16 \times 55.5} = 2.55 \times 10^4$$

By combining this with the ionization constant of water, we obtain $K_a = 2.25 \times 10^{-10}$ for

$$H \rightleftharpoons H^+ + e^-_{aq}$$

Thus we have a measure of the acidity of the simplest possible aqueous acid!

The reactions of e^-_{aq} with H^+ and H_2O to form atomic H are examples of a general reaction of e^-_{aq} with protonic acids:

$$e^-_{aq} + HX \quad \rightarrow \quad H + X^-$$

As one might expect, the rate constants for such reactions can be correlated with the acidities of the acids. Figure 9.3 shows a plot of log k versus pK for various aqueous acids; it can be seen that these quantities are linearly related.[4] Obviously the relatively low reactivity of e^-_{aq} with the water molecule (see Table 9.1) is related to the weak acidity of the water molecule.

---

[4] J. Rabani, *Adv. Chem. Ser.*, **50**, 242 (1965).

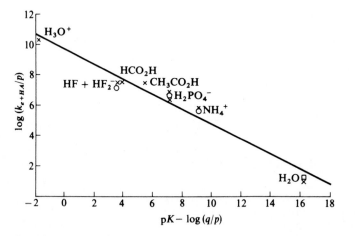

FIGURE 9.3

Plot of log k for e^-_{aq} + HX reactions versus pK_a for HX. The k and pK_a values have been slightly corrected for statistical effects. The crosses refer to radiation data, the circles to photochemical data, and the square to a study of H_2O in ethylenediamine. *(Reproduced with permission from J. Rabani, in "Solvated Electron," Advances in Chemistry Series, No. 50, p. 247, American Chemical Society, Washington, D.C., 1965.)*

In Table 9.2, rate constants for e^-_{aq} reactions can be compared with rate constants for the analogous atomic H reactions. Although atomic H is a powerful reducing agent, e^-_{aq} is stronger. This can be seen from the reduction potentials for these species.[5]

$$e^- + H^+ \quad \rightarrow \quad H \qquad E° = -2.3 \text{ V}$$

$$e^- \quad \rightarrow \quad e^-_{aq} \qquad E° = -2.8 \text{ V}$$

In general, the e^-_{aq} rate constants are greater than the corresponding atomic H rate constants, except in the case of reacting species which have no stable vacant orbitals to accept electrons. The e^-_{aq} reactions probably always involve the formation of intermediate species which correspond to the addition of a negative charge to the reacting species. High rate constants seem to be associated with the formation of stable intermediates, and relatively low rate constants with the formation of unstable intermediates. Thus the reaction with Zn^{2+} is relatively slow because of the instability of Zn^+, and Eu^{3+} reacts much more rapidly than Gd^{3+} because of the stability of Eu^{2+} compared to Gd^{2+}.

METAL-AMMONIA SOLUTIONS[6]

Physical Characteristics

In contrast to water, which reacts rapidly with alkali metals to form hydrogen and the alkali-metal hydroxides, liquid ammonia reversibly dissolves the alkali metals and other electropositive metals to form metastable solutions containing ammoniated electrons. Metal-ammonia solutions have been extensively studied because of their unusual composition and because of their usefulness in inorganic and organic synthesis. We shall first discuss some of the physical characteristics of these solutions which give us insight to the microscopic structure of the solutions.

It is significant that the metals which are soluble in liquid ammonia are the electropositive metals with aqueous reduction potentials more negative than -2.5 V. Obviously, the factors which cause a metal to have a very negative reduction potential (high solvation energy for the ion, low ionization potential, and low sublimation energy) are the same as those which cause it to have a high solubility in ammonia. This observation is consistent with the concept that metal-

---

[5] It should be remembered that in half-reactions of this type the symbol e^- stands for $[\frac{1}{2}H_2 - H^+]$ and does not represent an electron of any type.

[6] W. L. Jolly, *Prog. Inorg. Chem.,* **1,** 235 (1959); Solvated Electron, *Adv. Chem. Ser.,* **50,** 1–304 (1965); J. J. Lagowski and M. J. Sienko, eds., "Metal-Ammonia Solutions," Butterworth, London, 1970 (supplement to *Pure and Applied Chemistry*); W. L. Jolly, "Metal-Ammonia Solutions," Dowden, Hutchinson, and Ross, Stroudsburg, Pa., 1972; J. L. Dye, *Prog. Inorg. Chem.,* **32,** 327 (1984).

ammonia solutions are electrolytic, i.e., that they contain ammoniated metal ions and electrons.

Dilute solutions of metals in ammonia are bright blue; the color is due to the short wavelength tail of a very broad and intense absorption band with a peak at approximately 1500 nm (1.5 μm). This absorption spectrum is shown in Fig. 9.4. The fact that the spectra of the various metals are essentially identical proves that the absorption is due to a common species and strongly suggests that this species is the ammoniated electron. Further indication that the blue color is due to the ammoniated electron is found in the electrolytic behavior of metal-salt solutions. When a solution of a salt such as sodium bromide in liquid ammonia is electrolyzed with direct current using inert electrodes (e.g., platinum electrodes), a blue color appears in the region of the cathode. The electrode reaction may be written $e^- \rightarrow e^-_{am}$, where e^-_{am} stands for the ammoniated electron.

Metal-ammonia solutions are extremely good conductors of electricity; at all concentrations the equivalent conductances are greater than those found for any other known electrolyte in any known solvent. In Fig. 9.5 the equivalent conductances for Li, Na, and K are plotted against the negative logarithm of the molarity of the metal in the ammonia. In the very dilute region, the equivalent conductances approach a limiting value characteristic of the ions M^+_{am} and e^-_{am}. As the solutions are made more concentrated, the equivalent conductances decrease, as in the case of most electrolytic solutions, because of ionic association. However, at approximately 0.05 M the conductances reach a minimum value. Further increase in concentration causes the conductances to increase rapidly until, in the saturated solutions, conductances comparable to those of metals are reached. These concentrated, highly conducting solutions are not blue, like the relatively dilute, electrolytic solutions, but rather bronze-colored, like a molten metal. Indeed, these metallike solutions have been likened to molten "expanded metals" in which ammoniated metal ions are held together by electrons. Further evidence for such a description is found in the fact that the metallike concentration regions correspond to NH_3/metal ratios which are reasonable coordination numbers for the metal ions (e.g., in the case of Na, the metallike region ranges from $NH_3/Na \approx 5.5$ to $NH_3/Na \approx 10$).

The magnetic susceptibility of a very dilute metal-ammonia solution corresponds to that expected for a solution of ammoniated valence electrons with

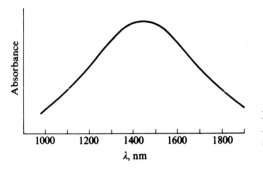

FIGURE 9.4
Absorption spectrum of a dilute (10^{-4} M) solution of a metal in liquid ammonia at $-70°C$.

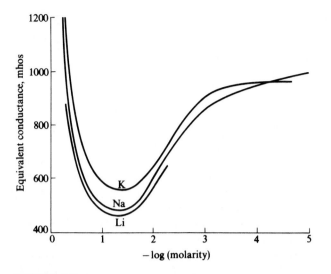

FIGURE 9.5
Equivalent conductances of metal-ammonia solutions at $-33°C$ versus the logarithm of dilution. *(Reproduced with permission from D. M. Yost and H. Russell, Jr., "Systematic Inorganic Chemistry," Prentice-Hall, Englewood Cliffs, N.J., 1946.)*

independent unpaired spins. In more concentrated solutions, the molar suscepti-bility decreases, as shown by the data in Table 9.3. Obviously spin pairing occurs in the more concentrated solutions.

To explain the various physical phenomena, a model has been proposed for metal-ammonia solutions. It is believed that the ammoniated cations and ammo-niated electrons are involved in aggregation equilibria of the following type:

$$M^+_{am} + e^-_{am} \rightleftharpoons (M^+_{am})(e^-_{am})$$

$$2M^+_{am} + 2e^-_{am} \rightleftharpoons (M^+_{am})_2(e^-_{am})_2$$

The species $(M^+_{am})(e^-_{am})$ and $(M^+_{am})_2(e^-_{am})_2$ are obviously neutral and non-conducting; their formation accounts for the decrease in equivalent conductance

TABLE 9.3
Molar magnetic susceptibility of potassium solutions at $-33°C$†

M	$\chi \times 10^6$
0.00341	1268
0.00406	1240
0.00812	974
0.0318	402
0.482	29.9

† S. Freed and Sugarman, *J. Chem. Phys.*, **11**, 354 (1943).

with increasing concentration in the concentration region around 10^{-2} M. The $(M^+_{am})_2(e^-_{am})_2$ species is believed to be diamagnetic; its formation thus accounts for the decrease in magnetic susceptibility with increasing concentration in the same concentration region. In very concentrated solutions (> 1 M), there is not enough ammonia to adequately solvate the electrons; the solution may then be described as $(M^+_{am})e^-$, i.e., as a molten expanded metal.

The Miscibility Gap and Metal Ammoniates

A striking feature of most metal-ammonia systems is the existence of liquid-liquid miscibility gaps. That is, it is possible to have two immiscible solutions of the same metal at equilibrium. The heavier phase is blue and less concentrated in metal than the less dense, bronze-colored phase. For example, consider the process of gradually adding ammonia to a saturated (10.8 m) solution of sodium in ammonia at $-47.5°C$. At first, the solution is simply diluted, but when a concentration of 3.95 m is reached, a second liquid phase of concentration 1.51 m begins to separate out. Continued addition of ammonia causes the dilute phase to grow at the expense of the concentrated phase until only the dilute phase remains. The phase diagram for sodium and ammonia is shown in Fig. 9.6.

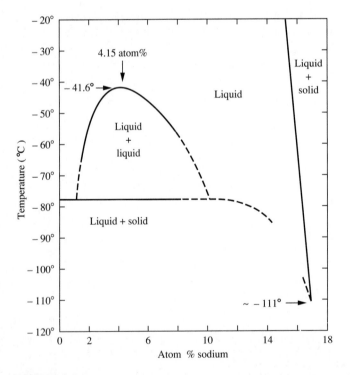

FIGURE 9.6
Sodium-ammonia phase diagram, showing the liquid-liquid miscibility gap.

The phase diagram for the lithium-ammonia system, which has been extensively studied, is shown in Fig. 9.7.[7] Note the deep eutectic at 20 mole percent lithium and 89 K. Considerable evidence indicates that a solid compound, $Li(NH_3)_4$, is formed at or near this point. In fact, there is evidence for three different phases of $Li(NH_3)_4$. Phase transitions, indicated by breaks in the thermal and magnetic properties, occur at 82 K and $\sim$ 25 K. Similar metal ammoniates, but of the general composition $M(NH_3)_6$, are formed by calcium, strontium, barium, europium, and ytterbium.

Reactions

The ammoniacal electron is a very reactive species, widely used as a reagent in synthetic chemistry.[8] We shall now briefly summarize the main types of reaction that the ammoniacal electron is known to undergo.

Many species react with metal-ammonia solutions simply by accepting electrons, without any bond breaking or extensive changes in geometry. Frequently the reactions serve as useful methods for the synthesis of reduced species. For example, manganate(VI), salts, hydronitrites, superoxides, and tetracyanonickelate(0) can be prepared as follows:

$$MnO_4^- + e^-_{am} \longrightarrow MnO_4^{2-}$$

$$NO_2^- + e^-_{am} \longrightarrow NO_2^{2-}$$

$$O_2 + e^-_{am} \longrightarrow O_2^-$$

$$Ni(CN)_4^{2-} + 2e^-_{am} \longrightarrow Ni(CN)_4^{4-}$$

A much larger class of metal-ammonia reactions is that in which the attacking electron causes a bond cleavage. For purposes of categorization, these reactions can be classified in terms of their overall stoichiometry, as shown in Fig. 9.8. The reaction paths of Fig. 9.8 are not necessarily indicative of the reaction mechanisms.

Addition of one electron usually goes according to general reaction I, in which one fragment of the bond cleavage effectively dimerizes:

$$NH_3 + e^-_{am} \rightarrow \tfrac{1}{2}H_2 + NH_2^-$$

$$NH_4^+ + e^-_{am} \rightarrow \tfrac{1}{2}H_2 + NH_3$$

$$GeH_4 + e^-_{am} \rightarrow \tfrac{1}{2}H_2 + GeH_3^-$$

$$C_6H_5I + e^-_{am} \rightarrow \tfrac{1}{2}C_6H_5{-}C_6H_5 + I^-$$

[7]   A. M. Stacy and M. J. Sienko, *Inorg. Chem.*, **21**, 2294 (1982).

[8]   J. Jander, "Anorganische und Allgemeine Chemie in Flüssigem Ammoniak," Interscience, New York, 1966; H. Smith, "Organic Reactions in Liquid Ammonia," Interscience, New York, 1963; W. L. Jolly, "Metal-Ammonia Solutions," Benchmark Papers in Inorganic Chemistry, Dowden, Hutchinson and Ross, Stroudsburg, Pa., 1972; D. Nicholls, "Inorganic Chemistry in Liquid Ammonia," pp. 172–205, Elsevier, Amsterdam, 1979.

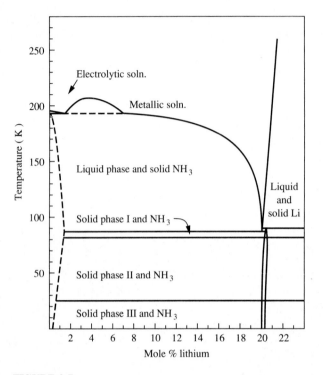

FIGURE 9.7
Lithium-ammonia phase diagram, showing the liquid-liquid miscibility gap and the stability regions for three phases of $Li(NH_3)_4$. [*Reproduced with permission from A. M. Stacy and M. J. Sienko, Inorg. Chem., 21, 2294 (1982).*]

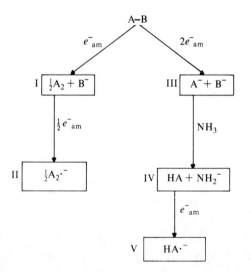

FIGURE 9.8
General reactions of the ammoniacal electron involving bond cleavage. The most important general reactions are I, III, and IV.

The reaction of the electron with ammonia is extremely slow in clean, cold metal-ammonia solutions; indeed, if it were not so, metal-ammonia solutions would not be as valuable as they are. However, the reaction can be catalyzed by certain high-surface-area transition-metal compounds (e.g., iron rust and platinum black), and then the reaction serves as a useful synthesis of alkali-metal amides. The reaction of the electron with the ammonium ion is extremely fast and serves as a convenient way to destroy excess metal after completing a reduction with a metal-ammonia solution.

An example of general reaction II is found in the reaction of iodobenzene with excess ammoniated electrons:

$$2C_6H_5I + 3e^-_{am} \longrightarrow C_6H_5-C_6H_5 \cdot^- + 2I^-$$

The $C_6H_5-C_6H_5 \cdot^-$ ion is a typical aromatic radical anion; such species are highly colored and are readily formed by the reaction of polycyclic aromatic hydrocarbons with alkali metals. The odd electron occupies a highly delocalized π orbital of the hydrocarbon.

General reaction III, involving the formation of two anions, generally occurs when two electrons are added to a molecule:

$$Mn_2(CO)_{10} + 2e^-_{am} \longrightarrow 2Mn(CO)_5^-$$
$$C_6H_5NHNH_2 + 2e^-_{am} \longrightarrow C_6H_5NH^- + NH_2^-$$
$$R_3Ge-GeR_3 + 2e^-_{am} \longrightarrow 2GeR_3^-$$

Frequently one of the anions formed in a two-electron addition undergoes ammonolysis, corresponding to general reaction IV:

$$N_2O + 2e^-_{am} + NH_3 \longrightarrow N_2 + OH^- + NH_2^-$$
$$NEt_4^+ + 2e^-_{am} + NH_3 \longrightarrow NEt_3 + C_2H_6 + NH_2^-$$
$$RBr + 2e^-_{am} + NH_3 \longrightarrow Br^- + RH + NH_2^-$$

It happens that the amide ion formed in each of the latter three reactions reacts with the starting material to give the following side reactions:

$$N_2O + 2NH_2^- \longrightarrow N_3^- + OH^- + NH_3$$
$$NEt_4^+ + NH_2^- \longrightarrow NEt_3 + C_2H_4 + NH_3$$
$$RBr + NH_2^- \longrightarrow RNH_2 + Br^-$$

General reaction V occurs in the reaction of aryl halides with the ammoniacal electron:

$$ArX + 3e^-_{am} + NH_3 \longrightarrow ArH \cdot^- + X^- + NH_2^-$$

Kinetics

The kinetics of some of the reactions of metal-ammonia solutions have been studied, and in some cases the mechanisms of the reactions have been established.[9,10] One type of reaction which has been fairly thoroughly studied is the reaction of a weak protonic acid with the ammoniacal electron,

$$HA + e^-_{am} \longrightarrow \tfrac{1}{2}H_2 + A^-$$

In cases where the weak acid cannot initially react to form a radical anion (e.g., when HA = water, ethanol, and urea), the following mechanism appears to hold[9,10]:

$$HA + NH_3 \underset{2}{\overset{1}{\rightleftharpoons}} NH_4^+ + A^-$$

$$NH_4^+ + e^-_{am} \overset{3}{\longrightarrow} \tfrac{1}{2}H_2 + NH_3$$

(9.5)

These reactions start out fairly rapidly with a simple first-order dependence on the weak acid and a zero-order dependence on sodium. As the reactions proceed, the rates drop markedly and become first-order in both the acid and sodium. The rates can be expressed quantitatively by the following rate law (derived from the above mechanism, assuming a steady state in NH_4^+):

$$-\frac{d[e^-_{am}]}{dt} = \frac{k_1[HA][e^-_{am}]}{(k_2/k_3)[A^-] + [e^-_{am}]}$$

ALKALI-METAL ANIONS AND ELECTRIDES[11]

Alkali metals also dissolve (to form relatively unstable solutions) in certain ethers, low-molecular-weight amines, and hexamethyl phosphoryl triamide. These solutions have properties which indicate that they contain species distinctly different from those found in metal-ammonia solutions. For example, they have optical spectra with two types of absorption bands. One of these (an infrared band) is metal-independent and is assigned to the solvated electron. The other band (V band) has a peak position which depends on the metal and is assigned to the metal anion species, M^-. Spectra for solutions of Cs, K, and Na in tetrahydrofuran are shown in Fig. 9.9. The ratio of the intensity of the V band to that of the infrared band depends on both solvent and metal. The less polar the solvent, or the lower the atomic weight of the metal, the larger the ratio of the V-band intensity to the infrared band intensity. Evidence for the existence of alkali-metal anions is found in the following results.

1. Solutions with high V-band–infrared-band intensity ratios are diamagnetic, as expected for a spin-paired ground state for M^-.

---

[9] W. L. Jolly, in "Metal-Ammonia Solutions," J. J. Lagowski and M. J. Sienko, eds., pp. 167–181, Butterworth, London, 1970 (supplement to *Pure and Applied Chemistry*); *Adv. Chem.*, **50**, 27 (1965).

[10] R. R. Dewald, *J. Phys. Chem.*, **79**, 3044 (1975); R. R. Dewald, R. L. Jones, and H. Boll, in "Electrons in Fluids," J. Jortner and N. R. Kestner, eds., pp. 473–478, Springer-Verlag, New York, 1973.

[11] J. L. Dye, *J. Chem. Educ.*, **54**, 332 (1977); *Prog. Inorg. Chem.*, **32**, 327 (1984); S. B. Dawes, D. L. Ward, O. Fussa-Rydel, R.-H. Huang, and J. L. Dye, *Inorg. Chem.*, **28**, 2132 (1989).

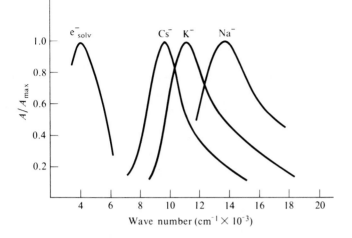

FIGURE 9.9
Spectra at 25°C of e^-_{solv}, Cs$^-$, K$^-$, and Na$^-$ in tetrahydrofuran. [*Reproduced with permission from J. L. Dye, Chem. Educ.,* **54,** *332 (1977).*]

2. The solvent, metal, and temperature dependence of the V band is similar to that of the charge-transfer-to-solvent band of I$^-$ and similar to that expected for M$^-$.

3. When solutions containing Na$^+$ ions are subjected to an electron pulse from a linear accelerator, the rate of formation of the V band is second-order in e^-_{solv}, consistent with the formation of Na$^-$.

4. The molar conductivity of sodium solutions decreases slowly with increasing concentration, as expected for a solute which forms ion pairs Na$^+$Na$^-$ with a very large anion.

5. Although sodium is insoluble in 1,2-dimethoxyethane, sodium-potassium alloy dissolves to give equimolar Na and K, presumably as K$^+$ and Na$^-$.

Alkali-metal cations form very stable complexes with complexing agents called "crown ethers" and "cryptands" that have structures such as the following:

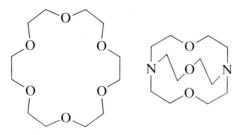

The addition of a crown ether or cryptand (abbreviated C) to ethylamine, tetrahy-drofuran, or polyether greatly increases the solubility of alkali metals by shifting the following equilibrium to the right:

$$2M(s) \;\rightleftharpoons\; M^+ + M^-$$

The net reaction for the dissolution then becomes

$$2M(s) + C \;\rightleftharpoons\; MC^+ + M^-$$

Actual crystals of sodides having the compositions $[NaC^+]Na^-$, $[KC^+]Na^-$, $[KC_2^+]Na^-$, $[RbC^+]Na^-$, $[RbC_2^+]Na^-$, $[CsC^+]Na^-$, and $[CsC_2^+]Na^-$ have been isolated from solutions either by slow evaporation of solvent or by cooling of the saturated solutions. In the case of $[NaC^+]Na^-$, the structure consists of essentially close-packed sodium-cryptand cations, with sodide ions in the octahedral holes. These holes have a radius of about 2.2 Å, which means that Na^- has about the same size as I^-. Obviously the -1 oxidation state of alkali metals is well established.

Crystallization from solutions containing 1:1 or 1:2 ratios of alkali metal to cryptand have yielded remarkable materials (*electrides*) in which the holes in a lattice of metal-cryptand cations are occupied by electrons. For example, the electride $[CsC_2^+]e^-$ consists of cesium cations, each of which is complexed by two cryptand molecules that pack in such a way that anionic sites of 4–5 Å are produced. The structure supports the conjecture that localized electrons are centered in the anionic sites and that they do not interact strongly with each other or with the cations.

PROBLEMS

9.1 Hydrated electrons can be generated by ultraviolet irradiation of the aqueous I^- ion:

$$I^- \;\xrightarrow{h\nu}\; I + e^-_{aq}$$

What is the overall net reaction for the ultraviolet irradiation of an aqueous HI solution?

*9.2 Which species would you expect to be more reactive toward e^-_{aq}: (*a*) ClO_4^- or ClO^-? (*b*) Cu^{2+} or Ni^{2+}? (*c*) $Fe(CN)_6^{4-}$ or $Fe(CN)_6^{3-}$?

9.3 If a sodium-ammonia solution were electrolyzed by using direct current and inert electrodes, what would be the electrode reactions?

*9.4 A solvated electron in H_2O or NH_3 is believed to occupy a cavity in the solvent, in which the protons of adjacent solvent molecules are oriented toward the electron in the cavity. Thus a solvated electron is situated in a potential well created by the polarization of solvent molecules. The absorption spectra are analogous to the $1s \rightarrow 2p$ transition of an H atom. On the basis of spectral information given in this chapter, what can you say about the relative sizes of the electron cavities in H_2O and NH_3?

9.5 Write equations for the net reactions which occur when the following materials are added to a sodium-ammonia solution: (*a*) methylgermane; (*b*) iodine; (*c*) diethyl sulfide.

*9.6 In the reaction of a weak acid with a metal-ammonia solution (see mechanism 9.5), it is conceivable that atomic hydrogen is formed, which then either dimerizes ($2H \rightarrow H_2$) or reacts by the following sequence of rapid reactions:

$$H + e^-_{am} \quad \rightarrow \quad H^-$$

$$H^- + NH_3 \quad \rightarrow \quad H_2 + NH_2^-$$

$$NH_2^- + HA \quad \rightarrow \quad NH_3 + A^-$$

In principle, could these two possibilities be distinguished kinetically?

9.7 Can you rationalize the shift in peak wavelength with a change in alkali metal in the absorption spectra of the M^- species (Fig. 9.9)?

CHAPTER
10

BORON HYDRIDES, THEIR DERIVATIVES, AND MAIN-GROUP CLUSTERS

For many years the formulas and structures of the boron hydrides defied rationalization by conventional concepts of bonding. These compounds are beautiful examples of nonclassical, multicenter bonding, and the fact that they can be readily explained by simple molecular orbital (MO) theory is another proof of the power of that theory. The formulas and structures of the higher, polyhedral boron hydrides have been shown to conform to a simple rule based on the number of cluster electrons. This rule also has a molecular orbital basis and is widely applicable to all kinds of "cluster" compounds, including various polyanions and transition-metal cluster compounds (to be discussed in Chap. 16).

SYNTHESES

Boron hydrides, or boranes, were first isolated and characterized by Alfred Stock and his coworkers in Germany. In the period from 1912 to 1930 they used vacuum-line techniques (most of which they devised themselves) to prepare B_2H_6, B_4H_{10}, B_5H_9, B_5H_{11}, B_6H_{10}, $B_{10}H_{14}$, and many derivatives of these compounds.[1] Further

---

[1] A. Stock, "Hydrides of Boron and Silicon," Cornell University Press, Ithaca, N.Y., 1933.

progress in the field was rather slow until, during World War II, Schlesinger's group at the University of Chicago developed easy methods for preparing $NaBH_4$ (sodium tetrahydroborate, usually called sodium borohydride), which is useful for the preparation of B_2H_6 (diborane).[2] Physical properties of some of the many boron hydrides which are now known are listed in Table 10.1.

The Synthesis of Boron Hydrides[3,4]

Diborane can be prepared by the reaction of a boron trihalide with a strong hydriding agent such as sodium tetrahydroborate or lithium aluminum hydride in an aprotic solvent. This synthesis is usually carried out with BF_3 and $NaBH_4$ in diglyme (the dimethyl ether of diethylene glycol):

$$3NaBH_4 + BF_3 \rightarrow 2B_2H_6 + 3NaF$$

A convenient laboratory synthesis involves the careful addition of $NaBH_4$ to concentrated H_2SO_4 or H_3PO_4:

$$2NaBH_4 + 2H_2SO_4 \rightarrow 2Na^+ + 2HSO_4^- + B_2H_6 + 2H_2$$

A synthetic method of potential industrial significance is the reduction of boric oxide with aluminum and hydrogen at high pressure in the presence of aluminum

---

[2] H.I. Schlesinger and H. C. Brown, *J. Am. Chem. Soc.,* **75**, 219 (1953). Also see H. C. Brown, *Science,* **210**, 485 (1980).

[3] E. L. Muetterties, ed., "The Chemistry of Boron and Its Compounds," Wiley, New York, 1967; E. L. Muetterties and W. H. Knoth, "Polyhedral Boranes," Dekker, New York, 1968.

[4] R. W. Parry and M. K. Walter, *Prep. Inorg. React.,* **5**, 45 (1968).

TABLE 10.1
The boron hydrides and some of their physical properties

	Melting point, °C	Boiling point, °C	ΔH_f°(at 25°C for gases), kcal mol^{-1}†
B_2H_6	−164.86	−92.84	7.5
B_4H_{10}	−120.8	16.1	13.8
B_5H_9	−46.75	58.4	15.0
B_5H_{11}	−123.2	65	22.2
B_6H_{10}	−63.2	108	19.6
B_6H_{12}	−83	80–90	
B_8H_{12}	~ −20		
B_8H_{18}			
B_9H_{15}	2.6	0.8 mm at 28°C	
$B_{10}H_{14}$	99.5	~ 213	2.8
$B_{10}H_{16}$			
$B_{18}H_{22}$	178.5		
iso-$B_{18}H_{22}$			
$B_{20}H_{16}$	199		

† 1 kcal = 4.1840 kJ

chloride catalyst:

$$B_2O_3 + 2Al + 3H_2 \xrightarrow{Al_2Cl_6} B_2H_6 + Al_2O_3$$

Several of the higher boron hydrides, B_4H_{10}, B_5H_9, B_5H_{11}, B_6H_{10}, and $B_{10}H_{14}$, can be prepared by the thermal decomposition of diborane. It is believed that the initial stages of the thermal decomposition involve the formation at low concentrations of high-energy intermediates such as BH_3 and B_3H_7.[5] For example, the following mechanism has been proposed for the formation of tetraborane:

$$B_2H_6 \rightleftharpoons 2BH_3$$

$$B_2H_6 + BH_3 \rightleftharpoons B_3H_7 + H_2$$

$$B_3H_7 + B_2H_6 \longrightarrow B_4H_{10} + BH_3$$

The same higher hydrides, as well as B_8H_{12}, B_9H_{15}, and $B_{10}H_{16}$, can be prepared by the appropriate treatment of lower hydrides with an electric discharge. For other preparative procedures, the review by Parry and Walter[4] should be consulted. The structures of several boron hydrides are illustrated in Fig. 10.1.

---

[5]   R. Greatrex, N. N. Greenwood, and S. M. Lucas, *J. Am. Chem. Soc.*, **111**, 8721 (1989).

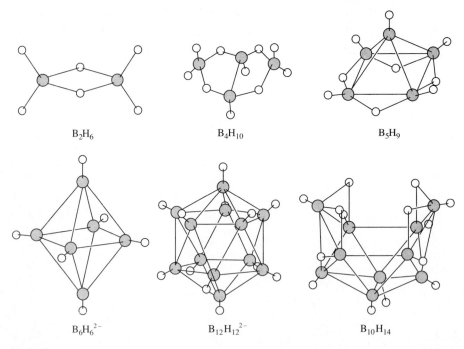

B_2H_6 B_4H_{10} B_5H_9

$B_6H_6{}^{2-}$ $B_{12}H_{12}{}^{2-}$ $B_{10}H_{14}$

FIGURE 10.1
Structures of some boron hydrides and borane anions. Boron atoms are represented by the shaded balls; hydrogen atoms by the smaller, unshaded balls.

The Synthesis of Borane Anions

In Chap. 7 we briefly discussed the synthesis of the tetrahydroborate ion, BH_4^-, and the octahydrotriborate ion, $B_3H_8^-$. A large number of higher borane anions are known; these have been prepared mainly by two general methods. The first method, the "BH condensation method," consists of the effective addition of BH groups to borane anions by treatment with diborane or other sources of BH groups. The general mechanism is believed to involve a sequence of steps in which BH_3 groups are added and H_2 molecules are lost. This sort of process is undoubtedly involved in the synthesis of $B_3H_8^-$ from BH_4^- and diborane:

$$BH_4^- + [BH_3] \rightleftharpoons B_2H_7^-$$

$$B_2H_7^- + [BH_3] \longrightarrow B_3H_8^- + H_2$$

Similar mechanisms have been postulated for the synthesis of the anions $B_nH_n^{2-}$, where n ranges from 6 to 12. The second method for preparing higher borane anions, specifically the $B_nH_n^{2-}$ ions, is the pyrolysis of salts of lower borane anions. The particular product obtained is highly dependent on the temperature, the cation, and the solvent (if any). Some useful conversions are indicated in the following:

$$(CH_3)_4NB_3H_8 \xrightarrow{\Delta} (CH_3)_3NBH_3 + [(CH_3)_4N]_2B_{10}H_{10} + [(CH_3)_4N]_2B_{12}H_{12}$$

$$CsB_3H_8 \xrightarrow{\Delta} Cs_2B_9H_9 + Cs_2B_{10}H_{10} + Cs_2B_{12}H_{12}$$

$$CsB_3H_8 \xrightarrow[\Delta]{\text{traces of ether}} Cs_2B_{12}H_{12}$$

$$(C_2H_5)_4NBH_4 \xrightarrow{\Delta} [(C_2H_5)_4N]_2B_{10}H_{10}$$

All the $B_nH_n^{2-}$ anions have boron frameworks corresponding to highly symmetric polyhedra. The octahedral and icosahedral structures of $B_6H_6^{2-}$ and $B_{12}H_{12}^{2-}$, respectively, are shown in Fig. 10.1. It should be noted that the boron atom frameworks of the neutral boranes in Fig. 10.1 are essentially polyhedral fragments. Indeed, the larger boranes and their derivatives are usually classified in terms of their structural relationships to polyhedra.[6] A "closo" borane (e.g., $B_6H_6^{2-}$ or $B_{12}H_{12}^{2-}$) has framework atoms at the vertices of a polyhedron which has all equilateral, or nearly equilateral, triangular faces (i.e., a deltahedron). A "nido" borane is formally derivable from a closo borane by the removal of one framework BH group and the addition of two protons and two hydrogen atoms to the resulting open face. Both B_5H_9 (formally derivable from $B_6H_6^{2-}$) and $B_{10}H_{14}$ (formally derivable from $B_{11}H_{11}^{2-}$) are nido boranes. An "arachno" borane is formally derivable from a closo compound by the removal of two framework atoms. Thus B_4H_{10} (formally derivable from $B_6H_6^{2-}$) is an arachno borane. The relatively rare "hypho" boranes are formally derivable from closo compounds by

---

[6] ' R. W. Rudolph, *Acc. Chem. Res.*, **9**, 446 (1976).

the removal of three framework atoms and have even more opened-out structures than the arachno boranes. The anion $B_5H_{12}^-$ (related to $B_8H_8^{2-}$) is an example of a hypho borane.[7]

The Synthesis of Carboranes

A boron-hydrogen compound in which carbon atoms occupy structural sites similar to those occupied by boron atoms is called a "carborane." The most important carborane synthesis is that of $1,2\text{-}B_{10}C_2H_{12}$, which is isoelectronic with the $B_{12}H_{12}^{2-}$ ion. The reactions involved in the synthesis follow:

$$B_{10}H_{14} + 2Et_2S \xrightarrow{n\text{-Pr}_2O} B_{10}H_{12}(Et_2S)_2 + H_2$$

$$B_{10}H_{12}(Et_2S)_2 + C_2H_2 \xrightarrow{n\text{-Pr}_2O} B_{10}C_2H_{12} + H_2 + 2Et_2S$$

The second step involves the insertion of the two carbon atoms of acetylene into the $B_{10}H_{12}$ framework, yielding the icosahedral $1, 2\text{-}B_{10}C_2H_{12}$ framework, with the carbon atoms occupying adjacent positions on the icosahedron. When this compound is heated, rearrangement occurs to the 1,7 and 1,12 isomers, in which the carbon atoms occupy positions separated by one and two boron atoms, respectively.

STRUCTURE AND BONDING[3,8]

Three-Center Bonding

The boron hydrides, borane anions, and carboranes have extraordinary structures. Some of these structures are shown in Fig. 10.1. It should be noted that the framework atoms of the larger molecules form the vertices of either a regular polyhedron or a polyhedral fragment. Compounds of this type have fascinated chemists not only because of their unusual structures and reactions but also because simple valence bond theory cannot account for their bonding. It is easy to understand why BH_3 is a reactive species; there are only six valence electrons in the compound, and it is impossible for the boron atom to achieve a complete valence octet. However, even the stable dimer, diborane, lacks enough electrons to permit conventional bonding as in ethane. The key to the bonding of diborane is found in the fact that two of the hydrogen atoms form bridges between the two boron atoms. These B—H—B bridges in diborane are examples of "three-center bonds," in which a bonding pair of electrons holds together three atoms. In our previous discussions of valence bond theory, we have considered only "two-center bonds," in which bonding electrons hold together two atoms. By including the three-center

[7]   R. J. Remmel, H. D. Johnson, I. S. Jaworiwsky, and S. G. Shore, *J. Am. Chem. Soc.*, **97**, 5395 (1975).

[8]   W. N. Lipscomb, "Boron Hydrides," Benjamin, New York, 1963.

bond concept in valence bond theory, we can write octet-satisfying structures for the boron hydrides and their derivatives. The B—H—B bridge is usually repre-sented by the structure B $\overset{\displaystyle H}{\diagup\diagdown}$ B . The only other type of three-center bond which is encountered in boron hydrides is the B—B—B bond, usually represented by the structure B $\overset{\displaystyle B}{\diagup\diagdown}$ B . In carboranes one must consider analogous B—B—C or B—C—C bonds.

Three-center bonding is easily accommodated by simple MO theory. In fact, in Chap. 4 we have already considered[9] the MOs in linear and triangular H_3^+, which are analogous to the MOs in the "open" B—H—B bond and the "closed" B—B—B bond. When two of the three atoms (the "end" atoms) do not interact (that is, $\beta_{13} = 0$), the energy levels are, in order of increasing energy, $\alpha + \sqrt{2}\beta, \alpha, \alpha - \sqrt{2}\beta$. When all three atoms interact equally with one another, the energy levels are $\alpha + 2\beta, \alpha - \beta, \alpha - \beta$. Because there must be a continuous change in the MO energy levels as the interaction between the end atoms (β_{13}) increases from zero to β_{12}, we can draw a correlation diagram, as shown in Fig. 10.2. It can be seen that, whether the bonding is closed and completely symmetric (as in the B—B—B bonding of $B_{12}H_{12}^{2-}$) or somewhere between open and closed (as in the B—H—B bonding of B_2H_6), the lowest MO, which accepts the two valence electrons, has definite bonding character.

To describe the bonding of a boron compound satisfactorily, the following criteria must be fulfilled:

1. The total number of valence electrons must equal twice the number of bonds, including both two- and three-center bonds.

---

[9] See Prob. 4.2 and its answer in the back of the book.

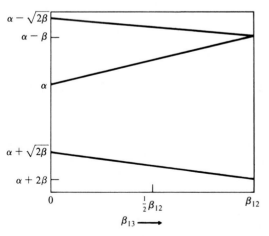

FIGURE 10.2
Energy levels for the three-center bond as a function of β_{13}/β_{12}, where $\beta_{12} = \beta_{23}$. The "open" bond prevails at $\beta_{13}/\beta_{12} = 0$, and the symmetric "closed" bond at $\beta_{13}/\beta_{12} = 1$.

2. Each boron atom must use four orbitals and must achieve a complete octet of valence electrons.

3. The bonding must be consistent with the observed structure of the molecule and with an approximately tetrahedral disposition of bonds to each boron atom.

It is useful to express the first two of these rules in the form of topological equations. For the case of a species of formula $B_bH_h{}^q$, where q is the net charge, let α represent the number of two-center bonds and β the number of three-center bonds. Rule 1 corresponds to

$$\alpha + \beta = \tfrac{1}{2}(3b + h - q)$$

Rule 2 corresponds to equating the total number of two- and three-center connections to the number of atomic orbitals, or

$$2\alpha + 3\beta = 4b + h$$

Solving for α and β, we obtain

$$\alpha = \tfrac{1}{2}(b + h - 3q) \tag{10.1}$$

$$\beta = b + q \tag{10.2}$$

If the numbers of two- and three-center bonds in the proposed bonding for a species $B_bH_h{}^q$ correspond to Eqs. 10.1 and 10.2, we can be sure that rules 1 and 2 are satisfied.[10] We shall presently describe, for several boranes, bonding assignments which are consistent with rules 1 to 3. A shorthand scheme for describing borane bonding, devised by Lipscomb, amount to listing four numbers (s, t, y, and x), defined as follows.[11]

$s \equiv$ the number of B—H—B bonds

$t \equiv$ the number of B—B—B bonds

$y \equiv$ the number of B—B bonds

$x \equiv$ the number of BH_2 groups plus twice the number of BH_3 groups

Hereafter we shall give the appropriate *styx* numbers when discussing the bonding in particular boranes.

First let us consider the tetrahydroborate ion, $BH_4{}^-$. The bonding in this simple species can be described by using classical valence bond theory because, according to Eq. 10.2, $\beta = 1 - 1 = 0$. Clearly rule 3 is satisfied because the species is tetrahedral and the boron uses four sp^3 hybrid orbitals.

---

[10] Lipscomb (ref. 8, chap. 2) has devised a similar, but more complicated set of topological equations.

[11] If we assume that each boron atom in a species $B_bH_h{}^q$ is bonded to at least one terminal hydrogen atom (by a B—H bond), then it can be easily shown that the *styx* parameters are related to α, β, and b as follows: $\alpha - b = x + y$; $\beta = s + t$.

Next we shall consider diborane, B_2H_6. From Eqs. 10.1 and 10.2 we obtain $\alpha = 4$ and $\beta = 2$, consistent with the bonding

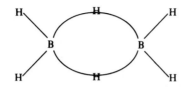

This bonding topology, corresponding to the *styx* numbers 2002, requires that the bridging hydrogens lie in a plane perpendicular to the plane of the other atoms, as shown in Fig. 10.1.

Now let us consider tetraborane, B_4H_{10}, for which we calculate $\alpha = 7$ and $\beta = 4$. The 4012 bonding shown below is consistent with these parameters and fits the observed structure of the molecule.

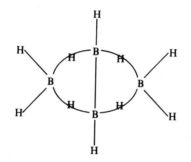

In the case of pentaborane-9, B_5H_9, $\alpha = 7$ and $\beta = 5$. The 4120 bonding assignment, which follows, is consistent with these parameters and with the structure shown in Fig. 10.1:

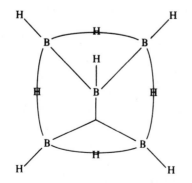

This molecule has the shape of a square pyramid. Inasmuch as the four basal boron atoms are actually equivalent, the bonding must be described as a resonance hybrid of the four bonding structures corresponding to permutation of the two B—B

bonds and the B—B—B bond. Pentaborane-9 is the simplest boron hydride which uses all the possible types of two- and three-center bonds.

Finally we shall discuss the bonding in the closo octahedral anion $B_6H_6^{2-}$ in terms of two- and three-center bonds. For this species, we calculate $\alpha = 9$ and $\beta = 4$. From Fig. 10.1 we see that each boron atom is bonded to a terminal hydrogen atom by a two-center B—H bond; hence the octahedral B_6 framework must be held together by three two-center bonds and four three-center bonds. One of many possible permutations of this 0430 bonding, consistent with approximately tetrahedral bonding to the boron atoms, is shown in the following structure:

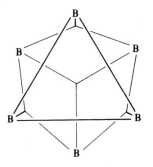

Obviously there are many resonance structures which one can write for this ion.

One of the intriguing aspects of the polyhedral $B_nH_n^{2-}$ ions is the fact that they all have a -2 charge. Neutral B_nH_n molecules would be consistent with valence bond theory: Each molecule would have nB—H bonds and nB—B—B bonds (that is, $\alpha = \beta = n$). The nonexistence of neutral molecules can be rationalized by the fact that the number of resonance structures that can be written for a $B_nH_n^{2-}$ ion is always greater than that for the corresponding B_nH_n neutral molecule.[12] Positive B_nH_n ions are unacceptable because they would not have enough valence electrons. Equations 10.1 and 10.2 yield the impossible result for $q > 0$ that the number of two-center bonds would be less than the number of B—H bonds and that the number of three-center bonds would be greater than the number of boron atoms. Finally if we rule out charges more negative than -2 (-4, -6, etc.) because of the destabilization caused by charge repulsion, we complete a rationalization of the charge on the $B_nH_n^{2-}$ ions.[12,13] To obtain a more rigorous explanation of the stability of the -2 charge of these species and to avoid the awkwardness of writing many resonance structures to explain the bonding in these species, we must go to simple MO theory.

---

[12]   I. R. Epstein and W. N. Lipscomb, *Inorg. Chem.*, **10**, 1921 (1971), app. II.

[13]   By simple valence bond considerations, it can be shown that q must equal -2 for the unknown trigonal bipyramidal $B_5H_5^q$. See Prob. 10.6.

MO Theory of Clusters[14]

Let us assume that we have n cluster atoms at the vertices of a regular deltahedron. The four valence orbitals on each vertex atom may be divided into one external sp hybrid orbital, two equivalent tangential p orbitals, and one internal radial sp hybrid orbital. The external orbital is used to form a σ bond to an external hydrogen atom or group. The tangential orbitals and internal orbital are used to bond the cluster atoms together. This orbital arrangement is shown in Fig. 10.3.

Pairwise interaction of the $2n$ tangential orbitals in the "surface" bonding of the polyhedron leads to n bonding and n antibonding orbitals. The n internal sp hybrid orbitals overlap at the center of the polyhedron to generate one strongly bonding orbital and $n - 1$ nonbonding or antibonding orbitals. Thus the cluster contains $n + 1$ bonding orbitals. Upon filling these bonding orbitals with electrons, we obtain a total of $2n + 2$ cluster electrons. Thus we have rationalized the so-called $2n + 2$ rule of polyhedral clusters.[15] Clearly any $B_nH_n{}^q$ species conforms to the rule if $q = -2$; the boron atoms contribute $2n$ electrons to the cluster, and the -2 charge accounts for the other two electrons.

It is not necessary that the n vertex atoms be boron atoms, or that they be identical. Thus the closo carboranes $C_2B_{n-2}H_n$ ($n = 5$ to 12), which are iso-electronic and isostructural with the $B_nH_n{}^{2-}$ ions, also follow the $2n + 2$ rule. Nido molecules, such as B_5H_9, B_6H_{10}, and $B_3C_3H_7$, have $2n + 4$ framework electrons; arachno molecules, such as B_4H_{10}, B_5H_{11}, and $B_7C_2H_{13}$, have $2n + 6$ framework electrons; and hypho molecules, such as $B_5H_{12}{}^-$ and $B_6H_{10}(PMe_3)_2$, have $2n + 8$ framework electrons. (In all cases n is the number of framework atoms.) It is significant that nido, arachno, and hypho molecules usually have as many framework electrons as the closo compounds to which they are structurally

---

[14] R. B. King and D. H. Rouvray, *J. Am. Chem. Soc.*, **99**, 7834 (1977); A. J. Stone, *Inorg. Chem.*, **20**, 563 (1981); A. J. Stone and M. J. Alderton, *Inorg. Chem.*, **21**, 2297 (1982).

[15] K. Wade, *Chem. Commun.*, 792 (1971); R. E. Williams, *Inorg. Chem.*, **10**, 210 (1971); R. W. Rudolph and W. R. Pretzer, *Inorg. Chem.*, **11**, 1974 (1972); R. W. Rudolph, *Acc. Chem. Res.*, **9**, 446 (1976).

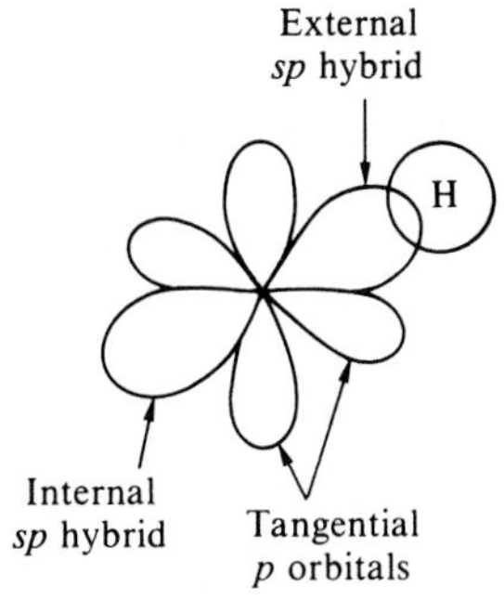

External
sp hybrid

H

Internal
sp hybrid

Tangential
p orbitals

FIGURE 10.3
The orbitals used by a polyhedral vertex atom in cluster bonding.

related. That is, the removal of vertices from an n-vertex closo polyhedron usually does not change the number of framework electrons; the resulting polyhedral fragment generally still has $2n + 2$ electrons (where n refers to the parent closo polyhedron).

Exceptions to the $2n + 2$ rule are found in boron halide clusters.[16] Although polyhedral ions of the type $B_nCl_n{}^{2-}$ and $B_nBr_n{}^{2-}$, which follow the $2n + 2$ rule, are well established, various neutral $2n$ clusters such as B_9Cl_9 and B_nBr_n ($n = 7$ to 10), and even a few examples of $2n + 1$ radical anions of the type $B_9X_9{}^-$ have been prepared and characterized. As far as has been determined, these species all have closo polyhedral structures. The stability of these highly electron-deficient species has been rationalized by assuming halogen π donation to the cluster MOs.[17]

METALLOCARBORANES AND METALLOBORANES[3,18]

Treatment of $1,2\text{-}B_{10}C_2H_{12}$ with ethoxide ion in ethanol at $70°C$ causes a degradation corresponding to the deletion of a B^+ ion from the molecule:

$$B_{10}C_2H_{12} + OC_2H_5{}^- + 2C_2H_5OH \quad \rightarrow \quad B_9C_2H_{12}{}^- + B(OC_2H_5)_3 + H_2$$

The $B_9C_2H_{12}{}^-$ ion can be deprotonated by a strong base to give the "open" $B_9C_2H_{11}{}^{2-}$ ion pictured in Fig. 10.4. The latter ion can react with various transition-metal ions or organo transition-metal ions to form complexes in which the metal atom completes the icosahedron. For example,

$$4B_9C_2H_{11}{}^{2-} + 3Co^{2+} \quad \rightarrow \quad Co + 2Co(B_9C_2H_{11})_2{}^-$$

The structure of the $Co(B_9C_2H_{11})_2{}^-$ ion is shown in Fig. 10.5. Each atom in each of the two icosahedra contributes three orbitals to the icosahedral framework bond-

---

[16] E. H. Wong and R. M. Kabbani, *Inorg. Chem.*, **19**, 451 (1980); N. A. Kutz and J. A. Morrison, *Inorg. Chem.*, **19**, 3295 (1980).

[17] For an MO discussion, see M. E. O'Neill and K. Wade, *Inorg. Chem.*, **21**, 461 (1982).

[18] M. F. Hawthorne and G. B. Dunks, *Science*, **178**, 462 (1972).

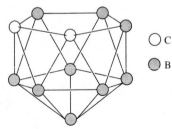

○ C

◉ B

FIGURE 10.4
Skeletal structure of the $B_9C_2H_{11}{}^{2-}$ ion. (*Reproduced with permission from E. L. Muetterties and W. H. Knoth, "Polyhedral Boranes," Dekker, New York, 1968.*)

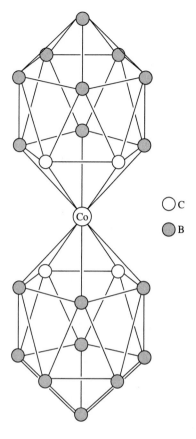

FIGURE 10.5

Skeletal structure of the $Co(B_9C_2H_{11})_2^-$ ion. *(Reproduced with permission from E. L. Muetterties and W. H. Knoth, "Polyhedral Boranes," Dekker, New York, 1968.)*

ing. Each icosaheadron receives 2 electrons from the Co^- group.[19] 18 electrons from nine BH groups, and 6 electrons from two CH groups. Thus 26 electrons are available per icosahedron, in agreement with the $2n + 2$ rule.

Reaction of B_5H_9 with $Fe(CO)_5$ at 220°C gives 10 to 20 percent yields of $B_4H_8Fe(CO)_3$, which has a B_5H_9 structure in which the apex BH group has been replaced by an $Fe(CO)_3$ group, as shown in Fig. 10.6a.[20] Reaction of a mixture of B_5H_9 and $Fe(CO)_5$ with $LiAlH_4$ gives ~ 1 percent yields of $B_3H_7Fe_2(CO)_6$ which, similarly, has a B_5H_9 structure in which the apex BH group and one of the basal BH groups have been replaced by $Fe(CO)_3$ groups, as shown in Fig.

---

[19] Transition-metal atoms in such cluster compounds acquire valence shells containing 18 electrons, corresponding to rare-gas configurations. The Co^- ion has 10 valence electrons; 8 more are obtained by sharing from other cluster atoms. the 6 framework orbitals provided by the cobalt atom require 12 electrons. Since 8 of these are provided by other atoms, 4 (or 2 per cluster) are provided by the Co^- itself.

[20] N. N. Greenwood, C. G. Savory, R. N. Grimes, L. G. Sneddon, A. Davison, and S. S. Wreford, *J. Chem. Soc. Chem. Commun.*, 718 (1974).

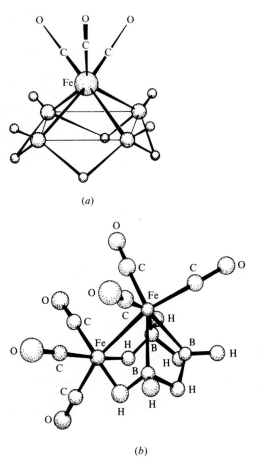

FIGURE 10.6
The structures of ferraboranes.
(a) $B_4H_8Fe(CO)_3$; (b) $B_3H_7Fe_2(CO)_6$. [*Reproduced with permission from N. N. Greenwood et al., J. Chem. Soc. Chem. Commun., 718 (1974), and from E. L. Andersen, K. J. Haller, and T. P. Fehlner, J. Am. Chem. Soc., **101**, 4390 (1979). Copyright 1979 American Chemical Society.*]

10.6*b*.[21] In these ferraboranes, each $Fe(CO)_3$ group contributes three orbitals and two electrons to the bonding (just as a BH group does); thus both compounds have the same number of framework electrons (14) as B_5H_9 and may be considered nido clusters. The $2n + 2$ rule and the relationships between framework structure and the number of framework electrons have been useful guides in the synthesis of these and other boron cluster compounds.[22] In Chap. 16 we shall discuss transition-metal cluster compounds in more detail.

OTHER MAIN-GROUP CLUSTER COMPOUNDS

The $2n + 2$ rule appears to be generally applicable to clusters containing main-group elements. In these clusters, atoms which have no substituents are assumed

21 E. L. Andersen, K. J. Haller, and T. P. Fehlner, *J. Am. Chem. Soc.*, **101**, 4390 (1979).

22 R. W. Rudolph, *Acc. Chem. Res.*, **9**, 446 (1976).

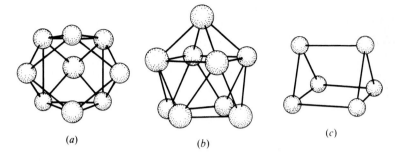

(a) (b) (c)

FIGURE 10.7
The structures of some nonmetal cluster ions. (a) The tricapped trigonal prism structure of Ge_9^{2-}.
(b) The monocapped archimedean antiprism structure of Ge_9^{4-}. Note that this structure has one
vertex less than the closo bicapped archimedean antiprism. (c) The trigonal prism structure of Te_6^{4+}.
[*Reproduced with permission from C. H. E. Belin, J. D. Corbett, and A. Cisar, J. Am. Chem. Soc.,
99, 7163 (1977), and from R. C. Burns et al., Inorg. Chem., **18**, 3086 (1979), Copyright 1977 and
1979 American Chemical Society.*]

to have one lone pair of electrons each, directed outward from the surfaces of the
clusters. For example, Ge_9^{2-} has a tricapped trigonal prism structure (Fig. 10.7a),
a closo structure corresponding to 20 framework electrons. On the other hand,
Ge_9^{4-} has a nido configuration based on a bicapped archimedean antiprism.[23]
(See Fig. 10.7b). The cation Te_6^{4+}, with 20 framework electrons, has a trigonal
prism structure[24] (Fig. 10.7c), which may be considered a hypho structure related
to the tricapped trigonal prism structure that would presumably be adopted by
hypothetical Te_9^{16+}. The rather complicated structure of the As_{11}^{3-} anion (with
36 framework electrons) has been rationalized by looking upon it as a 17-vertex
polyhedron from which 6 vertices have been deleted.[25]

"Magic Numbers"

Atomic clusters in the gas phase can be generated by focusing laser light on a solid
element.[26] In some cases this procedure yields cluster anions and in other cases
neutral clusters that can be photoionized to cationic species. The cluster ions
thus prepared can be examined by mass spectrometry, and clues to the cluster
geometries can be gleaned from the distribution of masses. The predominate
masses in the mass spectra are associated with particularly stable structures.
  For example, consider the mass spectrum of carbon clusters shown in Fig.
10.8.[27] The most remarkable feature of the spectrum is the prominence of one cluster,

23   C. H. E. Belin, J. D. Corbett, and A. Cisar, *J. Am. Chem. Soc.*, **99**, 7163 (1977).
24   R. C. Burns, R. J. Gillespie, W. C. Luk, and D. R. Slim, *Inorg. Chem.*, **18**, 3086 (1979).
25   C. H. E. Belin, *J. Chem. Soc.*, **102**, 6036 (1980).
26   M. A. Duncan and D. H. Rouvray, *Sci. Am.*, December 1989, pp. 110–115.
27   R. F. Curl and R. E. Smalley, *Science*, **242**, 1017 (1988); H. Kroto, *Chem. Brit.*, **26** (1), 40
(1990).

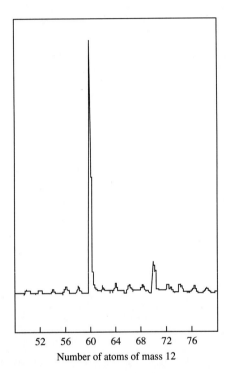

52   56   60   64   68   72   76
Number of atoms of mass 12

FIGURE 10.8
Mass spectrum of carbon clusters formed in the laser vaporization of graphite into a helium atmosphere. [*Reproduced with permission from H. Kroto, Chem. Brit., 26(1), 40 (1990).*]

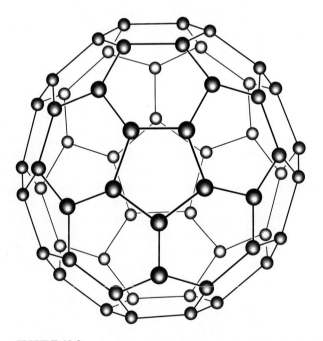

FIGURE 10.9
Proposed structure for C_{60}, sometimes referred to as Buckminsterfullerene.

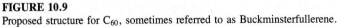

C_{60}. In a search for an explanation for this feature, it was noted that a very symmetric structure can be formed by a 60-atom carbon cluster. By placing atoms at the vertices of a truncated icosahedron as shown in Fig. 10.9, one obtains a pattern of bonds like the pattern of seams on a soccerball. A key to the stability of this unique structure is that the truncated icosahedron arranges all of the atoms uniformly on the surface of an imaginary sphere. Note that all of the carbon atoms in the proposed structure for C_{60} are identical; each atom is a bridge joining two hexagonal faces and one pentagonal face. In effect, C_{60} is a spherical analog of graphite. These conjectures have been confirmed[28] by the actual isolation and characterization of crystalline C_{60}.

For another example, consider the negative ion spectrum obtained by laser ablation of aluminum metal, shown in Fig. 10.10.[29] The most abundant aluminum

---

[28]   W. Krätschmer, L. D. Lamb, K. Fostiropoulos, and D. R. Huffman, *Nature*, **347**, 354 (1990).
[29]   R. L. Hettich, *J. Am. Chem. Soc.*, **111**, 8582 (1989).

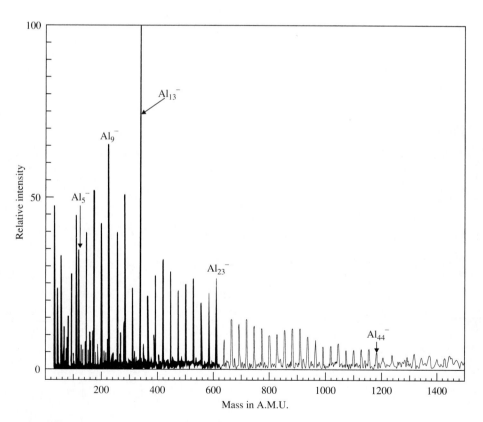

FIGURE 10.10
Negative ion spectrum obtained by laser ablation of aluminum metal. [*Reproduced with permission from R. L. Hettich, J. Am. Chem. Soc., 111, 8582 (1989).*]

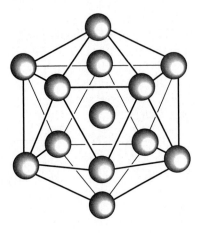

FIGURE 10.11
Proposed icosahedral structure for Al_{13}.

cluster anion is $Al_{13}{}^-$, a fact that can be rationalized by considering the cluster to be an icosahedral fragment of metallic aluminum. The atoms in metallic aluminum are packed exactly as are the atoms in the 13-atom icosahedral cluster pictured in Fig. 10.11. Note that in this cluster, one atom is a central atom, surrounded by 12 atoms at the corners of a regular icosahedron. Many other gaseous metal clusters have been studied, and the "magic numbers" associated with spikes in the mass spectra have generally been similarly rationalized in terms of exceptionally stable geometric structures.

PROBLEMS

10.1 Draw three possible structures for the $B_3H_8{}^-$ ion, showing any three-center bonds clearly. Which structure do you think is more plausible? Give the *styx* numbers for all three structures.

10.2 Draw a plausible structure for $B_3C_2H_5$. How many bonds of each type (B—H, B—B, C—H, B—B—B, B—H—B, B—B—C, B—C—C) are there?

10.3 In a polyhedral anion of formula $B_nH_n{}^{2-}$, how many B—B bonds and how many B—B—B bonds are there?

*10.4 At least two different mechanisms have been proposed for the isomerization of $1,2\text{-}B_{10}C_2H_{12}$. One involves the rotation of triangular faces of the icosahedron (e.g., a B_2C face). A second involves the conversion of two adjacent triangular faces into a square face, followed by conversion into two triangular faces:

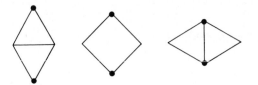

A cube octahedron intermediate, or activated complex, has been proposed for the latter mechanism:

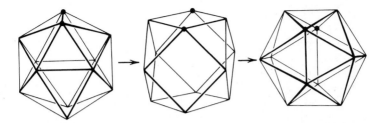

Show that the first mechanism can explain the formation of the 1,7 and 1,12 isomers, but that the second mechanism cannot account for the formation of the 1,12 isomer.

10.5 Sketch and label an MO energy level diagram for the bonding of the bridging hydrogens in B_2H_6. Assume that the boron atoms lie on the z axis and that the bridging hydrogen atoms lie on the x axis, and make the approximation that the bridging hydrogen atoms interact only with the $2p_x$ orbitals of the boron atoms.

*10.6 Using the valence bond method and the topological rules outlined on page 266, show that the only acceptable value of q for $B_5H_5{}^q$ is -2, corresponding to $2n + 2$ framework electrons.

10.7 Rationalize the fact that $C_2B_9H_{11}{}^{2-}$, formed by the reduction of polyhedral $C_2B_9H_{11}$, has an open, polyhedral fragment structure.

10.8 Propose topologically acceptable *styx* numbers for the species $B_5H_{12}{}^-$, B_6H_{10}, and B_6H_{12}.

10.9 Predict the structure of $As_6{}^{2-}$.

10.10 Explain the fact that the proton nmr spectrum of $(CCH_3)_6{}^{2+}$ shows two peaks with a 5:1 intensity ratio. Propose a plausible structure for the $(CH)_5{}^+$ ion.

10.11 Reaction of $[(CH_3)_2C_2B_4H_4]_2FeH_2$ with B_5H_9 gives, among other things, $(CH_3)_4C_4B_{11}H_{11}$. The structure shown below has been proposed for the C_4B_{11} cluster of this molecule. Rationalize this structure in terms of electron counting, and show how you count the cluster electrons.

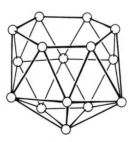

10.12 Rationalize the trigonal prismatic structure of $(RGe)_6$, where R is the bis-(trimethylsilyl)methyl group, (*a*) in terms of the $2n + 2$ rule, and (*b*) in terms of localized sp^3 bonding.

CHAPTER
11

THE
SOLID
STATE

Up to this point, practically all of the chemical species we have discussed have been molecules or ions—either in the gas or liquid phase, or in solution. We have said very little about solid compounds. In this chapter, in which we discuss the solid state, you will find that you are familiar with most of the principles used. However, you will find it necessary to employ a talent that you probably have not used much before: you will be required to visualize relatively complex three-dimensional structures. Some people can do this much more easily than others. Fortunately this ability "to think in three dimensions" can be improved by practice with structural models, and it is strongly recommended that you obtain access to a good model-building kit or at least a large set of balls. Then you will be able to assemble actual models of many of the structures illustrated in this chapter.

CLASSIFICATION OF BONDS AND CRYSTALS[1]

Bonding between atoms may be classified into the following types:

1. *Covalent.* This is the type of bonding which we have principally discussed thus far in this book. Covalent bonding is generally described in terms of electron pairs which are shared by the bonded atoms.

---

[1] A. F. Wells, "Structural Inorganic Chemistry," 4th ed., Oxford University Press, London, 1975; C. Kittel, "Introduction to Solid State Physics," 5th ed., Wiley, New York, 1976.

2. *Ionic.* This is the bonding caused by the electrostatic (coulombic) attraction between oppositely charged atoms or ions. In the limit of pure ionic bonding between atoms, each ion would be spherically symmetric and there would be no shared electrons.

3. *Van der Waals.* This bonding, sometimes referred to as London dispersion forces, corresponds to the relatively weak interaction which occurs between atoms and molecules caused by instantaneous dipole-induced dipole interactions. Even in species which have no permanent dipole moment, instantaneous dipoles arise because of momentary asymmetry in electron distribution. These instantaneous dipoles induce dipoles in adjacent species. In other words, the electrons in separate species tend to synchronize their movements to minimize electron-electron repulsion and to maximize electron-nucleus attraction. Van der Waals forces increase rapidly with molecular volume and with the number of polarizable electrons. They are the principal forces holding together the atoms in solid argon and are the principal *inter*molecular forces in solid CCl_4 and I_2.

4. *Metallic.* This type of bonding holds together regular lattices of electron-deficient atoms, or groups of atoms, such as in metals and alloys. The characteristic feature of metallic bonding is that the bonding electrons are delocalized over the entire crystal.

It might appear that the best way of classifying solids would be in terms of the types of bonds between the atoms, i.e., covalent, ionic, van der Waals, and metallic. However, bonds which are purely of one type or another are rare, and in most crystals there are bonds of more than one type. Figure 11.1 illustrates how covalent, ionic, and metallic bonding mix together even in simple compounds and shows that the bonding in most compounds cannot be cleanly classified into any one of these groups. It is more useful to classify solids in terms of their structural

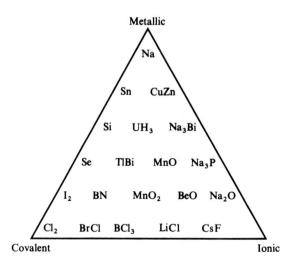

FIGURE 11.1
Classification of compounds according to the degree of covalent, ionic, and metallic character in the bonding.

features, and we shall now show how it is possible to roughly classify crystals in the following four categories: (1) crystals containing finite, discrete complexes, (2) crystals containing infinite one-dimensional complexes, (3) crystals containing infinite two-dimensional complexes, and (4) crystals consisting of infinite three-dimensional complexes.

Crystals Containing Finite Complexes

This category includes most solid compounds of the nonmetals in which the molecules are held together by van der Waals or dipole-dipole forces. In the case of atoms (for example, Ne, Ar, Kr, and Xe) and molecules with approximately spherical shapes (for example, HCl, H_2S, and SiF_4), the so-called close-packed structures and the body-centered cubic structure are generally found. The close-packed structures may be understood by reference to Fig. 11.2. When identical spheres are packed together as closely as possible on a plane surface, they are arranged as shown, with each sphere touching six others. An exactly similar layer of spheres can be laid on top of this layer, so that the spheres of the second layer rest in depressions of the first layer. When a third layer is placed on the second layer, there are two choices: The spheres may be placed in the depressions directly over the first-layer spheres or in the depressions which lie over depressions in the first layer. The two simplest sequences of layers may be indicated *ABAB* . . . (called "hexagonal close packing") or *ABCABC* . . . (called "cubic close packing" or "face-centered cubic packing"). In each case each sphere touches 12 others, and the fraction of space occupied by spheres is 0.7405. Sketches of spheres packed in these two ways are shown in Fig. 11.3.

The body-centered cubic structure is illustrated in Fig. 11.4. In the case of molecular crystals, this type of packing is less common, probably because the packing is slightly less efficient than in the close-packed structures. (The fraction of space occupied by spheres is 0.68.) Unfortunately there does not appear to be any simple rationale for predicting the type of packing in molecular crystals. Thus, although the *molecular* structures of Se_4N_4 and S_4N_4 are practically identical, the packing of Se_4N_4 is approximately body-centered cubic, whereas that of S_4N_4 is almost cubic close-packing.[2] Very subtle changes in van der Waals bonding and

---

[2] H. Bärnighausen, T. von Volkmann, and J. Jander, *Acta Crystallogr.*, **21**, 571 (1966).

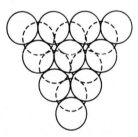

FIGURE 11.2
Two adjacent layers of close-packed spheres. Notice that there are two ways of placing a third layer. If the spheres in the third layer are directly over those of the first layer, the packing is hexagonal close packing. Otherwise the packing is cubic close packing.

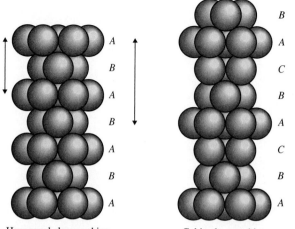

Hexagonal close packing   Cubic close packing

FIGURE 11.3
Arrangement of spheres in hexagonal close packing (left) and cubic close packing (right). *(Adapted with permission from L. Pauling, "The Nature of the Chemical Bond," 3d ed., Cornell University Press, Ithaca, N.Y., 1960.)*

electron-electron repulsions may be the determining factors in changes in crystal structure.

The packing of nonspherical or dipolar molecules generally deviates from the undistorted close-packed and body-centered cubic structures. The packing diagrams of two rather complicated molecules [an iron(III) complex of *N*-(2-phenylethyl)salicylaldimine and chloride,[3] and pentacarbonyl(tetraphosphorus trisulfide)molybdenum[4]] are shown in Figs. 11.5 and 11.6, respectively. Inasmuch as these molecules are not spherically symmetric, it is necessary to specify their angular orientations as well as their positions when describing the packing.

---

[3]   J. A. Bertrand, J. L. Breece, and P. G. Eller, *Inorg. Chem.,* **13**, 125 (1974).

[4]   A. W. Cordes, R. D. Joyner, R. D. Shores, and E. D. Dill, *Inorg. Chem.,* **13**, 132 (1974).

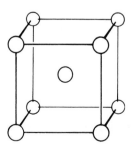

FIGURE 11.4
Body-centered cubic structure. This type of packing is found for the atoms in various metals and for the molecules in the crystals of many molecular compounds [e.g., SiF_4, $N_4(CH_2)_6$, and $MoAl_{12}$].

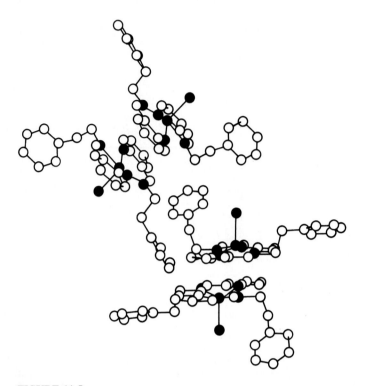

FIGURE 11.5
Packing diagram of Fe(SANE)$_2$Cl. [SANE = N-(2-phenylethyl)salicylaldimine.] The asymmetric units are linked into loose "dimers." Carbon atoms are represented by open circles; other atoms by filled circles. [*Reproduced with permission from J. A. Bertrand et al., Inorg. Chem., **13**, 125 (1974). Copyright 1974 American Chemical Society.*]

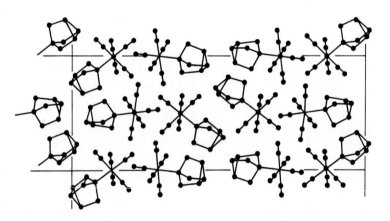

FIGURE 11.6
Projection of the crystal packing of Mo(CO)$_5$P$_4$S$_3$. [*Reproduced with permission from A. W. Cordes et al., Inorg. Chem., **13**, 132 (1974). Copyright 1974 American Chemical Society.*]

Crystals Containing Infinite One-Dimensional Complexes

The simplest examples of one-dimensional complexes in the solid state consist of infinite linear molecules held together in lattices by van der Waals bonds. The SiS_2 structure consists of SiS_4 tetrahedra which share edges to form long chains, as shown in Fig. 11.7. In palladium(II) chloride, $PdCl_2$, infinite chains are built of planar $PdCl_4$ groups sharing opposite edges, as shown in Fig. 11.8. In elemental tellurium, the atoms are joined in the form of infinite helical chains. Similar helical chains exist in the "metallic" form of selenium and the fibrous form of sulfur. The fibrous sulfur structure is illustrated in Fig. 11.9.

Many crystals contain anionic chains held together by cations. For example, the complex $KCu(CN)_2$ has a spiral polymeric anion in which each Cu(I) atom is bound to two carbon atoms and one nitrogen atom:

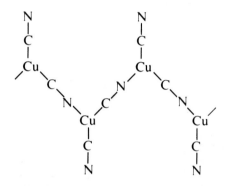

Silicate chain ions are found in the two large classes of minerals, the "pyroxenes," and the "amphiboles." The former [including enstatite, $MgSiO_3$; diopside,

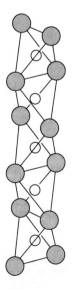

FIGURE 11.7
Structure of SiS_2. Small circles represent silicon atoms; large circles sulfur atoms.

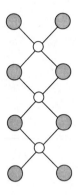

FIGURE 11.8
Structure of $PdCl_2$.

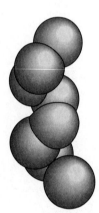

FIGURE 11.9
Structure of fibrous sulfur. *(Adapted with permission from L. Pauling, "The Nature of the Chemical Bond," 3d ed., Cornell University Press, Ithaca, N.Y., 1960.)*

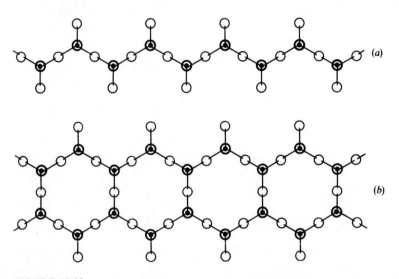

FIGURE 11.10
Silicon-oxygen anionic chains in pyroxenes (*a*) and amphiboles (*b*). *(Reproduced with permission from A. F. Wells, "Structural Inorganic Chemistry," 3d ed., Clarendon Press, Oxford, 1962.)*

$CaMg(SiO_3)_2$; jadeite, $NaAl(SiO_3)_2$; and spodumene, $LiAl(SiO_3)_2$] contain simple chains of tetrahedral SiO_4 groups connected by shared oxygen atoms, as shown in Fig. 11.10a. The latter [including tremolite, $Ca_2Mg_5(OH)_2(Si_4O_{11})_2$, and arfvedsonite, $Na_3Mg_4Al(OH)_2(Si_4O_{11})_2$] contain double chains in which one-half of the SiO_4 groups are joined to two adjacent groups by shared oxygen atoms, and one-half are joined to three adjacent groups by shared oxygen atoms, as shown in Fig. 11.10b.

Crystals Containing Infinite Two-Dimensional Complexes

Structures in which layers are held together by van der Waals or weak covalent bonds are found in elemental arsenic and graphite (Fig. 11.11) and in $CdCl_2$, MoS_2, $CrCl_3$, and HgI_2 (Fig. 11.12).

The graphite structure differs form the other layer structures in that the atoms of each layer are coplanar. The large distance between layers (3.35 Å) compared with the C—C distance of 1.42 Å in the layers indicates very weak bonding between layers, thus allowing the layers to slide easily over one another, giving graphite its valuable lubricating properties. In the common hexagonal graphite structure, the atoms of alternate layers lie above one another, and the repeat

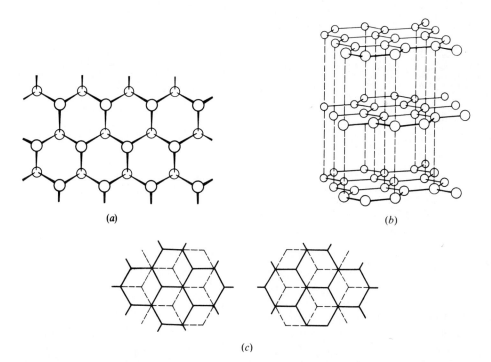

(a)

(b)

(c)

FIGURE 11.11
Layer structures of arsenic (a) and hexagonal graphite (b). (c) Relative alignments of adjacent layers in graphite. See text.

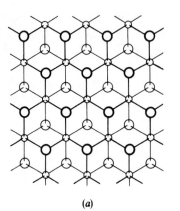

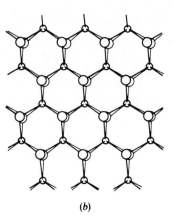

(a)   (b)

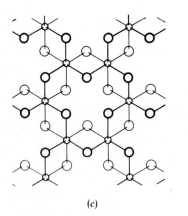

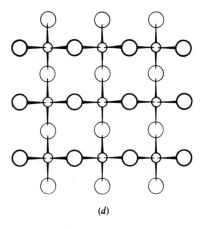

(c)   (d)

FIGURE 11.12
Structures of some simple MX_2 and MX_3 layers. Small circles represent metal atoms in the plane of the paper; large circles, the X atoms which lie in planes above and below that of the metal atoms. The layers are $CdCl_2$ (a), MoS_2 (b), $CrCl_3$ (c), and HgI_2 (d). (*Reproduced with permission from A. F. Wells, "Structural Inorganic Chemistry," 3d ed., Clarendon Press, Oxford, 1962.*)

distance perpendicular to the layers is 6.70 Å. In Fig. 11.11c, it can be seen that there are two different ways in which adjacent layers can be symmetrically aligned, excluding the more repulsive eclipsed alignment. If we give the layer drawn with solid lines the symbol A, and the other two dashed-line layers the symbols B and C, then hexagonal graphite can be represented by stacking the layers $ABABAB \ldots$, $ACACAC \ldots$, or $BCBCBC \ldots$, all of which are equivalent structures. A second form of graphite, rhombohedral graphite, has the stacking order $ABCABC \ldots$, in which every third layer is superposed. It is interesting that in hexagonal graphite there are two crystallographically different types of carbon atoms, whereas in rhombohedral graphite, all the atoms are equivalent.

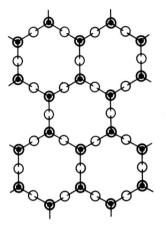

FIGURE 11.13
Structure of a silicon-aluminum-oxygen sheet. Oxygen atoms lying above the Si(Al) atoms (small black circles) are drawn more heavily. *(Reproduced with permission from A. F. Wells, "Structural Inorganic Chemistry," 3d ed., Clarendon Press, Oxford, 1962.)*

Structures with layers which are held together by hydrogen bonds are found in various hydroxy compounds, such as aluminum hydroxide, $Al(OH)_3$. The arrangement of Al and O atoms in this compound is similar to that of the Cr and Cl atoms in $CrCl_3$, shown in Fig. 11.12c; hydrogen bonds link together the oxygen atoms of adjacent layers. In boric acid (Fig. 8.6) hydrogen bonds hold together the atoms *within* each layer.

Examples of anionic layers held together by cations are found in certain aluminosilicates such as the hexagonal form of $CaAl_2Si_2O_8$. This compound contains silicon-aluminum-oxygen sheets of the type shown in Fig. 11.13. Such sheets are joined together in pairs by oxygen atoms to form double layers of empirical composition $(AlSiO_4)_n^{n-}$. These double layers are held together by calcium ions as shown in Fig. 11.14. The micas have related structures.

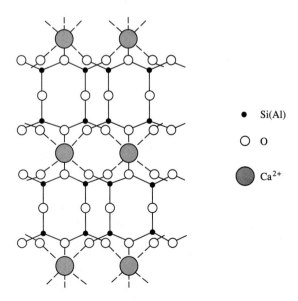

● Si(Al)

○ O

⬤ Ca^{2+}

FIGURE 11.14
Side view of the structure of hexagonal $CaAl_2Si_2O_8$, showing the double layers, interleaved with Ca^{2+} ions. *(Reproduced with permission from A. F. Wells, "Structural Inorganic Chemistry," 3d ed., Clarendon Press, Oxford, 1962.)*

Crystals Containing Infinite Three-Dimensional Complexes

Metals, which we shall discuss separately in Chap. 12, have three-dimensional frameworks which generally can be described in terms of one of the following types of atomic structures: hexagonal close packing, cubic close packing, and body-centered cubic.

A wide variety of three-dimensional framework structures have been observed for compounds. In the case of simple compounds having the general formula M_mX_x, where X is a relatively large atom, most observed structures consist of close-packed arrangements of the X atoms, with the M atoms occupying the holes. There are two types of holes in a close-packed structure: tetrahedral and octahedral. The hole formed by the following arrangement of spheres is a tetrahedral hole:

The hole formed by the following arrangement of spheres is an octahedral hole:

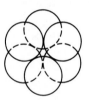

By examination of Fig. 11.2 it can be seen that the number of tetrahedral holes is equal to twice the number of close-packed spheres and that the number of octahedral holes is equal to the number of close-packed spheres.

In Table 11.1 are listed some common structures based on close packing. Let us consider the NaCl structure. The chlorine atoms are arranged in a cubic close-packed array, and all the octahedral holes are occupied by sodium atoms. The octahedral holes have a cubic close-packed arrangement, and so the NaCl lattice may be considered as two interpenetrating cubic close-packed lattices. The conventional representation of the NaCl lattice is shown in Fig. 11.15a; however, reference to Fig. 11.15b and c will help make obvious the close-packed arrangements of the two sets of equivalent atoms. Many MX compounds have the NaCl crystal structure. These include most of the alkali-metal halides and many metal oxides, sulfides, carbides, nitrides, and phosphides.

Five other structures based on close-packed X atoms are illustrated in Fig. 11.16. In fluorite (CaF_2), the calcium atoms have a cubic close-packed arrangement, with all the tetrahedral holes occupied by fluorine atoms. Each calcium atom has eight nearest-neighbor fluorine atoms at the corners of a cube, and each fluorine atom has four nearest-neighbor calcium atoms at the corners of a tetra-

TABLE 11.1
Structures based on close packing of X atoms

| Formula | Fraction of holes occupied by M atoms | | Type of close packing for X atoms | | Coordination number | |
	Tetrahedral	Octahedral	hcp	ccp	M	X
M_2X	1	0		F_2Ca (fluorite)	4	8
M_3X_2	$\frac{3}{4}$	0		Zn_3P_2 O_3Mn_2	4	6
MX	0	1	NiAs	NaCl	6	6
	$\frac{1}{2}$	0	ZnS (wurtzite)	ZnS (zinc blende)	4	4
M_2X_3	0	$\frac{2}{3}$	α-Al_2O_3 (corundum)		6	4
	$\frac{1}{3}$	0	β-Ga_2S_3	γ-Ga_2S_3	4	
MX_2	0	$\frac{1}{2}$	CdI_2 TiO_2 (rutile)	$CdCl_2$ TiO_2 (anatase)	6	3
	$\frac{1}{4}$	0	β-$ZnCl_2$	HgI_2 γ-$ZnCl_2$ SiS_2 OCu_2 α-$ZnCl_2$	4	2
MX_3	0	$\frac{1}{3}$	BiI_3	$CrCl_3$	6	2
	$\frac{1}{6}$	0	Al_2Br_6		4	2
MX_4	$\frac{1}{8}$	0	SnI_4		4	1
MX_6	0	$\frac{1}{6}$		α-WCl_6 UCl_6	6	1

hedron. A number of difluorides, dioxides, and disilicides have the fluorite structure. In the so-called anti-fluorite structure, the positions occupied by the cations and anions of the fluorite structure are interchanged. The anti-fluorite structure is adopted by most of the alkali-metal oxides and sulfides.

In nickel arsenide, the arsenic atoms have the hexagonal close-packed arrangement, with all the octahedral holes occupied by nickel atoms. Thus each nickel atom is octahedrally coordinated by arsenic atoms, and each arsenic atom

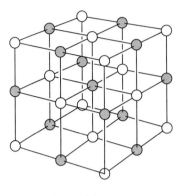

(a)

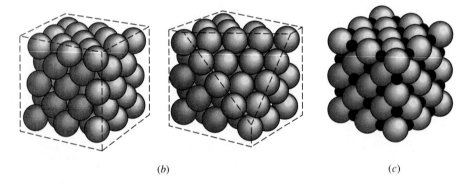

(b)   (c)

FIGURE 11.15
Conventional NaCl structure (a), cubic close packing of spheres (b), and another representation of the NaCl structure (c). [*From W. Barlow, Z. Kristallogr., 29, 433 (1898).*]

has six nickel neighbors at the corners of a trigonal prism. The NiAs structure is adopted by many MX compounds in which M is a transition metal and X is Sn, As, Sb, S, Se, or Te.

The rutile structure is named after one of the crystal forms of TiO_2. This structure can be obtained by systematic removal of one-half of the nickel atoms from nickel arsenide, followed by a slight distortion of the arsenic atoms from hexagonal close packing. The rutile structure consists of chains of MX_6 octahedra, in which each octahedron shares a pair of opposite edges and which are further linked by sharing corners to form a three-dimensional structure in which each M atom has six X neighbors and each X atom has three coplanar M neighbors. Many dioxides and difluorides have the rutile structure, and many compounds of the type ABX_4, A_2BX_6, etc., have rutile-like structures in which the metal atoms are either randomly or regularly distributed in the M sites.

Zinc sulfide is found in two crystal modifications, one called zinc blende (based on cubic close-packed sulfur atoms) and the other wurtzite (based on hexagonal close-packed sulfur atoms). In each structure, the zinc atoms occupy half of the tetrahedral holes, and both the zinc and sulfur atoms have four tetrahedral near-

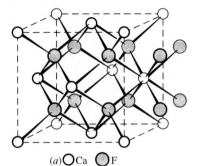

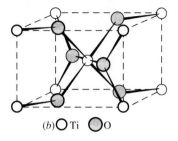

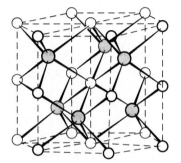

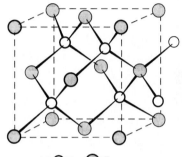

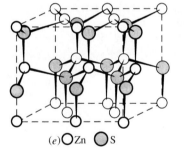

FIGURE 11.16
Structures of fluorite (CaF$_2$) (*a*), rutile (TiO$_2$) (*b*), nickel arsenide (NiAs) (*c*), zinc blende (ZnS) (*d*), and wurtzite (ZnS) (*e*). *(Reproduced with permission from A. F. Wells, "Structural Inorganic Chemistry," 3d ed., Clarendon Press, Oxford, 1962.)*

est neighbors. Although many MX compounds (such as ZnO, SiC, CuCl, AlP, and InAs) adopt wurtzite or zinc-blende structures, even more compounds having two or more different metal atoms are known which adopt structures related to the ZnS structures. Thus BeSiN$_2$ has a wurtzite-type structure, and Cu$_2$FeSnS$_4$ has a zinc-blende-like structure. Diamond has a zinc-blende-type structure, in which both the Zn and S sites are occupied by carbon atoms. Silicon, germanium, and gray tin also crystallize in the diamond structure. A rare hexagonal form of diamond

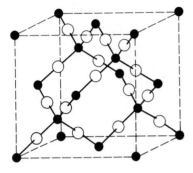

FIGURE 11.17
The β-cristobalite structure of SiO_2 idealized. In fact, the Si—O—Si bond angles are not 180°, but close to 144°. Compare with the zinc-blende structure. *(Reproduced with permission from A. F. Wells, "Structural Inorganic Chemistry," 3d ed., Clarendon Press, Oxford, 1962.)*

(wurtzite-type structure) was discovered in meteorites and has been synthesized in the laboratory.[5] Ordinary ice has a hexagonal structure similar to that of wurtzite, in which both the Zn and S sites are occupied by oxygen atoms. As shown in Fig. 8.3, the hydrogen atoms lie approximately on the lines connecting the oxygen atoms, 1.01 Å from one oxygen atom and 1.75 Å from the other. The short O—H bonds are randomly distributed, but with the restriction that at any given time only two H atoms are close to any one O atom. The high-temperature form of silica (cristobalite) has a structure like that of zinc blende, with silicon atoms at the Zn and S sites and oxygen atoms near the midpoints of the lines connecting the silicon atoms (Fig. 11.17).

It should be noted that Table 11.1 includes several structures which do not fit strictly into the category of infinite three-dimensional networks. For example, Al_2Br_6 consists essentially of a lattice of close-packed bromine atoms, with one-sixth of the tetrahedral holes occupied by aluminum atoms in such a way that adjacent pairs of tetrahedral holes are occupied. Thus, pairs of bromine atoms on tetrahedral edges are shared, and the structure can be looked upon as a packing of discrete molecules of the following type:

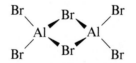

An important structure for compounds of the type MX, which does not correspond to a close packing for either the M or X atoms, is the cesium chloride structure, shown in Fig. 11.18. Each atom has eight nearest neighbors, as in the body-centered cubic structure.

Another important structure, for compounds of the type ABX_3, is the *perovskite* structure shown in Fig. 11.19. The "ideal" structure is cubic, with large A cations surrounded by 12 anions, and smaller B cations surrounded by 6 anions. The structures of many compounds (including the mineral perovskite itself,

---

[5]   R. E. Hanneman, H. M. Strong, and F. P. Bundy, *Science,* **155**, 955 (1967).

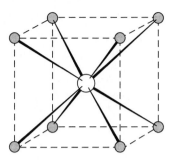

FIGURE 11.18
The CsCl structure.

CaTiO$_3$) are slightly distorted variants of the ideal structure. Some of these compounds are ferroelectric and can exist in two polarization states, analogous to those of KH$_2$PO$_4$, discussed on p. 219. In idealized lead titanate (PbTiO$_3$), for example, Ti^{4+} ions occupy the centers of the cubes, Pb^{2+} ions occupy the corners, and O^{2-} ions occupy the centers of the cube faces. In the actual crystal there are regions which possess a net dipole due to displacement up or down of the Ti^{4+} ions with respect to the other ions. If a large electric field is applied, all the regions (*domains*) can be lined up, and the polarization can be switched from up to down by reversing the applied field. Such materials can be used as random-access memories in integrated circuits and may replace magnetic memory systems in many applications.[6]

There are many ionic compounds of the type M$_m$X$_x$ in which M, X, or both M and X are replaced by complex ions. When the complex ions are approx-

---

[6] J. F. Scott, and C. A. Paz de Araujo, *Science,* **246,** 1400 (1989); J. M. Herbert, *Chem. Brit.,* 728 (September 1983).

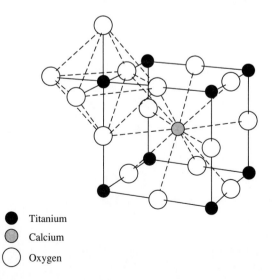

- ● Titanium
- ◉ Calcium
- ○ Oxygen

FIGURE 11.19
The idealized perovskite structure.

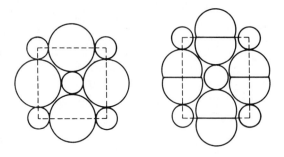

FIGURE 11.20
Sections through the structures of NaCl and CaC$_2$, showing the relationship of the packing. *(Reproduced with permission from A. F. Wells, "Structural Inorganic Chemistry," 3d ed., Clarendon Press, Oxford, 1962.)*

imately spherical, the spatial arrangement of the ions is often the same as in one of the symmetric structures which we have discussed. For example, $[Co(NH_3)_6]I_2$ and $[Mg(NH_3)_6]I_2$ adopt the fluorite (CaF$_2$) structure, and K$_2$PtCl$_6$ and K$_2$SnCl$_6$ adopt the anti-fluorite structure (i.e., a fluorite-like structure in which anions occupy the Ca^{2+} positions and cations occupy the F$^-$ positions). The structure of $[Ni(H_2O)_6][SnCl_6]$ is a slightly deformed version of that of CsCl. Polyatomic ions can attain effective spherical symmetry by rapid free rotation or random orientation. Such behavior is often found in salts containing the ions NH$_4^+$, NO$_3^-$, CN$^-$, and SH$^-$.

Salts containing polyatomic ions sometimes adopt structures similar to the corresponding simple M$_m$X$_x$ compounds but of lower symmetry because of the nonspherical shapes of the ions. For example, the structure of calcium carbide, CaC$_2$, is essentially that of NaCl, but the structure is distended in the direction along which the C$_2^{2-}$ ions are aligned, as shown in Fig. 11.20. Calcite, CaCO$_3$, also has a structure like that of NaCl but elongated along a threefold axis, as shown in Fig. 11.21. When one of the complex ions is very large and irregularly shaped, and the counter-ion is small, the structure is usually determined by the packing of the large ions, the counter-ions fitting in the interstices.

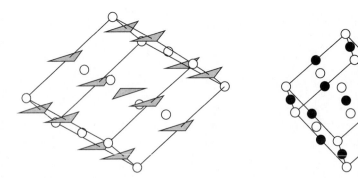

FIGURE 11.21
The CaCO$_3$ and NaCl structures similarly oriented. *(Reproduced with permission from A. F. Wells, "Structural Inorganic Chemistry," 3d ed., Clarendon Press, Oxford, 1962.)*

THE EFFECT OF RADIUS RATIO AND CHARGE ON STRUCTURE

There is no sharp dividing line between ionic and covalent bonding. The ionic character of a bond increases with increasing electronegativity difference between the atoms. Thus the bonding in solid NaCl is principally ionic, and that in diamond is completely covalent. However, in compounds such as ZnO and AgBr, the bonding is intermediate in character. An abrupt change in a physical property in a series of compounds has often been mistakenly ascribed to an abrupt change in the ionic character of the bonding. For example, consider the melting points of the fluorides of the second-row elements:

NaF	MgF_2	AlF_3	SiF_4	PF_5	SF_6
988°C	1266°C	1291°C (sublimes)	−90°C	−94°C	−50°C

The enormous change in melting points between AlF_3 and SiF_4 is not due to a marked change in ionic character (the electronegativities of Al and Si are 1.5 and 1.8, respectively) but rather to a change from an infinite lattice structure in which each aluminum is coordinated to six fluorine atoms to a lattice of discrete SiF_4 molecules held together by van der Waals bonds.

In most saline halides (i.e., halides with infinite three-dimensional ionic lattices), the number of halide ions coordinated to each metal ion is greater than the number of halide ions that could be bound to the metal ion in any conceivable discrete molecule. Generally a halide will exist in the form of molecules only when, in a given molecule, the number of halogen atoms bonded to the electropositive atom equals the maximum coordination number of the electropositive atom for that halogen. A halide MX_n will have a three-dimensional crystal lattice if it is thereby possible for the electropositive atom to achieve a coordination number greater than n. Obviously the radius ratio of the atoms is the critical factor. By simple geometric considerations, we may calculate the minimum possible values of the ratio r_M/r_X for various types of coordination of X atoms around M atoms. These results are given in Table 11.2. In principle, these data, when combined with atomic or ionic radii, should be useful for predicting coordination numbers.

TABLE 11.2
Allowed values of r_M/r_X for various types of coordination

Coordination number	Type of coordination	Minimum value of r_M/r_X
2	Linear	0
3	Triangular planar	0.155
4	Tetrahedral	0.225
4	Square Planar ⎫	
6	Octahedral ⎬	0.414
8	Square antiprismatic	0.645
8	CsCl structure	0.732
12	Cubooctahedral	1.000

However, such predictions are quite approximate because atomic and ionic radii are not well-defined, constant quantities. However, by using, for example, a consistent set of ionic radii such as those deduced by Pauling from crystal structure data (Table 11.3), it is possible to observe a correlation of the calculated radius ratios with the observed coordination numbers in crystals. The r_M/r_X values for metals in various oxide lattices and in various halides are listed with the corresponding observed coordination numbers for the metals in these compounds in Table 11.4. It can be seen that r_M/r_X values at the border lines between various coordination numbers are in approximate agreement with the theoretical r_M/r_X values of Table 11.2.

The Pauling ionic radii of Table 11.3 are useful for qualitative comparisons or rough quantitative estimations of interionic distances, but serious errors can be made if the values are used indiscriminately. One factor which influences the effective radius of an ion in a crystal is the coordination number. Other things being equal, an increase in coordination number causes the valence electrons to be shared among more bonds; hence the covalent character of the bonds is reduced and the effective ionic radius increases. For example, the interionic distance in TlCl is 3.15 Å when the crystal has the NaCl structure (coordination number 6)

TABLE 11.3
Pauling ionic radii (in angstroms)†

Ag^+	1.26	Fe^{2+}	0.76	Pb^{4+}	0.84
Al^{3+}	0.50	Fe^{3+}	0.64	P^{3-}	2.12
As^{3-}	2.22	Ga^+	1.13	P^{5+}	0.34
As^{5+}	0.47	Ga^{3+}	0.62	Pd^{2+}	0.86
Au^+	1.37	Ge^{2+}	0.93	Ra^{2+}	1.40
B^{3+}	0.20	Ge^{4+}	0.53	Rb^+	1.48
Ba^{2+}	1.35	H^-	1.40	S^{2-}	1.84
Be^{2+}	0.31	Hf^{4+}	0.81	Sb^{3-}	2.45
Br^-	1.95	Hg^{2+}	1.10	Sc^{3+}	0.81
C^{4-}	2.60	I^-	2.16	Se^{2-}	1.98
C^{4+}	0.15	In^+	1.32	Sr^{2+}	1.13
Ca^{2+}	0.99	In^{3+}	0.81	Sn^{2+}	1.12
Cd^{2+}	0.97	K^+	1.33	Sn^{4+}	0.71
Ce^{3+}	1.11	La^{3+}	1.15	Te^{2-}	2.21
Ce^{4+}	1.01	Li^+	0.60	Ti^{2+}	0.90
Cl^-	1.81	Lu^{3+}	0.93	Ti^{3+}	0.76
Co^{2+}	0.74	Mg^{2+}	0.65	Ti^{4+}	0.68
Co^{3+}	0.63	Mn^{2+}	0.80	Tl^+	1.40
Cr^{2+}	0.84	Mn^{3+}	0.66	Tl^{3+}	0.95
Cr^{3+}	0.69	Mo^{6+}	0.62	U^{3+}	1.11
Cr^{6+}	0.52	N^{3-}	1.71	U^{4+}	0.97
Cs^+	1.69	N^{5+}	0.11	V^{2+}	0.88
Cu^+	0.96	Na^+	0.95	V^{3+}	0.74
Cu^{2+}	0.70	NH_4^+	1.48	V^{4+}	0.60
Eu^{2+}	1.12	Ni^{2+}	0.72	Y^{3+}	0.93
Eu^{3+}	1.03	O^{2-}	1.40	Zn^{2+}	0.74
F^-	1.36	Pb^{2+}	1.20		

† L. Pauling, "The Nature of the Chemical Bond," 3d ed., Cornell University Press, Ithaca, N.Y., 1960.

TABLE 11.4
Radius ratios and coordination numbers in oxides and halides

Cation	r_M/r_O	Oxide lattices Commonly observed coordination nos.	Salt	r_M/r_X	Halides Coordination no.
B^{3+}	0.14	3, 4	MgI_2	0.30	6
Be^{2+}	0.22	4	$MgBr_2$	0.33	6
Si^{4+}	0.29	4	$MgCl_2$	0.36	6
Al^{3+}	0.36	4, 6	CaI_2	0.46	6
Ge^{4+}	0.38	4, 6	MgF_2	0.48	6
Li^+	0.43	4	$CaBr_2$	0.51	6
Mg^{2+}	0.46	6	SrI_2	0.52	>6
Ti^{4+}	0.48	6	$CaCl_2$	0.55	6
Zr^{4+}	0.57	6	$SrBr_2$	0.58	>6
Sc^{3+}	0.58	6	BaI_2	0.62	>6
Na^+	0.68	6	$SrCl_2$	0.62	8
Ca^{2+}	0.71	8	$BaBr_2$	0.69	>6
Ce^{4+}	0.72	8	CaF_2	0.73	8
K^+	0.95	9	$BaCl_2$	0.75	>6
Cs^+	1.21	12	SrF_2	0.83	8
			BaF_2	0.99	8

and 3.32 Å when it has the CsCl structure (coordination number 8). Of course, even though the interionic distance increases on going to the structure with a higher coordination number, the packing efficiency increases. Thus the density of TlCl is 6.37 g cm^{-3} in the NaCl structure and 7.07 g cm^{-3} in the CsCl structure. In the case of some transition-metal ions, another factor influencing ionic radius is the spin state: the greater the spin, the greater the ionic radius. The reason for this relationship between spin and effective ionic radius will be discussed in Chap. 18. Shannon[7] has prepared a table of effective ionic radii in which, for a given metal ion, separate values are listed for various coordination numbers and, when appropriate, for high- and low-spin states. Ionic radii from his compilation are given in Appendix F.

The radius ratio of two atoms is related to the electronegativity difference between the atoms. In general, the greater the electronegativity difference between M and X, the greater the value of r_M/r_X. Thus it is not surprising that Mooser and Pearson[8] observed that, if the average principal quantum number of the valence shells of M and X is held constant, a transition from tetrahedral coordination to octahedral coordination occurs when the electronegativity difference exceeds a particular value. In Fig. 11.22, the average principal quantum number is plotted against the electronegativity difference for a large number of compounds of type

---

[7]   R. D. Shannon, *Acta Crystallogr.*, **A32**, 751 (1976).

[8]   E. Mooser and W. B. Pearson, *Acta Crystallogr.*, **12**, 1015 (1959).

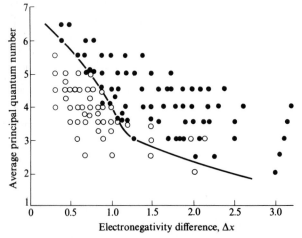

FIGURE 11.22
Average principal quantum number of valence shells versus electronegativity difference for MX compounds. Open circles correspond to tetrahedral structures; solid structures correspond to octahedral structures. [*From E. Mooser and W. B. Pearson, Acta Crystallogr.,* **12**, *1015 (1959).*]

MX. It can be seen that a fairly sharp boundary separates the tetrahedral structures (open circles) from the octahedral structures (solid circles).

Striking evidence that the effective radius of an ion depends on the nature of its counter-ion is given in Table 11.5, which lists the interionic distances for AgX and NaX compounds having the NaCl crystal structure. If we make the completely arbitrary assumption that the X^- radius is constant on going from AgX to NaX, then we conclude that Ag^+ is larger than Na^+ when $X^- = F^-$, but that Na^+ is larger than Ag^+ when $X^- = Cl^-, Br^-$, or PF_6^-. These results can be rationalized by recognizing that Ag^+ is a much "softer" ion than Na^+ and that it forms much more covalent (relatively short) bonds with polarizable ions such as Cl^-, Br^-, and PF_6^-.

Pauling[9] proposed several rules for predicting the types of anion-cation co-ordination in stable ionic lattices containing more than two kinds of ions. His first

---

[9]   L. Pauling, "The Nature of the Chemical Bond," 3d ed., Cornell University Press, Ithaca, N.Y., 1960.

TABLE 11.5
A comparison of Ag^+ and Na^+
in NaCl-type structures

Compound	M—X distance, Å
AgF	2.47
NaF	2.31
AgCl	2.77
NaCl	2.82
AgBr	2.89
NaBr	2.99
$AgPF_6$	3.76
$NaPF_6$	3.80

rule says, in effect, that the charge on a given anion should be canceled by its share of the charges on the cations to which it is coordinated. Thus an oxide ion will be stable if it is coordinated to any of the following combinations of cations:

Two tetrahedral Si^{4+} ions, as in many silicates $(2 = \frac{4}{4} + \frac{4}{4})$

One tetrahedral Si^{4+} ion and two octahedral Al^{3+} ions, as in topaz, $Al_2SiO_4F_2$ $(2 = \frac{4}{4} + \frac{3}{6} + \frac{3}{6})$

One tetrahedral Si^{4+} ion and three octahedral Mg^{2+} ions, as in olivine, Mg_2SiO_4 $(2 = \frac{4}{4} + \frac{2}{6} + \frac{2}{6} + \frac{2}{6})$

One tetrahedral Si^{4+} ion and two tetrahedral Be^{2+} ions, as in phenacite, Be_2SiO_4 $(2 = \frac{4}{4} + \frac{2}{4} + \frac{2}{4})$.

Pauling's second rule is concerned with the number of anions simultaneously coordinated to two different cations, i.e., with the sharing of polyhedron vertices, edges, and faces. The rule states that shared edges, and especially shared faces, decrease the stability of a structure. The effect is greatest for cations of high charge and low coordination number and is caused by the increased electrostatic repulsion between cations. In agreement with this rule, SiO_4 tetrahedra tend to share only vertices with one another and with other polyhedra. However, in compounds in which the bonding is relatively covalent and in which the atomic charges are lower, exceptions to this second rule occur. Thus we have seen, in Fig. 10.7, how SiS_4 tetrahedra share edges in SiS_2 to form infinite chains,

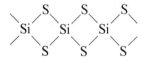

Pauling's third rule, which also has an electrostatic basis, states that in a crystal containing different cations, those with high charge and low coordination number tend not to occupy polyhedra which share vertices or edges with each other. The highly charged cations tend to be as far apart from one another as possible.

LATTICE ENERGY

The energy required to break up a crystal into infinitely separated ions is called the lattice energy U:

$$M_mX_x(s) \quad \longrightarrow \quad mM^{z_M}(g) + xX^{z_X}(g)$$

If the bonding in the crystal were completely ionic, the lattice energy could be calculated from the ionic charges and the geometric arrangement of the ions in the lattice. It is convenient to express this theoretical electrostatic lattice energy U_e as a function of the stoichiometric numbers m and x, the ionic charges z_M and z_X, the closest interionic distance r, and a geometric factor called the reduced

Madelung constant, M':

$$U_e = -M' \frac{z_M z_X (m + x) e^2}{2r}$$

In the case of a salt for which $m = x = 1$, the quantity $M' - 1$ is essentially a measure of the stabilization in the crystal lattice relative to that in a simple ion pair. ($M' = 1$ for an ion pair.) Values of M' for various crystal structures are given in Table 11.6. Values for many other structures are given in the literature.[10] It can be seen that M' has essentially the same magnitude (namely ~ 1.7) for all the crystal structures listed, showing that the lattice stabilization is of comparable magnitude for these structures.[11] Most texts quote the Madelung constants $\mathcal{M}$ or M, which are related to M' by the equations given in Table 11.6. When these Madelung constants are used, the theoretical electrostatic lattice energy must be calculated by the following equations:

$$U_e = -\mathcal{M} \frac{z_M z_X e^2}{r}$$

$$U_e = M \frac{e^2}{r}$$

Inasmuch as Madelung constant of type M contains the $z_M z_X$ factor, it can be used only for crystals having a particular charge type. The Madelung constants M' and $\mathcal{M}$ are more generally useful because they can be applied to crystals of any charge type. Thus $M' = \mathcal{M} = 1.7476$ for any compound of type $M^{n+} X^{n-}$ which has the sodium chloride structure (for example, MnO), whereas $M = 1.7476$ when $n = 1$ and $M = 6.9902$ when $n = 2$.

---

[10]   Q. C. Johnson and D. H. Templeton, *J. Chem. Phys.*, **34**, 2004 (1961).
[11]   D. H. Templeton, *J. Chem. Phys.*, **23**, 1826 (1955).

TABLE 11.6
Values of Madelung constants for several crystal structures

Crystal structure	Reduced Madelung constant, M'	Conventional Madelung constants	
		$\mathcal{M} = \left(\frac{m + x}{2}\right) M'$	$M = \frac{z_M z_X (m + x)}{2} M'$
NaCl	1.7476	1.7476	1.7476
CsCl	1.7627	1.7627	1.7627
ZnS (zinc blende)	1.6381	1.6381	6.5522
ZnS (wurtzite)	1.6413	1.6413	6.5653
CaF$_2$ (fluorite)	1.6796	2.5194	5.0388
TiO$_2$ (rutile)	1.6053	2.4080	19.264
Al$_2$O$_3$ (corundum)	1.6688	4.172	25.031

There is no experimental method known for directly measuring lattice energies. However, it is possible to obtain "experimental" values of lattice energies from appropriate thermodynamic data. Thus the lattice energy of a salt such as NaCl can be calculated as the sum of the energies of the following reactions, whose sum corresponds to the dissociation of the crystal into gaseous ions.

$$NaCl(s) \rightarrow Na(s) + \tfrac{1}{2}Cl_2(g) \qquad -\Delta H_f^\circ(NaCl) = 98.3 \text{ kcal mol}^{-1}$$

$$Na(s) \rightarrow Na(g) \qquad\qquad\qquad S(Na) = 25.6 \text{ kcal mol}^{-1}$$

$$\tfrac{1}{2}Cl_2(g) \rightarrow Cl(g) \qquad\qquad\qquad \tfrac{1}{2}D(Cl_2) = 29.1 \text{ kcal mol}^{-1}$$

$$Na(g) \rightarrow Na^+(g) + e^-(g) \qquad\qquad IE(Na) = 118.5 \text{ kcal mol}^{-1}$$

$$Cl(g) + e^-(g) \rightarrow Cl^-(g) \qquad\qquad -EA(Cl) = -83.4 \text{ kcal mol}^{-1}$$

$$U(NaCl) = -\Delta H_f^\circ(NaCl) + S(Na) + \tfrac{1}{2}D(Cl_2) + I(Na) - EA(Cl)$$
$$= 188.1 \text{ kcal mol}^{-1}$$

Each of the energy terms involved in the summation has been evaluated by a separate physical chemical method; hence we may consider the sum 188.1 kcal mol^{-1} as the experimental lattice energy of sodium chloride. (Similar calculations can be carried out for other salts using the heats of formation of solids given in Appendix E, the heats of formation of gaseous atoms in Table 3.12, the ionization energies given in Table 1.4, and the electron affinities in Appendix C.) It is interesting to compare the experimental value of U for NaCl with the theoretical electrostatic value. From the observed interionic distance in crystalline NaCl (2.814 Å) we calculate[12]

$$U_e = \frac{1.7476 \times (4.803 \times 10^{-10})^2 \times 1.4394 \times 10^{13}}{2.814 \times 10^{-8}} = 206.2 \text{ kcal mol}^{-1}$$

The discrepancy of 18.1 kcal mol^{-1} may be attributed to repulsions between the electron clouds of adjacent ions in the lattice. These repulsions may be approximately accounted for by the use of the relation

$$U = U_e\left(1 - \frac{\rho}{r}\right) \tag{11.1}$$

where ρ is a constant, approximately 0.31 Å. In the case of the alkali-metal halides, lattice energies calculated by this equation are accurate to about ± 2 percent.

In Table 11.7, experimental values of lattice energies and lattice energies calculated from Eq. 11.1 are listed for several alkaline-earth halides. It can be seen that the agreement is fairly good for the fluorides and poor for the iodides (especially for CaI_2, for which the experimental value is 13.2 percent greater than the calculated[13] value). The reason for the poor agreement is undoubtedly

[12]   The factor for converting ergs per molecule to kilocalories per mole is 1.4394×10^{13}.

[13]   T. E. Brackett and E. B. Brackett, *J. Phys. Chem.*, **69**, 3611 (1965).

TABLE 11.7
Comparison of experimental and calculated lattice energies†

Salt	U_{exp}, kcal mol^{-1}	U_{calcd}, kcal mol^{-1}	Δ
CaF$_2$	628.0	615.9	12.1
CaI$_2$	494.2	436.5	57.7
BaF$_2$	561.0	551.8	9.2
BaI$_2$	446.6	423.5	23.1

† T. E. Brackett and E. B. Brackett, *J. Phys. Chem.*, **69**, 3611 (1965).

the neglect, in the calculation, of the polarizabilities of the ions. A large ion, such as the iodide ion, is much more easily distorted in an electric field (i.e., more easily polarized) than a small ion, such as the fluoride ion. (Values of some ionic polarizabilities are given in Table 11.8.) The soft iodide ion, when coordinated to the small, doubly charged calcium ion, is strongly polarized, and the interaction of the induced moment with the charge of the calcium ion increases the bonding energy beyond that expected for spherically symmetric ions. The explanation of the enhanced stability of CaI$_2$ in terms of polarization is essentially equivalent to ascribing a large covalent contribution to the bonding. Thus we see that the estimation of lattice energies by Eq. 11.1 for relatively covalent compounds such as ZnS and InSb would be almost hopeless.

The reader may verify, from data in Appendix E, that ΔH_f° for the alkali-metal fluorides *decreases* from CsF to LiF, whereas ΔH_f° for the alkali-metal iodides *increases* from CsI to LiI. To understand the reason for these different

TABLE 11.8
Ionic polarizabilities†

Ion	α, Å^3	Ion	α, Å^3
Li$^+$	0.0	Cl$^-$	3.00
Na$^+$	0.21	Br$^-$	4.13
K$^+$	0.97	I$^-$	6.18
Rb$^+$	1.47	OH$^-$	2.0
Cs$^+$	2.36	NO$_3^-$	3.78
Mg^{2+}	0.18	IO$_3^-$	6.18
Ca^{2+}	0.7	ClO$_4^-$	4.16
Sr^{2+}	0.84	SO$_4^{2-}$	3.81
Ba^{2+}	1.63	O^{2-}	2.8
OH$_3^+$	1.2	S^{2-}	9.0
NH$_4^+$	1.7	HS$^-$	5.3
F$^-$	0.84		

† Polarizabilities calculated from molar refractions (in cubic centimeters) by the relation $\alpha = 0.396R \times 10^{-24}$, using mainly data of Böttcher from R. J. W. Le Fèvre, *Adv. Phys. Org. Chem.*, **3**, 1 (1965).

trends, it is helpful to consider the relation

$$\Delta H_f^\circ = S(M) + \tfrac{1}{2}D(X_2) + I(M) - EA(X) - U(MX)$$

The only terms on the right side of this equation that change on going from CsX to LiX are $S(M)$, $I(M)$, and $U(MX)$, each of which increases. Inasmuch as the trend in ΔH_f° is determined by the trend in $S(M) + I(M) - U(MX)$, whether or not ΔH_f° increases or decreases is determined by the relative rates of change of $S(M) + I(M)$ and $U(MX)$. Obviously the trend in ΔH_f° for the fluorides is established by the trend in $U(MF)$, which exceeds the trend in $S(M) + I(M)$, and the trend in ΔH_f° for the iodides is established by the trend in $S(M) + I(M)$, which exceeds the trend in $U(MI)$. The reason for the relatively small trend in $U(MI)$ is the fact that the interionic distance r is principally determined by the larger iodide ion. Hence the fractional change in r on going from CsI to LiI is relatively small, and the lattice energy does not increase markedly. On going from CsF to LiF, the fractional change in r is relatively great; consequently the lattice energy increases markedly.

APPLICATION OF THE ISOELECTRONIC PRINCIPLE

Compounds of elements which have similar electronegativities and polarizabilities often have the same type of structure if the average number of valence electrons per atom is the same. A large class of isostructural compounds, known as "Grimm-Sommerfeld" compounds, have diamond-like structures (wurtzite or zinc blende). These compounds have an average of four valence electrons per atom. Examples of such compounds containing group III–group V combinations are BN, AlP, GaAs, and InSb. Group II–group VI combinations are ZnSe and CdTe. Group I–group VII combinations are CuBr and AgI. Even some ternary compounds, such as $CuInTe_2$ and $ZnGeAs_2$, have diamond-like structures which can be rationalized on this basis.

In a binary compound of elements whose electronegativities are quite different, the more the electropositive element generally transfers its valence electrons to the more electronegative element. If there are not enough electrons transferred to give the anions noble-gas configurations, the anions may covalently bond with each other other to achieve a structure analogous to that of the corresponding isoelectronic element, conforming to the octet rule. This is the essence of the so-called Zintl concept of bonding.[14] For example, in MgB_2 the B^- ions form planar sheets having the graphite structure; in LiAs the arsenic atoms form anionic spiral chains as in elemental selenium, and in $CaSi_2$ the silicon atoms form anionic puckered layers as in elemental arsenic. (However, in CaC_2 the carbon atoms form diatomic C_2^{2-}, isoelectronic with N_2.) The compound NaTl has a structure with the thallium atoms bonded together in a three-dimensional

14   H. Schäfer, B. Eisenmann, and W. Müller, *Angew. Chem. Internat. Ed.*, **12**, 694 (1973).

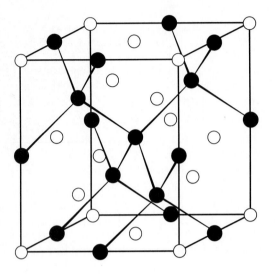

FIGURE 11.23
Unit cell of the NaTl phase. The Tl atoms (black balls) form a diamond-like lattice.

diamond-like lattice and the sodium atoms in the interstices, as shown in Figure 11.23. The Zintl concept can even be applied to compounds with more than two elements; in $Ba_7Ga_4Sb_9$, the gallium and antimony atoms form eight-membered rings, linked together in a layer lattice by weak Sb-Sb bonds.[15]

GLASSES

A glass is a material, formed by cooling from the normal liquid state, which has undergone no discontinuous change such as crystallization but which has become more rigid through an increase in viscosity. It is commonly assumed that a liquid has become a glass when its shear viscosity exceeds about 10^{13} poise. Glasses, like ordinary liquids, have crystallographic order only over a range of one or two interatomic spacings. A crude representation of the difference between a crystal and a corresponding glass is shown in Fig. 11.24. The x-ray diffraction pattern of a glass or liquid shows only one or two diffuse rings—completely different from the sharp rings of powdered crystals. Although the word "glass" is usually associated with silicates, ceramics, and related nonmetallic materials, it has been discovered that very fast cooling ($\sim 10^6$ deg s^{-1}) of certain molten alloys can yield metallic materials that are rigid and have liquidlike molecular structures. These metallic glasses have exceptional strength, corrosion resistance, and ease of magnetization.[16]

[15]   P. Alemany, S. Alvarez, and R. Hoffmann, *Inorg. Chem.*, **29**, 3070 (1990).

[16]   J. J. Gilman, *Science,* **208,** 856 (1980).

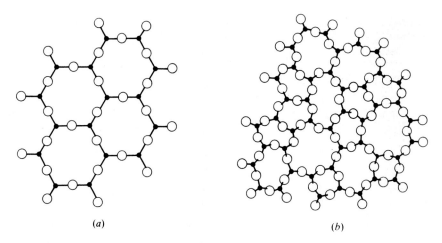

FIGURE 11.24
Schematic two-dimensional analogs of the regular structure of a crystal (*a*) and the semirandom network of a glass (*b*).

POLYMERS

Nearly all polymer chemistry to date has been based on the chemistry of carbon. The study of new polymers, with elements other than carbon as the backbone, is an area with many opportunities for novel synthetic methods.[17,18]

The most highly developed field of polymer chemistry based on an inorganic backbone is that of the silicones.[19] Silicones are the polymeric materials formed by the hydrolysis of alkyl-substituted silicon halides. The hydrolysis of dimethyldichlorosilane first generates cyclosiloxanes, which are then induced to undergo ring opening to give a linear high polymer.

$$(CH_3)_2SiCl_2 \xrightarrow{H_2O} \begin{matrix} (CH_3)_2Si-O-Si(CH_3)_2 \\ | \qquad\qquad | \\ O \qquad\qquad O \\ | \qquad\qquad | \\ (CH_3)_2Si-O-Si(CH_3)_2 \end{matrix} \xrightarrow[KOH]{\Delta} \left(\begin{matrix} CH_3 \\ | \\ O-Si-O \\ | \\ CH_3 \end{matrix} \right)_n$$

The molecular weight and degree of chain branching can be modified by appropriate inclusion of other silicon chlorides in the reaction mixture. Thus varying proportions of terminal groups and branching groups can be introduced in the final product:

---

[17] H. R. Allcock, *Chem. & Eng. News,* **63,** 22 (March 18, 1985).

[18] M. Zeldin, K. J. Wynne, and H. R. Allcock, "Inorganic and Organometallic Polymers," ACS Symposium Series 360, Washington, 1988.

[19] E. G. Rochow, "The Metalloids," Heath, Boston, 1966.

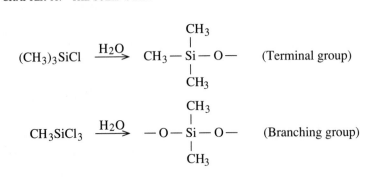

Silicones may be solids, rubbers, greases, or liquids; they are remarkably stable to heat and chemical attack; they have high dielectric strength and are water-repellent.

Attempts to prepare silicon-nitrogen analogs of the silicones by ammonolysis or aminolysis of organosilicon halides have been disappointing because of the relatively easy hydrolysis of the Si—N bonds. However, the Si—N compounds can serve as precursors of silicon nitride (Si_3N_4) ceramics. For example, polysilazanes can be prepared by the reaction of SiH_2Cl_2 with methylamine to give a product which on pyrolysis gives a ceramic material composed mainly of silicon nitride:

$$SiH_2Cl_2 \xrightarrow{CH_3NH_2} CH_3NH-[SiH_2-NCH_3]_x-H \xrightarrow{900°C} H_2 + Si_3N_4 + \cdots$$

When phosphorus pentachloride is heated with a suspension of ammonium chloride in an inert solvent, the ring compounds $(NPCl_2)_3$ and $(NPCl_2)_4$ are formed:

$$PCl_5 + NH_4Cl \xrightarrow{\Delta} \frac{1}{n}(NPCl_2)_n + 4HCl$$

When these materials are heated, they first melt and then are transformed into a rubbery material known as "inorganic rubber."

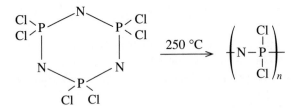

This polymer is insoluble in all solvents and is hydrolyzed to phosphoric acid, ammonium chloride, and hydrogen chloride in a moist atmosphere. However, it can be converted to useful polymers by replacement of the chlorine atoms with various groups. For example, treatment with a primary amine yields a polymer of the type

This, when heated, loses RNH_2 with formation of a highly cross-linked ceramic containing —NR— groups linking the chains.

PROBLEMS

11.1 From the heat of formation of ammonium chloride, NH_4Cl, and estimated lattice energies, calculate the gas-phase proton affinity of ammonia. (NH_4Cl has the CsCl structure, with $r_{N-Cl} = 3.347$ Å.)

11.2 From the heat of formation of BaO and other information, estimate the heat of the reaction $O(g) + 2e^-(g) \rightarrow O^{2-}(g)$. How reliable do you think your estimate is?

11.3 Estimate the heat of formation of $CaCl(s)$. Do you think $CaCl(s)$, if prepared, would be a stable compound? Explain.

11.4 In the compound MX_3, if all the M atoms are equivalent and all the X atoms are equivalent, what are the *possible* coordination numbers for each atom?

*11.5 Calculate the minimum radius ratio for trigonal prismatic sixfold coordination, as in MoS_2 (see Fig. 11.12*b*).

11.6 Give the formula of a beryllium silicon nitride (containing equal numbers of Be and Si atoms) that would be expected to have a diamond-like crystal structure.

*11.7 The second ionization potential of Mg is approximately twice the first, and the conversion of O^- to O^{2-} is endothermic, whereas the conversion of O to O^- is exothermic. Nevertheless we formulate MgO as $Mg^{2+}O^{2-}$ rather than as Mg^+O^-. Why? What simple experiment would show the latter formulation to be unrealistic?

11.8 Some salts can be prepared with either a CsCl structure or an NaCl structure. Which structure would you expect to be preferred at high pressures? Why?

*11.9 The cation-to-anion distances for several compounds having the NaCl structure are listed below

MgO	2.10 Å	MgS	2.60 Å	MgSe	2.73 Å
MnO	2.24 Å	MnS	2.59 Å	MnSe	2.73 Å

How can you explain these data? Calculate the radius of the S^{2-} ion from these data.

*11.10 Imagine a CsCl-type structure from which half of the M atoms are removed in such a way that each X atom has four tetrahedral cation neighbors. What is this well-known MX_2 structure?

*11.11 Boron nitride, BN, has a graphite-like layer structure in which the layers have an eclipsed alignment, with B and N atoms alternating both in the bonds of each layer and in the direction perpendicular to the layers. Can you rationalize the difference between this structure and that of graphite?

11.12 What is the C—C bond order in graphite? Is the C—C distance of graphite (1.42 Å) consistent with that of benzene (1.40 Å), in which the C—C bond order is 1.5?

*11.13 Calculate the fraction of space occupied by spheres packed in (a) the diamond structure and (b) the simple cubic structure (i.e., the NaCl-like structure in which all the atoms are identical). How can you rationalize the existence of elements with the diamond structure and the nonexistence of elements with the simple cubic structure?

11.14 For a compound MX_2 in which all the M atoms are tetrahedrally coordinated and equivalent, describe conceivable structures which are (a) one-dimensional (chain), (b) two-dimensional (layer), and (c) three-dimensional. Give a known example for each structure type.

11.15 All the alkaline-earth oxides except BeO have the NaCl structure. How and why does BeO differ?

11.16 Calculate the relative densities of a compound MX in the NaCl and CsCl structures, assuming a constant M—X interatomic distance.

11.17 Estimate the heat of formation of crystalline CsF_3.

*11.18 Predict the average number of silicon atoms per molecule in a silicone prepared from a mixture of $(CH_3)_3SiCl$, $(CH_3)_2SiCl_2$, and CH_3SiCl_3 in a 3:6:2 mole ratio.

CHAPTER 12

METALS AND METALLIC COMPOUNDS

Most of the elements in the periodic table are metals. In this chapter we will show how the physical properties of these elements, and of the compounds formed by the reaction of these elements with one another (alloys), can be systematized in terms of electron configurations. Then we will discuss the very important class of materials known as metallic compounds—i.e., compounds, containing nonmetallic elements, which have the physical properties (especially the electrical properties) of metals. An understanding of the material in the preceding chapter on the solid state is essential to an understanding of the topics in this chapter.

BONDING ENERGIES AND STRUCTURES

Metals have very characteristic physical properties, including the following: (1) high reflectivity (often called metallic luster); (2) high electrical conductivity, which decreases with increasing temperature; (3) high thermal conductivity; and (4) malleability and ductility. Many of a metal's properties, such as chemical reactivity, hardness, strength, melting point, and boiling point, can be correlated with the strength with which the atoms of the metal are held together. This bonding strength is most simply measured by the energy required to break up the metal into gaseous atoms, i.e., the atomization energy. As one might expect, metals with low atomization energies generally are soft and have low melting points, and metals with high atomization energies generally are hard and have high melting points. In Table 12.1 we have listed the heats of atomization and melting points of the metals in a periodic-table arrangement. In the case of the nontransition metals (groups I to III), the atomization energies increase from left to right. This trend continues for two or three positions into the transition series, in periods 3, 4, and 5. The trend is strong evidence that metallic bonding energy is directly

TABLE 12.1
Heats of atomization† (in kilocalories per mole at 25°C) and melting points (in degrees Celsius) of the metals

Li 38.1 180°	Be 77.5 1283°										
Na 25.6 97.5°	Mg 35.0 650°	Al 78.7 660°									
K 21.3 63.4°	Ca 42.6 850°	Sc 90.3 1539°	Ti 112.5 1725°	V 123 1730°	Cr 95 1900°	Mn 67.7 1247°	Fe 99.3 1535°	Co 101.7 1493°	Ni 102.9 1455°	Cu 80.7 1083°	Zn 31.2 420°
Rb 19.3 38.8°	Sr 39.5 770°	Y 101.0 1509°	Zr 144.5 1852°	Nb 175.2 2487°	Mo 157.6 2610°	Tc 158	Ru 155.8 2400°	Rh 133 1960°	Pd 90.0 1550°	Ag 68.1 961°	Cd 26.7 321°
Cs 18.2 28.7°	Ba 43.5 704°	La 103.1 920°	Hf 148.5 2300°	Ta 187 2997°	W 205.5 3380°	Re 185.4 3150°	Os 188.6 2700°	Ir 160.3 2454°	Pt 134.9 1769°	Au 88.0 1063°	Hg‡ 15.3 −38.9°

† L. Brewer, Lawrence Berkeley Laboratory Report LBL-3720 Rev., May 4, 1977.

‡ At melting point.

related to the number of valence electrons. Sodium, with a single $3s$ electron, can form only one electron-pair bond per atom. This bond is equally distributed in fractional bonds to the neighboring atoms in the lattice, and we can crudely represent the metal as a resonance hybrid of structures such as the following:

$$
\begin{array}{ccccc}
\text{Na} & \text{Na} & \quad -\text{Na} \quad \text{Na}- & \quad \text{Na}^- -\text{Na} & \\
| & | & & | & \quad \text{etc.} \\
\text{Na} & \text{Na} & \quad \text{Na} - \text{Na} & \quad \text{Na} \quad \text{Na}^+ &
\end{array}
$$

The magnesium atom, with the $3s^2$ electron configuration, can be "prepared" for covalent bonding by promotion to the $3s3p$ configuration, with two unpaired electrons. Thus a magnesium atom can distribute two bonds to its neighbors in the lattice. The energetics of the situation may be illustrated as follows:

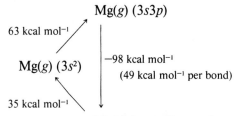

The aluminum atom, with the $3s^2 3p$ ground state, can be promoted to the $3s3p^2$ state, and it can form three bonds per atom in the metal lattice. Further increase in the number of bonds per atom occurs as we move from a group III metal such as scandium into the transition-metal series, where d electrons are involved in the bonding. However, because of the pairing of d electrons and the increased promotion energies, the number of bonds per atom eventually drops off.

We may now state the first rule of Engel and Brewer:[1] *The bonding energy of a metal or alloy depends on the average number of unpaired electrons per atom available for bonding.* Low-lying excited electron configurations with more unpaired electrons than the ground-state configuration may be important if the bonding energy from the additional electron-pair bonds compensates for the promotion energy.

The second rule of Engel and Brewer is concerned with the crystal structures of metals: *The crystal structure depends on the average number of s and p orbitals per atom involved in bonding,* i.e., upon the average number of unpaired s and p electrons in the atoms in their "prepared-for-bonding" state. When the number of bonding s,p electrons is less than or equal to 1.5, the body-centered cubic (bcc) structure is observed. When the number of bonding s,p electrons is between 1.7 and 2.1, the hexagonal close-packed (hcp) structure is observed. When the

[1] L. Brewer, *Science,* **161**, 115 (1968); in "Electronic Structure and Alloy Chemistry of the Transition Elements," P. A. Beck, Ed., pp. 221–235, Interscience, New York, 1963.

number of s,p electrons is in the range 2.5 to 3.2, the cubic close-packed (ccp) structure is observed. Of course, when the number of s,p electrons is near 4, a nonmetallic diamond-like structure is found.

A remarkable feature of the second rule is that it relates metal structure only to the number of unpaired s and p electrons in the prepared-for-bonding state. In fact one can restate the second rule in terms of the number of p electrons: 1, 2, 3, and 4 s,p electrons are equivalent to 0, 1, 2, and 3 p electrons. In either case, although the number of unpaired d electrons is relevant with respect to atomization energy, it is irrelevant with respect to structure. The dependence of structure on the number of p electrons probably follows from the fact that, compared to s orbitals and d orbitals, p orbitals are highly directional in character. However, there does not yet appear to be a simple explanation for the particular structures observed.

In Table 12.2 we have listed the observed crystal structures of the metals. As predicted by the second rule, the alkali metals all have the bcc structure. However, of the alkaline-earth metals, only beryllium and magnesium rigorously obey the rule by having only the hcp structure. The bcc structure shown by calcium, strontium, and barium can be explained by the general rule that the relative importance of d orbitals increases with atomic number. Thus, in these cases, excitation to an $(n - 1)dns$ state is energetically favored over excitation to an $nsnp$ state. Unfortunately, there does not seem to be any simple explanation for the ccp structures shown by calcium and strontium. The bcc and hcp structures shown by scandium, yttrium, and lanthanum (in contrast to the normal ccp structure of

TABLE 12.2
Crystal structures of metals†

The structures are listed in order of temperature stability, with the room-temperature structure lowest.

Li	Be										
bcc	hcp										
Na	Mg	Al									
bcc	hcp	ccp									
K	Ca	Sc	Ti	V	Cr	Mn	Fe	Co	Ni	Cu	Zn
bcc	bcc	bcc	bcc	bcc	bcc	bcc	bcc	ccp	ccp	ccp	hcp
	ccp	hcp	hcp			ccp	ccp	hcp			
						β	bcc				
						χ					
Rb	Sr	Y	Zr	Nb	Mo	Tc	Ru	Rh	Pd	Ag	Cd
bcc	bcc	bcc	bcc	bcc	bcc	hcp	hcp	ccp	ccp	ccp	hcp
	hcp	hcp	hcp								
	ccp										
Cs	Ba	La	Hf	Ta	W	Re	Os	Ir	Pt	Au	
bcc	bcc	bcc	bcc	bcc	bcc	hcp	hcp	ccp	ccp	ccp	
		ccp	hcp								
		hcp									

† L. Brewer, *Science*, **161**, 115 (1968).

aluminum) can be explained in terms of d-orbital participation, as in the cases of calcium, strontium, and barium. In fact, the bcc structure shown by the first four members of each transition series can be explained by assuming that each atom is prepared for bonding by achieving a $d^{v-1}s$ valence electron configuration (where v is the number of valence electrons).

As we move immediately to the right of chromium, molybdenum, and tungsten, the added electrons are generally put into p orbitals to maintain the maximum possible amount of d-orbital bonding. However, with continued addition of electrons, d electrons are paired and eventually the bonding is due only to s,p orbitals. In copper, silver, and gold, the prepared electronic state is d^8sp^2, corresponding to the ccp structure and five bonds per atom. When we reach zinc and cadmium, the d electrons are held too tightly to be easily promoted, and the prepared electronic state is $d^{10}sp$, corresponding to the hcp structure and two bonds per atom. It can be seen from Table 12.1 that the atomization energies are qualitatively in accord with these bonding descriptions.

A cubic close-packed lattice has four sets of parallel close-packed layers, perpendicular to the four body diagonals of the unit cell. This fact can be recognized by examination of Fig. 11.15b. On the other hand, a hexagonal close-packed lattice has only one set of close-packed layers. The ductility, malleability, and softness of a metal depends on the ease with which adjacent layers of atoms can glide over one another. Gliding takes place readily along close-packed planes, and since there are four times as many of these planes in ccp metals as in hcp metals, ccp metals are softer than hcp metals, other factors being equal. Thus we can rationalize the softness of copper, sliver, and gold compared to other metals, of ccp cobalt relative to hcp colbalt, and of ccp iron relative to the high-pressure hcp form of iron.

The Engel-Brewer rules can be readily applied to alloys.[2] The metals of the first half of a transition series, which have fewer than five d electrons in their prepared-for-bonding states, provide d-orbital sinks for electrons until the d^5 configuration is achieved. This fact and the second rule lead to the conclusion that a transition-metal alloy will have a bcc structure if the average number of valence electrons is less than 6.5, corresponding to the prepared electron configuration $d^5s^1p^{0.5}$. On this basis we predict the maximum solubilies of metals such as Re, Os, Ir, and Pt in the bcc phase of tungsten to be that mole fraction corresponding to 6.5 valence electrons per atom. The predicted and observed mole fractions are given in Table 12.3. Although the agreement is far from perfect, it shows that qualitative and even semiquantitative predictions can be made by using the Engel-Brewer rules.

Another, related, application of the rules is the prediction of the effect of small additions of alloying metals upon the relative stability of two crystal struc-

---

[2] L. Brewer, *Science*, **161**, 115 (1968); *Acta Metal.*, **15**, 553 (1967); L. Brewer and P. R. Wengert, *Metal. Trans.*, **4**, 83 (1973).

TABLE 12.3
**Maximum solubilities (in mole percentages)
of some metals in the bcc phase of tungsten**

Metal	Predicted solubility	Observed solubility
Re	50	35–43
Os	25	10–20
Ir	16.7	10–15
Pt	12.5	4–10

tures.[3] From Table 12.2 we see that Ti, Zr, and Hf can have either a bcc structure (corresponding to configuration d^3s) or an hcp structure (corresponding to configuration d^2sp). Addition of d-electron-rich metals to the right of these metals favors d-orbital bonding and thus stabilizes the bcc structure. The addition of metals containing no d electrons stabilizes the hcp structure because that structure will suffer less by a reduction in the amount of d-orbital bonding. All the experimental data available confirm these predictions of the effects of alloying on the relative stabilities of the bcc and hcp structures.

Brewer has shown that the alloying of d-electron-poor metals such as Zr, Nb, Ta, and Hf with d-electron-rich metals such as Re, Ru, Rh, Ir, Pt, and Au yields extremely stable alloys.[3] For example, if platinum is heated with ZrC (one of the most stable carbides known), the alloy $ZrPt_3$ and graphite are formed.

$$ZrC + 3Pt \xrightarrow{\Delta} ZrPt_3 + C$$

These interactions of "electron-deficient" and "electron-rich" transition metals may be considered an extension of Lewis acid-base reactions.

Hume-Rothery Compounds[4]

The importance of the average number of s,p valence electrons in the correlation of alloy composition with structure was recognized long ago by Hume-Rothery. The phase diagrams of a large class of binary alloys, e.g., those of copper and silver with the zinc family of metals or the aluminum family of metals, show a wide variety of phases. Three of these phases, called Hume-Rothery compounds, have structures which may be associated with certain electron/atom ratios. The first phase, having the β-brass structure (a bcc-type structure), corresponds to an s,p electron/atom ratio of 3:2. Examples of such phases are CuZn, AgCd, Cu_3Al, Cu_5Sn, and NiAl. (In these and the following examples, Cu and Ag must be considered as $d^{10}s$ metals, and Ni must be considered a d^{10} metal.) The second

---

[3] Brewer and Wengert, op. cit.

[4] H. J. Emeléus and J. S. Anderson, "Modern Aspects of inorganic Chemistry," 3d ed., pp. 507–510, Van Nostrand, New York, 1960.

phase, having the γ-brass structure (a complex cubic structure), corresponds to an s,p electron/atom ratio of 21:13. Examples are Cu_5Zn_8, Ag_5Hg_8, Cu_9Al_4, $Cu_{31}Sn_8$, and Ni_5Zn_{21}. The third phase, having the ϵ-brass structure (an hcp-type structure), corresponds to an s,p electron/atom ratio of 7:4. Examples are $CuZn_3$, $AgCd_3$, and Cu_3Sn.

It can be seen that the β-brass type of alloys and the ϵ-brass type of alloys are in accord with the Engel-Brewer rule relating s,p electrons to structure. However, the reader may question the apparently ad hoc assignment of one s,p electron to Cu, Ag, and Au, and no s,p electrons to transition elements such as Ni, Pt, Co, and Fe. Each of these metals, in the pure state, has a prepared-for-bonding electron configuration of d^5sp^2, d^6sp^2, d^7sp^2, or d^8sp^2, corresponding to the ccp structure. However, when any one of these metals is diluted sufficiently with a metal such as zinc (with prepared-for-bonding configuration $d^{10}sp$), the number of d orbitals available for bonding is reduced so much that there is no advantage in promotion to these excited states, and the metals then have, in effect, prepared-for-bonding electron configurations with more of the d orbitals filled. The limiting electron configurations for such metals are d^8, d^9, d^{10}, and $d^{10}s$.

To show the typical complexity of the phase diagrams of such systems, the Cu–Zn phase diagram is given in Fig 12.1. As can be seen from this diagram, there are other phases besides the Hume-Rothery phases in this system.

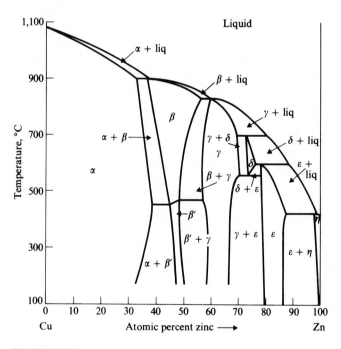

FIGURE 12.1
The Cu-Zn phase diagram. (*From L. V. Azároff, "Introduction to Solids," pp. 294–297. McGraw-Hill, New York, 1960.*)

BAND THEORY[5]

Electrical Conductivity

When two lithium atoms are brought together, the $2s$ atomic orbital levels are split into two levels, corresponding to the $2s\sigma$ bonding molecular orbital (MO) and the $2s\sigma^*$ antibonding MO, as shown in Fig. 12.2. If six lithium atoms were brought together to form a cluster, the atomic orbitals would be split into six MOs, ranging in character from completely bonding to completely antibonding, as shown in Fig. 12.3. When a very large number (say Avogadro's number) of lithium atoms are brought together to form a regular array, as in the normal bcc metal, the resulting energy levels are so closely spaced that they form an essentially continuous energy band, as shown in Fig. 12.4. The bottom of the band is a bonding level, and the top of the band is an antibonding level. The band is composed of as many levels as there are atoms, and each level can hold two electrons of opposite spin. Inasmuch as each lithium atom furnishes one valence electron, the band is exactly half-filled. Another way of depicting energy bands is in the form of a density-of-states diagram, as shown in Fig. 12.5. Here the number of electrons that can be accommodated in a narrow range of energy is plotted as a function of energy. Cross-hatching is used to indicate the filled portion of a band.

The individual levels of a band correspond to MOs which extend throughout the metal lattice. The valence electrons can be thought of as moving throughout the crystal as waves. Of course, in any isolated piece of metal, there are as many electrons moving in one direction as in another. For example, in a horizontal wire stretched from left to right, the number of electrons moving to the left is equal to the number of electrons moving to the right. This situation is depicted by the double density-of-states diagram of Fig. 12.6a, in which points above the

---

[5] C. A. Wert and R.M. Thomson, "Physics of Solids," McGraw-Hill, New York, 1964; C. Kittel, "Introduction to Solid State Physics," 5th ed., Wiley, New York, 1976.

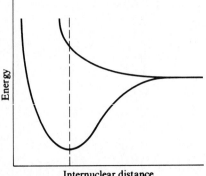

FIGURE 12.2
Interaction of the $2s$ orbitals of two lithium atoms. The dashed line indicates the internuclear distance in the Li_2 molecule.

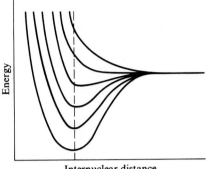

FIGURE 12.3
The molecular orbitals resulting from interaction of the 2s orbitals of six lithium atoms. The relative positions of the six levels depend on the arrangement of the atoms in the cluster.

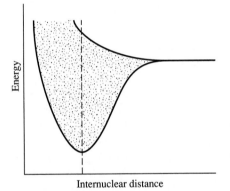

FIGURE 12.4
The valence band in lithium metal, shown as a function of the internuclear distance. The dashed line indicates the equilibrium internuclear distance in the actual metal.

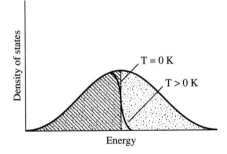

FIGURE 12.5
Density of states as a function of energy for the valence band of a metal such as lithium. The vertical line, which marks the boundary between the filled and empty parts of the band at 0 K, corresponds to the Fermi energy. At a finite temperature, this boundary is fuzzy, as shown.

horizontal axis refer to left-moving electrons, and points below the horizontal axis refer to right-moving electrons. If we apply an electric field to the wire, so that the left side is positively charged relative to the right side, the energy of the left-moving electrons is reduced and the energy of the right-moving electrons is increased, as shown by the diagram of Fig. 12.6b, in which the two bands are

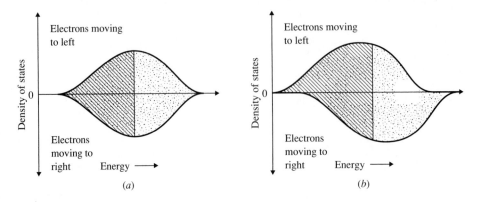

FIGURE 12.6
Density-of-states diagrams for electrons moving to the left (upper curves) and to the right (lower curves). (*a*) The diagram for a metal wire in the absence of an electric field. (*b*) The diagram for a metal wire subjected to an electric filed; $+$ at the left end, and $-$ at the right end.

shifted relative to one another. The electrons fill up the lowest available energy levels, resulting in more electrons moving to the left than to the right and a net electric current.

When the temperature of a metal is increased, lattice irregularities become more pronounced because of increased atomic vibrations. These irregularities scatter the electrons, thus reducing the electrical conductivity. This effect of temperature on conductivity may also be explained in terms of MO theory. In a highly regular metal, the electron orbitals extend for great distances through the lattice, and the electrons are highly mobile because of the delocalization. The introduction of irregularities increases the amount of localized bonding and hence decreases the conductivity.

The electrical conductivity of a filled band is zero because, in such a band, the number of electrons moving in one direction must always be equal to the number moving in the opposite direction. Even though application of an electric field can shift the relative positions of the left-moving and right-moving bands on the energy scale, because each band is full it is impossible for electrons to spill out of the higher-energy band into the lower-energy band. For this reason one might expect an alkaline-earth metal such as calcium, which has a filled valence *s* shell in the free atomic state, to be a nonconductor. However, at the interatomic distances found in this metal, the valence *s* and *p* bands overlap, as shown in Fig. 12.7. Consequently the density-of-states diagram is as shown in Fig. 12.8, and the metal exhibits typical metallic electrical conductivity.

An "experimental" density-of-states diagram for a metal can be obtained by photoelectron spectroscopy. The metal is prepared with a clean surface in an ultrahigh vacuum ($< 10^{-9}$ torr) and is irradiated with monochromatic photons. A plot of the number of emitted electrons per absorbed photon versus kinetic energy has approximately the shape of the density-of-states diagram, assuming that the yield of electrons is proportional to the electron population. Figure 12.9 shows the observed photoemission energy distributions for copper for five different photon

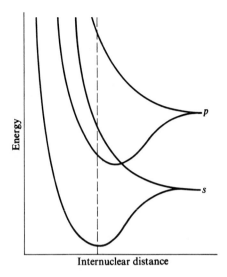

Energy

Internuclear distance

FIGURE 12.7
The valence band of an alkaline-earth metal such as calcium. The dashed line indicates the equilibrium internuclear distance in the actual metal.

energies.[6] The shapes of the curves differ because the relative cross sections of the electrons in the band are a function of photon energy and electron energy.

The phenomenon of *superconductivity* is discussed at the end of this chapter.

The density-of states diagram for an insulator looks like that in Fig. 12.10, in which the high-energy empty band corresponds to nonbonding or antibonding levels. Sometimes it is possible to convert such a material into a metallic conductor by the application of high pressure (for example, 100 kbars). From Fig. 12.11 it can be seen that, if the interatomic distance is sufficiently reduced by increased pressure, the bands will overlap, yielding a density-of-states diagram similar to that of calcium (Fig. 12.8). In Table 12.4 the electrical resistivities of the elements of the first long row of the periodic table are listed. The metals with partially filled *d* bands generally have higher resistivities than other metals such as K, Ca, Cu,

---

[6]   D. E. Eastman, in "Electron Spectroscopy," D. A. Shirley, ed., p. 487, North-Holland, Amsterdam, 1972.

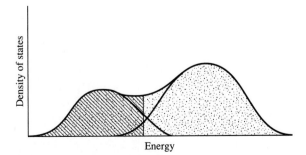

Density of states

Energy

FIGURE 12.8
Density of states as a function of energy for an alkaline-earth metal such as calcium.

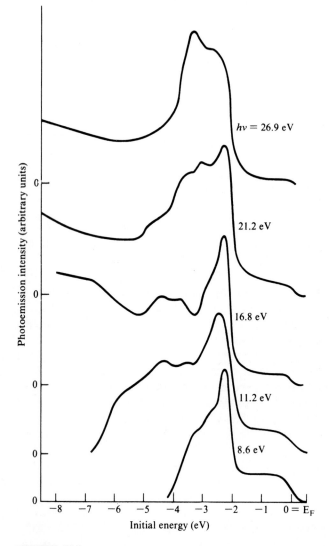

FIGURE 12.9

Photoemission energy distributions for Cu. *(Reproduced with permission from D. E. Eastman, in "Electron Spectroscopy," D. A. Shirley, ed., p. 487, North-Holland, Amsterdam, 1972.)*

and Zn. This greater resistivity is probably due to the fact that the *d* orbitals are somewhat "inner" orbitals and do not overlap as effectively as *s* and *p* orbitals. Consequently the valence electrons are relatively "localized" in narrow bands and do not hop from one atom to another as rapidly as they would if they occupied broader *s* or *p* bands. The very high resistivities of Ge and Se are expected in view of the nonresonating covalent structures of these elements. However, the low resistivity as As is remarkable in view of its structure (Fig. 11.11a). Apparently

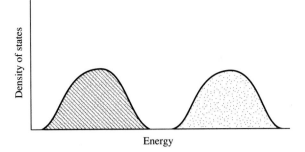

FIGURE 12.10
Density of states as a function of energy for an insulator.

the bonding valence band and a higher antibonding band are so broad that they overlap and permit metallic conductivity.

It is interesting to consider the trend in electrical conductivity of the tetrahedrally bonded elements of main group IV: C, Si, Ge, and Sn. Diamond is an insulator, silicon and germanium are semiconductors (corresponding to a narrow gap between the filled and empty bands), and tin is a good conductor. This trend is probably related to the decrease in the covalent bond energy between the atoms on going down the family. The stronger the bonds, the greater the separation between the filled bonding bands and the empty antibonding bands. In diamond, the band gap is high; in tin, the bands actually overlap and cause metallic behavior.

Ferromagnetism

Some substances have permanent magnetic moments even in the absence of applied magnetic fields and are called "ferromagnetic." Only a few elements are ferromagnetic: Fe, Co, Ni, and several lanthanides. A density-of-states diagram

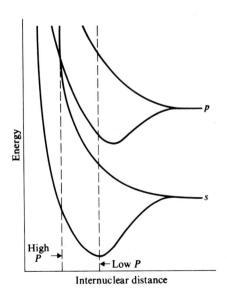

FIGURE 12.11
The valence bands of an insulator which can be converted into a metal by high pressure.

TABLE 12.4
Electrical resistivities at 22°C† of the elements of the first long row of the periodic table

Element	Resistivity, $\mu\Omega$ cm	Element	Resistivity, $\mu\Omega$ cm
K	7.19	Co	5.80
Ca	3.35	Ni	7.04
Sc	46.8	Cu	1.70
Ti	43.1	Zn	5.92
V	19.9	Ga	14.8
Cr	12.9	Ge	$\sim 10^8$
Mn	136	As	29
Fe	9.8	Se	$\sim 10^{11}$

† D. E. Gray, ed., "American Institute of Physics Handbook," 3d ed., McGraw-Hill, New York, 1972.

for the $3d$ and $4s$ shells of Fe, Co, or Ni would look something like that shown in Fig. 12.12. Here the upper diagram corresponds to electrons with their spins oriented one way (say "up"), and the lower diagram corresponds to electrons with their spins oriented the opposite way (say "down"). The remarkable thing about these metals is that the two bands are spontaneously displaced so that they are unequally filled. Thus more electrons are aligned one way than in the opposite way, and the metal has a permanent magnetic moment. The thermal energy of the crystal tends to misalign the electronic spins and to bring the density-of-states

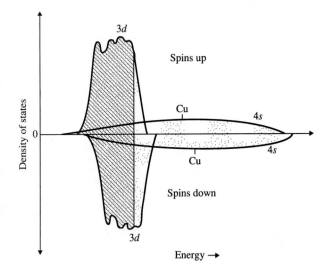

FIGURE 12.12
Density-of-states diagrams for a ferromagnetic transition metal. The upper bands correspond to electrons with spins "up"; the lower bands correspond to electrons with spins "down." The points marked Cu indicate the Fermi energy of copper, for which there is no relative displacement of the bands.

bands together. Obviously energy is required to displace the bands relative to one another as shown in Fig. 12.12. It is believed that the source of this energy is the exchange interaction between neighboring aligned spins. (See Chap. 1.) Raising the temperature of a ferromagnet causes the magnetization to decrease. When the temperature reaches the Curie temperature, at which the thermal energy of the crystal tending to bring the bands together equals the exchange energy, the magnetization is zero. At temperatures above the Curie temperature (1043, 1404, and 631 K for Fe, Co, and Ni, respectively) the metal is no longer ferromagnetic.

Metals which are ferromagnetic have high densities of states at the Fermi energy (i.e., the energy corresponding to the highest filled level in the band). Thus many electrons change their spin direction (and yield exchange energy) for a small displacement of the bands. For ordinary metals, the density of states at the Fermi energy is lower and the energy required to displace the bands is not compensated by the exchange energy. For example, in copper metal the levels are filled to a point where the density of states is very low (see Fig. 12.12); consequently this metal is not ferromagnetic. The reason for the high density of states in Fe, Co, and Ni is the fact that the $3d$ atomic orbitals are somewhat interior orbitals which do not overlap strongly with those on other atoms in the lattice; thus the $3d$ bands for these metals are narrow.

Magnetic data for alloys of the metals in the vicinity of Fe, Co, and Ni in the periodic table show that the number of aligned spins per atom, N_B, is a continuous function of the average number of electrons per atom.[5] In Fig. 12.13 it can be seen that the maximum number of aligned spins per atom is 2.4, a value achieved by an Fe–Co alloy. For some unknown reason, when the constituent metals of the alloy differ in atomic number by more than two units, the points deviate markedly from the curve.

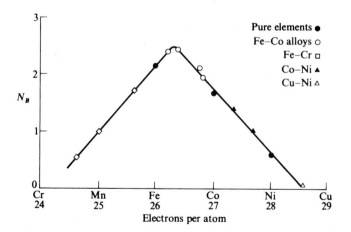

FIGURE 12.13
The number of aligned spins per atom as a function of the average number of electrons per atom. *(From C. A. Wert and R. M. Thomson, "Physics of Solids," McGraw-Hill, New York, 1964.)*

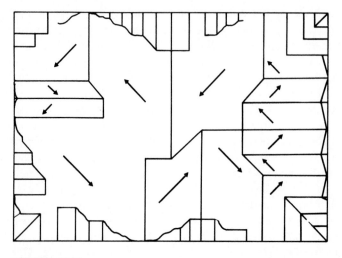

FIGURE 12.14
A ferromagnetic domain pattern on a single crystal face of nickel. (*Adapted from C. Kittel, "Introduction to Solid State Physics," 5th ed., p. 485, Wiley, New York, 1976.*)

On an ultramicroscopic scale, all the electronic spins of a ferromagnetic material are aligned in the same direction. However, in different regions called "domains" (with linear dimensions in the range $\sim 10^{-4}$ to $\sim 10^{-1}$ cm), the directions of magnetization can be different. Thus the net magnetic moment of a macroscopic ferromagnetic crystal can be near zero because of cancelation of the moments of the individual magnetic domains. A pattern of domains with approximately zero resultant magnetic moment is shown in Fig. 12.14. The net magnetic moment of a sample can increase under the influence of an applied magnetic field by two independent processes: (1) In weak applied fields, the

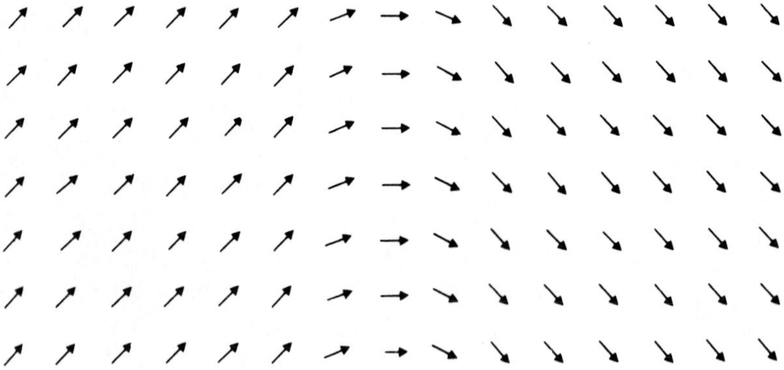

FIGURE 12.15
Schematic diagram of the boundary between two magnetic domains. Actual boundaries are hundreds of atoms thick. By the gradual turning of the spins within the boundary, the boundary can move in response to an applied magnetic field.

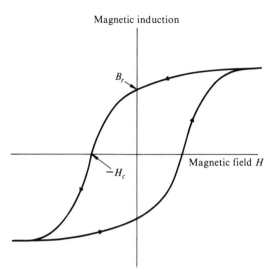

Magnetic induction

B_r

$-H_c$

Magnetic field H

FIGURE 12.16

A magnetic hysteresis loop. The arrows indicate the direction of the change in magnetic field. The coercive force, H_c, is the field required to bring the induction to zero; the remanence, B_r, is the value of the magnetic induction at zero applied field.

volumes of domains favorably oriented in the field can increase at the expense of adjacent domains unfavorably oriented. (2) In strong fields, the direction of magnetization within a domain can rotate toward alignment with the field. The first process involves the shifting of the boundaries between domains. The change in spin direction does not occur suddenly across a sharp boundary, but rather gradually over a boundary hundreds of atoms thick, as shown schematically in Fig. 12.15. Because the shifting of domain boundaries is not a completely reversible process, the magnetization of a ferromagnetic material shows hysteresis, as seen in the magnetization curve of Fig. 12.16. The coercive force is the field H_c required to reduce the induction to zero. It may range from 20,000 gauss in a high-stability Fe-Pt magnet to 0.004 gauss in a Supermalloy pulse transformer.

METALLIC COMPOUNDS

In crystalline sodium chloride, the two valence bands (attributable to the sodium $3s$ and chlorine $3p$ orbitals, respectively) are widely separated in energy, as expected from the great difference in electronegativity between sodium and chlorine. The lower-energy, chlorine, band is filled; the higher-energy, sodium, band is empty. Hence sodium chloride is an insulator. Titanium carbide (TiC) has the sodium chloride–type structure, but it differs from sodium chloride in two important respects. First, the titanium and carbon atoms are not extremely different in electronegativity. Second, the titanium atoms are much bigger than the carbon atoms, thus allowing adjacent titanium atoms to essentially touch one another, with considerable overlap of their valence orbitals. Hence it is reasonable to expect the titanium and carbon bands to overlap, and titanium carbide to be metallic. Indeed, this and numerous other binary compounds of the early transition metals and electronegative elements do show metallic character, as evidenced by their high electrical conductivity with a negative temperature coefficient. These include some very hard, high-melting, and chemically inert borides, carbides, nitrides, oxides, and sulfides. These compounds may be looked upon as roughly three-dimensional cationic metal lattices, held together by metallic bonding, with interspersed anions. In most cases the compounds are markedly variable in composition. When the metal atoms are in a cubic close-packed array, the nonmetal atoms fill all the octahedral holes, giving compositions such as MC and MN, with the NaCl structure. However, when the metal atoms are hexagonally close-packed, only one of each pair of octahedral holes directly above and below each other on opposite sides of each close-packed layer is ever occupied by a nonmetal atom—hence stoichiometries such as V_2C, Nb_2C, Ta_2C, Mo_2C, W_2C, and Mo_2N. Intermediate stoichiometries are found when the M atom stacking alternates (e.g., V_4C_3) or when there are ordered defect structures (e.g., V_8C_7 and V_6C_5).

One of the highest-melting substances known is the reaction product of a 4:1 mixture of TaC and ZrC, which melts at 4215 K. Metallic carbides are chemically very inert; they do not react with water, they are oxidized by air only at extremely high temperatures, and they dissolve (slowly) only in reagents such as concentrated nitric and hydrofluoric acids. The hardness, high melting points, and inertness of tungsten and tantalum carbides have led to their use as high-speed cutting tools. Many metallic compounds become superconducting at low temperatures; this topic is discussed at the end of this chapter. A few properties of some metallic carbides and nitrides are listed in Table 12.5.

Most metallic carbides are prepared by the powder metallurgical reaction between the metal or metal oxide and carbon.[7] An intimate pulverized mixture of the reactants is heated in an induction or resistance furnace, usually to a temperature of at least 1500°C. For example, Cr_3C_2 can be prepared at 1800°C by the following reactions:

$$3Cr + 2C \longrightarrow Cr_3C_2$$

$$3Cr_2O_3 + 13C \longrightarrow 2Cr_3C_2 + 9CO$$

Metallic nitrides which are thermodynamically stable with respect to the elements can be prepared by reaction of the metals with nitrogen or ammonia.[8,9] For example,

$$Ti + \tfrac{1}{2}N_2 \xrightarrow{1200°C} TiN$$

$$4Fe + 2NH_3 \xrightarrow{500°C} 2Fe_2N + 3H_2$$

Some nitrides can be prepared by reaction of a metal halide with ammonia, thus[8,9]

$$CrCl_3 + 4NH_3 \xrightarrow{800°C} CrN + 3NH_4Cl$$

[7] S. Windisch and H. Nowotny, in "Preparative Methods in Solid State Chemistry," P. Hagenmuller, ed., pp. 533–562, Academic, New York, 1972.

[8] G. Brauer, "Handbook of Preparative Inorganic Chemistry," vol. 2, Academic, New York, 1965.

[9] R. Kieffer and P. Ettmayer, *High Temp. High Pressures*, **6**, 253 (1974).

TABLE 12.5
Properties of some metallic carbides and nitrides†

Compound	Density, g cm^{-3}	Melting point, °C	Electrical resistivity, $\mu\Omega$ cm
TiC	4.93	2940	68
VC	5.36	2684	60
Cr$_3$C$_2$	6.68	1810	75
ZrC	6.46	3420	42
NbC	7.78	3613	35
Mo$_2$C	9.18	2520	71
MoC	9.15	2600	
HfC	12.3	3820	37
TaC	14.48	3825	25
WC	15.7	2720	19
ThC	10.64	2625	25
ThC$_2$	8.65	2655	30
UC	13.63	2650	40
UC$_2$	11.86	~2500	90
TiN	5.43	2950	25
VN	6.10	2350	85
CrN	6.14	1080d	640
ZrN	7.3	2980	21
NbN	8.47	2630d	78
Mo$_2$N	9.46	790d	
HfN	14.0	3330	33
ϵ-TaN	14.3	2950d	128
δ-TaN	15.6	2950d	
ThN	11.9	2820	20
UN	14.4	2800	176

† "Kirk-Othmer Encyclopedia of Chemical Technology," 3d ed., vol. 4, pp. 490–505, Wiley-Interscience, New York; R. Kieffer and P. Ettmayer, *High Temp. High Pressures*, **6**, 253 (1974).

Rhenium trioxide is a red solid with metallic luster and a structure in which each Re atom is octahedrally coordinated by oxygen atoms, as shown in Fig. 12.17. ReO$_3$ has a fairly high conductivity with a negative temperature coefficient; clearly the single valence electron of the Re(VI) is delocalized in a partially filled conduction band of the crystal. This conduction band is probably formed by the overlap of rhenium $d\pi$ and oxygen $d\pi$ orbitals throughout the ReO$_3$ framework.

Tungsten trioxide has a structure very similar to that of ReO$_3$, but of course W(VI) has no valence electrons, and so WO$_3$ is an insulator. However, the *tungsten bronzes* are a class of metallic, nonstoichiometric compounds of general formula M$_x$WO$_3$, where $0 < x < 1$ and where M can be an alkali or alkaline earth metal, lead, thallium, copper, silver, or a lanthanide.[10] Sodium tungsten bronzes, Na$_x$WO$_3$ ($0.32 < x < 0.93$), can be made by the high-temperature reduction of

[10] E. Banks, and A. Wold, *Prep. Inorg. React.*, **4**, 237 (1968).

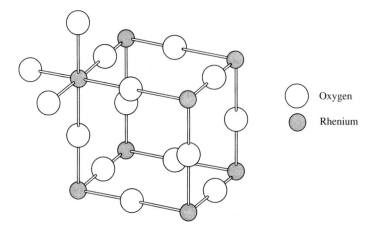

Oxygen

Rhenium

FIGURE 12.17
The ReO_3 structure. This structure is like the perovskite structure, but with the central atom missing.

sodium tungstate, Na_2WO_4, with a wide variety of reducing agents. The structures correspond essentially to M-deficient perovskite lattices in which the M atoms have released their valence electrons to the tungsten-oxygen conduction band. Some of the properties of sodium tungsten bronzes are listed in Table 12.6.

It is possible to prepare materials of the type $WO_{3-x}F_x$ by replacing some of the oxygen atoms of WO_3 by fluorine atoms.[11] The materials for which $0.17 < x < 0.66$ are analogous to tungsten bronzes; the extra electrons introduced by

---

[11]   A. W. Sleight, *Inorg. Chem.*, **8**, 1764 (1969); C. E. Derrington et al., *Inorg. Chem.*, **17**, 977 (1978).

TABLE 12.6
Properties of Na_xWO_3 crystals†

x	Color	Phase	Lattice parameter,‡ a_0, Å
0.37	Blue	Tetragonal	
0.40	Blue	Tetragonal	
0.44	Blue	Cubic	3.821
0.56	Violet	Cubic	3.831
0.58	Violet	Cubic	3.832
0.65	Red	Cubic	3.838
0.70	Orange	Cubic	3.842
0.81	Yellow	Cubic	3.851

† E. Banks and A. Wold, *Prep. Inorg. React.*, **4**, 237 (1968).
‡ The lattice parameter of a cubic crystal is the "repeat distance," i.e., the side dimension of the unit cell.

the fluorine atoms partially fill the tungsten-oxygen-fluorine conduction band and cause the materials to be good conductors.

"One-Dimensional" Metallic Compounds

Many compounds are known which contain extended chains of metal atoms or chains of metal clusters. These structural features are often found in transition-metal halides or chalcogenides with low nonmetal/metal ratios.[12] Thus $NaMo_4O_6$, formed by heating a mixture of Na_2MoO_4, MoO_2, and Mo,[13] contains anionic chains of the repeating unit $Mo_4O_6^-$. These chains contain Mo_6 octahedra sharing opposite edges, with oxygen atoms forming bridges between molybdenum atoms in the same and adjacent chains, as shown in Fig. 12.18. Similar chains of octahedral metal clusters exist in the compounds Sc_5Cl_8, which is better represented by the formula $(ScCl_2^+)$ $(Sc_4Cl_6^-)$, and Gd_2Cl_3.[12]

When liquid mercury is treated with a solution of AsF_5 in SO_2, golden crystals of Hg_3AsF_6 are formed.[14] This compound contains infinite chains of mercury atoms in a lattice of AsF_6^- ions, as shown in Fig. 12.19. The Hg—Hg distance and the spacing of the AsF_6^- ions are not commensurate with the formula Hg_3AsF_6, but rather with the composition $HG_{2.86}AsF_6$. This anomaly can be explained either by assuming that the compound is indeed nonstoichiometric (with an average oxidation state of mercury of $1/2.86 = 0.35$) or by assuming that

---

[12] For a review on halides of this type, see J. D. Corbett, *Acc. Chem. Res.*, **14**, 239 (1981).

[13] C. C. Torardi and R. E. McCarley, *J. Am. Chem. Soc.*, **101**, 3963 (1979).

[14] D. Brown, B. D. Cutforth, C. G. Davies, R. J. Gillespie, P. R. Ireland, and J. E. Vekris, *Can. J. Chem.*, **52**, 791 (1974).

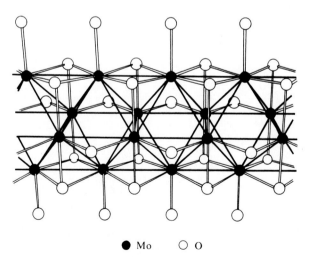

FIGURE 12.18
View of one cluster chain in $NaMo_4O_6$. [*Reproduced with permission from C. C. Torardi and R. McCarley, J. Am. Chem. Soc.*, **101**, *3963 (1979). Copyright 1979 American Chemical Society.*]

● Mo   ○ O

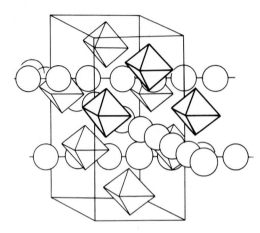

FIGURE 12.19
A view of the Hg_3AsF_6 structure, showing chains of Hg atoms (circles) running through the lattice of AsF_6^- ions (octahedra). [*Reproduced from D. Brown, B. D. Cutforth, C. G. Davis, R. J. Gillespie, P. R. Ireland, and J. E. Vekris, Can. J. Chem.,* **52**, *791 (1974).*]

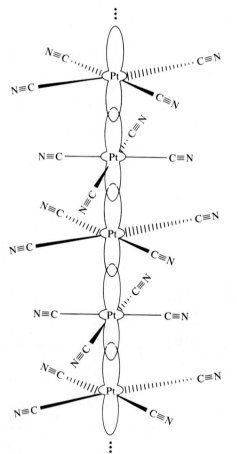

FIGURE 12.20
The stacking of planar $Pt(CN)_4$ groups in partially oxidized $K_2Pt(CN)_4 \cdot 3H_2O$ complexes. Note the overlapping of the d_{z^2} orbitals and the staggering of the cyano groups. [*Reproduced with permission from J. S. Miller and A. J. Epstein, Prog. Inorg. Chem.,* **20**, *1 (1976).*]

TABLE 12.7
Properties of some tetracyanoplatinates†

Complex	Pt oxidation state	Pt—Pt distance, Å	Color	Conductivity, Ω^{-1} cm^{-1}
Pt metal	0	2.775	Metallic	9.4×10^4
$K_2Pt(CN)_4 \cdot 3H_2O$	+2	3.50	White	5×10^{-7}
$K_2Pt(CN)_4Br_{0.3} \cdot 3H_2O$	+2.3	2.89	Bronze	$4 - 830$
$K_2Pt(CN)_4Cl_{0.32} \cdot 3H_2O$	+2.32	2.88	Bronze	~ 200
$K_{1.75}Pt(CN)_4 \cdot 1.5H_2O$	+2.25	2.96	Bronze	~ 80
$Cs_2Pt(CN)_4(HF_2)_{0.39}$	+2.39	2.83	Gold	

† J. S. Miller and A. J. Epstein, *Prog. Inorg. Chem.*, **20**, 1 (1976); G. D. Stucky, A. J. Schultz, and J. M. Williams, *Ann. Rev. Mater. Sci.*, **7**, 301 (1977).

the compound is stoichiometric (Hg_3AsF_6), with vacancies randomly distributed among 4.7 percent of the AsF_6^- sites in the lattice.[15]

An interesting class of one-dimensional conducting solids can be prepared by partial oxidation of the tetracyanoplatinate complex, $Pt(CN)_4^{2-}$, either electrolytically or with chlorine or bromine.[16] These compounds contain square planar $Pt(CN)_4^{n-}$ ions stacked such that the platinum $5d_{z^2}$ orbitals overlap, as shown in Fig. 12.20. The formulas and properties of some of these compounds are given in Table 12.7. The unoxidized platinum(II) compound, $K_2Pt(CN)_4 \cdot 3H_2O$, is a nonconductor of electricity. Its platinum oxidation state is integral ($+2$), and the Pt—Pt separation is so great that there is no significant bonding between $Pt(CN)_4$ groups. However, in the oxidized compounds the Pt—Pt separations are almost as short as in platinum metal itself, and the compounds are electrically conducting. The electrical conductivity arises from partial filling of one-dimensional bands.

Polysulfur nitride, $(SN)_x$, is remarkable because it is a polymeric metal which contains no metal atoms.[17] It is prepared by the spontaneous polymerization of S_2N_2 at temperatures between 0°C and room temperature, whereby shiny metallic fiber bundles are formed. The crystals contain parallel chains of alternating S and N atoms with approximately coplanar atoms in the following configuration:

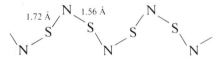

Although all the sulfur atoms and all the nitrogen atoms in these chains are stereochemically equivalent, there are two markedly different N—S distances.

15 A. J. Schultz, J. M. Williams, N. D. Miro, A. G. MacDiarmid, and A. J. Heeger, *Inorg. Chem.*, **17**, 646 (1978).

16 J. S. Miller and A. J. Epstein, *Prog. Inorg. Chem.*, **20**, 1 (1976).

17 M. M. Labes, P. Love, and L. F. Nichols, *Chem. Rev.*, **79**, 1 (1976).

A valence bond representation of the chains requires that they be considered resonance hybrids:

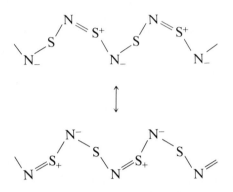

The conductivity of $(SN)_x$ is about $4 \times 10^3 \ \Omega^{-1} \ cm^{-1}$ at room temperature, increasing by a factor of 200 to 250 on cooling to liquid helium temperature. In 1975, it was found that $(SN)_x$ becomes superconducting below 0.26 K.[18] This discovery stimulated both extensive work on sulfur-nitrogen compounds and attempts to prepare other superconducting compounds of nonmetals.

"Two-Dimensional" Metallic Compounds

The electrical resistivity of graphite (see Fig. 11.11b) in the direction perpendicular to the layers is 0.3 to 0.5 Ω cm at 50 K and 0.15 to 0.25 Ω cm at room temperature. The magnitude and temperature coefficient of this resistivity is characteristic of a semiconductor. However, in a direction parallel to the layers, the resistivity is much lower: 2 $\mu\Omega$ cm at 10 K and 40 $\mu\Omega$ cm at room temperature. This latter resistivity is almost as low as that found for typical metals, and graphite is often referred to as a semimetal.[19] The semimetallic behavior of graphite can be explained by the density-of-states diagram. The Fermi level occurs at a low minimum in the band, in the region where a filled band and an empty band barely overlap. Because of the low density of states at the Fermi surface, the application of an electric field does not cause as much net electron flow as in the case of a typical metal, in which the Fermi surface is situated in a region with a high density of states.

The layers of the graphite structure are so widely separated and so weakly held together that many molecules and ions can penetrate the "galleries" between layers and form so-called intercalates or layered interstitial compounds. Most of

[18] R. L. Greene, G. B. Street, and L. J. Suter, *Phys. Rev. Lett.*, **34**, 577 (1975). See the last part of this chapter for a discussion of superconductivity.

[19] A. K. Holliday, G. Hughes, and S. M. Walker, in "Comprehensive Inorganic Chemistry," vol. 1, pp. 1259–1294, Pergamon, Oxford, 1973.

the stable graphite intercalates are essentially salts in which the graphite layers have had electrons either removed or added. Changing the number of valence electrons in graphite causes a shift of the Fermi surface away from the density-of-states minimum and a consequent increase in electrical conductivity.

Among the first layer-structure intercalates to be discovered were the graphite-alkali metal compounds, which are readily prepared by direct reaction of the metals with graphite. The most metal-rich compositions which have been obtained are C_8M for potassium, rubidium, and cesium, C_6Li for lithium, and $C_{64}Na$ for sodium.[19] Each metal atom contributes its valence electron to the conduction band of the graphite network.

Generally any given type of graphite intercalate can be prepared in a series of different stoichiometries, each stoichiometry corresponding to a "stage" for that set of compounds. The stage is defined as the reciprocal of the fraction of the galleries occupied. For example, a stage 1 intercalate has all of the galleries occupied, a stage 2 intercalate has one-half of the galleries occupied, etc., as indicated in Fig. 12.21. In the stage 1 potassium compound (C_8K), the potassium atoms are arranged in a triangular net, with the metal atoms centered between hexagonal C_6 rings. In all the higher-stage potassium compounds, every third potassium atom is missing, producing a hexagonal net and a general formula $C_{12n}K$, where n is the stage number.[20] It should be noted that the offset stacking pattern of ordinary graphite ($ABABA\cdots$) is not maintained in the carbon layers on either side of an occupied gallery: the two layers adjacent to an occupied gallery

---

[20] N. N. Greenwood and A. Earnshaw, "Chemistry of the Elements," p. 316, Pergamon, Oxford, 1984.

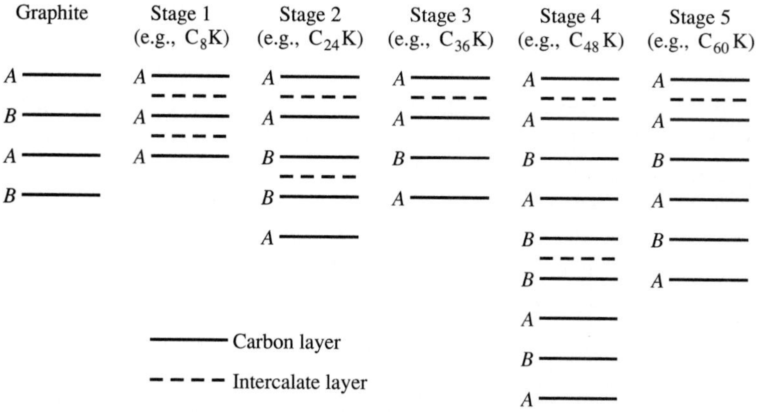

FIGURE 12.21
The stacking of carbon and intercalate layers in various stages of graphite intercalation compounds. The *A*s and *B*s indicate the carbon stacking patterns.

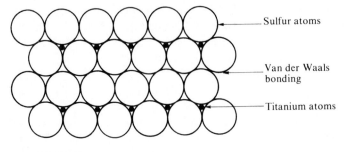

Sulfur atoms

Van der Waals bonding

Titanium atoms

FIGURE 12.22
Layer structure of TiS_2. Atoms intercalate between the layers held together by van der Waals bonding.

always have their carbon atoms superimposed. As expected, these layers are more widely separated (about 5.5 Å) than those in pure graphite (3.35 Å).

A much larger number of graphite intercalates are known in which the carbon layers are believed to be positively charged. The most thoroughly studied of these are graphite bisulfates, formed by treatment of graphite with a mixture of sulfuric and nitric acids.[19] The most bisulfate-rich composition achieved is $C_{24}^+HSO_4^- \cdot 2H_2SO_4$. Although many graphite intercalates are known in which it appears that neutral molecules have intercalated, recent work indicates that in most (if not all) of these cases the graphite is oxidized. Thus, although AsF_5 reacts with graphite to form a highly conducting compound of empirical formula C_8AsF_5, the actual intercalated species appear to be AsF_6^- and AsF_3.[21] That is, the graphite is oxidized according to the following equation:

$$3AsF_5 + 24C \quad \rightarrow \quad C_{24}^{2+}(AsF_6^-)_2 \cdot AsF_3$$

Intercalation will not occur with species which do not have oxidizing power themselves unless an oxidizing agent is present during the reaction. Thus graphite is not intercalated by GeF_4 alone but is intercalated by GeF_4-fluorine mixtures to yield, at the intercalation limit, $C_{12}GeF_{5-6}$, which is in equilibrium with gaseous fluorine.[22]

Titanium disulfide, in common with several other transition-metal disulfides, has a layered structure (Fig. 12.22). Because it is a metallic conductor and can be readily intercalated with metallic lithium, TiS_2 has received considerable attention as a possible electrode material in rechargeable batteries.[23] If TiS_2 is made one electrode in an electrolytic cell with a second electrode of lithium metal and a solution of a lithium salt in an aprotic polar solvent such as dioxolane as electrolyte, the reactions which occur spontaneously in the shorted cell are as follows.

[21] E. M. McCarron and N. Bartlett, *J. Chem. Soc. Chem. Commun.*, 404 (1980); N. Bartlett, B. McQuillan, and A. S. Robertson, *Mater. Res. Bull.*, **13**, 1259 (1978).

[22] E. M. McCarron, Y. J. Grannec, and N. Bartlett, *J. Chem. Soc. Chem. Commun.*, 890 (1980).

[23] M. S. Whittingham and R. R. Chianelli, *J. Chem. Educ.*, **57** 569 (1980); D. W. Murphy and P. A. Christian, *Science*, **205**, 651 (1979).

$$\text{Li} \longrightarrow \text{Li}^+ + e^- \qquad \text{at anode}$$

$$\text{Li}^+ + \frac{1}{x}\text{TiS}_2 + e^- \longrightarrow \frac{1}{x}\text{Li}_x\text{TiS}_2 \qquad \text{at cathode}$$

$$\text{Li} + \frac{1}{x}\text{TiS}_2 \longrightarrow \frac{1}{x}\text{Li}_x\text{TiS}_2 \qquad \text{overall reaction}$$

The process is reversible; by applying a reverse potential to the cell, the TiS_2 can be deintercalated. Because of the low equivalent weight of lithium, a battery of this type would have possible application in electric automobiles.

SUPERCONDUCTIVITY[24,25]

In 1911 H. Kamerlingh Onnes,[26] while studying the electrical resistance of mercury metal at very low temperatures, discovered that when the temperature was lowered below a critical temperature (T_c) of 4.2 K, the resistivity dropped to an immeasurably small value (see Fig. 12.23). Soon this phenomenon of supercon-

---

[24] A. W. Sleight, *Science*, **242**, 1519 (1988).

[25] A. B. Ellis, *J. Chem. Educ.*, **64**, 836 (1987).

[26] H. K. Onnes, *Akad. van Wetenschappen (Amsterdam)*, **14**, 113, 818 (1911).

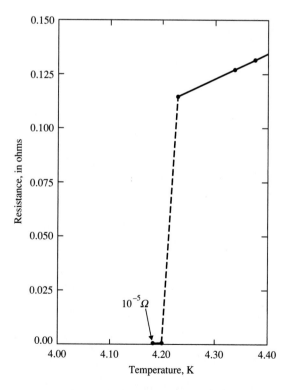

FIGURE 12.23

Electrical resistance of a sample of mercury as a function of absolute temperature. These data of Kamerlingh Onnes were the first evidence for the phenomenon of superconductivity.

TABLE 12.8
Superconductivity transition temperatures of the elements†

Element	T_c, K
Al	1.140
Ti	0.39
V	5.38
Zn	0.875
Ga	1.091
Zr	0.546
Nb	9.50
Mo	0.92
Tc	7.77
Ru	0.51
Cd	0.56
In	3.404
Sn(w)	3.722
La(ccp)	6.00
Lu	0.1
Hf	0.12
Ta	4.483
W	0.012
Re	1.4
Os	0.655
Ir	0.14
Hg(α)	4.153
Tl	2.39
Pb	7.193
Th	1.368
Pa	1.4

† C. Kittel, "Introduction to Solid State Physics," 5th ed., Wiley, New York, 1976.

ductivity was observed for about two dozen other metals, for which the critical temperatures are given in Table 12.8. The highest T_c value for a pure metal is that observed for niobium, 9.50 K. It has also been shown that certain alloys and metallic *compounds* exhibit superconductivity, often with transition temperatures considerably higher then those found for pure metals (see Table 12.9). It should be noted that even $(SN)_x$, a polymer containing no metal atoms, becomes superconducting below 0.26 K.

For many years a major obstacle to the use of superconductors was their low critical temperatures, generally below the boiling point of helium (4.3 K) or hydrogen (20.4 K). This obstacle was removed soon after Bednorz and Müller[27] found, in 1986, that the compound $Ba_xLa_{2-x}CuO_4$ has a superconductivity critical

[27]   J. G. Bednorz and K. A. Müller, *Z. Phys. B*, **64**, 189 (1986).

TABLE 12.9
Superconductivity transition
temperatures of some compounds

Compound	T_c, K
$(SN)_x$	0.26
TiO	1
TiC	1.15
WC	1.28
ϵ-TaN	1.8
Mo_2C	2.78
Ti_2Co	3.44
TiN	4.8
Mo_2N	5.0
VN	7.5
ZrN	9
TaC	9.7
La_3In	10.4
NbC	11.1
$LiTi_2O_4$	13
NbN	16.0
V_3Ga	16.5
V_3Si	17.1
Nb_3Al	17.5
δ-TaN	17.8
Nb_3Sn	18.05
Nb_3Ge	23.2
$(La, Ba)_2CuO_4$	35
$YBa_2Cu_3O_7$	95
$Tl_2Ba_2Ca_2Cu_3O_{10}$	120

temperature of about 35 K. Their discovery spurred intense synthetic activity, and in 1987 Chu and Wu[28] reported the preparation of $YBa_2Cu_3O_{7-x}$ ($x \leq 0.5$), with a 95 K critical temperature (18° above the boiling point of nitrogen). Although this material can be prepared by heating a mixture of Y_2O_3, CuO, and $BaCO_3$ to about 950°, a better product, consisting of finer and more densely sintered particles, can be prepared by heating an intimate mixture of precursors not containing the difficult-to-decompose $BaCO_3$. Thus a relatively high-density form of $YBa_2Cu_3O_{7-x}$ can be prepared by heating the hydrolysis products of a mixture of $Y(OCHMe_2)_3$, $Ba(OCHMe_2)_2$, and $CuNBu_2$.[29]

The structure of $YBa_2Cu_3O_{7-x}$, shown in Fig. 12.24, is similar to the perovskite structure shown in Fig. 11.19. The perovskite structure (typified by $CaTiO_3$) has a cubic unit cell with a calcium atom at the center, a titanium atom

[28] M. K. Wu, Jr., J. R. Ashburn, C. J. Torng, P. H. Hor, R. L. Meng, L. Gao, Z. J. Huang, Y. Q. Wang, and C. W. Chu, *Phys. Rev. Lett.*, **58**, 908 (1987).

[29] H. S. Horowitz et al., *Science*, **243**, 66 (1989).

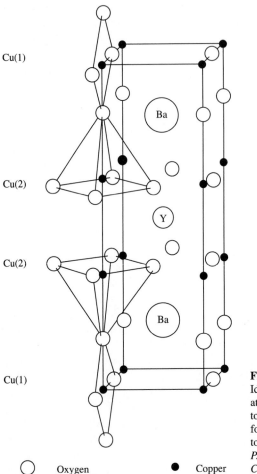

Cu(1)

Cu(2)

Cu(2)

Cu(1)

○ Oxygen   ● Copper

FIGURE 12.24
Idealized structure of $YBa_2Cu_3O_7$. Oxygen atoms can be randomly removed from the top and bottom edges of the unit cells to form $YBa_2Cu_3O_{7-x}$, with x ranging from 0 to 1. [*Reproduced with permission from P. P. Edwards, M. R. Harrison, and R. Jones, Chem. Brit., 23, 962 (1987).*]

at each corner, and an oxygen atom at the middle of each edge. The lattice may be looked upon as a close-packed array of oxygen and calcium atoms, with titanium atoms in one-quarter of the octahedral holes. The $YBa_2Cu_3O_{7-x}$ unit cell is essentially a group of three adjacent perovskite unit cells, with an yttrium and two barium atoms replacing the calcium atoms and copper atoms replacing the titanium atoms. If all the oxygen positions were occupied, the formula would be $YBa_2Cu_3O_9$, and the average oxidation state of the copper would be 11/3 or 3.67. Such a composition would be extremely unstable toward loss of oxygen because of the strong oxidizing power of +4 copper, and thus one can rationalize the oxygen vacancies in the lattice. (Notice that the top and bottom planes of the cell contain only two oxygen atoms each, and that there are no oxygen atoms whatever in the horizontal plane passing through the yttrium atom.)

There are two structurally distinct sites for the copper atoms of $YBa_2Cu_3O_{7-x}$: Cu(1) has a square planar coordination of oxide ions, whereas

Cu(2) is located near the base of a square pyramid of oxide ions. The five-coordinate Cu(2) is displaced about 0.3 Å from the plane of the oxide ions. This distortion, giving a dimpled CuO_2 plane, may be of importance to the superconductivity properties of the material.[30]

There are difficult problems associated with the practical application of high-temperature superconductors. When fashioned into wires for magnets or electrical transmission, these superconductors usually have a low critical current density (the maximum current a superconductor can carry before it loses its superconductivity). However, in thin films, critical current densities of several million amperes per square centimeter—enough for microelectronic applications—have been achieved, and early applications of these superconductors will probably involve such systems.

PROBLEMS

*12.1 The compression of a transition metal by application of pressure improves the overlap of the valence d orbitals and increases the bonding ability of the d orbitals relative to that of the s and p orbitals. Predict the effect of pressure on the relative stabilities of the bcc and hcp structures of zirconium and the bcc and ccp phases of iron.

12.2 Why would you expect an increase in temperature to stabilize a bcc structure relative to an hcp structure instead of vice versa?

*12.3 How would you expect the electrical conductivity of a solid-solution alloy (e.g., an alloy of Cu and Au) to vary with composition?

12.4 What are the structure types (bcc, hcp, ccp, or diamond-like) of the following alloys: CuBe, Ag_5Cd, AgMg, Ag_5In_3, Cu_3Ge, FeAl, and InSb?

12.5 Give the empirical formulas (or approximate composition ranges) that might be expected for five solid phases in the Ag-Al phase diagram. For each phase, predict the crystal structure type.

12.6 Arrange the following materials in order of hardness: Cs, Na, C(diamond), W, Al, Fe, Mg.

12.7 Can a tantalum crucible be used in the melting of (a) Sr, (b) Re, (c) Pt?

*12.8 Why are none of the metals of the second and third transition series ferromagnetic?

*12.9 The densities of the lanthanide metals change gradually from 6.19 g cm^{-3} for La to 9.84 for Lu, except for Eu and Yb, which have low densities of 5.24 and 6.98, respectively. Explain.

12.10 What species probably occupy the galleries in the graphite intercalate $C_{12}GeF_{5-6}$?

*12.11 Using the information that in C_8K the C—C bond distance is 1.42 Å and that the metal atoms form regular triangular nets between the carbon layers, calculate the distance between K atoms in each plane of K atoms in C_8K.

---

[30] For a discussion of the factors affecting the critical temperature of cuprate superconductors, see M.-H. Whangbo and C. C. Torardi, *Science*, **249**, 1143 (1990).

12.12 Why does the Pt—Pt distance decrease on oxidizing $K_2Pt(CN)_4 \cdot 3H_2O$ to $K_2Pt(CN)_4Br_{0.3} \cdot 3H_2O$?

*12.13 What fraction of the energy band formed by the overlap of the $p\pi$ orbitals of the S and N atoms of $(SN)_x$ is filled with electrons?

12.14 Without referring to any data, list the following metals in order of increasing melting point K, Cr, Co, W, Li, Zn.

12.15 Without referring to any data, indicate the structure types (bcc, hcp, ccp, or diamond-like) of the following metals: Ge, Rb, Cu, Cr, Mg, Al, Cd, Pd.

12.16 Although iron metal is ferromagnetic, a typical iron nail does not act as a magnet (e.g., it does not attract iron filings). Explain.

CHAPTER
13

SEMICONDUCTORS[1]

There is no question that semiconductors are important solid-state materials. Almost everyone has heard of them and knows something about their technical application. In this chapter we will briefly describe the electrical properties of semiconductors (here your understanding of the solid state and band theory will be valuable), and we will discuss the chemistry involved in the manufacture of semiconductors and in various semiconductor devices.

A semiconductor is a crystal with a narrow energy gap between a filled valence band and a conduction band. If the crystal were cooled to absolute zero, the conduction band would be empty and the material would be a perfect nonconductor. However, at ordinary temperatures some electrons are thermally excited from the valence band to the conduction band, enough to give the material an electrical conductivity between that of a metal and that of an insulator. (Typical room-temperature *resistivities* of metals, semiconductors, and insulators are 5, 5×10^7, and $10^{23} \mu\Omega$ cm, respectively.) The conductivity of a semiconductor is proportional to the number of electrons in the conduction band, which in turn is proportional to the Boltzmann factor $e^{-E_g/RT}$, where E_g is the energy gap. Consequently the conductivity of a semiconductor increases exponentially with increasing temperature.

The semiconducting material of greatest commercial importance is elemental silicon. If one were able to prepare a perfectly pure crystal of silicon, the small

---

[1] C. A. Wert and R. M. Thomson, "Physics of Solids," McGraw-Hill, New York, 1964; L. V. Azaroff, "Introduction to Solids," McGraw-Hill, New York, 1960; C. Kittel, "Introduction to Solid State Physics," 5th ed., Wiley, New York, 1976.

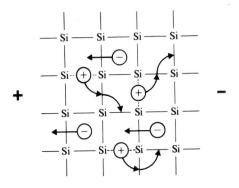

FIGURE 13.1
Schematic drawing of silicon, showing the negative-electron and positive-hole current carriers which give rise to a small intrinsic conductivity in the pure element.

number of electrons in the conduction band at any given temperature would equal the number of positive "holes" in the valence band. Both conduction electrons and holes contribute to electrical conductivity. The electrons travel through the interstices of the lattice, and the holes jump from one bond to another, as crudely illustrated in Fig. 13.1. The countercurrent motion of conduction electrons and valence holes in an applied field constitutes the "intrinsic" semiconductivity of the crystal. Generally even highly purified silicon contains enough impurity atoms to increase the semiconductivity far above the hypothetical intrinsic semiconductivity. For example, let us suppose that a minute fraction of the silicon atoms are randomly replaced by atoms containing more than four valence electrons each, such as arsenic atoms. Because the crystal must remain essentially electrically neutral, each arsenic atom contributes an electron which can enter the conduction band. The principal conducting species in the sample of silicon would be the negative electrons, and such a material is called an *n*-type (for "negative") semiconductor. A schematic illustration of this type of semiconductor and the corresponding density-of-states diagram are shown in Fig. 13.2. At absolute zero, the extra electrons introduced by the arsenic atoms would be localized around the arsenic atoms (which have $+1$ formal charges), but at ordinary temperatures, many

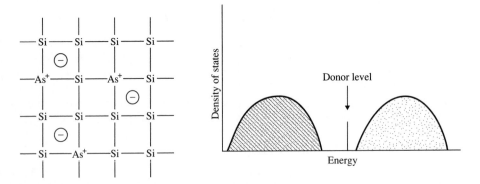

FIGURE 13.2
Schematic drawing of *n*-type silicon, and the corresponding density-of-states diagram.

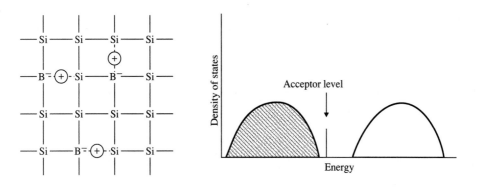

FIGURE 13.3
Schematic drawing of *p*-type silicon, and the corresponding density-of-states diagram.

of them would be excited to the conduction band. The energy gap between the donor level and the conduction band is the energy required to break an interstitial electron completely away from an arsenic atom.

Now suppose that a very small fraction of the atoms in a crystal of silicon were replaced by atoms containing *fewer* than four valence electrons each, such as boron atoms. The principal conducting species in the material would be positive valence holes, as shown in Fig. 13.3, and such a material is called a *p*-type (for "positive") semiconductor. At absolute zero, the positive holes introduced by the impurity boron atoms would be localized around the boron atoms (which would have negative formal charges), whereas at finite temperatures many of these holes would be excited so that they would be free to hop among the silicon-silicon bonds and thus to carry electric current. In the density-of-states diagram of Fig. 13.3, the energy gap between the valence band and the acceptor level is the minimum energy required to force a normal valence electron to neutralize a hole next to a boron atom, so that the boron atom is bonded by single bonds to four silicon atoms. In effect, holes are excited to the left into the valence band.

PREPARATION OF SEMICONDUCTORS

Silicon which is to be used in semiconductors must first be highly purified. Crude silicon ($\sim$ 98 percent pure) is generally produced by heating quartz rock and carbon above 1800°C in an electric-arc furnace.[2]

$$SiO_2 + C \xrightarrow{\Delta} CO_2 + Si$$

This silicon has about 20 million times more impurity than can be tolerated. It can be purified by converting it to trichlorosilane with hydrogen chloride, carefully

---

[2] *Chem. Eng. News,* 46 (Nov. 30, 1970).

distilling the trichlorosilane, and then decomposing it to give purified silicon. The chemical reactions involved are

$$3HCl + Si \;\underset{1100°C}{\overset{350°C}{\rightleftharpoons}}\; SiHCl_3 + H_2$$

If necessary, this silicon can then be further purified by zone refining.[3] In this process a rod of the element is melted near one end, and, by moving the furnace, the short melted zone is moved slowly to the other end of the rod. Because impurities are more soluble in the melt than in the solid, they concentrate in the melt and are carried to one end of the rod. After the process is repeated several times, the impure end is removed.

Sometimes semiconductors can be purified by gas-phase thermal transport processes.[4] For example, germanium can be purified by placing it in one end of an evacuated tube with a small amount of iodine and then heating the tube so that the end containing the impure germanium is at 500°C and the other end is at 350°C. The iodine reacts with the germanium to form vapors of GeI_4 and GeI_2. Soon a steady state is reached in which the following reaction takes place continuously, as written, in the high-temperature zone, and the reverse reaction takes place in the initially empty, low-temperature zone.

$$Ge(s) + GeI_4(g) \;\underset{350°C}{\overset{500°C}{\rightleftharpoons}}\; 2GeI_2(g)$$

Thus germanium is transported from one end of the tube to the other, leaving impurities behind.

Purified silicon and germanium crystals can be converted into n-type or p-type semiconductors by high-temperature diffusion of the appropriate impurity elements ("dopants") into the crystals. Arsenic and phosphorus are often used for n-type semiconductors, and boron and aluminum are often used for p-type semiconductors.

Elements from main groups III and V form 1:1 semiconducting compounds of the type GaAs that are isoelectronic with and similar in structure to silicon and germanium. These compounds are often prepared as thin films on other semiconductor substrates by deposition from the vapor phase, employing high-temperature reactions such as[5]

$$Ga(CH_3)_3 + AsH_3 \;\overset{580°}{\longrightarrow}\; GaAs + 3CH_4$$

The main advantage of gallium arsenide over silicon is the greater ease with which electrons move through it. Gallium arsenide circuits are faster at a given power

[3]   W. G. Pfann, "Zone Melting," Wiley, New York, 1966.

[4]   H. Schäfer, "Chemical Transport Reactions," Academic, New York, 1963; also in "Preparative Methods in Solid State Chemistry," P. Hagenmuller, ed., p. 251, Academic, New York, 1972.

[5]   E. Yablonovitch, *Science,* **246,** 347 (1989).

level than are silicon circuits. This high electron mobility is valuable in high-speed computer circuits and is important for high-frequency, low-noise operation, as in the detection of television and microwave signals.[6]

BAND GAPS

The band gap of a semiconductor is the energy difference between the top of the valence band and the bottom of the conduction band. Values of the band gap in various substances are listed in Table 13.1.

The binding energies associated with impurities are much smaller than the band gaps. For example, the energy required to excite an electron away from a phosphorus atom impurity in silicon to the conduction band is about 0.044 eV. Similarly, the energy required to excite a hole from an aluminum atom impurity in silicon to the valence band is about 0.057 eV. These excitation energies may be compared with the band gap, 1.1 eV.

A major advantage of gallium arsenide over silicon is the ease with which its band gap can be changed. The gap is larger in gallium arsenide than in silicon, but it can be narrowed or widened by appropriate substitution with other elements. If aluminum is substituted for gallium, a much wider band gap is obtained, and partial substitutions produce gaps proportional to the fraction of aluminum. Other valuable materials are formed by substituting some indium for gallium, some phosphorus for arsenic, or both at the same time.

---

[6]   M. H. Brodsky, *Sci. Am.*, **262**(2), 68 (1990).

TABLE 13.1
Band gaps of some semiconductors

Compound	E_g, at 0 K, eV	Compound	E_g, at 0 K, eV
α-Sn	0.0	CdTe	1.6
InSb	0.2	Se	1.8†
PbTe	0.2	Cu_2O	2.2
Te	0.3	InN	2.4
PbS	0.3	CdS	2.6
InAs	0.4	ZnSe	2.8
ZnSb	0.6	GaP	2.88
Ge	0.7	ZnO	3.4
GaSb	0.8	$SrTiO_3$	3.4
Si	1.1	ZnS	3.9
InP	1.3	AlN	4.6
GaAs	1.5	Diamond	5.4
		BP	6.0

† 300 K.

DEFECT SEMICONDUCTORS[7]

Many compounds are semiconductors because they are nonstoichiometric. For example, when compounds such as NaCl, KCl, LiH, and δ-TiO are subjected to high-energy radiation or are heated with an excess of their constituent metals, the compounds become deficient in the electronegative elements and their compositions may be represented by the general formula MY_{1-x}, where x is a small fraction. The crystal lattice of such a compound has anion vacancies, each of which is usually occupied by an electron. Such electron-occupied holes are called "F centers." The electron of an F center can be thermally excited to a conduction band, thus giving rise to n-type semiconduction.

Another class of n-type semiconductor are those compounds which contain excess interstitial metal atoms and whose compositions correspond to the formula $M_{1+x}Y$. The compounds ZnO, CdO, Cr_2O_3, and Fe_2O_3 show this type of structural

---

[7] H. J. Eméleus and J. S. Anderson, "Modern Aspects of Inorganic Chemistry." 3d ed., Van Nostrand, New York, 1960.

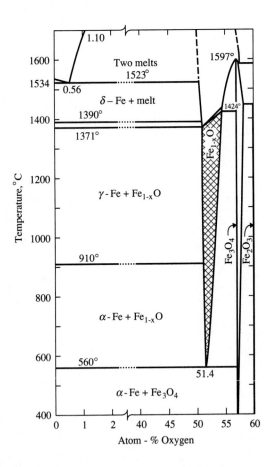

FIGURE 13.4
The iron-oxygen phase diagram. Notice that $Fe_{1-x}O$ is unstable with respect to disproportionation below 560°. However, room-temperature metastable $Fe_{1-x}O$ can be prepared by rapid quenching of the high-temperature phase. (*Adapted from Gmelin-Durrer, "Metallurgie des Eisens," Band 1b, p. 59, Verlag Chemie, Weinheim, 1964.*)

defect. The interstitial metal atoms are readily ionized, allowing their valence electrons to enter a conduction band and leaving interstitial metal ions. When a defect oxide of this type is heated in oxygen, its room-temperature conductivity decreases because of the loss of some of the interstitial metal atoms by oxidation.

Compounds which are deficient in metal ions, of general formula $M_{1-x}Y$, are p-type semiconductors. This type of semiconductor can be found in Cu_2O, FeO, NiO, δ-TiO, CuI, and FeS. Electrical neutrality is maintained in these compounds, in spite of cation vacancies, by the presence of metal ions of higher oxidation state. Thus the composition $Fe_{0.95}O$ is more informatively represented by the formula $Fe^{II}_{0.85}Fe^{III}_{0.10}O$. (See Fig. 13.4 for the iron-oxygen phase diagram.) Electrical conductivity is achieved by the hopping of valence electrons from lower-oxidation-state metal ions to higher-oxidation-state metal ions. However, because the Fe(III) ion concentration is much lower than the Fe(II) ion concentration, the effective migrating unit is a positive hole in a metal ion and the conductivity is of the p type. The energy gap for this type of semiconductor corresponds to the energy required to move a positive hole away from the vicinity of a cation vacancy, where it is electrostatically held. When a defect oxide of this type is heated in oxygen, its room-temperature conductivity increases because of oxidation of some of the metal ions and the consequent increase in positive-hole concentration.

CONTROLLED-VALENCE SEMICONDUCTORS[8]

Defect semiconductors are generally difficult to obtain with large deviations from stoichiometry. Hence only limited variations in electrical properties are possible. In addition, compositions are difficult to reproduce exactly. These difficulties can be overcome by the use of "controlled-valence" semiconductors, first prepared by Verwey and his coworkers.

For example, consider the material formed by heating an intimate mixture of NiO and a small amount of Li_2O in air at 1200°C. Oxygen is absorbed, and a single phase of composition $Li_xNi_{1-x}O$ is formed:

$$\frac{x}{2}Li_2O + (1 - x)NiO + \frac{x}{4}O_2 \quad \rightarrow \quad Li_xNi_{1-x}O$$

Inasmuch as Li^+ and Ni^{2+} ions have similar radii, Li^+ ions can substitute for Ni^{2+} ions in the lattice. One Ni^{3+} ion is formed for every Li^+ ion introduced to preserve electrical neutrality. The resulting compound can be represented more precisely by the expanded formula $Li_xNi^{III}_xNi^{II}_{1-2x}O$. The positive holes in the Ni^{2+} ions (i.e., the Ni^{3+} ions) can move from one nickel ion to another and thus give rise to p-type semiconduction. By simply controlling the ratio of lithium to nickel, the electrical conductivity can be varied at will. The conductivity of pure

---

[8] W. D. Johnston, *J. Chem. Educ.*, **36**, 605 (1959); P. E. Snyder, *Chem. Eng. News.* 102 (Mar. 13, 1961).

NiO is about $10^{-10}\ \Omega^{-1}\ cm^{-1}$, whereas the compound in which 10 percent of the nickel atoms have been replaced by lithium atoms has a conductivity of about $1\ \Omega^{-1}\ cm^{-1}$.

Mixed-valence systems can also be obtained with compounds having the perovskite structure such as barium titanate, $BaTiO_3$. (This compound consists of a ccp lattice of Ba^{2+} and O^{2-} ions in which the octahedral holes surrounded exclusively by O^{2-} ions are occupied by Ti^{4+} ions.) If a small number of the barium ions are replaced by $+3$ ions such as lanthanum ions, a corresponding number of Ti^{4+} ions must be reduced to Ti^{3+}, forming the system $La_xBa_{1-x}Ti^{III}{}_xTi^{IV}{}_{1-x}O_3$. This material exhibits n-type semiconductivity.

Compounds of the type $Li_xMn_{1-x}O$ are semiconducting, but they cannot be made by heating a mixture of Li_2O and MnO in air, because MnO is easily oxidized in air at high temperatures. In this case, the reaction is carried out in a sealed container, and the oxygen is introduced in the form of lithium peroxide:

$$\frac{x}{2}Li_2O_2 + (1-x)MnO \quad \rightarrow \quad Li_xMn_{1-x}O$$

APPLICATIONS

Semiconductors have innumerable practical applications which contribute to our high standard of living. We shall describe just four applications: the photovoltaic cell, the photoelectrolytic cell, the rectifier, and the insulated-gate field-effect transistor. Typically these devices involve a semiconductor crystal which is p-type in one part and n-type in another. The boundary region is called a p-n junction and has interesting electrical properties. Such a crystal can be made, for example, by allowing small amounts of boron to diffuse into one side of a slice of n-type silicon at high temperatures. If properly doped, the crystal will then be p-type on one side and n-type on the other, with a region within the slice where the two types of silicon meet: the p-n junction.

At the p-n junction, a few electrons are transferred from the n-type semiconductor to the p-type semiconductor, causing a partial depletion of charge carriers in both phases near the junction. This charge transfer equalizes the Fermi levels and causes band bending (i.e., an electric-field gradient) near the junction, as shown in Fig. 13.5.

The Photovoltaic Cell[9]

Let us suppose that a p-n junction is irradiated with light. If the photon energy equals or exceeds the band gap, electron-hole pairs will be formed in the irradiated region, as indicated in Fig. 13.5. Because of the band bending, the electrons will migrate toward the bulk of the n-type semiconductor and the holes will migrate

[9] Wert and Thomson, ref. 1.

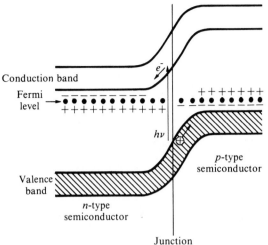

FIGURE 13.5
Energy level diagram for a *p-n* junction, showing band bending and the creation of an electron-hole pair upon absorption of a photon with energy equal to or greater than the band gap.

toward the bulk of the *p*-type semiconductor. Thus the *n*-type region will achieve a negative potential relative to the *p*-type region. If the *n*-type and *p*-type regions are electrically connected to an external circuit, a current will flow as long as the junction is irradiated. Obviously such a crystal can be used as a kind of battery whose energy is derived from light. One of the aims of solid-state chemists is to prepare efficient cells of this type economically enough to permit their use in harnessing solar energy.

The reverse of a photovoltaic cell is a light-emitting diode (LED). Voltage applied across a *p-n* junction injects excess populations of holes and electrons into the boundary. When an electron and a hole annihilate one another, the band-gap energy is released as a photon. Many millions of aluminum gallium arsenide LEDs (with the familiar red glow) are made every year.

The Photoelectrolytic Cell

When an *n*-type semiconductor is immersed in a solution containing a suitable redox couple A^+/A, a few electrons near the surface of the semiconductor react with the oxidizing agent of the couple, A^+, and thus deplete the semiconductor surface region of charge carriers and cause a bending of the bands as shown in Fig. 13.6. Absorption at the interface of a photon with energy greater than the band gap promotes an electron to the conduction band and creates a hole in the valence band. Because of the band bending, the electron and hole spontaneously separate; the electron e^- moves toward the bulk of the semiconductor, and the hole (h^+) remains at the interface. If the reduction potential of the solution is low enough, the electron-transfer reaction

$$A + h^+ \quad \rightarrow \quad A^+$$

occurs at the interface. The excited electron can move through a wire connected to the semiconductor to a nonphotoactive electrode where an oxidized species,

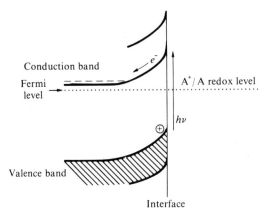

FIGURE 13.6

Energy level diagram for an *n*-type semiconductor in contact with a solution containing the A^+/A redox couple. Photon absorption creates an electron-hole pair.

B^+, can be reduced:

$$B^+ + e^- \quad \rightarrow \quad B$$

This combination of *n*-type semiconductor and inert electrode immersed in the solution comprises a photoelectrochemical cell[10] in which light promotes the reaction

$$A + B^+ \quad \rightarrow \quad A^+ + B$$

If the A/A^+ and B^+/B couples are H_2O/O_2 and H^+/H_2 couples,

$$H_2O \quad \rightleftharpoons \quad \tfrac{1}{2}O_2 + 2H^+ + 2e^-$$

$$2H^+ + 2e^- \quad \rightleftharpoons \quad H_2$$

the net cell reaction will be the photoelectrolysis of water to H_2 and O_2:

$$H_2O \quad \rightarrow \quad H_2 + \tfrac{1}{2}O_2$$

In fact, this reaction has been carried out using photoelectrolytic cells involving the semiconductor $SrTiO_3$. Because the band gap of $SrTiO_3$ is about 3.4 V, it is necessary to use ultraviolet light with such cells. The ultraviolet light is converted fairly efficiently (about 25 percent) to storable energy in the form of hydrogen and oxygen. Unfortunately, because sunlight consists mainly of visible light, the device has very low efficiency (about 1 percent) for converting solar energy into hydrogen and oxygen. Practical solar photoelectrolysis requires the development of semiconductors that can use visible light efficiently and that have long-term stability under photoelectrolysis conditions. Some fluorine-doped oxide semiconduc-

---

[10] A. J. Bard, *Science,* **207,** 139 (1980); M. S. Wrighton, *Ace, Chem. Res.,* **12,** 303 (1979); M. S. Wrighton, *Chem. Eng. News,* 29 (Sept. 3, 1979). Also see "Photoeffects at Semiconductor-Electrolyte Interfaces," A.C.S. Symposium Series, No. 146, American Chemical Society, Washington, D.C., 1981.

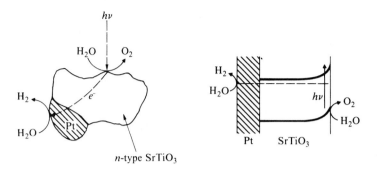

FIGURE 13.7
Representation of SrTiO$_3$-Pt particle and the corresponding energy level diagram. When an aqueous suspension of such particles is irradiated with ultraviolet light, water is electrolyzed into hydrogen and oxygen.

tors show promise for such applications. Thus $TiO_{2-x}F_x$ (where $x \approx 0.002$)[11] and $WO_{3-x}F_x$ (where $x \approx 0.01$)[12] are n-type semiconductors which have photoelectrolytic properties superior to those of the corresponding defect semiconductors, TiO_{2-x} and WO_{3-x}.

The concepts learned from photoelectrolytic cells with semiconductor electrodes have been used to design relatively economical systems in which aqueous suspensions of semiconductor particles are used for similar purposes. In a typical system, n-type SrTiO$_3$ powder with platinum dispersed on the surface of the particles has been used.[10] Each particle is essentially a short-circuited photoelectrolytic cell in which the semiconductor electrode and counter-electrode have been brought into contact. (See Fig. 13.7.) Irradiation of such particle suspensions still involves the electron/hole formation and surface oxidation and reduction reactions found in cells, but without external current flow. The Pt-SrTiO$_3$ system, when irradiated with ultraviolet light, has been shown to effect the decomposition of water into hydrogen and oxygen, the hydrogen forming at the platinum surface and the oxygen forming at the SrTiO$_3$ surface. Unfortunately, just as in the corresponding photoelectrolytic cell process, the system has a low solar efficiency.

The Rectifier

Let us suppose that a voltage is applied across a p-n junction so that the p-type region is positive relative to the n-type region. In the p-type region, holes will migrate toward the junction, and in the n-type region electrons will migrate toward the junction. At the junction, the holes and electrons will be essentially

---

[11] S. N. Subbarao, Y. H. Yun, R. Kershaw, K. Dwight, and A. Wold, *Inorg. Chem.*, **18,** 488 (1979).

[12] C. E. Derrington, W. S. Godek, C. A. Castro, and A. Wold, *Inorg. Chem.*, **17,** 977 (1978).

annihilated. That is, the migrating electrons of the *n*-type region will drop down into the valence-shell vacancies of the *p*-type region. The migration of holes and electrons can continue indefinitely; i.e., a current will flow as long as the voltage is applied. Now let us suppose that the voltage is reversed so that the *p*-type region is negative relative to the *n*-type region. In the *p*-type region, holes will migrate away from the junction, and in the *n*-type region, electrons will migrate away from the junction. Thus, in the region of the junction, the current carriers will be depleted and no current can flow. Actually, current can flow under these conditions if the voltage is high enough. It is necessary to supply enough energy to excite an electron from the valence band of the *p*-type region all the way up to the conduction band so that the electron can cross the barrier into the *n*-type region. Obviously a *p-n* junction can serve as a rectifier for converting alternating current into direct current, as long as the voltage is not too high.

The Insulated-Gate Field-Effect Transistor[13]

This type of transistor, also called a "metal oxide semiconductor" (MOS) transistor, can be understood by reference to Fig. 13.8. When the gate voltage (V_g) is zero, no current flows from the source to the drain because these *n*-type regions are insulated from each other by *p*-type silicon. The excess electrons are indicated by dots, and the holes are indicated by circles. When a positive voltage is applied to the gate, holes are driven away from the interface of the silicon dioxide and the silicon under the gate. With a sufficiently high gate voltage, a very thin region of the silicon under the gate is changed from *p* type to *n* type. This inverted region is called an *n* channel and provides a conduction path for electrons between the source and the drain. Thus the gate voltage controls the magnitude of the

---

[13] W. C. Hittinger, *Sci. Am.*, **229**, 48 (August 1973).

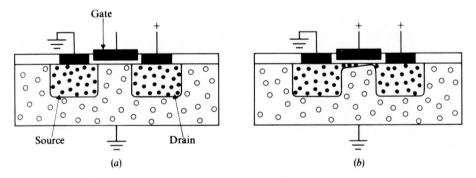

(a)

(b)

FIGURE 13.8
Cross section of an enhancement-type MOS transistor. The black regions correspond to Al metal, the white regions to SiO_2, the regions with black dots to *n*-type Si, and the regions with open circles to *p*-type Si. (a) The source and drain are insulated from each other. (b) Sufficiently high voltage has been applied to the gate to create an *n*-type channel between the source and the drain.

current. Devices of this type can be made microscopically small and allow the miniaturization of integrated circuits. MOS circuits are widely used in computers and pocket calculators.

AMORPHOUS SEMICONDUCTORS[14]

When crystalline silicon is doped with phosphorus atoms, the geometric constraints of the crystal force the phosphorus atoms to occupy silicon sites so that they are obliged to bond tetrahedrally. Thus the phosphorus atom acquires a positive formal charge, and the extra valence electron is held weakly nearby at an energy level slightly below that of the bottom of the conduction band. However, when an amorphous or glassy sample of silicon is doped with phosphorus, the situation is different. There are no geometric constraints, and the amorphous silicon has a high concentration of "dangling bonds" which can be used to bond to the entering phosphorus atoms. Thus by breaking only one Si—Si bond and using one dangling bond, a phosphorus atom can bond trigonally, as shown below.

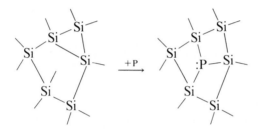

Thus phosphorus doping of ordinary amorphous silicon does not alter the conductivity. However, if the doping is carried out in the presence of hydrogen (either by exposure to hydrogen gas during the doping or by using amorphous silicon formed by the decomposition of silane), it is possible to form an *n*-type semiconductor. It has been suggested that the hydrogen ties up the dangling bonds as SiH groups and thereby forces the impurity atoms to bond tetrahedrally. It appears that the economic breakthrough necessary to achieve the large-scale generation of electric power from solar energy may come about through the use of relatively cheap amorphous silicon rather than through the development of cheaper methods of making crystalline silicon devices.

Chalcogenide glasses, such as compounds of selenium or tellurium and arsenic, are photoconductive and thus find application in xerography and television. Doping is not necessary to achieve photoconductivity in these materials; the pure materials contain positively and negatively charged sites which can trap electrons and holes, respectively. The positively charged sites are believed to be trivalent chalcogen atoms at branching points in the chalcogen chains. The negatively

---

charged sites are believed to be monovalent chalcogen atoms which terminate chains and side chains.[15] These two sites are illustrated below.

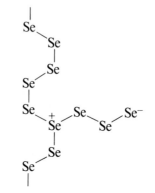

Optical excitation promotes an electron to the conduction band and creates a hole in the valence band. The electron is trapped near a positive site at an energy level slightly below the bottom of the conduction band, and the hole is trapped near a negative site at an energy level slightly above the top of the valence band. This situation constitutes the semipermanent photoconductive state.

PROBLEMS

13.1 Estimate the longest wavelength of light which can generate a current in a silicon photovoltaic cell.

*13.2 Predict the effect on the electrical conductivity of adding iron to an $Li_xNi_{1-x}O$ semiconductor (to form $Li_xFe_yNi_{1-x-y}O$). Explain.

13.3 Indicate the type of semiconduction (n or p) shown by the following materials: (a) Zn-doped GaAs, (b) $In_{1+x}As_{1-x}$, where $x \ll 1$, (c) x-ray-irradiated KCl, (d) $Li_{0.05}Cu_{0.95}O$, (e) $WO_{2.999}$.

13.4 Categorize the following materials, on the basis of their electrical conductivities, as metallic, semiconducting or insulating: Bi, AlF_3, CdS, CuZn, $NaHCO_3$, InP, CeS, Os, $Na_{0.5}WO_3$, P_4.

13.5 Compare the electrical conductivities of germanium crystals containing (a) 0.1 ppm of As, (b) 0.1 ppm of As and 0.05 ppm of Al, (c) 0.1 ppm of As and 0.1 ppm of Al, (d) 1 ppm of Al.

*13.6 Using a simple thermodynamic argument, explain why, in the thermal transport of germanium by iodine, the germanium is transported from the higher temperature zone to the lower temperature zone, rather than vice versa.

*13.7 Laser emission can occur at a p-n junction if a dc voltage is applied across the junction. In the case of gallium arsenide, the light is emitted in the near infrared

[15] M. Kastner, D. Adler, and H. Frizsche, *Phys. Rev. Lett.*, **37**, 1504 (1976).

at about 8400 Å, with the exact wavelength tunable by variation of temperature or pressure. State and explain the direction of the wavelength shift with increasing pressure. Would you expect the GaP laser to emit at a shorter or longer wavelength than the GaAs laser? Explain.

13.8 What magnitude band gap would be suitable for a photoelectrolytic cell designed to operate efficiently in sunlight at a wavelength of $\sim$ 5500 Å?

*13.9 Without reference to any data, arrange the following materials in order of increasing band gap; CdS, CdSe, InSb, GaSb, diamond, α-Sn.

*13.10 When an electrode of p-type indium phosphide in an acidic aqueous solution of V^{3+} is irradiated with sunlight, the V^{3+} is reduced to V^{2+} with 11.5 percent solar conversion efficiency. Sketch the valence and conduction bands and acceptor level in the region of the electrode surface and indicate the mechanism of the photoelectrolytic process.

CHAPTER
14

INTRODUCTION TO TRANSITION-METAL CHEMISTRY

The remaining chapters in this book cover the chemistry of transition-metal compounds. By transition metal we mean any one of the elements that has a partly filled d shell in one of its commonly occurring oxidation states. The "inner" transition metals, which are characterized by partly filled f shells, will only occasionally be discussed. The transition-metal compounds have certain characteristic properties which make it convenient, if not necessary, to study them separately from the compounds of the main-group elements. We have already pointed out, in Chap. 12, that the transition metals are hard, strong, high-melting metals which form alloys with one another and with other metals and which have high electrical and thermal conductivities. Other characteristic properties which we shall later discuss are (1) each transition metal generally has several readily accessible oxidation states; (2) the compounds are often colored; (3) many of the compounds are paramagnetic; (4) the metal ions are often obtained in the form of complexes with a wide variety of ligands; and (5) the compounds often have catalytic properties.

OXIDATION STATES

The multiplicity of oxidation states shown by most transition metals is due to the ability of these elements to lose a variable number of d electrons. For example, in aqueous solution, vanadium exists in the $+2$, $+3$, $+4$, and $+5$ oxidation states, corresponding to the electron configurations d^3, d^2, d^1, and d^0, respectively. Consequently the aqueous chemistry of vanadium consists, to a large extent, of the

356

oxidation-reduction reactions of vanadium. These types of reactions may be understood and predicted with the aid of the following reduction-potential diagram.

$$VO_2^+ \xrightarrow{1.0} VO^{2+} \xrightarrow{0.36} V^{3+} \xrightarrow{-0.25} V^{2+} \xrightarrow{-1.2} V$$

Yellow   Blue   Green   Violet

The fact that the vanadium reduction potentials gradually become more negative from left to right in this diagram indicates that the species V^{2+}, V^{3+}, and VO^{2+} are stable with respect to disproportionation. A striking demonstration can be carried out by the addition of excess zinc pellets to a solution of $(VO_2)_2SO_4$ in dilute sulfuric acid. The originally yellow solution successively turns green, blue, green, and, finally, violet as the vanadium is reduced stepwise from VO_2^+ to V^{2+}. The Fe^{3+}, Fe^{2+} potential is 0.77 V; consequently the addition of Fe^{2+} to an acidic solution of VO_2^+ reduces the vanadium to VO^{2+}, and no further. The Sn^{2+}, Sn potential is -0.14 V; thus tin reduces VO_2^+ cleanly to V^{3+}. The Zn^{2+}, Zn potential is -0.76 V, and, as we have indicated, zinc reduces VO_2^+ to V^{2+}.

The reduction-potential diagrams given in Chap. 5 can be likewise used to interpret the oxidation-reduction reactions of other transition metals.

COLORS

The colors of transition-metal compounds are usually attributable to electronic transitions involving d orbitals. These transitions are of two main types. In one type, the "d-d transition," an electron is transferred between orbitals that have predominantly metal d-orbital character. In this type of transition, there is little shift of electron density from any atom to another. In the other type of transition, the "charge-transfer transition," an electron is transferred from a molecular orbital centered mainly on the ligands (or a ligand) to one centered mainly on the metal atom, or vice versa. In this type of transition, the atomic charges in the initial and final states are appreciably different. As a rough rule of thumb, d-d transitions give pale colors, and charge-transfer transitions give dark colors.

MAGNETISM

A substance which has no unpaired electrons is "diamagnetic"; i.e., it is repelled by a magnetic field. On the other hand, a substance with unpaired electrons is generally "paramagnetic" and shows the opposite behavior: When it is placed in a magnetic field gradient, it is attracted toward the region of greater field strength with a force proportional to the field strength times the field gradient. An applied magnetic field causes the electron spins to become somewhat ordered so that they are oriented favorably with respect to the field. In a "ferromagnetic" substance at low temperature, there is a spontaneous cooperative interaction between spins of the unpaired electrons such that they are all aligned and the substance has a high net magnetic moment. In an "antiferromagnetic" substance at low temperature, half of the electron spins are aligned one way and half are aligned the opposite way, so that the substance not only has no magnetic moment but also does not exhibit paramagnetic behavior in a magnetic field. The cooperative spin alignments of both ferromagnetic and antiferromagnetic substances become disordered as the temperature is increased, and at sufficiently high temperatures (i.e., above the Curie temperature of a ferromagnetic substance, and above the Néel temperature of an antiferromagnetic substance), the substances become paramagnetic.

In our study of electron configurations in Chap. 1, we saw that free transition-metal ions frequently have unpaired d electrons. Many compounds contain these ions with electron configurations which are relatively unperturbed, at least with respect to the number of unpaired d electrons. Hence many transition-metal compounds show one of the types of magnetism just described.

COMPLEXES

Transition metals have a much more diverse and interesting chemistry than, say, the alkali metals. In the case of transition metals in high oxidation states, this difference arises partly because the highly charged ions can electrostatically strongly bind a wide variety of negative or polar ligands. In the case of transition metals in low oxidation states, the difference arises because of the presence of electrons in the metal d orbitals. These d electrons can become involved in the bonding and permit the formation of complexes which cannot be formed in the case of the "hard" alkali-metal ions. Thus, much of the chemistry of the transition metals is the chemistry of complexes, or coordination compounds.

The term "coordination compound" is usually applied to any compound containing molecules or ions in which metal atoms are surrounded by other atoms or groups. Thus both $CH_3Mn(CO)_5$ and $K_3[Fe(CN)_6]$ are considered coordination compounds. Metal-containing molecules and ions are often called "complexes" or, in the case of ions, "complex ions." In fact, the term "complex ion" is even applied to ions in which the central atom is a nonmetal, such as BCl_4^- and PF_6^-. However, neutral nonmetallic species such as $SiBr_4$ and SF_6 are usually simply referred to as molecules rather than as coordination compounds or complexes.

CATALYSIS

Most industrially important reactions are carried out with the aid of catalysts, and the catalysts are usually transition metals or transition-metal compounds. For example, the Haber process for the synthesis of ammonia from nitrogen and hydrogen involves an iron–iron oxide catalyst. The oxidation of ammonia to nitric oxide (which is used in the synthesis of nitric aid) is catalyzed by a platinum or platinum-rhodium catalyst. The Ziegler-Natta polymerization of ethylene and propene is catalyzed by a mixture of $TiCl_4$ and $Al(C_2H_5)_3$. Many biologically important reactions take place with the aid of proteins in which transition-metal atoms play important roles. For example, a cobalt atom lies at the heart of the vitamin B_{12} coenzymes. Iron atoms are importantly involved in the hemoglobin of blood and in the ferredoxins of the photosynthetic process. Both molybdenum and iron are contained in nitrogen-fixing enzymes.

The transition-metal atoms in catalysts act as sites for holding the reacting species in conformations favorable for reaction and act as sources of, or sinks for, electrons. These functions are obviously aided by the abilities of transition metals to form complexes and to undergo oxidation-reduction reactions.

COORDINATION GEOMETRY

In the following paragraphs we shall discuss the commonly observed geometrical arrangements of ligand donor atoms in complexes. For convenience, we shall classify these structures in terms of the coordination numbers of the metal atoms.

Coordination Number 2

Although this coordination number is found in high-temperature, gas-phase species such as $MgCl_2(g)$, it is rather unusual for ordinary, stable complexes. It is found principally in some of the complexes of the ions Cu^+, Ag^+, Au^+, and Hg^{2+}, each of which has a ground-state d^{10} electron configuration. Typical examples of such complexes are $Cu(NH_3)_2{}^+$, $AgCl_2{}^-$, $Au(CN)_2{}^-$, and $HgCl_2$. All such complexes are linear; i.e., the ligand-metal-ligand bond angle is 180°. As a rough approximation, we may describe the bonding as a consequence of the overlap of σ orbitals of the ligands with sp hybrid orbitals of the metal atom. However, a metal d orbital is probably involved in the bonding to some extent.[1] Let us assume that the bonds lie on the z axis of the metal atom and that a small amount of the non-bonding electron density of the metal d_{z^2} orbital is shifted to the metal s hybrid orbital. In this case the metal orbitals used in bonding would not be simple sp_2 hybrid orbitals but rather hybrid orbitals with 50 percent p_z character, a small amount of d_{z^2} character, and the remainder s character. A pair of electrons would occupy a hybrid nonbonding metal orbital, principally of d_{z^2} character with a little s character. This type of hybridization would shift some nonbonding electron density away from the regions between the metal and the ligands and thus would be energetically favored.

Coordination Number 3

This coordination number is rare among metal complexes. Most crystalline compounds of stoichiometry MX_3 have structures in which the coordination number of M is greater than 3. For example, $CrCl_3$ has an infinite layer lattice in which each Cr atom is coordinated to six Cl atoms (see Fig. 11.12c). In $CsCuCl_3$ each Cu atom is coordinated to four Cl atoms in infinite anionic chains: $-Cl-CuCl_2-Cl-CuCl_2$. And $AuCl_3$ really exists as planar Au_2Cl_6 molecules

---

[1] L. E. Orgel, "An Introduction to Transition-Metal Chemistry: Ligand-Field Theory," 2d ed., pp. 69–71, Wiley, New York, 1966.

in which each Au atom is bonded to two bridging Cl atoms and two terminal Cl atoms. Coordination number 3 is occasionally encountered in complexes with extremely bulky ligands and, in the case of a few d^{10} metal complexes, even with ligands that are not very bulky. Some well-established examples are $KCu(CN)_2$ (see page 283), $[(CH_3)_3S^+][HgI_3^-]$, $Cr\{N[Si(CH_3)_3]_2\}_3$, $Fe\{N[Si(CH_3)_3]_2\}_3$, $\{Cu[SC(NH_2)_2]_3\}Cl$, $[Cu(SPPh_3)_3]ClO_4$, and $Pt(PPh_3)_3$. In all these cases the metal atom and the three directly coordinated ligand atoms are coplanar.

Coordination Number 4

This is a very important coordination number, for which two limiting configurations are commonly observed, "tetrahedral" and "square planar." In Chap. 3 it was pointed out that 4-coordinate non-transition-element complexes in which the central atom has no lone-pair electrons (such as $BeCl_4^{2-}$, BF_4^-, $ZnCl_4^{2-}$, $SnCl_4$, and AlF_4^-) are almost always tetrahedral. The same geometry is also generally found for 4-coordinate transition-element complexes in which the central atom or ion does *not* have a d^8 electron configuration. Thus species such as $Co(CO)_4^-$, $Ni(CO)_4$, $FeCl_4^-$, $CoCl_4^{2-}$, MnO_4^-, VO_4^{3-}, and FeO_4^{2-} are tetrahedral. When the central atom has a d^8 configuration, one sometimes finds tetrahedral geometry, but, more commonly, square planar geometry. Examples of d^8 tetrahedral complexes are those with "weak ligand fields" such as $NiCl_4^{2-}$, $NiBr_4^{2-}$, $NiCl_3OPPh_3$, and Co(I) and Ni(II) complexes that have ligands with bulky substituents. Examples of d^8 square planar complexes are those with "strong ligand fields" such as $IrCl(CO)(PEt_3)_2$, $Ni(CN)_4^{2-}$, $PdCl_4^{2-}$, $Pt(NH_2CH_2CH_2NH_2)_2^{2+}$, AgF_4^-, Au_2Cl_6, and $[Rh(CO)_2Cl]_2$. The last two complexes have bridging chlorine atoms which are shared by the two 4-coordinate metal atoms. The distinction between weak and strong ligand fields is discussed in Chapter 15.

Coordination Number 5

The structures of 5-coordinate complexes lie between two limiting geometries: trigonal bipyramidal and square pyramidal. These limiting structures are not markedly different, as can be seen by examination of Fig. 14.1. The conversion of one structure into the other requires a relatively slight distortion. In Fig. 14.2, seven different 5-coordinate structures, including two nonmetallic compounds, are illustrated to show the gradual transition from $CdCl_5^{3-}$ (which has almost the ideal trigonal bipyramid structure) to $Ni(CN)_5^{3-}$ (which, in one of its forms, has almost the ideal square pyramid structure). Several of the intermediate structures are almost equally well described as distorted trigonal bipyramids or as distorted square pyramids. To eliminate such ambiguity, Muetterties and Guggenberger[2] proposed the specification of dihedral angles of the polyhe-

[2] E. L. Muetterties and L. J. Guggenberger, *J. Am. Chem. Soc.*, **96**, 1748 (1974).

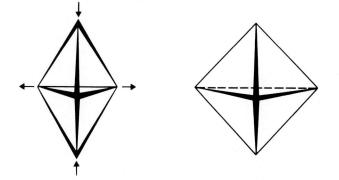

FIGURE 14.1
The distortion of a trigonal bipyramid to a square pyramid. [*Reproduced with permission from E. L. Muetterties and L. J. Guggenberger, J. Am. Chem. Soc., 96, 1748 (1974). Copyright 1974 American Chemical Society.*]

dron as an objective way to describe the shape of a 5-coordinate compound. Dramatic proof of the fact that the various conformations from trigonal bipyramidal to square pyramidal have comparable energies is provided by the compound $[Cr(NH_2CH_2CH_2NH_2)_3][Ni(CN)_5] \cdot 1.5H_2O$, which contains two types of $Ni(CN)_5^{3-}$ ions: One has the geometry of a regular square pyramid, and the other has a geometry practically midway between that of a trigonal bipyramid and a square pyramid.[3]

---

[3] K. N. Raymond, P. W. R. Corfield, and J. A. Ibers, *Inorg. Chem.*, **7**, 1362 (1968).

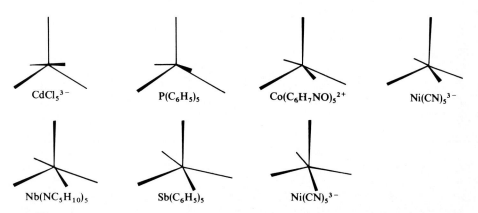

FIGURE 14.2
The 5-coordinate structures found in seven different species. The structures range from almost ideal trigonal bipyramidal to almost ideal square pyramidal. [*Reproduced with permission from E. L. Muetterties and L. J. Guggenberger, J. Am. Chem. Soc., 96, 1748 (1974). Copyright 1974 American Chemical Society.*]

Further evidence for a small energy difference between trigonal bipyramidal and square pyramidal configurations is the ease with which the axial and equatorial ligands of trigonal bipyramidal complexes are scrambled. These ligands do maintain their stereochemical integrity long enough to allow structure determinations by diffraction methods and vibrational spectroscopic studies, but in some cases the axial and equatorial sites exchange at a rate such that these sites cannot be distinguished by nmr (which has a time scale[4] typically around 10^{-3} s). For example, the ^{13}C nmr spectrum of $Fe(CO)_5$ shows only one resonance.[5] It is generally believed that the scrambling of the axial and equatorial sites occurs by the so-called Berry mechanism, which has already been discussed in terms of the stereochemical nonrigidity of PF_5 (Chap. 3).[6] In this mechanism, the square planar configuration is believed to be an intermediate or activated complex. Stereochemical nonrigidity, with consequent easy scrambling of ligands, is a relatively common feature of discrete complexes which have low activation energy routes between alternative geometries.[7]

Caution must be used when interpreting stoichiometric data as evidence for coordination number 5. Thus Cs_3CoCl_5 contains the anions $CoCl_4^{2-}$ and Cl^-, Tl_2AlF_5 contains infinite chain anions of the type $-F-AlF_4-F-AlF_4-$, and $Co(NH_2CH_2CH_2NHCH_2CH_2NH_2)Cl_2$ contains the 6-coordinate ions $Co(NH_2CH_2CH_2NHCH_2CH_2NH_2)_2^{2+}$ and the 4-coordinate ions $CoCl_4^{2-}$.

Coordination Number 6

This is the commonest and most important coordination number for transition-metal complexes. The geometry usually corresponds to six coordinated atoms at the corners of an octahedron or a distorted octahedron. The regular octahedron, shown in the following sketch,

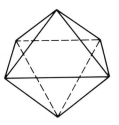

---

[4] We loosely refer to the minimum lifetime a species must have in order that a characteristic nmr spectrum be observed (related to the nmr frequency) as the nmr "time scale."

[5] F. A. Cotton, A. Danti, J. S. Waugh, and R. W. Fessenden, *J. Chem. Phys.*, **29**, 1427 (1958).

[6] For a historical survey of early work on stereochemical nonrigidity, see F. A. Cotton, *J. Organomet. Chem.*, **100**, 29 (1975).

[7] J. P. Jesson and E. L. Muetterties, in "Dynamic Nuclear Magnetic Resonance Spectroscopy," L. Jackman and F. A. Cotton, eds., p. 253, Academic, New York, 1975.

is a highly symmetric polyhedron. The reader can probably recognize the existence of six twofold rotational axes, four threefold rotational axes, three fourfold rotational axes, three mirror planes of one type, six of another type, and a center of symmetry. (See Chap. 2 for a discussion of symmetry elements.) One common type of distortion of the octahedron, "tetragonal" distortion, involves either elongation or contraction of the octahedron along one of the fourfold symmetry axes, as shown below.

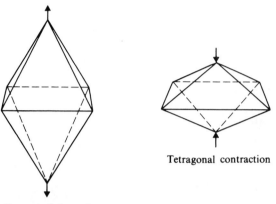

Tetragonal contraction

Tetragonal elongation

Another type of distortion, "trigonal" distortion, involves either elongation or contraction along one of the threefold symmetry axes, as shown below, to form

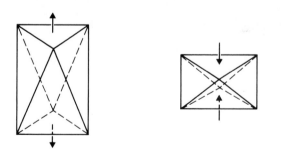

trigonal antiprisms. Very rarely one finds a 6-coordinate complex with trigonal prismatic geometry:

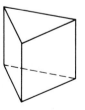

This geometry is unusual probably because repulsions between the coordinated atoms are greater than in the antiprismatic geometry which would be obtained if one rotated one triangular face 60° relative to the other.

It should be noted that, although a hexacoordinate complex of formula MA_6 can have a regular octahedral geometry (O_h symmetry), complexes of the type MA_5B, MA_4B_2, etc., cannot have O_h symmetry. In such mixed-ligand complexes, the ligands generally are not even at the corners of regular octahedra. Nevertheless, we usually refer to such complexes as octahedral complexes. Only when distortions from octahedral symmetry are great do we describe the structures differently. For example, the hexacoordinate iron compound $H_2Fe(CO)_4$ has C_{2v} symmetry, and, although it has approximate octahedral structure, it can also be described as having the structure of a bicapped tetrahedron, in which the hydride ligands are positioned over two faces of an $Fe(CO)_4$ tetrahedron. Of course, similar remarks regarding the loss of symmetry in complexes with more than one kind of ligand apply to other coordination numbers as well.

Coordination Number 7

Coordination numbers higher than 6 in discrete complexes are most commonly found for second- and third-row transition metals, lanthanides, and actinides. Coordination number 7 is relatively uncommon; it has been found with three limiting geometries: the pentagonal bipyramid, the monocapped octahedron, and the trigonal prism with one of its rectangular faces capped.[8] Examples of pentagonal bipyramidal geometry are ReF_7, $V(CN)_7^{4-}$, $Mo(CN)_7^{5-}$, $UO_2F_5^{3-}$, and $NbOF_6^{3-}$. Examples of capped octahedra are $Mo(CO)_3(PEt_3)_2Cl_2$ and $W(CO)_4Br_3^-$, and examples of capped trigonal prisms are NbF_7^{2-} and $Mo(CNR)_7^{2+}$.

Coordination Number 8

The geometries that have been observed for 8-coordinate complexes are illustrated in Fig. 14.3. Cubic (O_h) coordination is almost never found in discrete complexes. Two exceptions are the complex anions in Na_3PaF_8 and $[Et_4N]_4[U(NCS)_8]$, in which the coordinated fluoride ions and nitrogen atoms, respectively, surround the actinide ions at the corners of a cube.[9] Eight-coordinate complexes more often have the D_{4d} geometry of a square antiprism, which is obtained by rotating one face of a cube 45° relative to the opposite face. Another commonly found geometry is that of the D_{2d} dodecahedron, which may be looked upon as a combination of an elongated tetrahedron and a flattened tetrahedron. The square antiprismatic and dodecahedral configurations have very similar energies, and, in the case of octacyano complexes such as $Mo(CN)_8^{3- \text{ or } 4-}$ and $W(CN)_8^{3- \text{ or } 4-}$, either geometry can be obtained in solids by appropriate choice of the cation. For example, the $Mo(CN)_8^{3-}$ ion is square antiprismatic in $Na_3Mo(CN)_8 \cdot 4H_2O$ and dodecahedral in $[N(n\text{-}C_4H_9)_4]_3Mo(CN)_8$.

---

[8]   M. G. B. Drew, *Prog. Inorg. Chem.*, **23**, 67 (1977).

[9]   D. Brown, J. F. Easey, and C. E. F. Rickard, *J. Chem. Soc. A*, 1161 (1969); R. Countryman and W. S. McDonald, *J. Inorg. Nucl. Chem.*, **33**, 2213 (1971).

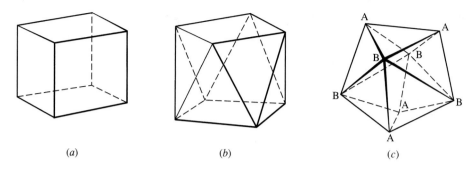

(a) *(b)* *(c)*

FIGURE 14.3

The cube (*a*), the square antiprism (*b*), and the dodecahedron (*c*). Note that, whereas all the vertices of the cube and the square antiprism are equivalent, the dodecahedron contains two types of vertices (A and B).

Higher Coordination Numbers

Nine-coordinate complexes, such as $Nd(H_2O)_9^{3+}$ and ReH_9^{2-}, have geometries corresponding to a tricapped trigonal prism, as illustrated in Fig. 14.4. It appears that there are no examples of the capped square antiprism geometry in discrete ML_9 complexes.[10] Coordination numbers 10 to 12 are occasionally found in complexes of lanthanide and actinide elements but, in view of their rarity, are not very important. Twelve-coordination generally involves the formation of a distorted icosahedral geometry, either with six bidentate ligands, as in $Ce(NO_3)_6^{2-}$, or with four tridentate ligands, as in $Zr(BH_4)_4$, $Hf(BH_4)_4$, $Np(BH_4)_4$, and $Pu(BH_4)_4$. The structure of these borohydrides is illustrated in Fig. 14.5.

The Relative Prevalence of Coordination Numbers 4,6,8, and 12

Coordination numbers of 4,6,8, and 12 are more common than coordination numbers of 5,7,9,10, and 11, both in the solid state and in discrete complexes. It is

---

[10] L. J. Guggenberger and E. L. Muetterties, *J. Am. Chem. Soc.*, **98**, 7221 (1976).

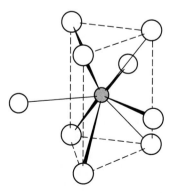

FIGURE 14.4

The geometry of a 9-coordinate complex such as ReH_9^{2-}. The ligands are at the vertices of a trigonal prism which has been capped on each of its rectangular faces.

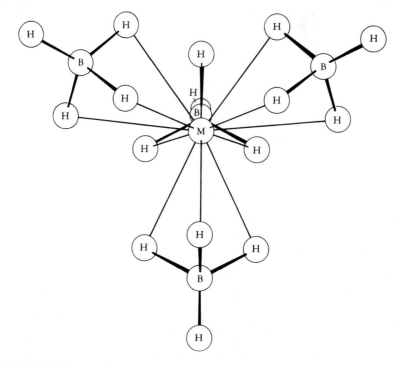

FIGURE 14.5
The structure of the borohydrides of zirconium, hafnium, neptunium, and plutonium: $M(BH_4)_4$. The BH$_4$ groups are tetrahedrally arranged, and the 12 coordinated hydrogen atoms are at the corners of a distorted icosahedron.

easy to understand why these coordination numbers are common in solids, inasmuch as close-packed atoms have a coordination number of 12 and form both tetrahedral and octahedral holes, and the fluorite and cesium chloride structures involve coordination number 8. However, it is not as obvious why these same coordination numbers are also relatively prevalent in discrete ML_n complexes. If the M—L bonds are assumed to be ionic (i.e., "bond energies" inversely proportional to the M—L distance), the preference for the common coordination numbers can be shown to be a simple matter of energetics. In Table 14.1 are given the M—L distances, d, corresponding to various coordination numbers, n, assuming that the ligand atoms are spheres of unit radius that touch each other and the atom M. The tabulated quantities n/d are thus proportional to the total bond energies of the ML_n complexes, and the quantities $\Delta(n/d)/\Delta n$ measure the increases in total bond energy accompanying unit increases in coordination number. It can be seen that, if a complex with a coordination number of 3, 5, or 7 were to increase its coordination number by 1, or if a complex with a coordination number of 10 were to increase its coordination number by 2, there would be a relatively large gain in bond energy that might more than compensate for the increase in interligand repulsion. On the other hand, if a complex with a coordination number of 5, 7, or 9 were to decrease its coordination number by 1, there would be a relatively

TABLE 14.1
Parameters proportional to bond distance and total bond energy in
ML_n complexes

Coordination no., n	Optimum figure	d	n/d	$\dfrac{\Delta(n/d)}{\Delta n}$
3	Triangle	1.155	2.60	
				0.67
4	Tetrahedron	1.225	3.27	
				0.27
5	Trigonal bipyramid†	1.414	3.54	
				0.70
6	Octahedron	1.414	4.24	
				0.16
7	Monocapped octahedron	1.592	4.40	
				0.46
8	Square antiprism	1.645	4.86	
				0.34
9	Tricapped trigonal prism	1.732	5.20	
				0.21
10	Bicapped square antiprism	1.850	5.41	
				0.45
12	Icosahedron	1.902	6.31	

† The tetragonal pyramid has the same value of d, etc.

small loss in bond energy that might be more than compensated by the decrease in interligand repulsion. Thus the $\Delta(n/d)/\Delta n$ values in Table 14.1 show that complexes with coordination numbers other than 4,6,8, and 12 would be expected to lose or gain ligands relatively easily and thus to be relatively rare.

LIGANDS

The ligands of a complex may be classified by the way in which they are co-ordinated to the central metal atom. A "monodentate" ligand is one which is attached to the metal atom by a bond from only one atom (the donor atom) of the ligand. Ligands such as H^-, F^-, Cl^-, O^{2-}, $\overset{*}{P}R_3$, $H_2\overset{*}{O}$, $\overset{*}{C}H_3{}^-$, $\overset{*}{O}R^-$, and $\overset{*}{C}O$ are monodentate. (The asterisks indicate the donor atoms of the polyatomic lig-ands.) A "polydentate" ligand (bidentate, tridentate, etc.) is one which can be attached to the metal atom by bonds from two or more donor atoms. Examples of bidentate ligands are ethylenediamine, $NH_2CH_2CH_2NH_2$ (which can coordinate through two nitrogen atoms and is often represented by the symbol "en"), and o-phenylenebis(dimethylarsine),

(which can coordinate through two arsenic atoms and is often represented by the symbol "diars").

Complexes of Monodentate Ligands

Simple metal cations in aqueous solution, usually represented by formulas such as Mn^{2+}, Fe^{3+}, and Zn^{2+}, are actually hydrated species which are better represented by formulas which indicate how many water molecules are directly coordinated to the metal atoms, e.g., $M_n(H_2O)_6^{2+}$, $Fe(H_2O)_6^{3+}$, and $Zn(H_2O)_4^{2+}$.[11] Treatment of such aqueous species with solutions containing appropriate monodentate ligands such as NH_3 and Br^- can cause displacement of water molecules by the ligands:

$$Zn(H_2O)_4^{2+} + 4NH_3 \rightleftharpoons Zn(NH_3)_4^{2+} + 4H_2O$$

$$Fe(H_2O)_6^{3+} + Br^- \rightleftharpoons Fe(H_2O)_5Br^{2+} + H_2O$$

Abbreviated equations, such as the following, are often written, but it should always be remembered that water molecules play an important part in the reactions.

$$Zn^{2+} + 4NH_3 \rightleftharpoons Zn(NH_3)_4^{2+}$$

$$Fe^{3+} + Br^- \rightleftharpoons FeBr^{2+}$$

When complexation of a metal ion takes place, it is often possible to detect the formation of a series of complexes, corresponding to the stepwise coordination of ligands to the metal ion. For example, in the case of the nickel ion, six different ammonia complexes are formed. The relative concentrations of the six complexes depend on the concentration of free ammonia and can be calculated from the equilibrium constants of the following reactions.[12]

$$Ni^{2+} + NH_3 \rightleftharpoons NiNH_3^{2+} \qquad \log K = 2.72$$

$$NiNH_3^{2+} + NH_3 \rightleftharpoons Ni(NH_3)_2^{2+} \qquad \log K = 2.17$$

$$Ni(NH_3)_2^{2+} + NH_3 \rightleftharpoons Ni(NH_3)_3^{2+} \qquad \log K = 1.66$$

$$Ni(NH_3)_3^{2+} + NH_3 \rightleftharpoons Ni(NH_3)_4^{2+} \qquad \log K = 1.12$$

$$Ni(NH_3)_4^{2+} + NH_3 \rightleftharpoons Ni(NH_3)_5^{2+} \qquad \log K = 0.67$$

$$Ni(NH_3)_5^{2+} + NH_3 \rightleftharpoons Ni(NH_3)_6^{2+} \qquad \log K = 0.03$$

It can be seen that the $\log K$ values decrease steadily as the number of coordinated ammonia molecules increases. This sort of trend is fairly generally observed and, in this case, is probably due to the steady buildup of negative charge on the nickel atom as ammonia molecules are coordinated. However, consider the following complexation constants for the chloride complexes of mercuric ion.[13]

$$Hg^{2+} + Cl^- \rightleftharpoons HgCl^+ \qquad \log K = 6.74$$

$$HgCl^+ + Cl^- \rightleftharpoons HgCl_2 \qquad \log K = 6.48$$

$$HgCl_2 + Cl^- \rightleftharpoons HgCl_3^- \qquad \log K = 0.9$$

$$HgCl_3^- + Cl^- \rightleftharpoons HgCl_4^{2-} \qquad \log K = 1.0$$

---

[11]   For many non-transition-metal ions, e.g., Ag^+, Zn^{2+}, Cd^{2+}, Hg^{2+}, and Ga^{3+}, the coordination number in aqueous solution is not known with certainty. See F. Basolo and R. G. Pearson, "Mechanisms of Inorganic Reactions," 2d ed., p. 153, Wiley, New York, 1967.

[12]   A. E. Martell and R. M. Smith, "Critical Stability Constants," vols. 1–4, Plenum, New York, 1974–1977; the data for Ni^{2+} and NH_3 are for 25°C and zero ionic strength.

[13]   Ibid.; the data for Hg^{2+} and Cl^- are for 25°C and ionic strength 0.5.

In this case, the driving force for complexation drops off enormously after the disubstituted complex forms. The break in the sequence of log K values at $HgCl_2$ is probably associated with a change in geometry. The species $HgCl_2$ is linear, whereas $HgCl_4^{2-}$ is tetrahedral. The data illustrate the strong tendency for d^{10} metal ions to form linear, 2-coordinate complexes.

All of the complexation reactions discussed in the preceding paragraph take place rapidly—essentially instantaneously for most purposes. These complexes are often referred to as *labile* complexes to indicate that the ligand exchange reactions are rapid. However, there are many so-called *inert* coordination compounds that undergo ligand exchange reactions relatively slowly and which cannot be prepared as readily as the nickel-ammonia complexes and the mercuric-chloride complexes. For example, cobalt(III) complexes are usually synthesized by the oxidation of aqueous cobalt(II) with oxygen or hydrogen peroxide in the presence of the ligands. Thus when air is bubbled vigorously through a solution containing cobalt(II), ammonia, ammonium ion, and suspended activated charcoal, good yields of the hexaamminecobalt(III) ion are obtained:

$$4Co(NH_3)_6^{2+} + O_2 + 4NH_4^+ \rightarrow 4Co(NH_3)_6^{3+} + 2H_2O + 4NH_3$$

A method for preparation of the remarkable complex $Ru(NH_3)_5N_2^{2+}$, in which the ruthenium atom is coordinated to five ammonia molecules and to one end of a molecule of nitrogen, is schematically represented as follows.

$$\text{``RuCl}_3 \cdot 3H_2O\text{''} \xrightarrow[\text{boil}]{\text{Zn, NH}_3, \text{NH}_4^+} Ru(NH_3)_6^{2+}$$

$$\text{``RuCl}_3 \cdot 3H_2O\text{''} \xrightarrow{Cl_2} Ru(NH_3)_5Cl^{2+} \xrightarrow{Zn} Ru(NH_3)_5H_2O^{2+} \xrightarrow{N_2} Ru(NH_3)_5N_2^{2+}$$

The hexacoordinate chromium(0) complex $Cr(CO)_5PMe_3$ can be prepared by the following sequence.

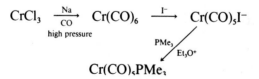

$$CrCl_3 \xrightarrow[\substack{CO \\ \text{high pressure}}]{Na} Cr(CO)_6 \xrightarrow{I^-} Cr(CO)_5I^-$$

$$\xrightarrow[\substack{PMe_3 \\ Et_3O^+}]{} Cr(CO)_5PMe_3$$

This route has the advantage of giving only the monosubstituted phosphine complex, $Cr(CO)_5PMe_3$. Direct reaction of PMe_3 and $Cr(CO)_6$ gives mono-, di-, and trisubstitution and is more difficult to control.

Frequently monodentate ligands are simultaneously coordinated to more than one metal atom and act as bridges between the metal atoms. (Obviously a complex with bridging ligands must be "polynuclear"; i.e., it must contain more than one metal center.) For example, when the pH of a VO_4^{3-} solution is reduced to a value below ~ 13, various polyvanadate ions form, in which tetrahedral VO_4 groups form chains by sharing oxygen atoms. The simplest of these species is the divanadate ion, $V_2O_7^{4-}$:

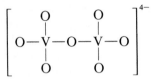

Figure 14.6 shows the principal vanadate species that exist in equilibrium at various pH values and total vanadium concentrations.

Hydrolysis of the ferric ion not only involves the formation of $FeOH^{2+}$ and $Fe(OH)_2^+$, but also at least one dinuclear species:

$$2Fe^{3+} + 2OH^- \rightleftharpoons Fe_2(OH)_2^{4+} \qquad K \approx 10^{25}$$
$$\text{or } Fe_2O^{4+} + H_2O$$

Notice that the reaction quotient is $(\text{complex})(Fe^{3+})^{-2}(OH^-)^{-2}$ whether the complex is formulated as $Fe_2(OH)_2^{4+}$ or Fe_2O^{4+}. Hence aqueous equilibrium data

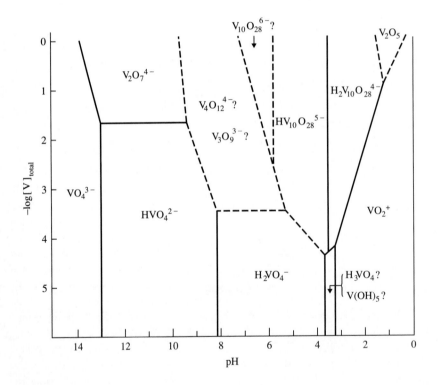

FIGURE 14.6
The principal vanadate species at various pH values and total vanadium concentrations. It should be emphasized that any point on this plot corresponds to the coexistence of several vanadium species at equilibrium. [*Reproduced with permission from M. T. Pope and B. W. Dale, Quart, Revs. 22, 527 (1968).*]

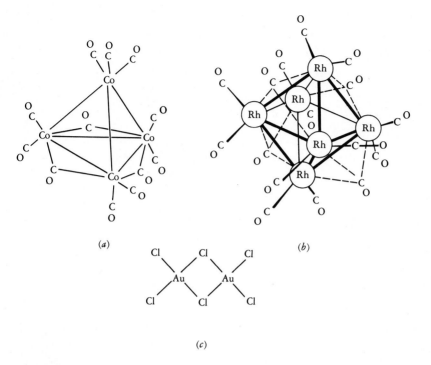

FIGURE 14.7
Some complexes with bridging monodentate ligands. (*a*) $Co_4(CO)_{12}$; (*b*) $Rh_6(CO)_{16}$; (*c*) Au_2Cl_6.

cannot be used to determine the nature of the bridging in this complex. Although the complex may consist of two hydroxide ions bridging two $Fe(H_2O)_4^{3+}$ groups, magnetic and spectroscopic studies[14] suggest that it probably consists of an oxide ion bridging between two $Fe(H_2O)_5^{3+}$ groups. Other examples of complexes containing bridging monodentate ligands, shown in Fig. 14.7, are $Co_4(CO)_{12}$ (which contains three CO groups that bridge across Co—Co bonds), $Rh_6(CO)_{16}$ (in which four CO groups occupy triply bridging positions over four of the triangular faces of the octahedron), and Au_2Cl_6 (with two bridging chloride ions). More unusual examples, in which ligands occupy central positions in metal clusters, are shown in Fig. 14.8. In the volatile complex $Be_4O(O_2CCH_3)_6$, a central oxygen atom is surrounded tetrahedrally by four beryllium atoms, and in $HCo_6(CO)_{15}^-$, the hydrogen atom is situated in the center of the metal octahedron.[15]

The remarkable complex $Tl_2Pt(CN)_4$, pictured in Fig. 14.9, is a rare example in which a metal ion (Tl^+) acts as a ligand.

[14]   J. M. Knudsen et al., *Acta Chem. Scand. Ser. A*, **29**, 833 (1975).
[15]   D. W. Hart et al., *Angew. Chem. Int. Ed.*, **18**, 80 (1979).

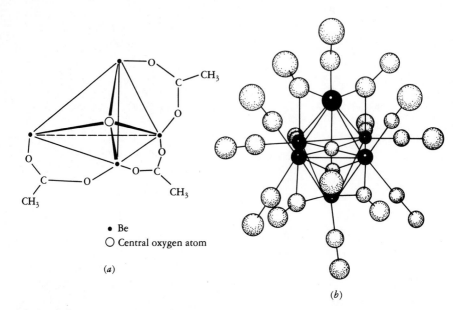

• Be
○ Central oxygen atom

(a)

(b)

FIGURE 14.8
Complexes with atoms inside metal clusters. (a) $Be_4O(O_2CCH_3)_6$, with only three of the acetate groups shown, for clarity; (b) $HCo_6(CO)_{15}^-$, with an "internal" hydrogen atom.

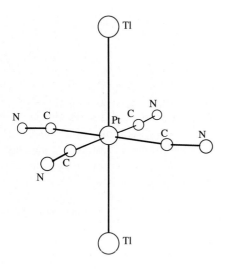

FIGURE 14.9
Structure of the $Tl_2Pt(CN)_4$ complex. [*Reproduced with permission from J. K. Nagle, A. L. Balch, and M. M. Olmstead, J. Am. Chem. Soc., **110**, 319 (1988).*]

Polydentate Ligands

Some polydentate ligands, and the abbreviations used to represent them in formulas, are listed in Table 14.2. A number of simple ligands, not listed in this table, can be either monodentate or bidentate; these include SO_4^{2-}, CO_3^{2-}, NO_3^-, and $CH_3CO_2^-$, which can coordinate through one or two oxygen atoms. The complex formed by the coordination of a polydentate ligand to a metal atom is called a "chelate complex"; several such complexes are illustrated schematically in Fig. 14.10.

The term "chelate effect" refers to the apparent enhanced stability of a chelate complex compared to that of a complex with similar ligand donor groups but no rings. For example, consider the conversion of $Cd(NH_2CH_3)_4^{2+}$ to $Cden_2^{2+}$:

$$Cd(NH_2CH_3)_4^{2+} + 2en \quad \rightleftharpoons \quad Cden_2^{2+} 4CH_3NH_2$$

Because the equilibrium constant[12] for this reaction is 1.2×10^4, we might conclude that the reaction intrinsically has a large driving force and that it provides evidence for the chelate effect. However, in a sense, this large driving force is fictitious, because the magnitude of the equilibrium constant depends on the arbitrary choice of the standard state for the solutes (conventionally, the 1 m solution). It has been pointed out that, if the standard state for solutes is taken to be the hypothetical mole fraction unity state, the driving force for chelation reactions of this type largely disappears.[16] In the case of an aqueous solution, unit molality corresponds to a mole fraction of $1/55.5$; hence for the cited reaction, the equilibrium constant becomes $1.2 \times 10^4 \times (1/55.5)^2$, or 3.9. In other words, chelation reactions proceed primarily because they are generally carried out in dilute solutions; their driving force increases with increasing dilution because of the mass action effect. It is significant that the enthalpies of chelation reactions of the type we are discussing (in which the nature of the ligand donor groups remains approximately constant) are generally small. This fact again emphasizes that the apparent free-energy change is mainly determined by $T\Delta S^\circ$, the value of which depends on the choice of standard state.[17]

To summarize: In the case of the following two reactions, whose reaction quotients have different dimensions,

$$Cd(NH_2CH_3)_4^{2+} \quad \rightleftharpoons \quad Cd^{2+} + 4CH_3NH_2$$

$$Cden_2^{2+} \quad \rightleftharpoons \quad Cd^{2+} + 2en$$

the driving forces cannot be meaningfully compared unless the concentrations of CH_3NH_2 and en are clearly specified.

β-Diketones, e.g., acetylacetone, can be readily deprotonated to form mononegative anions with delocalized π electrons. By chelation with a metal

---

[16] A. W. Adamson, *J. Am. Chem. Soc.*, **76**, 1578 (1954).

[17] For a change in molar volume from V_1 to V_2, $\Delta S^\circ = R \ln (V_2/V_1)$.

TABLE 14.2
Some polydentate ligands and their abbreviations

Name and abbreviation	Formula
Ethylenediamine, en	$H_2NCH_2CH_2NH_2$
2,2'-Bipyridine, bipy	
1,10-Phenanthroline, phen	
o-Phenylenebis(dimethylarsine), diars	
1,2-Bis(diphenylphosphino)ethane, diphos	$\begin{array}{l} CH_2-P(C_6H_5)_2 \\ \mid \\ CH_2-P(C_6H_5)_2 \end{array}$
Acetylacetonate, acac$^-$	$\overset{O}{\overset{\|}{CH_3C}}CH_2\overset{O^-}{\overset{\mid}{C}}CH_3$
Hexafluoroacetylacetonate, hfa$^-$	$\overset{O}{\overset{\|}{CF_3C}}CH_2\overset{O^-}{\overset{\mid}{C}}CF_3$
Salicylaldiminate, sal$^-$	
8-quinolinate, "oxinate"	

(*continued*)

TABLE 14.2 (*continued*)

Name and abbreviation	Formula
Oxalate, ox^{2-}	$^-O_2CCO_2^-$
Terpyridine, terpy (tridentate)	
Diethylenetriamine, dien (tridentate)	
Triethylenetetramine, tren (tetradentate)	$N(CH_2CH_2NH_2)_3$
Nitrilotriacetate, NTA^{3-} (tetradentate)	$N(CH_2CO_2)_3{}^{3-}$
Ethylenediaminetetraacetate, EDTA^{4-} (hexadentate)	
Diethylenetriaminepentaacetate, DTPA^{5-} (octadentate)	
3,4,3-LICAM-C (octadentate)	

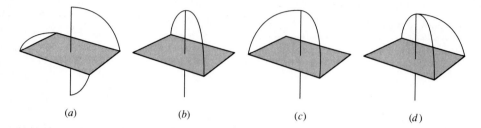

(a)   (b)   (c)   (d)

FIGURE 14.10

Schematic representations of chelates. (a) A complex with three bidentate ligands, such as $Co(en)_3^{3+}$; (b) meridional chelation of a tridentate ligand; (c) facial chelation of a tridentate ligand; (d) the chelation of a tetradentate ligand such as NTA^{3-}. Notice that, in all these cases, adjacent donor atoms of a ligand are coordinated cis to one another.

ion, very stable six-membered rings can be formed by these ligands:

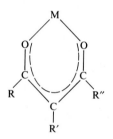

When the coordination number of the metal ion is twice its charge, it is possible to prepare neutral β-diketonates, such as $Be(acac)_2$ and $Cr(hfa)_3$, which are volatile and soluble in organic solvents. Bidentate ligands frequently act as bridges between metal atoms. For example, the addition of acetate to a solution of a chromium(II) salt precipitates $Cr_2(OAc)_4 \cdot 2H_2O$, in which the two chromium atoms are linked together by the four bridging acetate ions:

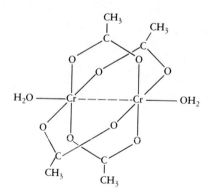

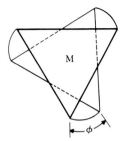

FIGURE 14.11
View down the threefold axis of a tris(bidentate ligand) complex.

This is one member of a large family of compounds (differing in the metal atoms, the carboxylate groups, and the axial monodentate ligands) with metal-metal bonding.[18]

Most hexacoordinate complexes with three bidentate ligands have structures which are intermediate between trigonal antiprismatic and trigonal prismatic. That is, the twist angle (defined in Fig. 14.11) generally lies between 0°, corresponding to the trigonal prism, and 60°, corresponding to the trigonal antiprism. Various factors are believed to be important in determining the geometry in such complexes. One important factor is the "bite" distance of the ligand, i.e., the distance between the two donor atoms of the same chelate.[19,20] As the ratio of the bite distance to the metal–donor atom distance decreases, the twist angle tends to decrease. (See Fig. 14.12.) This correlation of twist angle with bite distance can be rationalized in terms of ligand-ligand repulsion energies. A set of tris-chelate complexes having rather distinctive properties (represented as open squares in Fig. 14.12) are those with dithiolate ligands of the following type (or the selenium analogs):

Some of these complexes have essentially perfect trigonal prismatic geometry ($\phi = 0°$). Special electronic factors are probably responsible for these unusual structures.[21]

Multidentate ligands are useful for complexing large metal ions that have typical coordination numbers greater than 6. For example, the lanthanide ions are

[18] For further discussion, see pp. 400–402.
[19] D. L. Kepert, *Inorg. Chem.*, **11**, 1561 (1972); *Prog. Inorg. Chem.*, **23**, 1 (1977).
[20] A. Avdeef and J. P. Fackler, Jr., *Inorg. Chem.*, **14**, 2002 (1975).
[21] M. Cowie and M. J. Bennett, *Inorg. Chem.*, **15**, 1595 (1976).

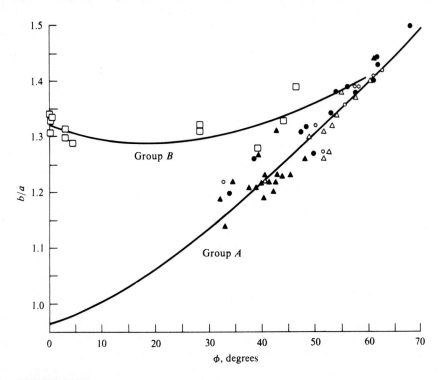

FIGURE 14.12

Plot of the ratio of the ligand bite distance to the metal–donor atom distance versus the twist angle. Group *A*: nitrogen ligands (△), oxygen ligands (○), sulfur and selenium ligands (▲), acac-type ligands (●). Group *B*: dithiolate ligands and Se analogs (□). [*Adapted with permission from A. Avdeef and J. P. Fackler, Inorg. Chem., 14, 2002 (1975). Copyright 1975 American Chemical Society*].

strongly complexed by EDTA, the structure of which is shown in Table 14.2. The lanthanides can be separated from one another by eluting them from a column of ion exchange resin with a solution of the triammonium salt of EDTA. The pertinent equilibria can be expressed approximately as

$$Ln^{3+}(resin) + 3NH_4^+ + EDTA \cdot H^{3-} + xH_2O$$
$$\rightleftharpoons [Ln(H_2O)_x EDTA \cdot H] + 3NH_4^+(resin)$$

The affinity of the Ln^{3+} ions for the resin increases very slightly with increasing atomic weight—not enough to effect a good separation by elution with ammonium ions alone. However, the complexation by EDTA increases markedly on going from Ce^{3+} to Lu^{3+}, because of the lanthanide contraction and the consequent increase in electrostatic bond energy between the metal ion and the donor atoms of EDTA. Consequently the elution causes the lanthanides to emerge from the column in the reverse order of atomic weights. The ligands diethylenetriaminepentaacetate and 3,4,3-LICAM-C, both pictured at the bottom of Table 14.2, have been found to form strong complexes with the 8-coordinate plutonium(IV) ion.[22] Plutonium is highly toxic, and rapid removal from the body is important in the case of accidental ingestion. Both chelating agents can be used for this purpose, but 3,4,3-LICAM-C, which coordinates through the eight deprotonated hydroxy groups of the catechol groups, is particularly effective.

Pi Complexes

Many olefins, aromatic species, and other π-bonded systems can function as ligands in which the π-bonded atoms coordinate to metal atoms. Ethylene, benzene, and allyl anion ($C_3H_5^-$), the cyclopentadienide ion ($C_5H_5^-$), and the cycloheptatriene (tropylium) cation ($C_7H_7^+$) are important examples of such ligands. For example, the trichloro(ethylene)platinate(II) ion is readily synthesized by displacing one of the chloride ions from $PtCl_4^{2-}$ by ethylene:

$$PtCl_4^{2-} + C_2H_4 \longrightarrow Pt(C_2H_4)Cl_3^- + Cl^-$$

The π-allyl complex, $C_3H_5Mn(CO)_4$ is prepared by the following sequence:

$$Mn_2(CO)_{10} \xrightarrow{Na} Mn(CO)_5^- \xrightarrow{CH_2=CHCH_2Cl} CH_2=CH-CH_2Mn(CO)_5$$

$$\pi-C_3H_5Mn(CO)_4 \xleftarrow[-CO]{\Delta \text{ or } h\nu}$$

Note that the conversion of the intermediate σ-allyl complex (which has only one Mn—C bond) to the π-allyl complex is accompanied by the loss of a CO molecule from the complex. Bis(cyclopentadienyl) iron ("ferrocene") can be prepared by treating an iron(II) halide with the cyclopentadienide ion:

$$C_5H_6 \xrightarrow[\text{"diglyme" ether}]{KOH} C_5H_5^- \xrightarrow{FeCl_2} Fe(C_5H_5)_2$$

The structures of the three π-bonded organometallic compounds just discussed are shown in Fig. 14.13.

Pi complexes differ from chelate complexes in two important respects: (1) The coordinated atoms of the ligand are coplanar and bonded together as a chain or ring, with the metal atom situated on one side of the plane; (2) the number of metal-ligand bonding electron pairs provided by the ligand (or the *effective* number of coordination sites occupied by the ligand) is less than the number of coordinated atoms. Thus ethylene in effect occupies one of the square planar coordination sites in $Pt(C_2H_4)Cl_3^-$; the π-allyl group occupies two of the octahedral sites in π-$C_3H_5Mn(CO)_4$, and the two $C_5H_5^-$ ions in $Fe(C_5H_5)_2$ in effect occupy six

[22] T. J. McMurry and K. N. Raymond, "Chelation," *McGraw-Hill Encyclopedia of Science and Technology*, p. 127, 1987.

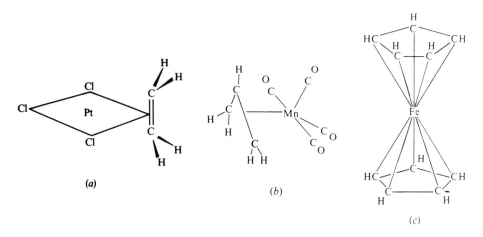

FIGURE 14.13
Structures of three π-bonded organometallic compounds. (a) $Pt(C_2H_4)Cl_3^-$; (b) $C_3H_5Mn(CO)_4$; (c) $Fe(C_5H_5)_2$ ferrocene.

octahedral coordination sites. The details of the bonding in such compounds will be discussed in later chapters.

Even hetero π-bonded species can form π complexes with transition metals.[23] For example, pyrrole can be deprotonated to form the pyrrolide anion, analogous to the $C_5H_5^-$ ion. The pyrrolide anion reacts with a metal carbonyl halide to form a σ complex which may then be converted via loss of two CO ligands into the corresponding π complex, as shown in the following scheme.

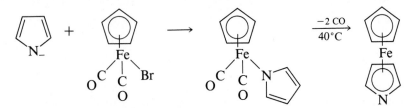

We have seen examples in which the allyl group is attached to a metal atom by either one or three carbon atoms and in which the pyrrolide ion is attached by either one or five atoms. There are many other examples of variations in the mode of bonding of π ligands. It is interesting that the $C_7H_7^+$ ligand can coordinate either seven, five, three, or one carbon atom to a metal atom, as in the following examples:

[23] K. H. Pannell, B. L. Kalsotra, and C. Parkanyi, *J. Heterocyclic Chem.*, **15**, 1057 (1978). A review on heterocyclic π complexes.

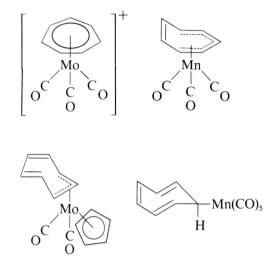

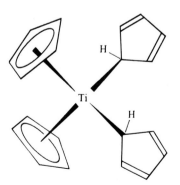

Cotton[24] devised the "hapto" nomenclature to describe in a simple way the various possible modes of bonding between metals and π systems. Based on this nomenclature, the C_7H_7-metal coordination in $C_7H_7Mo(CO)_3{}^+$ is called "heptahapto" (η^7) coordination, that in $C_7H_7Mn(CO)_3$ is called "pentahapto" (η^5) coordination, etc. The formulas for these complexes can be modified to express these modes of bonding as follows: $(\eta^7\text{-}C_7H_7)Mo(CO)_3{}^+$, $(\eta^5\text{-}C_7H_7)Mn(CO)_3$, etc. In $Ti(C_5H_5)_4$, two of the cyclopentadienyl groups are pentahapto and two are monohapto, as shown in Fig. 14.14. The proton nmr spectrum of this compound, shown in Fig. 14.15, undergoes marked changes as the temperature is changed.[25] At temperatures around $-30°C$ two peaks, corresponding to the $\eta^5\text{-}C_5H_5$ and $\eta^1\text{-}C_5H_5$ groups, are seen. The fact that only one peak is observed for the monohapto groups indicates that an intramolecular rearrangement of each monohapto

---

24 F. A. Cotton, *J. Am. Chem. Soc.*, **90**, 6230 (1968).

25 J. L. Calderon, F. A. Cotton, B. G. DeBoer, and J. Takats, *J. Am. Chem. Soc.*, **92**, 3801 (1970).

FIGURE 14.14
Structure of $Ti(C_5H_5)_4$. Notice that two of the rings are σ-bonded, through one carbon atom (monohapto), and that two of the rings are π-bonded, through five carbon atoms (pentahapto).

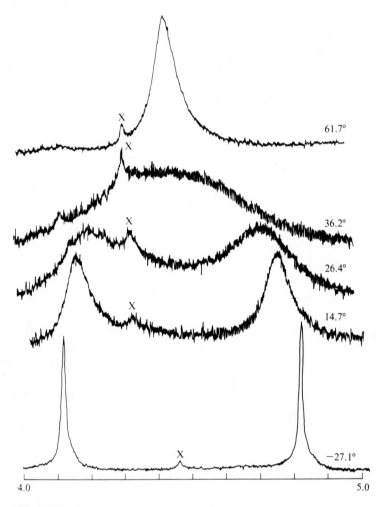

FIGURE 14.15
The pmr spectrum of $Ti(C_5H_5)_4$ at several temperatures. The X corresponds to $(C_5H_5)_3$ TiCl impurity. [*Reproduced with permission from J. L. Calderon, F. A. Cotton, B. G. DeBoer, and J. Takats, J. Am. Chem. Soc., 92, 3801 (1970). Copyright 1970 American Chemical Society.*]

ring occurs rapidly enough to make the protons equivalent on the time scale of the nmr experiment (roughly, 10^{-3} s). If the sample temperature is raised above 40°C, the two nmr peaks merge to form a single band. This behavior is ascribed to an increase in the rate of a scrambling process in which the two types of rings interchange their roles, so as to make all the protons equivalent on the nmr time scale.

ISOMERISM

Around the turn of the century, Alfred Werner explained many of the chemical and physical properties of a large number of complexes by means of his revolutionary

TABLE 14.3
Properties of platinum(IV) complexes of ammonia and chloride

Old formula	Relative molar conductivity of aqueous solution	No. of isomers known	Werner's formula
$PtCl_4 \cdot 6NH_3$	523	1	$[Pt(NH_3)_6]Cl_4$
$PtCl_4 \cdot 5NH_3$	404	1	$[Pt(NH_3)_5Cl]Cl_3$
$PtCl_4 \cdot 4NH_3$	228	2	$[Pt(NH_3)_4Cl_2]Cl_2$
$PtCl_4 \cdot 3NH_3$	97	2	$[Pt(NH_3)_3Cl_3]Cl$
$PtCl_4 \cdot 2NH_3$	0	2	$[Pt(NH_3)_2Cl_4]$
$PtCl_4 \cdot KCl \cdot NH_3$	108	1	$K[Pt(NH_3)Cl_5]$
$PtCl_4 \cdot 2KCl$	256	1	$K_2[PtCl_6]$

coordination theory.[26] The type of information at his disposal is typified by the data in Table 14.3. Werner proposed that, in each of these seven compounds, the platinum +4 ion is strongly coordinated to six ligands and that, in those cases in which the resulting complex has a net charge, the counter-ions (Cl^- or K^+) are not directly coordinated to the platinum. The formulas given in the fourth column, in which the formulas of the complexes are written inside brackets, readily explain the relative electrical conductivity data of the second column. As expected, the molar conductivity increases with the number of ions in the formula and is zero for the neutral complex $Pt(NH_3)_2Cl_4$.

Cis-Trans Isomerism

The number of isomers, given in column 3 of Table 14.3, was explained by Werner by postulating that the six ligands of each complex are attached to the platinum atom at the vertices of an imaginary octahedron around the platinum atom. Thus, in the complex $Pt(NH_3)_4Cl_2{}^{2+}$, the two chlorine atoms can be either at adjacent (cis) positions of the octahedron, as in the following representations:

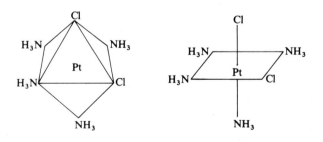

---

[26] A. Werner, *Z. Anorg. Chem.*, **3**, 267(1893); A. Werner, "Neuere Anschauungen auf der Gebiete der Anorganischen Chemie," Friedrich Vieweg, Brunswick, Germany, 1905; G. B. Kauffman, "Classics in Coordination Chemistry, Part I, The Selected Papers of Alfred Werner," Dover, New York, 1968.

or at opposite (trans) positions, as in the following representations:

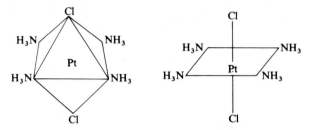

In a completely analogous way the isomerism of $Pt(NH_3)_2Cl_4$ can be explained in terms of cis and trans configurations of the coordinated NH_3 molecules.

One of the isomers of $Pt(NH_3)_3Cl_3{}^+$ has the structure in which the three NH_3 molecules are cis to one another and the three Cl atoms are cis to one another,

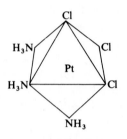

and the other isomer has the structure in which two NH_3 molecules are trans to one another and two Cl atoms are trans to one another,

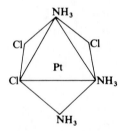

The former, "facially" coordinated, structure has one threefold rotational axis of symmetry, whereas the latter, "meridionally" coordinated, structure has one twofold rotational axis of symmetry.

Cis-trans isomerism is also found in square planar complexes. For example, two types of $Pt(NH_3)_2Cl_2$ are known, having the following structures:

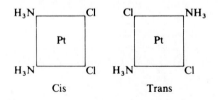

Optical Isomerism

A complex is optically active if its structure cannot be superimposed on its mirror image. For example, consider the complex cis-$Co(en)_2(NO_2)_2{}^+$, which can exist in either of the following configurations:

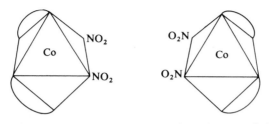

It should be noted that each structure is the mirror image of the other and that the structures are not superimposable. Consequently they are optical isomers. The usual method for the synthesis of cis-$Co(en)_2(NO_2)_2{}^+$ yields a product containing equal amounts of these isomers, i.e., a racemic mixture which does not rotate polarized light. The isomers cannot be separated by the usual techniques for separating other types of isomers (e.g., cis and trans isomers) because the optical isomers have essentially the same physical[27] and chemical properties, at least in systems which contain no other optically active species. However, optical isomers do interact differently with optically active species, and such specific interactions form the basis of methods for separating optical isomers. For example, if a solution containing the isomers of $Co(en)_2(NO_2)_2{}^+$ is treated with potassium antimonyl-$(+)$-tartrate, the $[(-)$-cis-$Co(en)_2(NO_2)_2][(+)$- $SbOC_4H_4O_6]$ combination (which is only slightly soluble) precipitates as crystals, leaving the $(+)$ -cis-$Co(en)_2(NO_2)_2{}^+$ isomer in solution.

Linkage Isomerism

Some ligands, called "ambidentate" ligands,[28] can bond to a metal atom through more than one type of donor atom. This flexibility leads to "linkage isomerism," first discovered by Jörgensen in 1894.[29] He characterized the nitrito and nitro isomers of the complex having the formula $Co(NH_3)_5NO_2{}^{2+}$. In the red nitrito-pentaamminecobalt(III) ion, the nitrite ion is coordinated through one of its oxygen atoms; thus, $Co-O-N{\diagdown}{}_O$. In the yellow nitropentaam-

minecobalt(III) ion, the nitrite ion is coordinated through its nitrogen atom; thus,

$Co-N{\overset{O}{\underset{O}{}}}$. The nitrito isomer is prepared by treatment of $Co(NH_3)_5H_2O^{3+}$ with

---

27 Of course, optical isomers rotate the plane of polarized light in opposite directions.

28 A. H. Norbury and A. I. P. Sinha, *Quart. Rev.*, **24**, 69 (1970).

29 S. M. Jörgensen, *Z. Anorg. Chem.*, **5**, 169 (1894).

a cold, buffered HNO_2–NO_2^- solution. The reaction is believed to proceed by the following mechanism:[30]

$$Co(NH_3)_5H_2O^{3+} + OH^- \xrightarrow{\text{fast}} Co(NH_3)_5OH^{2+} + H_2O$$

$$2HNO_2 \xrightarrow{\text{fast}} N_2O_3 + H_2O$$

$$Co(NH_3)_5OH^{2+} + N_2O_3 \longrightarrow \left[\begin{array}{c} (NH_3)_5Co-O---H \\ | \quad\quad | \\ O-N---ONO \end{array} \right]$$

$$\downarrow$$

$$(NH_3)_5Co-ONO^{2+} + HONO$$

The nitrito complex, when heated, rearranges by an intramolecular process to the more stable nitro complex. The nitrite ion can coordinate in three other ways, as shown below.

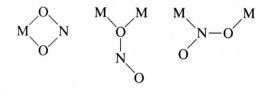

All three of these types of coordination are found in the complex $Ni_3(3\text{-}$ methylpyridine$)_6(NO_2)_6$, which has the following structure.[31]

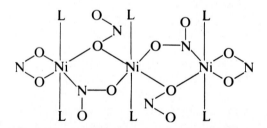

Another ambidentate ligand, the thiocyanate ion (SCN^-), can bond to metal atoms through the sulfur atom or the nitrogen atom. Soft metal ions such as Pd^{2+} and Hg^{2+} usually form sulfur-bonded thiocyanate complexes, whereas relatively hard metal ions such as Cr^{3+} and Fe^{2+} form nitrogen-bonded thiocyanate complexes. In a few cases, e.g., Pdbipy-$(NCS)_2$ and Pdbipy $(SCN)_2$, linkage isomers have been prepared. The nature of the bonding in such isomers can be readily determined by infrared spectroscopy. The C—S stretching vibration falls in the

[30] J. L. Burmeister and F. Basolo, *Prep. Inorg. React.*, **5**, 1 (1968).

[31] D. M. L. Goodgame, M. A. Hitchman, and D. F. Marsham, *J. Chem. Soc. (A)*, 259 (1971).

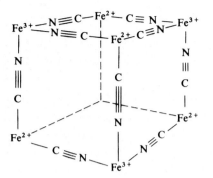

FIGURE 14.16
One-eighth of the unit cell of prussian (Turnbull's) blue. For simplicity, the alkali-metal ions and water molecules are omitted. Each iron atom is octahedrally coordinated.

range 690 to 720 cm^{-1} for M—SCN bonding, and in the range 780 to 860 cm^{-1} for M—NCS bonding.

In principle, the cyanide ion, CN$^-$, is an ambidentate ligand, but there are no known stable discrete complexes with metal–NC bonding.[32] However, there are many examples of solids in which the cyanide ion bridges between metal ions, the carbon atom to one metal ion and the nitrogen atom to another. For example, practically all chemists know that treatment of aqueous Fe^{3+} with hexacyanoferrate(II), Fe(CN)$_6$$^{4-}$, yields a blue precipitate called "prussian blue" and that treatment of aqueous Fe^{2+} with hexacyanoferrate(III), Fe(CN)$_6$$^{3-}$, yields a blue precipitate called "Turnbull's blue." These precipitates are actually identical; the empirical formula is Fe$_4$[Fe(CN)$_6$]$_3 \cdot x$H$_2$O(x = 14 to 16), i.e., "ferric ferrocyanide." The structure (shown in part in Fig. 14.16) consists of a simple cubic array of alternating Fe(II) and Fe(III) atoms, with the cyanide ions bridging thus: Fe(II)—CN—Fe(III). Of course, a perfect cubic array of this type would have equal numbers of Fe(II) and Fe(III) atoms, inconsistent with the formula. X-ray and neutron diffraction data[33] indicate that one-fourth of the Fe^{II}(CN)$_6$$^{4-}$ groups are randomly vacant. Six molecules of water are coordinated to Fe(III) at empty nitrogen sites; thus the average composition of each Fe(III) coordination unit is Fe^{III}(NC)$_{4.5}$(OH$_2$)$_{1.5}$. Approximately eight additional water molecules are present either as isolated molecules at the centers of the unit cell octants or as water molecules connected by hydrogen bonds to the coordinated ones.

In stable transition-metal carbonyls, the carbonyl group is always attached through the carbon atom (at least whenever CO functions as a monodentate ligand). However, in studies[34] of the cocondensation of Au atoms with a large excess of CO at 6 to 10 K, spectroscopic data suggest the formation of isocarbonyl(carbonyl)gold, COAuCO. The normal isomer, OCAuCO, was formed when Au and CO were cocondensed in rare-gas matrices.

---

[32] Metastable Co(CN)$_5$NC^{3-} and Cr(H$_2$O)$_5$NC^{2+} have been prepared. See J. Halpern and S. Nakamura, *J. Am. Chem. Soc.*, **87**, 3002 (1965); J. P. Birk and J. H. Espenson, *J. Am. Chem. Soc.*, **90**, 1153 (1968).

[33] A. Ludi et al., *Inorg. Chem.*, **16**, 2704 (1977); **19**, 956 (1980).

[34] D. McIntosh and G. A. Ozin, *Inorg. Chem.*, **16**, 51 (1977).

STRUCTURAL REORGANIZATION

The fact that it is possible to isolate and to characterize structural isomers of the types that we have discussed (cis-trans, optical, and linkage) indicates that the bonds between the metal atoms and ligands in these isomers are not rapidly broken and re-formed under ordinary conditions. If the ligand-to-metal bonds did dissociate rapidly, equilibrium would be rapidly achieved and it would be possible to isolate only the thermodynamically stable isomer (or mixture, in the case of optical isomers). Rapid interconversion of isomers is found in many complexes. For example, aluminum(III) forms a trisacetylacetonato complex, $Al(C_5H_7O_2)_3$, which undoubtedly consists of a mixture of $(+)$ and $(-)$ optical isomers. However, because of the rapid interconversion of these isomers, nobody has thus far succeeded in separating them. Obviously the prediction of the relative kinetic stabilities of coordination compounds is an important problem, one which we shall take up in Chap. 19.

PROBLEMS

*14.1 Plot, on one graph, separate curves showing the concentrations of Hg^{2+}, $HgCl^+$, etc., as functions of log (Cl^-) for an aqueous system $0.1\ M$ in mercury(II). What can you conclude about the relative stabilities of the complexes?

*14.2 In crystalline AgO, half of the silver atoms are linearly coordinated to two nearest-neighbor oxygen atoms and half are coordinated to four nearest-neighbor oxygen atoms in a square planar configuration. Explain.

14.3 Predict the structures of the following discrete complexes: OsO_4, $VOCl_3$, $Pt(NH_3)_4^{2+}$, $Ag(NH_3)_2^+$, $Pt(PPh_3)_4$, Nb_2Cl_{10}, $Cr_2O_7^{2-}$, Ru_4F_{20}.

14.4 Draw all the possible isomers (including optical) for each of the following complexes. (Be careful not to draw the same structure twice.) (a) $Co(en)_2(H_2O)Cl^{2+}$, (b) $Co(NH_3)_3(H_2O)ClBr^+$, (c) $Pt(NH_3)(NH_2OH)(NO_2)(C_5H_5N)^+$.

14.5 Which type of octahedral distortion would you expect in $Co(en)_3^{3+}$? In trans-$Co(NH_3)_4Cl_2^+$?

*14.6 The ^{13}C nmr spectrum of an aqueous solution of $Mo(CN)_8^{4-}$ shows only one peak. What may one conclude about the structure of the aqueous complex ion?

14.7 How many isomers are possible for the complexes (a) MX_4Y_2 and MX_3Y_3, assuming trigonal prism geometry, (b) MX_6Y_2, assuming square antiprism geometry, and (c) MX_6Y_2, assuming dodecahedron geometry?

*14.8 Suggest experimental techniques for determining whether $C_7H_7Mo(CO)_2C_5H_5$ has the structure $(\eta^7\text{-}C_7H_7)Mo(CO)_2(\eta^1\text{-}C_5H_5)$ or $(\eta^3\text{-}C_7H_7)Mo(CO)_2(\eta^5\text{-}C_5H_5)$, aside from an x-ray structure determination. Explain how you would interpret the experimental data.

*14.9 Platinum(II) forms a 4-coordinate complex with one molecule of mesostilbenediamine,

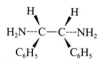

and one molecule of isobutylenediamine.

$$\text{H}_2\text{N}-\overset{\overset{\displaystyle \text{CH}_3}{\displaystyle |}}{\underset{\underset{\displaystyle \text{CH}_3}{\displaystyle |}}{\text{C}}}-\text{CH}_2\text{NH}_2$$

It is possible to resolve the complex into optically active isomers. Show that this result proves that the complex does not have tetrahedral geometry and that the result is consistent with square planar geometry.

14.10 Sketch three different trigonal bipyramidal structures for $(\text{Me}_2\text{PCH}_2\text{CH}_2\text{PMe}_2)_2\text{FeCO}$ and identify the optical isomers. Do the same for square pyramidal.

CHAPTER

15

THE
BONDING IN
COORDINATION
COMPOUNDS

Chemists have found that two approaches are particularly valuable for describing the bonding in transition-metal complexes: the "electrostatic ligand-field" approach and the molecular orbital (MO) approach. These methods are fundamentally quite different; the first assumes essentially pure ionic interactions[1] between the central metal ion and the surrounding ligand atoms or anions, and the second assumes the formation of covalent bonds between the central metal atom and surrounding ligands. Yet, as we shall see, these apparently contradictory methods lead to electronic energy level diagrams which are remarkably similar. Although the MO approach gives a more realistic and more widely applicable picture of bonding, the qualitative results of the simpler electrostatic ligand-field theory are completely adequate for many purposes.

In this chapter we describe both approaches, using only a few examples to illustrate their application. In subsequent chapters we shall consider in detail various physical and chemical properties of transition-metal compounds, using one or the other of these bonding approaches as a basis for systemization.

---

[1] The pure electrostatic approach, first expounded by Bethe in 1929, is sometimes referred to as "crystal-field theory," and the expression "ligand-field theory" is sometimes reserved for a modification in which certain physically unrealistic features of the model are corrected for. However, the expression "crystal-field theory" has misleading connotations, and "electrostatic ligand-field theory" is much more descriptive.

ELECTROSTATIC LIGAND-FIELD THEORY

Let us consider the bonding in transition-metal complexes from a very simple electrostatic point of view. We may consider an octahedral complex as a metal ion surrounded by either six negative ions or six dipolar groups with their negative ends pointing toward the metal ion. These ligands occupy the vertices of an imaginary octahedron. It is completely arbitrary how we align the cartesian coordinates of the metal ion with respect to the six ligands; for convenience, we suppose that the ligands lie exactly on the x, y, and z axes of the metal atom, as shown in Fig. 15.1. In view of the spatial configurations of the five metal d orbitals (see Fig. 1.9), which are degenerate in the free atom or ion, it is obvious that these d orbitals are no longer degenerate in the octahedral electric field of the ligands. In the complex, the lobes of the d_{xy}, d_{xz}, and d_{yz} orbitals are directed *between* the ligands, and the lobes of the d_{z^2} and $d_{x^2-y^2}$ orbitals are directed *toward* the ligands. Thus the five d orbitals are split by the octahedral field into two groups: a triply degenerate group composed of the d_{xy}, d_{xz}, and d_{yz} orbitals (which are labeled by the group-theoretical symbol t_{2g}) and a doubly degenerate group composed of the d_{z^2} and $d_{x^2-y^2}$ orbitals (labeled by the symbol e_g). From Fig. 1.9, it should be obvious that, in an octahedral environment, the t_{2g} orbitals are equivalent, but it is not obvious that the e_g orbitals are equivalent; however, this equivalence becomes apparent when one recognizes that the d_{z^2} orbital is a kind of hybrid of the $d_{z^2-x^2}$ and $d_{z^2-y^2}$ functions.

The d electrons of the metal ion would obviously prefer to occupy the t_{2g} set of orbitals than the e_g set of orbitals. That is, the e_g energy level lies above the t_{2g} energy level. Figure 15.2 is a plot of the energies of the d orbitals of a metal atom as a function of the distribution of six negative charges on a spherical shell surrounding the atom. The left side of the diagram corresponds to a uniform distribution of the six charges on the spherical shell; the right side corresponds to placing the negative charges at octahedral positions. Because the total energy of the d-orbital system is independent of the way the negative charges are distributed on the sphere, the "center of gravity" of the energy levels remains constant. Hence the e_g level is raised 1.5 times as much as the t_{2g} level is lowered. If we call

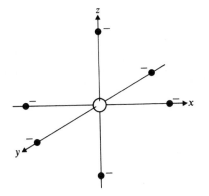

FIGURE 15.1
Octahedral arrangement of negative ligands around a metal ion.

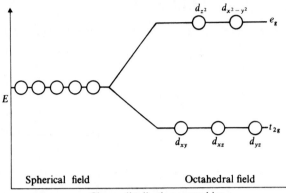

FIGURE 15.2
Splitting of d-orbital energies in an octahedral field.

the total octahedral-field splitting Δ_o, then the relative energies of the e_g and t_{2g} levels are $\frac{3}{5}\Delta_o$ and $-\frac{2}{5}\Delta_o$, respectively.[2]

Let us now consider a tetrahedral complex. For convenience, we shall consider the ligands to occupy tetrahedral corners of a cube positioned so that the cartesian coordinates of the metal atom pass through the centers of the cube faces, as shown in Fig. 15.3. The lobes of the d_{xy}, d_{xz}, and d_{yz} orbitals are directed toward the midpoints of the cube edges, at an angle of 35.3° with respect to the ligands. The lobes of the d_{z^2} and $d_{x^2-y^2}$ orbitals are directed toward the cube faces and bisect the angles between pairs of ligands, thus making an angle of 54.7° with the ligands. Calculations show that electrons in the d_{z^2} and $d_{x^2-y^2}$ orbitals are repelled less by the ligands than are electrons in the d_{xy}, d_{xz}, and d_{yz} orbitals. As a result, we obtain a splitting of the d-orbital energies into two

---

[2]   In some literature, Δ_o is referred to as $10Dq$.

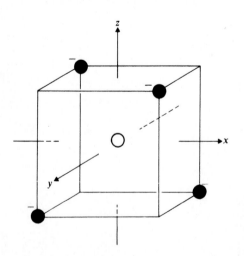

FIGURE 15.3
Tetrahedral arrangment of negative ligands around a metal ion.

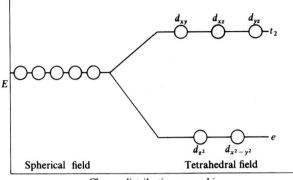

FIGURE 15.4
Splitting of d-orbital energies in a tetrahedral field.

levels, as shown in Fig. 15.4. The energy of the triply degenerate t_2 set is raised $\frac{2}{5}\Delta_t$, and the energy of the double degenerate e set is lowered $\frac{3}{5}\Delta_t$. Calculations show that, if the metal-ligand distances are maintained constant, $\Delta_t = \frac{4}{9}\Delta_o$; this relation is qualitatively predictable from the fact that the total negative charge of four ligands is less than that of six ligands.

It is interesting to consider the effect on the d-orbital energy levels of a tetragonal elongation of an octahedral complex. If we pull out the ligands on the z axis and push in the ligands on the x and y axes so as to maintain the total coulomb energy constant, the d_{z^2} and $d_{x^2-y^2}$ orbitals are no longer equivalent, and their energies diverge, as shown in Fig. 15.5. Similarly, the

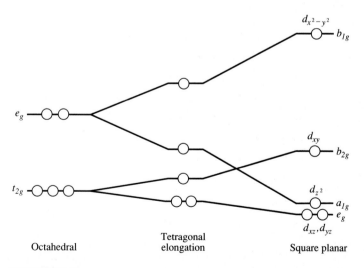

FIGURE 15.5
Splitting of d-orbital energies in a tetragonally distorted octahedral field and in a square planar field.

d_{xy} orbital is no longer equivalent to the d_{xz} and d_{yz} orbitals, and its energy is raised relative to that of the latter orbitals. In the limit of a square planar configuration, reached when the axial ligands have been completely withdrawn, the $d_{x^2-y^2}$ orbital definitely has the highest energy, but the relative energies of the other orbitals depend on quantitative properties of the metal ion and ligands. Theoretical calculations and experimental data indicate that the energy of the d_{z^2} orbital becomes comparable to that of the d_{xz} and d_{yz} orbitals in square complexes of Co(II), Ni(II), and Cu(II), as shown in Fig. 15.5.

It should be pointed out that the separations between the energy levels in Figs. 15.2, 15.4, and 15.5 are generally of the order of only a few electronvolts. These energy differences are relatively small compared to typical total interaction energies between positively charged metal ions and negatively charged ligands. For example, the energy required to dissociate completely the coordinated water molecules from the metal ion in a complex such as $Fe(H_2O)_6^{3+}$ (in the gas phase) is of the order of magnitude of 40 eV.[3]

MOLECULAR ORBITAL THEORY

We have seen that, by application of electrostatic ligand-field theory, we predict that the d-orbital energies of a transition-metal ion are split into two or more levels when the ion is subjected to a nonspherical electrostatic field such as it would experience in a coordination compound. Before discussing the various chemical ramifications of such orbital splitting, it will be useful to consider an entirely different approach to the bonding in transition-metal compounds, viz., MO theory, to show that the MO approach leads essentially to the same qualitative results. The MO approach is more realistic than the purely electrostatic approach and can explain several phenomena that cannot be explained by the electrostatic approach.

Let us apply simple MO theory to the σ bonding of a regular octahedral complex. The metal atom has, in its valence shell, five d orbitals, an s orbital, and three p orbitals. Each of the six equivalent ligand atoms contributes an atomic orbital, or a hybrid atomic orbital, to the σ-bonding system. We are concerned with how these 15 orbitals combine to form MOs. The combination of ligand orbitals which has the same symmetry as, and can interact with, the metal s orbital to yield a bonding MO and an antibonding MO is represented by the following group orbital wave function,[4]

$$\phi(a_{1g}) = \frac{1}{\sqrt{6}}(\sigma_1 + \sigma_2 + \sigma_3 + \sigma_4 + \sigma_5 + \sigma_6)$$

---

[3]   F. Basolo and R. G. Pearson, "Mechanisms of Inorganic Reactions," 2d ed., pp. 80–85, Wiley, New York, 1967.

[4]   The coefficient $1/\sqrt{6}$ is the "normalization constant," the value of which is determined by the requirement that the wave function be normalized. It has been assumed that $S = 0$. The coefficients of the other wave functions discussed in this paragraph are similarly calculated normalization constants.

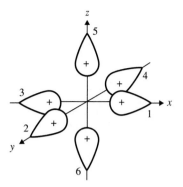

FIGURE 15.6
Combination of ligand σ orbitals which can interact with the metal s orbital.

where σ_1, σ_2, etc., stand for the ligand atomic orbital wave functions. This ligand combination is pictured in Fig. 15.6. The ligand-orbital combination which can interact with the metal d_{z^2} orbital is represented by the following function,

$$\phi(e_g) = \frac{1}{2\sqrt{3}}(2\sigma_5 + 2\sigma_6 - \sigma_1 - \sigma_2 - \sigma_3 - \sigma_4)$$

and is pictured in Fig. 15.7. The ligand-orbital combination which can interact with the metal $d_{x^2-y^2}$ orbital is represented as follows,

$$\phi(e_g) = \tfrac{1}{2}(\sigma_1 - \sigma_2 + \sigma_3 - \sigma_4)$$

and is pictured in Fig. 15.8. The combinations suitable for interacting with the metal p orbitals are given in the following functions:

$$\phi(t_{1u}) = \frac{1}{\sqrt{2}}(\sigma_1 - \sigma_3) \qquad \text{for } p_x$$

$$\phi(t_{1u}) = \frac{1}{\sqrt{2}}(\sigma_2 - \sigma_4) \qquad \text{for } p_y$$

$$\phi(t_{1u}) = \frac{1}{\sqrt{2}}(\sigma_5 - \sigma_6) \qquad \text{for } p_z$$

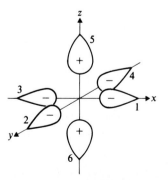

FIGURE 15.7
Combination of ligand σ orbitals which can interact with the metal d_{z^2} orbital.

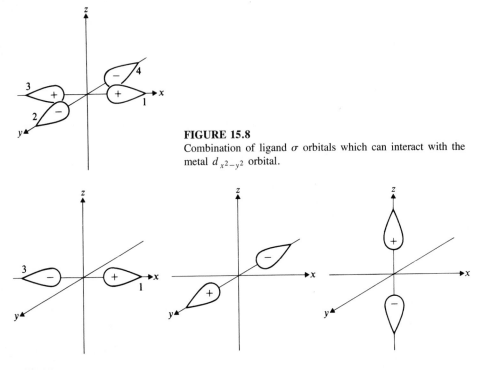

FIGURE 15.8
Combination of ligand σ orbitals which can interact with the metal $d_{x^2-y^2}$ orbital.

FIGURE 15.9
Combinations of ligand σ orbitals which can interact with the metal p orbitals.

These combinations are shown in Fig. 15.9. The metal d_{xz}, d_{xy}, and d_{yz} orbitals have no net overlap with the ligand σ orbitals, and therefore these d orbitals, which have t_{2g} symmetry, are nonbonding. The energy level diagram for the MOs which arise from the interactions which we have just described is shown in Fig. 15.10. The transition-metal valence s and p orbitals lie above the valence d orbitals in this diagram (rather than below, as in the free metal atom) because in a complex the metal atom generally has a positive charge, and the positive charge causes the relative energies of the orbitals to approach those of a hydrogen-like atom.[5] Thus in a metal ion from the first transition series, the energies of the $4s$ and $4p$ orbitals lie above those of the $3d$ orbitals. Notice that the separation between the t_{2g} and $e_g{}^*$ levels is identified with Δ_o. Each ligand contributes 2 electrons to the MOs; these 12 electrons can be used to fill the bonding a_{1g}, t_{1u}, and e_g MOs. The metal ion contributes as many valence electrons as it possesses. Thus an octahedral complex of a d^1 metal ion, such as Ti(H$_2$O)$_6{}^{3+}$, has one electron in one of the nonbonding t_{2g} orbitals. An octahedral complex of a d^8 metal ion, such as Ni(H$_2$O)$_6{}^{2+}$, has six electrons in the t_{2g} orbitals and two unpaired electrons in the $e_g{}^*$ orbitals.

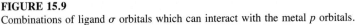

[5]   See p. 13.

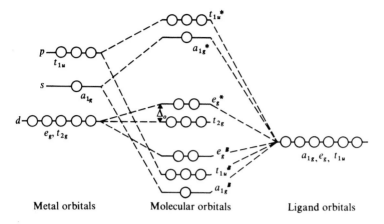

FIGURE 15.10
Molecular orbital energy level diagram for an octahedral complex with only σ bonding.

Notice that the descriptions of the d electrons are analogous to those using the energy level diagram of Fig. 15.2 based on electrostatic ligand-field theory. However, the MO treatment of the bonding has advantages over the corresponding electrostatic ligand-field treatment. The MO diagram shows the importance and significance of the σ bonding due to the electron pairs on the six ligands. We can ascribe various degrees of metal and ligand character to the MOs. Thus the e_g^B, t_{1u}^B, and a_{1g}^B orbitals have largely ligand lone-pair character but have acquired some of the character of the metal orbitals from which they are derived. On the other hand, the e_g^*, t_{1u}^*, and a_{1g}^* orbitals have mainly the characters of the corresponding metal orbitals but are somewhat delocalized onto the ligands. Only the nonbonding t_{2g} orbital has pure metal d orbital character. It is obvious from Fig. 15.10 that various electronic transitions, involving the e_g^B, t_{1u}^B, a_{1g}^B, a_{1g}^*, and t_{1u}^* orbitals, are possible in addition to those implied by the simplified diagram in which only the t_{2g} and e_g^* orbitals are shown. Of course, it should be recognized that even Fig. 15.10 is usually an approximation because it was constructed assuming the complete absence of π bonding.

Pi Bonding

The bonding of H^- is entirely σ-donor bonding, and that of NH_3 and CH_3^- is principally σ-donor bonding. Therefore the bonding of such ligands to metal atoms can be adequately described by an energy level diagram like that in Fig. 15.10, in which π bonding is ignored. However, ligands such as Cl^-, O^{2-}, and CO_3^{2-}, when engaged in σ bonding to metal atoms, have electrons in essentially nonbonding $p\pi$ orbitals of the donor atoms which can interact with appropriate $d\pi$ orbitals of the metal atoms. Similarly, ligands such as CN^-, CO, and NO^+ have empty antibonding π^* orbitals which can interact with metal $d\pi$ orbitals. In fact, it is believed that ligands such as CF_3^- and PF_3 also have empty orbitals which can interact with metal $d\pi$ orbitals. In the case of CF_3^-, it has been proposed that the empty orbitals are the antibonding σ^* orbitals of the C—F bonds, and in the case of PF_3, it has been proposed that the empty orbitals are the valence-shell d orbitals of phosphorus and the σ^* orbitals of the P—F bonds. All the π-type interactions which we have described are schematically indicated in the following sketches.

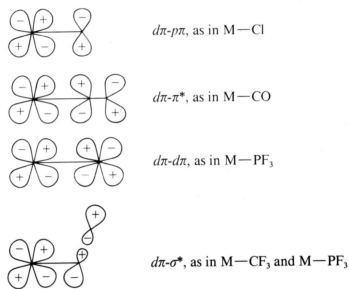

$d\pi$-$p\pi$, as in M—Cl

$d\pi$-π^*, as in M—CO

$d\pi$-$d\pi$, as in M—PF_3

$d\pi$-σ^*, as in M—CF_3 and M—PF_3

In an ML_6 complex in which each ligand donor atom has two π orbitals that can interact with the metal orbitals, the ligand π orbitals may be combined into four group orbitals of symmetry t_{1g}, t_{2g}, t_{1u}, and t_{2u}. The t_{1g} and t_{2u} combinations are nonbonding because there are no metal orbitals having these symmetries, and the t_{1u} combinations are essentially nonbonding because the metal p orbitals, which have this symmetry, are engaged in strong σ bonding with the ligands. However, the t_{2g} combinations can interact significantly with the metal t_{2g} orbitals. Two types of interaction are possible, differing as to whether the ligand π orbitals are filled and of lower energy than the metal t_{2g} orbitals, or empty and of higher energy than the metal t_{2g} orbitals. The first type of interaction, such as found in $MnCl_6^{4-}$, is illustrated in the energy level diagram of Fig. 15.11. Notice that, in this case, the $p\pi$-$d\pi$ interaction causes a decrease in Δ_o. The second type of interaction, such as found in $Fe(CN)_6^{4-}$, is illustrated in the energy level diagram of Fig. 15.12. In this case, the $d\pi \rightarrow p\pi$ back bonding causes an increase in Δ_o. Such back bonding serves to transfer electrons from the metal atom (which would be excessively negative in the absence of back bonding) to the ligand. The combination of σ donor bonding and π back bonding involved in the coordination of species such as CN^-, CO, N_2, etc., is often referred to as "synergic bonding." The σ basicity of N_2 is much too low to account for the strength of the many transition-metal–N_2 bonds that are known. However, the combination of σ and π bonding is remarkably strong, perhaps because it does not require a large net interatomic transfer of charge.

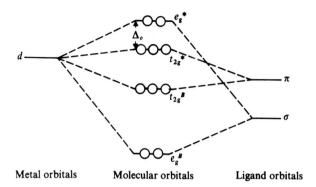

FIGURE 15.11
Energy level diagram showing the interaction of the $t_{2g}d$ orbitals with filled π orbitals of the ligands. The nonbonding ligand π orbitals are not shown. Notice that the interaction tends to decrease Δ_o.

The bonding of a transition-metal atom to an olefin can be similarly described in terms of a combination of a σ olefin $\rightarrow$ metal bond and a π metal $\rightarrow$ olefin bond, as indicated in the following sketches:

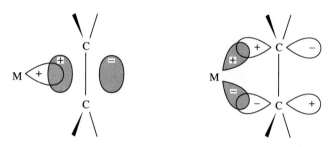

It is also acceptable to assume that the two metal d orbitals involved in the bonding

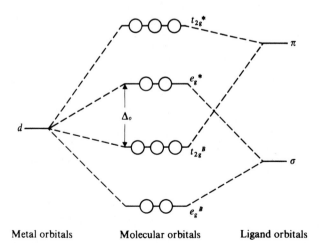

FIGURE 15.12
Energy level diagram showing the interaction of the $t_{2g}d$ orbitals with empty π orbitals of the ligands. The ligand π-orbital combinations of other symmetries are not shown. Notice that the interaction tends to increase Δ_o.

form two equivalent hybrid orbitals and to describe the bonding as analogous to that in cyclopropane:

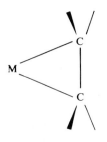

The Metal-Metal Bond[6]

Metal-metal bonding occurs in many transition-metal binuclear complexes in which the metal atoms are close enough to allow interaction of their d orbitals. It is helpful to consider first the bonding in a bare M_2 molecule. If the M—M bond axis is taken as the z axis, the d orbitals will be split more or less as indicated in the first energy level diagram of Fig. 15.13. The d_{z^2} orbitals on the atoms will combine to form a bonding σ MO and an antibonding σ^* MO. The d_{xz} and d_{yz} orbitals, which are physically indistinguishable, will combine to form two degenerate bonding π MOs and two degenerate antibonding π^* MOs. The $d_{x^2-y^2}$ and d_{xy} orbitals on the two atoms overlap each other to form δ MOs, i.e., MOs each of which has two nodal planes containing the M—M axis. Therefore degenerate pairs of bonding δ MOs and antibonding δ^* MOs will form. Because orbital overlap increases in the order $\delta < \pi < \sigma$, the bonding-antibonding separation increases in the same order, as indicated in Fig. 15.13.

When ligands are introduced along the $\pm x$ and $\pm y$ axes, such that these axes are in the eclipsed configuration (D_{4h} symmetry[7]), the $\delta_{x^2-y^2}$ orbitals are strongly destabilized by σ interactions with the ligands. Although the relative positions of the orbitals depend on the ligands and metals, an energy level diagram like the second one in Fig. 15.13 is expected. Note that the bonding M—L orbitals are not shown in the figure. Complexes with two d^4 metal atoms, such as $Re_2Cl_8^{2-}$ and $Mo_2Cl_8^{4-}$, generally have D_{4h} geometry and have short quadruple M≡M bonds, corresponding to the electron configuration $\sigma^2\pi^4\delta^2$. The eclipsed (D_{4h}) geometry of these complexes is strong experimental evidence for quadruple bonding. If the complexes had the sterically preferred, staggered (D_{4d}) geometry, the δ bonds would be broken and the metal atoms would be only triple-bonded.

The δ-bonding electrons are not very tightly bound, and it is possible to prepare metal-metal bonded compounds in which the δ orbitals are empty. Examples

---

[6]   W. C. Trogler, *J. Chem. Educ.*, **57**, 424 (1980); F. A. Cotton and R. A. Walton, "Multiple Bonds Between Metal Atoms," Wiley, New York, 1982.

[7]   See Chap. 2 for a discussion of symmetry and group-theoretical symbols such as D_{4h} and D_{4d}.

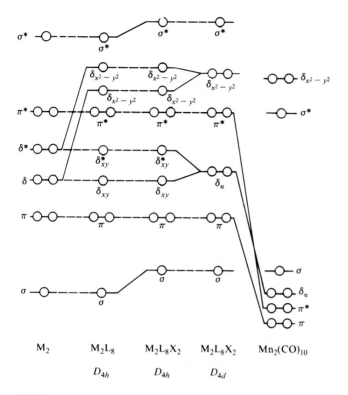

FIGURE 15.13
Energy level diagrams for the orbitals derived from the metal d orbitals of dinuclear metal complexes, showing the effects of the ligands. The bonding M—L and M—X orbitals are not shown.

of such M≡M $\sigma^2\pi^4$ triple bonding are found in the complexes $Mo_2(HPO_4)_4^{2-}$ and $Re_2(C_3H_5)_4$. Because the antibonding δ^* orbital is not far above the bonding δ-orbital, electrons may be placed in it to give the $\sigma^2\pi^4\delta^2\delta^{*2}$ configuration. This kind of triple bond, in which the δ bonding cancels the δ^* antibonding, is found in $Re_2Cl_4(PR_3)_4$. Even further addition of antibonding electrons can take place, each antibonding pair causing a decrease in bond order. In $Rh_2(O_2CR)_4$, for example, there are 14 metal valence electrons and only a single bond, $\sigma^2\pi^4\delta^2\delta^{*2}\pi^{*4}$.

In complexes of the type $M_2L_8X_2$, in which X ligands are introduced along the z axis, the σ donor orbitals of the X ligands interact with the σ and σ^* M—M orbitals. Two bonding M—X orbitals are formed, and the energies of the σ and σ^* M—M orbitals are raised, as shown in the third energy level diagram of Fig. 15.13. To the extent that the σ M—M orbital becomes involved in M—X bonding, the M—M bond is weakened. Thus Cr_2^{4+} complexes without axial ligands have very short Cr—Cr bonds (< 2.0 Å), whereas those with axial ligands have relatively long Cr—Cr bonds (2.2 to 2.5 Å). In fact, the shorter the Cr—X_{axial} distance in the latter compounds, the longer the Cr—Cr distance.

In staggered $M_2L_8X_2$ complexes (D_{4d} symmetry), the σ and π orbitals are the same as in the corresponding eclipsed complexes. However, the two $\delta_{x^2-y^2}$

orbitals (antibonding M—L orbitals) become equivalent, and the two d_{xy} orbitals also become equivalent and essentially nonbonding as indicated in the fourth energy level diagram of Fig. 15.13. This set of MOs can easily accommodate 10 d electrons, but in the case of a species such as $Mn_2(CO)_{10}$, which has a D_{4d} structure and 14 d electrons, it would apparently be necessary to put four electrons into strongly antibonding π^* orbitals. We can rationalize the existence of $Mn_2(CO)_{10}$ by recognizing that the strong π-acceptor CO ligands lower the energies of the d orbitals with appropriate symmetry, especially the π, π^*, and δ_n orbitals. Thus we obtain the fifth energy level diagram of Fig. 15.13, which can accommodate all 14 d electrons of $Mn_2(CO)_{10}$ in low-lying MOs, corresponding to an Mn—Mn single bond.

An important class of triply bonded species are those of the type $X_3M\equiv MX_3$, exemplified by $Mo_2(NMe_2)_6$ and $W_2(NEt_2)_4Cl_2$. In these compounds the trigonally disposed ligands raise the energy of both δ orbitals, leaving only σ and π bonding. In the absence of δ bonding, the repulsions between ligands on the two metal atoms lead to a staggered ligand configuration for all $X_3M\equiv MX_3$ molecules.

WEAK- AND STRONG-FIELD COMPLEXES

An octahedral complex having 4, 5, 6, or 7 d electrons can have either of two ground-state electron configurations. In a weak field, the maximum number of electrons are unpaired, as indicated by the following energy level diagrams:

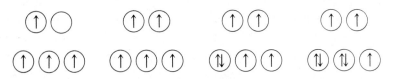

These configurations are referred to as "high-spin." In a strong field, the energy gap Δ_o is greater than the energy required to pair electrons in the same orbital and the t_{2g} level is filled as far as possible, as shown by the following diagrams:

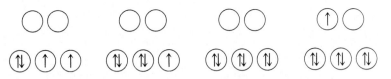

These configurations are referred to as "low-spin."

Many of the physical and chemical properties of low-spin complexes are markedly different from those of high-spin complexes. For example, the effective radius of the metal atom, the magnetic properties of the complex, the redox reactivity of the complex, the substitution lability of the complex, the geometric distortions of the complex, the optical absorption spectrum of the complex, and the strengths of the metal-ligand bonds are all factors which are a function of whether or not the complex is low-spin or high-spin. In succeeding chapters, we shall discuss these factors in relation to electron configurations.

In tetrahedral complexes, Δ_t is usually not great enough to give a low-spin configuration; that is, almost all tetrahedral complexes are "weak field." Exceptions are found among some organometallic compounds, e.g., $Co(Nor)_4$, in which cobalt is bonded to four 1-norbornyl groups.[8]

PROBLEMS

15.1 Sketch the d-orbital energy level diagrams for complexes of the following type: (*a*) 2-coordinate, linear; (*b*) 5-coordinate, trigonal bipyramidal; (*c*) 5-coordinate, square pyramidal, with the metal atom in the basal plane.

15.2 Rationalize the fact that Δ_o increases in the order $CrCl_6{}^{3-}$, $Cr(NH_3)_6{}^{3+}$, $Cr(CN)_6{}^{3-}$.

15.3 Rationalize the fact that Δ_o increases in the order $Co(H_2O)_6{}^{2+}$, $Co(H_2O)_6{}^{3+}$, $Rh(H_2O)_6{}^{3+}$.

15.4 Assuming that $d\pi \rightarrow p\pi$ back bonding occurs to the fullest possible extent in $Fe(CN)_6{}^{4-}$, use valence bond theory to estimate the average bond orders of the Fe—C and C—N bonds. What are the formal charges of the atoms?

15.5 Show that the complex $Pt(C_2H_4)Cl_3{}^-$, shown in Fig. 14.13*a*, can be looked upon as a square planar complex or, somewhat unrealistically, as a trigonal bipyramidal complex.

15.6 Which electron configuration of a d^5 transition-metal complex, the high-spin or low-spin configuration, would you expect to be favored by the application of high pressure?

*15.7 Why are the $ReCl_4$ groups of the complex $Re_2Cl_8{}^{2-}$ in the eclipsed configuration rather than staggered?

*15.8 Explain why $IrCl_6{}^{2-}$ has only one unpaired electron. Also explain the fact that the esr spectrum of $IrCl_6{}^{2-}$ is split into a multiplet by the nuclear moments of the chlorine atoms, showing that the electron spin density is delocalized onto the chlorine atoms.

15.9 Sketch the combination of chlorine p orbitals in the xy plane of $MnCl_6{}^{4-}$ which has the symmetry of the manganese d_{xy} orbital.

15.10 Would you expect the metal-metal bond to lengthen, shorten, or remain about the same on going from (*a*) $Mo_2(SO_4)_4{}^{4-}$, to $Mo_2(SO_4)_4{}^{3-}$ and (*b*) $Tc_2Cl_8{}^{3-}$ to $Tc_2Cl_8{}^{2-}$?

*15.11 The gas-phase Mo_2 molecule is believed to have a sextuple bond. How can this be, in view of the d-orbital energy level diagram in Fig. 15.13?

---

[8] E. K. Byrne, D. S. Richeson, and K. H. Theopold, *J. Chem. Soc., Chem. Commun.*, 1491 (1986).

CHAPTER

16

THE
18-ELECTRON
RULE

Now that we have discussed the theoretical aspects of the bonding in transition-metal coordination compounds, it is appropriate to introduce simple electron-counting rules, which can be justified using molecular orbital theory, for predicting the structures and formulas of these compounds. First we will discuss the 18-electron rule as applied to relatively simple complexes. Then we will show that the concept of oxidation state, applied to transition-metal compounds, is not always as straightforward as one might think. Finally we will apply a combination of the $2n + 2$ rule (covered in Chap. 10) and the 18-electron rule to metal cluster compounds.

In Chap. 3 we discussed the basis and application of the Lewis octet theory. The theory ascribes special stability to an electron configuration in which each non-transition-element atom achieves a number of valence electrons equal to that of a rare-gas atom, i.e., 8 valence electrons. Because transition-metal atoms have five d orbitals in addition to one s and three p orbitals in their valence shells, extension of the Lewis theory to transition-metal compounds corresponds to ascribing special stability to an electron configuration in which each transition-metal atom achieves 18 valence electrons. Thus we have the 18-electron rule (alias the rare-gas, nine orbital, or effective atomic number rule). This rule is not rigorously followed by some transition-metal compounds; in fact, Mitchell and Parish[1] pointed out that transition-metal compounds fall into three groups.

---

[1] P. R. Mitchell and R. V. Parish, *J. Chem. Educ.*, **46**, 811 (1969).

The first group consists of compounds with weak ligand fields. The 18-electron rule plays no part in determining the electron configurations of these compounds. Thus, in the case of octahedral complexes with low Δ values, there is essentially no restriction (except the Pauli restriction) on the number of electrons which can occupy the nonbonding t_{2g} and weakly antibonding $e_g{}^*$ orbitals. *In principal, the number of valence electrons can range from 12 to 22.* Some examples of such complexes are given in Table 16.1. It can be seen that all these examples involve first-row transition elements in low or medium oxidation states, with no strongly back-bonding ligands such as CO, olefins, or arenes. Notice that, although $Mn(CN)_6{}^{3-}$ is low-spin, it does not conform to the 18-electron rule.

The second group consists of compounds with relatively high Δ values but with ligands which do not engage in strong back bonding. In octahedral complexes of this type, there is essentially no restriction on the number of nonbonding t_{2g} electrons, but because of the high energy of the $e_g{}^*$ orbitals, electrons are forbidden to occupy these orbitals. *Therefore the number of valence electrons can range only from 12 to 18.* Some examples are given in Table 16.2. It can be seen that all the examples are second- and third-row transition-metal complexes, which have higher Δ values than the corresponding first-row metal complexes. There are no well-established complexes of these heavier metals with more than 18 valence electrons.

The third group consists of compounds with high Δ values and ligands which strongly back-bond. These compounds fairly rigorously conform to the 18-electron rule. In octahedral complexes of this type, the t_{2g} orbitals are *bonding*, and therefore it is energetically favorable for these orbitals to be completely filled. Obviously the 18-electron rule is most useful as a predictive guide for compounds of this category, particularly for π-bonded organometallic and carbonyl compounds. Some simple examples are given in Table 16.3.

Complexes with d^8 metal atoms sometimes have 18 valence electrons and sometimes only 16 valence electrons. The 18-electron complexes are found when

TABLE 16.1
Octahedral complexes with relatively low Δ values

Complex	No. of d electrons provided by metal ion	Total no. of valence electrons on metal atom
$TiF_6{}^{2-}$	0	12
$VCl_6{}^{2-}$	1	13
$V(C_2O_4)_3{}^{3-}$	2	14
$Cr(NCS)_6{}^{3-}$	3	15
$Mn(CN)_6{}^{3-}$	4	16
$Fe(C_2O_4)_3{}^{3-}$	5	17
$Fe(H_2O)_6{}^{2+}$	6	18
$Co(H_2O)_6{}^{2+}$	7	19
$Ni(en)_3{}^{2+}$	8	20
$Cu(NH_3)_6{}^{2+}$	9	21
$Zn(en)_3{}^{2+}$	10	22

TABLE 16.2
Complexes with high Δ values but with no strong back bonding†

Complex	No. of d electrons provided by metal ion	Total no. of valence electrons on metal atom
$ZrF_6{}^{2-}$	0	12
$ZrF_7{}^{3-}$	0	14
$Zr(C_2O_4)_4{}^{4-}$	0	16
$WCl_6{}^{-}$	1	13
$TcF_6{}^{2-}$	3	15
$OsCl_6{}^{2-}$	4	16
$W(CN)_8{}^{3-}$	1	17
$W(CN)_8{}^{4-}$	2	18
PtF_6	4	16
$PtF_6{}^{-}$	5	17
$PtF_6{}^{2-}$	6	18
$PtCl_4{}^{2-}$	8	16

† P. R. Mitchell and R. V. Parish, *J. Chem, Educ.*, **46**, 811 (1969).

the ligands are strongly back-bonding and can remove much of the electron density contributed to the metal atom by σ bonding. Thus one finds complexes such as $Fe(CO)_5$, $Fe(CNR)_5$, and $Pt(SnCl_3)_5{}^{3-}$. The 16-electron complexes are found with ligands which do not back-bond as strongly and cannot remove much electron density from the metal atom. Examples are $AuCl_4{}^{-}$, $PdCl_4{}^{2-}$, and $Ni(C_4H_7N_2O_2)_2$ [bis(dimethylglyoximato)nickel(II)]. The cyanide complexes of nickel(II) are a borderline case; both $Ni(CN)_4{}^{2-}$ and $Ni(CN)_5{}^{3-}$ are known.

The 18-electron rule is so valuable that a detailed discussion of its application is warranted. In the following paragraphs we shall show how one can rationalize the structure and bonding of a series of compounds of various degrees

TABLE 16.3
Complexes with back bonding which conform to the 18-electron rule

Complex	No. of d electrons provided by metal ion or atom	Total no. of valence electrons on metal atom
$V(CO)_6{}^{-}$	6	18
$Mo(CO)_3(PF_3)_3$	6	18
$HMn(CO)_5$	7	18
$Ni(CN)_5{}^{3-}$	8	18
$Fe(CO)_5$	8	18
$CH_3Co(CO)_4$	9	18
$Co(CO)_4{}^{-}$	10	18
$Ni(CNR)_4$	10	18

of complexity. First let us consider a simple example, $Cr(CO)_6$. The chromium atom furnishes 6 valence electrons, and the six carbon monoxide ligands furnish 12 electrons.

$$OC—\overset{\overset{\displaystyle O}{\overset{\displaystyle \|}{C}}}{\underset{\underset{\displaystyle C}{\underset{\displaystyle \|}{O}}}{\underset{}{Cr}}}\overset{\displaystyle {}_{C}{}^{\displaystyle \nearrow O}}{—CO}$$

Cr:	6
6 CO:	12
	18

In $Mn_2(CO)_{10}$, two $Mn(CO)_5$ groups are joined by an Mn—Mn bond. The formation of this bond effectively adds one valence electron to each manganese atom:

Mn:	7
5 CO:	10
Mn—Mn bond	1
	18

The Mn—Mn bond of $Mn_2(CO)_{10}$ can be easily cleaved by treatment with an alkali metal in an ether to give the $Mn(CO)_5^-$ ion. We may look upon this as a reduction of the Mn(0) to Mn(−I).

Mn^-:	8
5 CO:	10
	18

The $Mn(CO)_5^-$ ion reacts with alkyl halides to give organic derivatives such as $CH_3Mn(CO)_5$. The CH_3 group may be considered as a one-electron donor.

$$CH_3—\overset{\overset{\displaystyle O}{\overset{\displaystyle \|}{C}}}{\underset{\underset{\displaystyle C}{\underset{\displaystyle \|}{O}}}{\underset{}{Mn}}}\overset{\displaystyle {}_{C}{}^{\displaystyle \nearrow O}}{—CO}$$

Mn:	7
5 CO:	10
CH_3:	1
	18

Alternatively, we may consider the compound as a complex of Mn(I) and a CH_3^- ion:

Mn^+:	6
5 CO:	10
CH_3^-:	2
	18

In $Fe_2(CO)_9$, we have bridging carbon monoxide ligands (which in effect contribute one electron to each metal atom) and an Fe—Fe single bond.

Fe:	8
3 terminal CO:	6
3 bridging CO:	3
Fe—Fe bond :	1
	18

Ferrocene, $Fe(C_5H_5)_2$, may be looked upon as a complex of Fe(II) with two $C_5H_5^-$ ions (each with six π electrons) or as a complex of Fe(0) with two C_5H_5 radicals (each with five π electrons).

Fe^{2+} :	6
$2C_5H_5^-$:	12
	18

Fe:	8
$2C_5H_5$:	10
	18

It is convenient to consider bis(benzene)chromium as a complex of Cr(0) and benzene ligands:

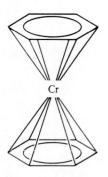

Cr:	6
$2C_6H_6$:	12
	18

The reaction of $Mo(CO)_6$ with cycloheptatriene yields the complex $(\eta^6\text{-}C_7H_8)Mo(CO)_3$, in which three essentially localized double bonds are coordinated to the molybdenum atom.

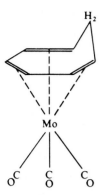

Mo:	6
C_7H_8 :	6
3 CO:	6
	18

When this compound is treated with the triphenylcarbonium ion, $C(C_6H_5)_3{}^+$, a hydride ion is abstracted, yielding a cationic complex, $(\eta^7\text{-}C_7H_7)Mo(CO)_3{}^+$ containing the planar, aromatic $C_7H_7{}^+$ ring:

Mo:	6
$C_7H_7{}^+$:	6
3 CO:	6
	18

Consider the $Re_3Cl_{12}{}^{3-}$ ion, illustrated in Fig. 16.1. A molecular orbital treatment of the bonding shows that the Re—Re bonds are double bonds. Indeed, such double bonding must also be assumed in order to obtain agreement with the 18-electron rule.

Re^{3+} :	4
$5Cl^-$:	10
2Re≡Re bonds :	4
	18

Figure 16.2 shows a skeletal representation of a molybdenum complex containing the cycloheptatrienyl group, C_7H_7, and the 3,5-dimethylpyrazolylborato anion, $H_2B(C_5H_7N_2)_2{}^-$. To rationalize this structure in terms of the 18-electron rule, it is necessary to assume that the C_7H_7 ring is bound in *trihapto* fashion and that the H atom near the molybdenum atom is engaged in a three-center B—H—Mo bond.

Mo^+ :	5
2 N donor atoms:	4
B—H—Mo :	2
2 CO:	4
trihapto C_7H_7 :	3
	18

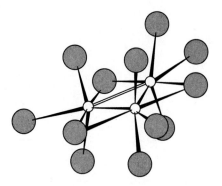

FIGURE 16.1
Structure of the $Re_3Cl_{12}{}^{3-}$ ion. [*Reproduced with permission from J. A. Bertrand, F. A. Cotton, and W. A. Dollase, J. Am. Chem. Soc., 85, 1349 (1963). Copyright 1963 American Chemical Society.*]

Exceptions to the 18-electron rule are known, even for compounds with strongly back-bonding ligands. However, most of the exceptional compounds are readily converted to compounds which follow the rule, and, in a sense, this reactivity can be looked upon as conformity to the rule. For example, both $V(CO)_6$ and $Fe(C_5H_5)_2{}^+$ have 17 valence electrons and are easily reduced to $V(CO)_6{}^-$ and $Fe(C_5H_5)_2$, respectively. The tetrakis(phosphine)platinum(O) complexes, $Pt(PR_3)_4$, have a tendency to lose PR_3, giving the 16- and 14-electron species, $Pt(PR_3)_3$ and $Pt(PR_3)_2$. This dissociation is most extensive when R is a very bulky group. Thus when R = *tert*-butyl or cyclohexyl, the $Pt(PR_3)_2$ species can actually be isolated.[2]

---

[2]   T. Yoshida and S. Otsuka, *Inorg. Syn.,***19**, 101 (1979).

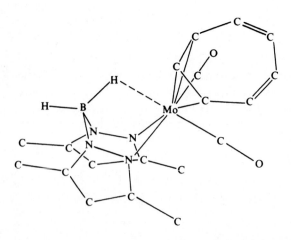

FIGURE 16.2
Skeletal representation of $[H_2B(3,5-$ dimethylpyrazolyl)$_2]$-$Mo(CO)_2C_7H_7$. [*Reproduced with permission from F. A. Cotton et al., J. Chem. Soc. Chem. Commun., 777, (1972).*]

OXIDATION-STATE AMBIGUITY

Nitrosyls

The following reactions illustrate four methods for the preparation of transition-metal nitrosyls (i.e., complexes containing the NO group as a ligand).

$$Fe_2(CO)_9 + 4NO \rightarrow 2Fe(NO)_2(CO)_2 + 5CO$$

$$Ir(PPh_3)_2(CO)Cl + NO^+ \rightarrow [Ir(PPh_3)_2(CO)(NO)Cl]^+$$

$$CoI_2 + Co + 4NO \rightarrow [Co(NO)_2I]_2$$

$$Ni(CN)_4^{2-} + NO^- \rightarrow Ni(CN)_3NO^{2-} + CN^-$$

In many nitrosyl complexes, the metal—N—O bond angle is very close to 180°, and it is convenient to assume that the ligand is the $:N{=}O:^+$ ion, which is isoelectronic with CO. In other complexes, the metal—N—O bond angle is around 130°, and it is reasonable to assume that the ligand is $:N{=}O:^-$, isoelectronic with O_2. However, these assumptions are somewhat arbitrary. For example, one may look upon *trans*-Co(diars)$_2$(NCS)NO$^+$ as a complex of Co(III) and NO$^-$,

$$L_5\overset{2+}{\overset{..}{Co}} + {}^{-}{:}N{=}\overset{..}{O} \longrightarrow L_5\overset{+}{Co}{-}N\overset{..}{\underset{\diagdown}{}}_{\overset{..}{O}{:}}$$

or as a complex of Co(I) and NO$^+$,

$$L_5\overset{..}{Co}{:} + {:}N{\equiv}O{:}^+ \longrightarrow L_5\overset{+}{Co}{-}N\overset{..}{\underset{\diagdown}{}}_{\overset{..}{O}{:}}$$

In the former case, NO$^-$ acts as a σ donor; in the latter case, NO$^+$ acts as a σ acceptor. In either case, it is easy to predict the bent configuration for the Co—NO linkage; a linear configuration would force an extra pair of electrons onto the cobalt atom, in violation of the 18-electron rule. The closely related complex, Co(diars)$_2$NO^{2+}, may be looked upon as a complex of Co(I) and NO$^+$,

$$L_4\overset{..}{Co}{}^+ + {:}N{\equiv}O{:}^+ \rightarrow L_4\overset{..}{Co}{-}\overset{+}{N}{\equiv}O{:}^+$$

or as a complex of Co(III) and NO$^-$,

$$L_4Co^{3+} + {}^{-}{:}N{=}\overset{..}{O} \rightarrow L_4\overset{..}{Co}{-}\overset{+}{N}{\equiv}O{:}^+$$

The linear geometry is readily predicted if one assumes that the cobalt atom must have 18 valence electrons. A bent configuration would remove a pair of electrons from the valence shell of the cobalt atom, leaving only 16 electrons. Of course, considerable back bonding occurs,

$$\text{L}_4\ddot{\text{C}}\text{o}-\overset{+}{\text{N}}\equiv\text{O:}^+ \quad \leftrightarrow \quad \text{L}_4\overset{+}{\text{C}}\text{o}=\overset{+}{\text{N}}=\overset{..}{\text{O}:}$$

but this interaction does not remove electrons from the formal valence shell of the cobalt atom.

The $Co(diars)_2NO^{2+}$ ion has a trigonal bipyramidal arrangement of donor atoms[3] (shown in Fig. 16.3), with the linear nitrosyl group occupying an equatorial position. The fact that the nitrosyl group occupies an equatorial position, rather than an axial position, can be explained on the basis that $d\pi$ back bonding is favored at an equatorial position relative to an axial position.[4] It is reasonable to assume that a linear nitrosyl group, which, in the absence of back bonding, has two atoms with $+1$ formal charges, would engage in stronger back bonding than an organoarsine ligand. Therefore one predicts that the stable structure would be that with an equatorial nitrosyl group.

When $Co(diars)_2NO^{2+}$ is treated with thiocyanate ion, *trans*-$Co(diars)_2$-$(NCS)NO^+$ is formed. The addition of the electron pair of the NCS^- ion to the cobalt atom would cause the 18-electron rule to be violated if no electronic rearrangement took place. Actually, the addition causes an electron pair to transfer from the cobalt to a nonbonding orbital of the nitrogen atom, thus enforcing the bent nitrosyl configuration:

$$SCN^- + [\text{L}_4Co=N=O]^{2+} \longrightarrow \left[(SCN)\text{L}_4Co-\overset{..}{N}\underset{\diagdown O}{} \right]^+$$

Enemark and Feltham[3] have called this a "stereochemical control of valence."

The 5-coordinate compound $IrCl_2(NO)(PPh_3)_2$, pictured in Fig. 16.4, is interesting because, although it is isoelectronic with $Co(diars)_2NO^{2+}$, it has an entirely different structure.[5] We have already pointed out that the $Co(diars)_2NO^{2+}$

---

[3] J. H. Enemark and R. D. Feltham, *Proc. Natl. Acad. Sci, USA*, **69**, 3534 (1972).

[4] The reason for this preference is discussed in Chap. 18, p. 474.

[5] D. M. P. Mingos and J. A. Ibers, *Inorg. Chem.*, **10**, 1035 (1971).

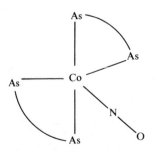

FIGURE 16.3
Structure of the $Co(diars)_2NO^{2+}$ ion. The C and H atoms have been omitted for clarity.

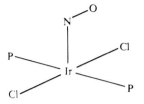

FIGURE 16.4
Structure of $IrCl_2(NO)(PPh_3)_2$. The C and H atoms have been omitted for clarity.

structure corresponds to a d^8 complex of Co^+ and NO^+, with 18 valence electrons. Although most d^8 complexes have only 16 valence electrons, 18 valence electrons are allowed in this case because the ligands engage in strong back bonding. The analogous structure for $IrCl_2(NO)(PPh_3)_2$ is unstable because most of the ligands do not engage in strong back bonding. To compensate for the buildup of negative charge on the iridium atom, a pair of electrons is transferred to the NO^+ ion, converting in into an NO^- ion with a bent coordination. The compound may be looked upon as a d^6 complex of iridium(III).

The strength of the N—O bond and the charge on the nitrogen atom in a metal nitrosyl complex would be expected to depend on whether the NO coordination is bent or linear, and, in the case of linear coordination, on the degree of $d\pi \rightarrow \pi^*$ back bonding. On the basis of valence bond structures for linear coordination $\left(\overset{-}{M}-\overset{+}{N}\equiv\overset{..}{O}:^+ \longleftrightarrow M=\overset{+}{N}= \overset{..}{O}:\right)$ and bent coordination $\left(\begin{matrix} M-\overset{..}{N} \\ \diagdown \\ :\overset{..}{O}: \end{matrix}\right)$ one would expect the N—O bond to be weaker in the bent complexes and in those complexes with strong back bonding. Strong back bonding should be favored in complexes of metal atoms with relatively high electron density. The nitrogen atom charge would be expected to be lower (more negative) in the bent complexes. The N—O bond strength can be measured by the N—O stretching frequency and the N—O bond distance, and the nitrogen atom charge can be measured by the nitrogen $1s$ electron binding energy. In Table 16.4 these quantities are tabulated for various transition-metal nitrosyl complexes.[6] As expected, the low stretching frequencies (corresponding to weak bonds) are found for the bent nitrosyls and for the linear nitrosyls of metals in very low oxidation states. The tabulated stretching frequencies, which range from 1550 to 1939 cm^{-1}, may be compared with the frequencies for free NO^+ (2200 cm^{-1}), free NO (1840 cm^{-1}), and organic nitroso compounds, RNO (~ 1550 cm^{-1}). The low nitrogen $1s$ binding energies (corresponding to relatively negative nitrogen atom charges) are found for the bent nitrosyls (as expected) and for the linear nitrosyls of metals in very low oxidation states. The correlation with oxidation state is reasonable inasmuch as one would expect the charge on the nitrogen atom to be affected by the charge on the adjacent metal atom. The N—O bond-distance data for the bent nitrosyls are inconclusive, but the data for the linear nitrosyls are consistent with weaker bonds for the nitrosyls of metals in low oxidation states.

---

6   P. Finn and W. L. Jolly, *Inorg. Chem.*, **11**, 893 (1972).

TABLE 16.4
Properties of some transition-metal nitrosyl complexes

Complexes with linearly coordinated NO groups	Oxidation state of metal, assuming NO^+	NO stretching frequency, cm^{-1}	$N\ 1s\ E_B$, eV	r_{N-O}, Å
$Na_2[Fe(NO)(CN)_5]$	Fe(II)	1939	403.3	1.13
trans-$[FeCl(NO)(diars)_2](ClO_4)_2$	Fe(II)	1865	402.9	
$[Ir(NO)_2(PPh_3)_2]PF_6$	Ir($-$I)	1740	400.2	1.21
$K_3[Mn(NO)(CN)_5]$	Mn(I)	1725	399.7	1.21
$Rh(NO)(PPh_3)_3$	Rh($-$I)	1650	400.8	
$K_3[Cr(NO)(CN)_5]$	Cr(I)	1645	400.7	1.21

Complexes with bent NO groups	Oxidation state of metal, assuming NO^-	NO stretching frequency, cm^{-1}	$N\ 1s\ E_B$, eV	r_{N-O}, Å
$[Co(NO)(NH_3)_5]Cl_2$	Co(III)	1620	400.7	1.15
$RhI_2(NO)(PPh_2CH_3)_2$	Rh(III)	1628	400.3	~1.23
trans-$[CoCl(NO)(diars)_2]Cl$	Co(III)	1550	400.5	

Bipyridyl Complexes[7]

Bipyridyl (bipy) acts as a bidentate ligand toward many metal atoms.

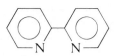

The free molecule reacts in ether solutions with alkali metals to form the anionic species bipy$^-$ (paramagnetic) and bipy^{2-} (diamagnetic), in which the extra electrons presumably occupy the π^* orbitals of the molecule. Transition-metal complexes of bipyridyl can similarly be reduced. For example, $Cr(bipy)_3^{2+}$ can be reduced to $Cr(bipy)_3^+$, $Cr(bipy)_3$, $Cr(bipy)_3^-$, $Cr(bipy)_3^{2-}$, $Cr(bipy)_3^{3-}$, and even $Cr(bipy)_3^{6-}$, as indicated by the formulas in Table 16.5. In all these species, there is ambiguity as to the oxidation state of the chromium. Thus $Cr(bipy)_3$ could be formulated as a Cr(III) complex of three bipy$^-$ anions, as a Cr(0) complex of three bipy ligands, or as several other possible combinations. The magnetic data of Table 16.5 help to rule out some of the possibilities. Inasmuch as $Cr(bipy)_3$ is diamagnetic, it is reasonable to describe it as a d^6 chromium(0) complex in which much of the electron density of the metal has been delocalized onto the ligands by $d\pi \rightarrow \pi^*$ back bonding. In the anionic complexes, the unpaired electron density is probably principally in the π^* orbitals of the bipyridyl groups.

[7]   W. R. McWhinnie and J. D. Miller, *Adv. Inorg. Chem. Radiochem.*, **12**, 135 (1969).

TABLE 16.5
Magnetic moments of chromium bipyridyl complexes†

Compound	μ_{eff}
$Cr(bipy)_3I_2$	2.9
$Cr(bipy)_3I$	2.07
$Cr(bipy)_3$	0.0
$Li[Cr(bipy)_3] \cdot 4THF$	1.83
$Na_2[Cr(bipy)_3] \cdot 7THF$	~2.85
$Na_3[Cr(bipy)_3] \cdot 7THF$	~3.85
$Ca_3[Cr(bipy)_3] \cdot 7NH_3$	~2.46

† W. R. McWhinnie and J. D. Miller, *Adv. Inorg. Chem. Radiochem.*, **12**, 135 (1969).

Metal Dithienes[8]

The reaction of metallic nickel with sulfur and diphenylacetylene yields the planar 4-coordinate complex indicated in the following reaction:

$$Ni + 4S + 2C_2Ph_2 \longrightarrow$$

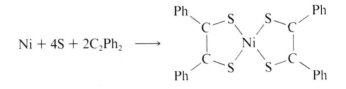

This complex can be reduced to a monoanion, $Ni(S_2C_2Ph_2)_2{}^-$, and a dianion, $Ni(S_2C_2Ph_2)_2{}^{2-}$, which have essentially the same structure. There is ambiguity in the assignment of oxidation states to the nickel atoms in these complexes. For example, the neutral complex can be formulated (1) as a nickel(0) complex of two dithioketone ligands,

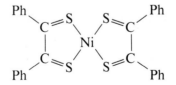

(2) as a nickel(II) complex of a dithioketone and a dithiolate anion,

8   G. N. Schrauzer, *Acc. Chem. Res.*, **2**, 72 (1969).

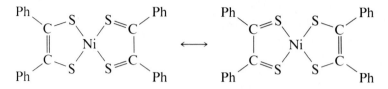

or (3) as a nickel(IV) complex of two dithiolate anions,

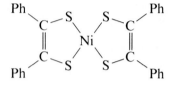

As a kind of noncommittal compromise, these compounds have been labeled "dithiene" complexes.

Some experimental evidence favors considering all three species as Ni(II) species, on which basis the reduction is formulated as follows:

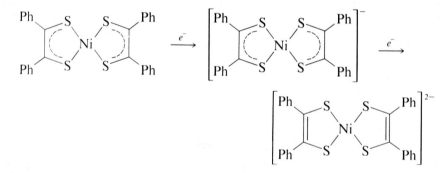

On successive reduction of the neutral complex, the stretching frequency of the C=C bond increases and that of the C=S bond decreases. The infrared and visible absorption spectra of the −2 complex are similar to those of organosubstituted ethylenedithiolates. The nickel $2p_{3/2}$ binding energies for all three species are practically the same (852.9, 852.2, and 852.8 eV, respectively), corresponding to similar charges on the nickel atom in all three species.[9]

POLYHEDRAL TRANSITION-METAL CLUSTER COMPOUNDS

Compounds containing metal-metal bonds are called metal cluster compounds. We have already considered three such complexes [$Mn_2(CO)_{10}$, $Fe_2(CO)_9$, and $Re_3Cl_{12}^{3-}$] and have shown that they conform to the 18-electron rule. Clusters of

---

[9] S. O. Grim, L. J. Matienzo, and W. E. Swartz, *J. Am. Chem. Soc.*, **94**, 5116 (1972).

more than three metal atoms generally consist of polyhedral arrangements of the metal atoms and thus are part of the fascinating and rapidly growing branch of chemistry which we can call "polyhedral cluster chemistry." In Chap. 10 we discussed the bonding in molecules containing polyhedral clusters of non-transition-metal atoms (e.g., boranes, carboranes, and species such as $Ge_9{}^{2-}$) and molecules containing clusters of both transition-metal and nonmetal atoms (e.g., metallo-carboranes). All such compounds, as well as those containing clusters of only transition-metal atoms, can be treated as one large family of compounds.

When applying the 18-electron rule to polyhedral clusters containing transition-metal atoms, it is usually convenient to treat the entire cluster as a unit. Of course, we assume that each transition-metal atom conforms to the 18-electron rule, and therefore we write, for a given metal atom,

$$18 = v + l + (6 - c) \tag{16.1}$$

Here v is the number of valence electrons of the free metal atom, l is the number of electrons donated by the ligands attached to the metal atom, and c is the net number of electrons contributed by the metal atom to the cluster bonds. It is assumed that three of the metal orbitals are engaged in bonding to other cluster atoms; hence six electrons are involved in bonding a given transition-metal atom to the other cluster atoms. Of these six electrons, c are contributed by the metal atom and $6 - c$ are contributed by the other cluster atoms. Thus we obtain the $6 - c$ term in Eq. 16.1. By rearrangement of Eq. 16.1 we obtain

$$c = v + l - 12$$

The total number of electrons involved in cluster bonding is the sum of the values of c for the transition-metal atoms and the number of electrons contributed by any non-transition-element atoms in the cluster. The number of cluster electrons, C, is given by the expression

$$C = V + L + m - 12t - q \tag{16.2}$$

where V is the total number of valence electrons of the transition elements in the cluster

L is the total number of ligand electrons donated to the transition-element atoms

m is the total number of electrons contributed to the cluster by non-transition-element cluster atoms

t is the number of transition-element atoms on the surface of the cluster

q is the charge on the entire cluster molecule

Calculation of the number of cluster electrons using Eq. 16.2 has the advantage that it is unnecessary to assign ligands to particular metal atoms. Thus the calculation is straightforward when there are ligands that bridge two or three metal atoms or even when the disposition of the ligands is unknown.

The $2n + 2$ rule, introduced in Chap. 10, has been applied with a fair record of success to all kinds of polyhedral cluster compounds. Let us review the rule.

If the number of cluster electrons equals $2n + 2$, where n is the number of cluster atoms, the cluster is expected to have the closo geometry of the corresponding n-vertexed regular polyhedron.[10] If $C = 2n + 4$, the cluster is expected to have a nido geometry, corresponding to an $(n + 1)$-vertexed regular polyhedron lacking one vertex. If $C = 2n + 6$ or $C = 2n + 8$, the cluster is expected to have an arachno or hypho geometry, corresponding to a regular polyhedron lacking two or three vertices, respectively. If $C < 2n + 2$, the cluster is expected to be a capped regular polyhedron. $C = 2n$ corresponds to monocapping, $C = 2n - 2$ corresponds to bicapping, etc. It should be noted that, according to the rules, the capping of a closo cluster compound and the removal of cluster atoms (and their associated ligands) from a closo compound to form nido, arachno, or hypho compounds are processes which involve no change in the number of cluster electrons. Hence each step in these conversions involves the addition or removal of a cluster atom which has three empty orbitals.

For simplicity, we shall refer to the various rules which correlate the number of cluster electrons with cluster geometry as the "$2n + 2$ rule," even though $2n + 2$ cluster electron systems strictly correspond to closo structures only. Now let us consider the application of the rule to various cluster compounds.

The trigonal bipyramidal complex $Os_5(CO)_{16}$, with 12 cluster electrons, is a closo cluster.[11]

$$
\begin{aligned}
V &= 5 \times 8 = &40 \\
L &= 16 \times 2 = &\underline{32} \\
& &72 \\
-12t &= -12 \times 5 = &\underline{-60} \\
C &= &12 = 2n + 2
\end{aligned}
$$

Formal *removal* of two $Os(CO)_2$ groups from this complex gives the triangular arachno complex $Os_3(CO)_{12}$.

$$
\begin{aligned}
V &= 3 \times 8 = &24 \\
L &= 12 \times 2 = &\underline{24} \\
& &48 \\
-12t &= -12 \times 3 = &\underline{-36} \\
C &= &12 = 2n + 6
\end{aligned}
$$

Formal *addition* of an $Os(CO)_2$ group to $Os_5(CO)_{16}$ gives $Os_6(CO)_{18}$, in which the six osmium atoms do not have an octahedral configuration (as one might offhand predict), but rather the configuration of a capped trigonal bipyramid, as shown in Fig. 16.5.[12]

---

[10]   The "regular," or closo, polyhedra include, but do not appear to be restricted to, triangulated polyhedra such as the trigonal bipyramid, octahedron, pentagonal bipyramid, dodecahedron, tricapped trigonal prism, bicapped square antiprism, and icosahedron.

[11]   C. R. Eady, B. F. G. Johnson, J. Lewis, B. E. Reichert, and G. M. Sheldrick, *J. Chem. Soc. Chem. Commun.*, **271** (1976).

[12]   R. Mason, K. M. Thomas, and D. M. P. Mingos, *J. Am. Chem. Soc.*, **95** 3802 (1973).

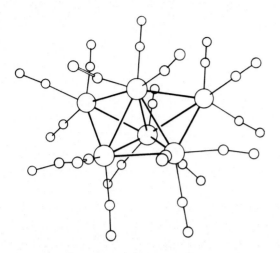

FIGURE 16.5
Structure of $Os_6(CO)_{18}$, showing the monocapped trigonal bipyramidal geometry of the Os_6 cluster.

$$V = 6 \times 8 = 48$$
$$L = 18 \times 2 = \underline{36}$$
$$84$$
$$-12t = -12 \times 6 = \underline{-72}$$
$$C = 12 = 2n$$

The tetrahedral complex $(\eta^5\text{-}C_5H_5)RhFe_3(CO)_{11}$,[13] with 12 cluster electrons, should be looked upon as a nido derivative of a trigonal bipyramidal cluster.

$$V = 3 \times 8 + 9 = 33$$
$$L = 11 \times 2 + 5 = \underline{27}$$
$$60$$
$$-12t = -12 \times 4 = \underline{-48}$$
$$C = 12 = 2n + 4$$

The complex $Re_4(CO)_{16}^{2-}$ has an open "butterfly" configuration, as shown in Fig. 16.6.[14] It can be looked upon as an arachno derivative of an octahedral structure, in accord with its 14 cluster electrons.

$$V = 4 \times 7 = 28$$
$$L = 16 \times 2 = 32$$
$$-q = \underline{2}$$
$$62$$
$$-12t = -12 \times 4 = \underline{-48}$$
$$C = 14 = 2n + 6$$

The complex $Ru_6(CO)_{17}C$ has a carbon atom situated in the center of an octahedral cage of metal atoms.[15] By assuming that the central carbon atom con-

[13] M. R. Churchill and M. V. Veidis, *J.Chem. Soc. Chem. Commun.*, 1470 (1970).

[14] M. R. Churchill and R. Bau, *Inorg. Chem.*, **7**, 2606 (1968).

[15] A. Sirigu, M. Bianchi, and E. Benedetti, *J. Chem. Soc. Chem. Commun.*, 546 (1969).

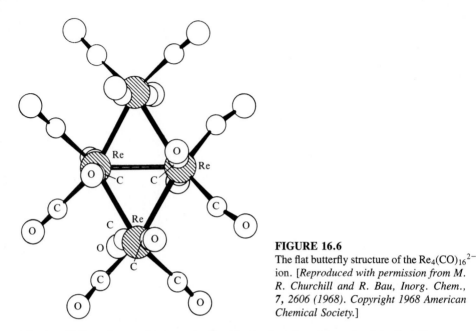

FIGURE 16.6
The flat butterfly structure of the $Re_4(CO)_{16}^{2-}$ ion. [*Reproduced with permission from M. R. Churchill and R. Bau, Inorg. Chem., 7, 2606 (1968). Copyright 1968 American Chemical Society.*]

tributes all its valence electrons to the cluster bonding, we calculate that there are 14 cluster electrons, in accord with the octahedral structure.

$$
\begin{aligned}
V = 6 \times 8 &= 48 \\
L = 17 \times 2 &= 34 \\
m &= \underline{4} \\
&86 \\
-12t = -12 \times 6 &= \underline{-72} \\
C &= 14 = 2n + 2
\end{aligned}
$$

The complex $Fe_5(CO)_{15}C$ has a square pyramidal arrangement of metal atoms, with a carbon atom situated slightly below the center of the square base.[16] The structure appears to be a nido derivative of the iron analog of the $Ru_6(CO)_{17}C$ complex. Indeed, the number of cluster electrons conforms to this interpretation.

$$
\begin{aligned}
V = 5 \times 8 &= 40 \\
L = 15 \times 2 &= 30 \\
m &= \underline{4} \\
&74 \\
-12t = -12 \times 5 &= \underline{-60} \\
C &= 14 = 2n + 4
\end{aligned}
$$

It has been proposed that $C_2B_3H_5Fe(CO)_3$ has the octahedral structure shown in Fig. 16.7.[17] Clearly this is a closo structure, with 14 cluster electrons.

---

[16] E. H. Braye, L. F. Dahl, W. Hubel, and D. L. Wampler, *J. Am. Chem. Soc.,* **84**, 4633 (1962).

[17] R. N. Grimes, *Ann. N.Y. Acad. Sci.,* from Symposium on New Horizons in Organometallic Chemistry, August 1973.

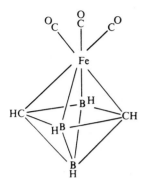

FIGURE 16.7
Proposed structure of $C_2B_3H_5Fe(CO)_3$. (*From R. N. Grimes, Ann. N.Y. Acad. Sci., Symposium on New Horizons in Organometallic Chemistry, August 1973.*)

$$
\begin{aligned}
V &= 8 \\
L = 3 \times 2 &= 6 \\
m\begin{cases} C: & 2 \times 3 = 6 \\ B: & 3 \times 2 = \underline{6} \end{cases} & \\
& 26 \\
-12t &= \underline{-12} \\
C &= 14 = 2n + 2
\end{aligned}
$$

The bis(acetylene) complex $(C_2Ph_2)_2Fe_3(CO)_8$, whose structure is shown in Fig. 16.8, is an example of a compound containing a pentagonal bipyramidal cluster.[18] In agreement with the rule, the cluster contains 16 electrons:

$$
\begin{aligned}
V = 3 \times 8 &= 24 \\
L = 8 \times 2 &= 16 \\
m = 4 \times 3 &= \underline{12} \\
& 52 \\
-12t = -12 \times 3 &= \underline{-36} \\
C &= 16 = 2n + 2
\end{aligned}
$$

The complex $Fe_2(CO)_6(HOCC(CH_3)C(CH_3)COH)$ has a nido structure based

---

[18] R. P. Dodge and V. Schomaker, *J. Organomet. Chem.*, **3**, 274 (1965).

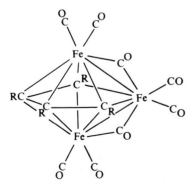

FIGURE 16.8
Structure of $(C_2Ph_2)_2Fe_3(CO)_8$.

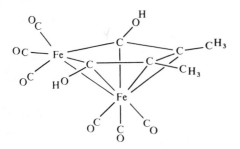

FIGURE 16.9
Structure of $Fe_2(CO)_6[C_6H_6(OH)_2]$.

on a pentagonal bipyramid,[19] as shown in Fig. 16.9, and accordingly has 16 cluster electrons.

$$V = 2 \times 8 = \quad 16$$
$$L = 6 \times 2 = \quad 12$$
$$m = 4 \times 3 = \underline{\quad 12}$$
$$40$$
$$-12t = -12 \times 2 = \underline{-24}$$
$$C = \quad 16 = 2n + 4$$

The anion $Co_8(CO)_{18}C^{2-}$ contains a carbon atom at the center of a square anti-prismatic cluster of cobalt atoms.[20] The closo geometry is consistent with the 18 cluster electrons of the complex. The metallocarborane $[(CH_3)_3P]_2Pt(CMe)_2(BH)_6$ contains a trigonal prismatic cluster of six boron atoms, with the rectangular faces capped with two carbon atoms and a platinum atom.[21] This closo geometry is consistent with the 20 cluster electrons. Twelve of the rhodium atoms in the anion $Rh_{13}(CO)_{24}H_3^{2-}$ are situated at the corners of a truncated hexagonal bipyramid, with the thirteenth rhodium atom in the center, as shown in Fig. 16.10.[22] Thus the atoms have a hexagonally close-packed arrangement. If we assume that the central rhodium atom contributes its 9 valence electrons to the cluster bonding, we calculate that there are 26 cluster electrons, consistent with the closo geometry.

Some Difficult Cases

Ingenuity is needed to demonstrate that certain cluster compounds follow the $2n + 2$ rule. For example, the complex $Ru_4Cl(CO)_{13}^-$ has a folded butterfly arrangement of four Ru atoms, with a dihedral angle of $91°$ between the two Ru_3 planes and a chlorine atom bridging the terminal Ru atoms:[23]

---

[19]   A. A. Hock and O. S. Mills, *Acta Crystallogr.,* **14**, 139 (1961).

[20]   V. G. Albano, P. Chini, G. Ciani, M. Sansoni, D. Strumolo, B. T. Heaton, and S. Martinengo, *J. Am. Chem. Soc.,* **98**, 5027 (1976).

[21]   A. J. Welch, *J. Chem. Soc. Dalton Trans.,* 225 (1976).

[22]   P. Chini, G. Longoni, and V. G. Albano, *Adv. Organomet, Chem.,* **14**, 285 (1976).

[23]   G. R. Steinmetz, A. D. Harley, and G. L. Geoffroy, *Inorg. Chem.,* **19**, 2985 (1980).

Assuming that the chlorine atom is a cluster atom and that it contributes 5 electrons to the cluster, we calculate a total of 16 cluster electrons. This result may be rationalized by looking upon the complex as an arachno derivative of a pentagonal bipyramidal cluster. (The chlorine atom and the two "hinge" Ru atoms occupy equatorial sites.) Alternatively, if we look upon the chlorine atom as a bridging *ligand* (contributing 3 electrons to the cluster), we calculate 14 cluster electrons. This result may be rationalized by looking upon the complex as an arachno derivative of an octahedral cluster. Thus the interpretation of this complex differs from that of $Fe_4N(CO)_{12}^-$, which has a similar geometry (a dihedral angle of 101°) but only 12 cluster electrons.[24] The latter complex obviously contains a closo trigonal bipyramidal cluster, with the nitrogen atom occupying an equatorial site.

---

[24] M. Tachikawa et al., *J. Am. Chem. Soc.*, **102**, 6648 (1980).

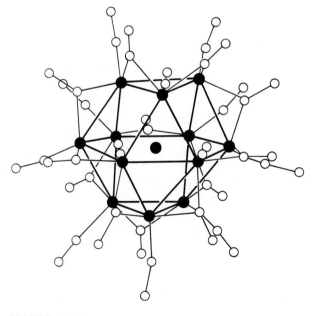

FIGURE 16.10
Structure of the $Rh_{13}(CO)_{24}H_3^{2-}$ ion. Notice that the metal atoms are part of an hcp lattice. [*Reproduced with permission from P. Chini, G. Longoni, and V. G. Albano, Adv. Organomet. Chem., 14, 285 (1976).*]

The complex $HOs_4(CO)_{13}^-$ has a tetrahedral arrangement of metal atoms and 12 cluster electrons, corresponding to a nido cluster based on a trigonal bipyramid.[25] On the other hand, the corresponding iron complex, $HFe_4(CO)_{13}^-$, has a butterfly structure with one of the CO groups bridging the terminal iron atoms of the butterfly.[26]

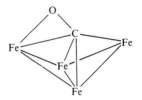

The bridging CO group is attached to one iron atom by the carbon atom and to the other iron atom by both the carbon and oxygen atoms. Presumably this CO group acts as a 4-electron donor; thus the complex has 14 cluster electrons and should be considered an arachno derivative of an octahedral cluster.

Numerous other examples of unsymmetric bridging CO groups (or "semibridging" CO groups) are known. A remarkable example is found in the cluster $(\eta^5\text{-}C_5H_5)_3Nb_3(CO)_7$, in which one of the seven CO groups acts as an $\eta^2\text{-}(\mu_3\text{-}C, \mu_2\text{-}O)$ bridge facing a nearly equilateral Nb_3 triangle.[27]

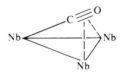

Presumably this unusual CO group acts as a 6-electron donor ligand, giving an electron pair to each of the niobium atoms. The number of cluster electrons is thus 12, corresponding to an arachno derivative of a trigonal bipyramidal cluster.

The complex $Ni_5(CO)_{12}^{2-}$ has a trigonal bipyramidal structure with 16 cluster electrons.[28] Perhaps one should look upon this as having an arachno cluster derived from a pentagonal bipyramidal geometry (two of the equatorial vertices corresponding to axial sites of the pentagonal bipyramid).

The complex $Rh_6(CO)_{15}C^{2-}$ contains a trigonal prismatic cluster of metal atoms with a central carbon atom.[29] The number of cluster electrons, 18, suggests

[25] P. A. Dawson, B. F. G. Johnson, J. Lewis, D. A. Kaner, and P. R. Raithby, *J. Chem. Soc. Chem. Commun.*, 961 (1980).

[26] M. Manassero, M. Sansoni, and G. Longoni, *J. Chem. Soc. Chem. Commun.*, 919 (1976).

[27] W. A. Herrmann et al., *J. Am. Chem. Soc.*, **103**, 1692 (1981).

[28] G. Longoni, P. Chini, L. D. Lower, and L. F. Dahl, *J. Am. Chem. Soc.*, **97**, 5034 (1975).

[29] V. G. Albano, M. Sansoni, P. Chini, and S. Martinengo, *J. Chem. Soc. Dalton Trans.*, 651 (1973).

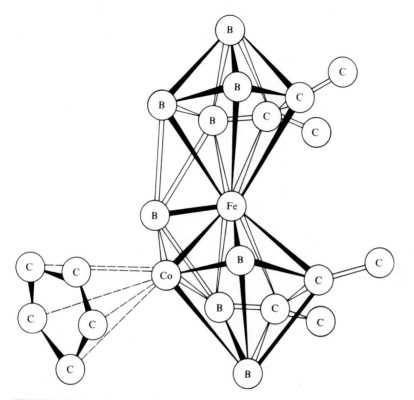

FIGURE 16.11

Structure of $(\eta^5\text{-}C_5H_5)CoFe(CMe)_4(BH)_8$. [*Reproduced with permission from R. B. King, E. K. Nishimura, and K. S. RaghuVeer, Inorg. Chem., **19**, 2478 (1980). Copyright 1980 American Chemical Society.*]

that the cluster is an arachno derivative of an eight-vertex regular polyhedral cluster. This interpretation is reasonable if we recognize that a trigonal prism is essentially a square antiprism with two vertices missing.

The nature of the bonding in $\eta^5\text{-}C_5H_5CoFe(CMe)_4(BH)_8$, pictured in Fig. 16.11, has been the subject of considerable discussion.[30] The feature causing difficulty is the "wedging" BH group which, as drawn in Fig. 16.11, simultaneously caps two pentagonal bipyramidal clusters. The situation is quite rational, however, if we consider the unique BH group to be simply a ligand of the iron atom, to which it indeed is rather closely attached. The iron atom achieves 18 valence electrons by accepting 2 electrons from the BH group and 4 from each of the two clusters. Since the iron atom uses three orbitals in its bonding to each cluster, it

[30] W. M. Maxwell, E. Sinn, and R. N. Grimes, *J. Am. Chem. Soc.*, **98**, 3490 (1976); R. B. King, E. K. Nishimura, and K. S. RaghuVeer, *Inorg. Chem.*, **19**, 2478 (1980).

must donate 2 electrons to each cluster. It is then an easy matter to show that each cluster has 16 cluster electrons, conforming to the closo pentagonal bipyramidal structures.

Equation 16.2 can be used to calculate the number of cluster electrons in a compound containing transition-metal atoms which are expected to conform to the 18-electron rule. Thus it is applicable to transition-metal clusters with π-acceptor ligands such as carbonyl groups, olefins, and cyclopentadienyl groups. However, cluster complexes such as $Mo_6Cl_8^{4+}$ and $Nb_6Cl_{12}^{2+}$, in which the metal atoms are not required to have 18 valence electrons, are best treated on an individual basis by simple MO theory[31] or ligand-field theory. The bonding in such complexes, involving metal atoms with oxidation states of $+2$ or higher, can be analyzed almost entirely in terms of the metal valence d orbitals.

The complex $Mo_6Cl_8^{4+}$, pictured in Fig. 16.12, consists of an octahedron of molybdenum atoms with a chlorine atom over each face. Let us assume that each Mo d_{z^2} orbital is oriented along the fourfold symmetry axis and that the lobes of the d_{xy} orbital are directed toward the four chlorine atoms which are essentially coplanar with the Mo atom. It is then reasonable to assign the d_{z^2}, d_{xz}, and d_{yz} orbitals to cluster bonding and to assume that the d_{xy} orbital is strongly antibonding because of repulsions by the four chlorine atoms. The $d_{x^2-y^2}$ orbitals of the six Mo atoms will then overlap in the regions over the Mo—Mo edges of the octahedral cluster. One combination of these six $d_{x^2-y^2}$ orbitals, pictured in Fig. 16.13, is strongly antibonding. The other five combinations are bonding, or at least nonbonding, and would be expected to hold 10 valence electrons. These 10 electrons, plus the 14 inner cluster electrons corresponding to the $2n + 2$ rule, make a total of 24 cluster electrons expected. In fact, if we strip the chloride ions from $Mo_6Cl_8^{4+}$, we are left with Mo_6^{12+}, which does contain 24 valence electrons.

The complex $Nb_6Cl_{12}^{2+}$, pictured in Fig. 16.14, consists of an octahedron of niobium atoms with a chlorine atom over each edge. Let us assume that each d_{z^2} orbital is aligned along the fourfold symmetry axis and that the lobes of the $d_{x^2-y^2}$ orbital are directed toward the four neighboring chlorine atoms. The

[31]   F. A. Cotton and T. E. Haas, *Inorg. Chem.*, **3**, 10 (1964).

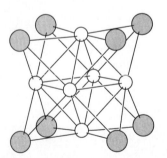

FIGURE 16.12
Structure of the $Mo_6Cl_8^{4+}$ complex.

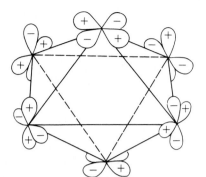

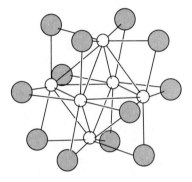

FIGURE 16.13
The a_{2g} combination of the $d_{x^2-y^2}$ orbitals of $Mo_6Cl_8^{4+}$, corresponding to a strongly antibonding MO.

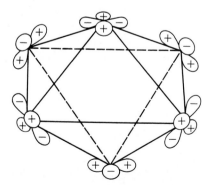

FIGURE 16.14
Structure of the $Nb_6Cl_{12}^{2+}$ complex.

d_{z^2}, d_{xz}, and d_{yz} orbitals are assigned to cluster bonding, and the $d_{x^2-y^2}$ orbital is strongly antibonding. The d_{xy} orbitals of the six Nb atoms then engage in rather poor three-center overlap in the regions over the triangular faces of the octahedral cluster. Five combinations of these six orbitals are probably antibonding and would not be expected to hold valence electrons. However, one combination, pictured in Fig. 16.15, is weakly bonding and might be expected to hold a pair of valence electrons. This pair of electrons, plus the 14 inner cluster electrons, make a total of 16 cluster electrons expected. In fact the Nb_6^{14+} cluster does have 16 valence

FIGURE 16.15
The a_{2u} combination of the d_{xy} orbitals of $Nb_6Cl_{12}^{2+}$, corresponding to a weakly bonding MO.

electrons, 2 of which are weakly held, as shown by the ready oxidation of the complex:[32]

$$Nb_6Cl_{12}^{2+} \xrightarrow{-e^-} Nb_6Cl_{12}^{3+} \xrightarrow{-e^-} Nb_6Cl_{12}^{4+}$$

Applications

The relations between cluster geometry and the number of cluster electrons have permitted the systematization and prediction of the structures of many unusual compounds. But perhaps even more important, these concepts have been successfully used to design syntheses of new types of compounds. Thus, recognition of analogies, such as that between $(\pi$-$C_5H_5)Co$ and BH (each contributes two electrons and three orbitals to polyhedral bonding) and that between $(\pi$-$C_5H_5)Ni$ and CH (each contributed three electrons and three orbitals) and recognition of the structural effects of oxidation and reduction reactions have led to the preparation of many novel metallocarboranes.[33,34]

High-Nuclearity Clusters

Many of the higher polyhedral cluster compounds do not follow the $2n+2$ rule. For example, most (if not all) eight-vertex clusters have closo structures (either square antiprismatic or dodecahedral) even when they do not have the prescribed 18 cluster electrons. Examples of exceptions to the $2n + 2$ rule are $(\eta^5$-$C_5H_5Co)_4B_4H_4$, with 16 cluster electrons, and $(\eta^5$-$C_5H_5Ni)_4B_4H_4$, with 20 cluster electrons.[35] Although both of these clusters have dodecahedral geometries, they have significant structural differences which are probably related to the electron deficiency of the cobalt compound and the electron surplus of the nickel compound. In the cobalt compound, the metal atoms occupy the high-coordinate dodecahedral sites (the B vertices of Fig. 14.3c), whereas, in the nickel compound, the metal atoms occupy the low-coordinate sites (the A vertices of Fig. 14.3c). In the nickel compound, the nickel atoms are essentially paired, with very short Ni—Ni distances. It seems likely that localized metal-metal bonding is involved in this electron-rich compound.

A 12-vertex cluster with 28 cluster electrons is predicted to have a nido structure, corresponding to a 13-vertex closo polyhedron with a missing vertex. The structure of the 28-electron metallocarborane $(\eta^5$-$C_5H_5Co)_2C_4B_6H_{10}$, shown in Fig. 16.16a, essentially fits this description. However, the structure of the 28-electron compound $(\eta^5$-$C_5H_5Co)(CCH_3)_4B_7H_6OC_2H_5$, shown in Fig. 16.16$b$, is

[32]   F. W. Koknat and R. E. McCarley, *Inorg. Chem.*, **13**, 295 (1974).

[33]   K. Wade, *Inorg. Nucl. Chem. Lett.*, **8**, 559, 563 (1972); *J. Chem. Soc. Chem. Commun.*, 792, (1971).

[34]   R. N. Grimes, *Acc. Chem. Res.*, **11**, 421 (1978).

[35]   J. R. Bowser, A. Bonny, J. R. Pipal, and R. N. Grimes, *J. Am. Chem. Soc.*, **101**, 6229 (1979).

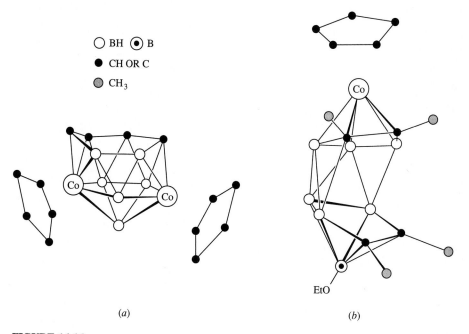

(a)   *(b)*

FIGURE 16.16

Structures of two 12-vertex complexes with 28 cluster electrons. (*a*) $(\eta^5\text{-}C_5H_5)_2Co_2C_4B_6H_{10}$; (*b*) $(\eta^5\text{-}C_5H_5)Co(CMe)_4B_7H_6OEt$. [*Reproduced with permission from J. R. Pipal and R. N. Grimes, J. Am. Chem. Soc., **100**, 3083 (1978). Copyright 1978 American Chemical Society.*]

completely different.[36] The molecule resembles a severely distorted icosahedron whose two halves have been partially separated, forming a large opening on one side. Although the $2n + 2$ rule is successful in this case to the limited extent of predicting a noncloso structure, no completely satisfactory explanation of this unusual structure has been found.

Mingos[37] has shown that the number of cluster electrons in many high-nuclearity metal cluster carbonyl compounds can be correlated with the ratio of the number of carbonyl groups to the number of surface, or outer, metal atoms. The compounds appear to fall into three groups, depending on the magnitude of this ratio.

When the carbonyl-to-surface-atom ratio is greater than about 2.0, the $2n + 2$ rule is applicable. Examples of this type of cluster, including $Rh_{13}(CO)_{24}H_3^{2-}$ (pictured in Fig. 16.10), are listed in Table 16.6.

When the carbonyl-to-surface-atom ratio falls in the range 1.3 to 2.0, the number of cluster electrons is generally 24, as shown by the data in Table 16.7. The structure of the largest cluster in this group, $Rh_{22}(CO)_{37}^{4-}$, is shown in Fig. 16.17.

[36] J. R. Pipal and R. N. Grimes, *J. Am. Chem. Soc.*, **100**, 3083 (1978).

[37] D. M. P. Mingos, *J. Chem. Soc. Chem. Commun.*, 1352 (1985).

TABLE 16.6
Geometries and numbers of cluster electrons for high-nuclearity clusters that conform to the $2n + 2$ rule.†

Compound	Structure	"Surface" metal atoms, t	Cluster electrons, C
$Rh_{13}(CO)_{24}H_3^{2-}$	hcp anticuboctahedron	12	26
$Rh_{15}(CO)_{30}^{3-}$	bcc deltahedron	14	30
$Ru_6C(CO)_{17}$	Octahedron‡	6	14
$Rh_{10}S(CO)_{22}^{2-}$	Bicapped square antiprism‡	10	22
$Rh_{12}Sb(CO)_{27}^{3-}$	Icosahedron‡	12	26

† D. M. P. Mingos, *J. Chem. Soc. Chem. Commun.*, 1352 (1985).

‡ With nonmetal internal atoms.

TABLE 16.7
High-nuclearity clusters with $1.3 < n_{co}/t \le 2.0$ and $C = 24$†

Compound	Structure	"Surface" metal atoms, t
$Rh_{14}(CO)_{25}^{4-}$	bcc	13
$Rh_{14}(CO)_{26}^{2-}$	bcc	13
$Rh_{15}(CO)_{27}^{3-}$	bcc/hcp	14
$Rh_{17}(CO)_{30}^{3-}$	hcp	16
$Rh_{22}(CO)_{37}^{4-}$	ccp/hcp	21

† D. M. P. Mingos, *J. Chem. Soc. Chem. Commun.*, 1352 (1985).

Another cluster, $Pt_{24}(CO)_{30}^{2-}$, with a carbonyl-to-surface-atom ratio of 1.30, has 26 cluster electrons and thus does not conform.

When the carbonyl-to-surface-atom ratio is less than 1.3, the shell of surface metal atoms and the attached groups function essentially as a set of ligands for the internal metal atoms. The "internal cluster" thus acquires a total valence electron

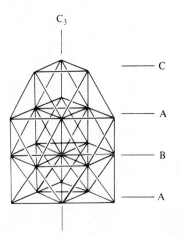

FIGURE 16.17
Skeletal structure of $Rh_{22}(CO)_{37}^{4-}$. [*Reproduced with permission from S. Martinengo, G. Ciani, and A. Sironi, J. Am. Chem. Soc.,* **102,** *7564 (1980). Copyright 1980 American Chemical Society.*]

TABLE 16.8
Valence electron counts and molecular analogs of "internal" metal atom clusters in high-nuclearity clusters†

No. of internal metal atoms	Valence electrons	Molecular analog
1	18	$Mo(CO)_6$
2	34	$Mn_2(CO)_{10}$
3 (triangular)	48	$Os_3(CO)_{12}$
3 (linear)	50	$OsRe_2(CO)_{14}$
4 (tetrahedral)	60	$Ir_4(CO)_{12}$
6 (octahedral)	86	$Rh_6(CO)_{16}$

†D. M. P. Mingos, *J. Chem. Soc. Chem. Commun.*, 1352 (1985).

TABLE 16.9
Geometries and numbers of cluster electrons for high-nuclearity clusters with $n_{co}/t < 1.3$†

Compound	"Surface" metal atoms, t	"Internal" metal atoms	Cluster electrons, C
$Au_9(PPh_3)_8^+$	8	1	18
$Pt_{19}(CO)_{22}^{4-}$	17	2	34
$Pt_{26}(CO)_{32}^{2-}$	23	3	50
$Ni_{38}Pt_6(CO)_{48}H_{6-n}^{n-}$	38	6	86

† D. M. P. Mingos, *J. Chem. Soc. Chem. Commun.*, 1352 (1985).

count equal to the number of cluster electrons of the overall cluster and equal to the number of metal valence electrons in a molecular carbonyl analog of the internal cluster. The number of electrons expected for various numbers of internal metal atoms, and molecular analogs of the "internal clusters" are given in Table 16.8. Examples of high-nuclearity cluster compounds that fall in this category are listed in Table 16.9.

Obviously the rules designed to rationalize polyhedral clusters must be used with discretion. Our present treatment of the bonding in these systems is much like the use which chemists made of rudimentary periodic tables in the nineteenth century. We can hope that soon our understanding of these systems will improve and that the empirical rules will be refined or replaced with more reliable rules.

PROBLEMS

16.1 Propose a formula and structure for an iron carbonyl complex of cyclobutadiene. Show that the 18-electron rule and the polyhedral cluster rules are obeyed.

16.2 Propose a structure for the acetylene complex $(C_6H_5)_2C_2Fe_3(CO)_9$ which satisfies the 18-electron and polyhedral cluster rules.

16.3 Rationalize the bonding in the molecule $H_6Cu_6(PPh_3)_6$, which contains an octahedral Cu_6 cluster.

16.4 Suggest a synthetic method to cap the rings of ferrocene to produce two pentagonal bipyramidal clusters joined by a common iron atom. Show that the 18-electron and polyhedral cluster rules would be satisfied in the product.

16.5 Which of the following complexes are likely to be unstable: $ZrCl_4$, diamagnetic $NiCl_4^{2-}$, $Ti(CO)_4^{4+}$, $Cd(CO)_3$, $Fe_3(CO)_{12}$, AuF_5^{2-}? Why?

16.6 Rationalize the relative magnitudes of the following reduction potentials.

$$e^- + Fe(CN)_6^{3+} = Fe(CN)_6^{4-} \qquad E° = 0.36 \text{ V}$$

$$e^- + Fe(H_2O)_6^{3+} = Fe(H_2O)_6^{2+} \qquad E° = 0.77 \text{ V}$$

$$e^- + Fe(phen)_3^{3+} = Fe(phen)_3^{2+} \qquad E° = 1.14 \text{ V}$$

16.7 Predict the structures of $Fe(CO)_4CN^-$, $Mn(CO)_4NO$, $Co(NO)_3$, and $[Fe(NO)_2Br]_2$.

16.8 For the "brown ring" species, $Fe(H_2O)_5NO^{2+}$, the N—O stretching frequency is 1745 cm^{-1} and $\mu_{eff} = 3.9$. Describe the bonding in this complex.

16.9 Show how oxidation-state ambiguities could arise in complexes of the following dianion:

16.10 Nickelocene, $Ni(C_5H_5)_2$, has a structure analogous that that of ferrocene, but the Ni—C distance is about 0.16 Å longer than the Fe—C distance. Can you rationalize the bonding in nickelocene with the 18-electron rule?

16.11 How many carbonyl ligands would you expect to find in the following complexes?

(a) $H_2Ru_4(CO)_x$, with a tetrahedral cluster

(b) $H_3Os_4(CO)_xI$, with a folded butterfly arrangement of Os atoms, and an I atom bridging the terminal Os atoms

(c) $HFe_5N(CO)_x$, with a square pyramidal arrangement of Fe atoms, and an N atom in the middle of the basal plane

16.12 What net charges, if any, would you expect on the following complexes?

(a) $RhFe_4C(CO)_{14}^{x\pm}$ (square pyramidal cluster with a C atom in the basal plane)

(b) $Rh_{10}As(CO)_{22}^{x\pm}$ (bicapped square antiprism with a central As atom)

(c) $(\eta^5\text{-}C_5H_5)_4Ni_4B_5H_5^{x\pm}$ (square antiprism with one square face capped)

16.13 What kinds of cluster structures would you expect in the following complexes?

(a) $Rh_7(CO)_{16}^{3-}$;

(b) $Fe_3S_2(CO)_9$;

(c) $Fe_4(CH)(CO)_{12}H$;

(d) $Ru_4(CO)_{12}(MeC\equiv CMe)$

16.14 Both $Ru_4(CO)_{12}H_4$ (12 cluster electrons) and $Re_4(CO)_{12}H_4$ (8 cluster electrons) have tetrahedral structures. However, the hydrogen atoms of the ruthenium complex occupy μ_2 edge-bridging sites, and those of the rhenium complex occupy μ_3 face-bridging sites. Explain the difference in the structures.

16.15 A compound of empirical formula $FeBC_7O_3H_5$ (formula determined by high-resolution mass spectroscopy) was obtained by irradiation at 360 nm of a solution of $C_4H_4Fe(CO)_3$ and B_5H_9 in ether. Suggest a structure and indicate how the $2n + 2$ cluster electron rule is satisfied.

16.16 Rationalize the change in structure of the M_3C_3 framework on going from $Fe_3(CO)_9(PhCCPh)$ (framework A) to $Os_3(CO)_{10}(PhCCPh)$ (framework B).

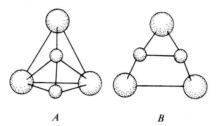

A   *B*

16.17 Rationalize the cluster structure of $Fe_2Ni(CO)_6(C_2Ph_2)(\eta^5-C_5H_5)^-$, shown in Fig. 16.18, in terms of the $2n + 2$ rule.

16.18 Some cluster compounds are best described as condensed polyhedra in which two polyhedra are joined through a common vertex, edge, or face. Mingos[38] has noted

---

[38] D. M. P. Mingos, *J. Chem. Soc. Chem. Commun.*, 706 (1983).

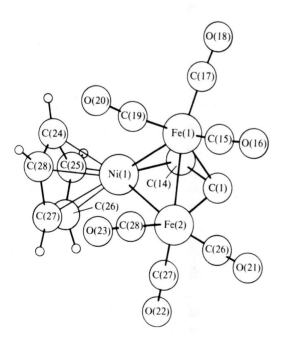

FIGURE 16.18
Structure of $Fe_2Ni(CO)_6-(C_2Ph_2)(\eta^5-C_5H_5)^-$. (See Prob. 16.17.) It is significant that the two Fe—C(1) distances are quite different. The phenyl groups are omitted for clarity. [*Reproduced with permission from M. I. Bruce, J. R. Rodgers, M. R. Snow, and F. S. Wong, J. Chem. Soc. Chem. Commun., 1285 (1980).*]

that in such cases the calculated number of cluster electrons is equal to the sum of the characteristic number of cluster electrons for the parent polyhedra minus the number of "cluster" electrons characteristic of the atom, pair of atoms, or face of atoms common to both polyhedra (6, 10, and 12 electrons, respectively). Show that this generalization aids rationalization of the structures of $Os_7(CO)_{21}$ (capped octahedron), $Ru_{10}C_2(CO)_{24}^{2-}$ (two octahedra sharing an edge), and $Os_8(CO)_{22}H^-$ (a capped trigonal bipyramid and a tetrahedron sharing an edge).

CHAPTER 17

ELECTRONIC SPECTRA OF COORDINATION COMPOUNDS

Most compounds absorb light somewhere in the spectral region between 200 and 1000 nm. These transitions correspond to excitation of the molecules to higher *electronic* states; therefore spectra in this region are often called electronic absorption spectra. The "visible" wavelength region ranges from about 400 nm (violet) to about 750 nm (red), and the "ultraviolet" spectral region corresponds to wavelengths less than about 400 nm. (See Table 17.1.)

The energy of a spectral transition is related to the frequency of the light by the relation $\triangle E = h\nu$, where h is Planck's constant and ν is the frequency.

TABLE 17.1
Approximate wavelengths and wave numbers corresponding to various colors

Color	Approximate wavelength, nm	Approximate wave number, cm^{-1}	
Deep blue (violet)	400	25,000	Approximate short-wavelength limit of human vision
Blue	450	22,200	
Blue-green	490	20,400	
Green	530	18,900	
Yellow	580	17,200	
Orange	620	16,100	
Red	700	14,300	
Deep red	750	13,300	Approximate long-wavelength limit of human vision

Generally light is characterized by its wavelength, λ, or its wave number, $\bar{\nu}$. These quantities are related to the frequency as follows: $\bar{\nu} = 1/\lambda = \nu/c$, where c is the velocity of light. The absorbance, or optical density, of a solution is defined by the equation

$$A = \log(I_o/I)$$

where I_o and I are the intensities of the incident and transmitted light, respectively. A plot of absorbance vs. wavelength or wave number, such as that shown in Fig. 17.1 for a $KMnO_4$ solution, is an *absorption spectrum*. Electronic transitions of this type, in which electrons are excited to higher, empty orbitals, often give rise to excited vibrational states of the molecule or ion. Such combined vibrational-electronic excitation can sometimes cause "fine structure" in the spectrum, as seen in Fig. 17.1. More often, however, the vibrational structure cannot be resolved and merely causes the spectrum to appear as a broadened band.

The human retina contains three kinds of cones, sensitive to the primary colors red, green, and blue. Light containing essentially equal proportions of these three colors is white; a combination of red and green is yellow, a combination of red and blue is magenta (purple), and a combination of green and blue[1] is cyan (blue-green). The color that one perceives for a particular solution is determined by the relative proportions of the primary colors transmitted by the solution. Thus the solution corresponding to the spectrum of Fig. 17.1, which absorbs in the middle (green region) of the visible spectrum, transmits red and blue light and appears magenta (purple).

---

[1] The color that most people associate with blue (that of aqueous Cu^{2+}) is really blue-green. True blue is better represented by the $Cu(NH_3)_4^{2+}$ ion.

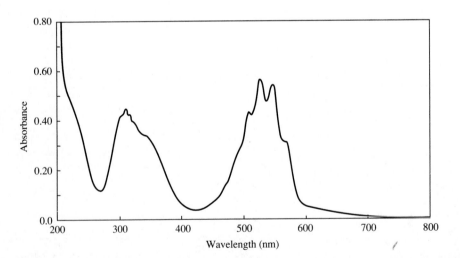

FIGURE 17.1
Absorption spectrum of an aqueous solution of $KMnO_4$.

Probably the most clear-cut evidence for the electronic energy levels of transition-metal compounds is provided by the electronic spectra of these compounds. Most of this chapter will be concerned with so-called ligand-field spectra, which can be rationalized using simple electrostatic ligand-field splitting diagrams. However, we will also discuss charge-transfer spectra and photoelectron spectra, which require energy level diagrams derivable from molecular orbital (MO) theory for their rationalization.

LIGAND-FIELD SPECTRA[2-4]

First we shall discuss electronic transitions between d orbitals for octahedral and tetrahedral complexes. These transitions are often referred to as d-d or ligand-field transitions. When such transitions occur in complexes with centers of symmetry, such as regular octahedral complexes, the intensities of the absorption bands are low. However, relatively strong absorption bands are obtained for d-d transitions in complexes which lack centers of symmetry, such as cis complexes of the type MA_4B_2 and tetrahedral complexes. The change in intensity of d-d transitions on going from octahedral to tetrahedral complexes can be demonstrated by adding excess concentrated hydrochloric acid to an aqueous solution of a Co(II) salt. The color changes from the pale red of $Co(H_2O)_6^{2+}$ to the intense blue of $CoCl_4^{2-}$.

Consider a d^1 octahedral complex, for which one observes a single absorption band corresponding to the process $t_{2g}^1 e_g^{*0} \rightarrow t_{2g}^0 e_g^{*1}$. This transition is represented in spectroscopic notation by the symbols $^2T_{2g} \rightarrow ^2E_g$, where the left superscripts indicate the spin multiplicities of the states. (The spin multiplicity is $2S + 1$, where S is the total spin of the state.) In the case of a d^1 tetrahedral complex, which has an energy level diagram that is essentially the inverse of that for a d^1 octahedral complex, the notation for the transition is $^2E \rightarrow ^2T_2$. The ligand-field transition of a d^9 complex may be treated in essentially the same way as that of a d^1 complex if one recognizes that a d^9 configuration corresponds to a positive hole in a filled d^{10} configuration. The electrostatic behavior of a hole is the opposite of that of an electron, and so for a d^9 octahedral complex one simply inverts the d^1 octahedral energy level diagram. The electronic transition is $^2E_g \rightarrow ^2T_{2g}$ (the same as that for a d^1 tetrahedral complex, except for the g subscripts). For a d^9 tetrahedral complex, the transition is $^2T_2 \rightarrow ^2E$. In general, a d^n octahedral complex has electronic states analogous to those of a d^{10-n} tetrahedral complex, and vice versa.

If a d^5 metal ion is subjected to an octahedral or tetrahedral field which is weak enough not to cause electron pairing, so that each d orbital has one electron, the metal ion will be essentially spherically symmetric, like a d^0 or d^{10} ion. Removal of a d electron from such a complex, to form a weak-field

[2]   B. N. Figgis, "Introduction to Ligand Fields," Interscience, New York, 1966.

[3]   C. K. Jørgensen, "Modern Aspects of Ligand Field Theory," pp. 343–353, American Elsevier, New York, 1971.

[4]   R. S. Drago, "Physical Methods in Chemistry," pp. 359–410, Saunders, Philadelphia, 1977.

d^4 complex, is analogous to removal of an electron from a d^{10} complex, and addition of a d electron, to form a weak-field d^6 complex, is analogous to addition of an electron to a d^0 complex. Thus a d^6 weak-field octahedral complex undergoes a $^5T_{2g} \rightarrow {}^5E_g$ transition analogous to the $^2T_{2g} \rightarrow {}^2E_g$ transition of a d^1 octahedral complex. In general, weak-field d^{n+5} complexes have transitions analogous to those of the corresponding d^n complexes.

Thus far, we have discussed one-electron and pseudo-one-electron systems. In such cases there is only one conceivable ligand-field transition, and only one ligand-field band is observed. The situation is much more complicated when more than one electron must be considered. That is, the energy level diagrams shown in Figs. 15.2, 15.4, and 15.10 are oversimplified. In general, there are several possible electronic states corresponding to a particular distribution of two or more electrons in d-orbital levels. These states arise because of the interaction of the electrons with one another. The states have different energies, causing relatively complicated spectra in some cases. If the ligand field of the complex were so strong as to make interelectronic interactions negligible by comparison, then only one d-d transition would be observed for a complex, for example, $t_{2g}{}^2 e_g{}^{*0} \rightarrow t_{2g}{}^1 e_g{}^{*1}$. Actually, of course, the electrons interact appreciably, and the energy level for each MO configuration is split into various terms. If we were able to decrease gradually the ligand-field strength to zero, we would observe that each term would gradually approach an energy corresponding to one of the states of the free ion. Thus the states of the complex can be correlated with those of the free metal ion. In Fig. 17.2 the correlation diagram for a d^2 octahedral complex is shown. Notice that, no matter how strong or weak the field, the ground state is always $^3T_{1g}$. Now, one of the rules of spectroscopy is that a transition involving a change in multiplicity is forbidden. Therefore, inasmuch as the ground state of the complex is a triplet state, the only allowed transitions are those to other triplet states. Notice that in Fig. 17.2 only the energy levels for triplet states are drawn with solid lines. The allowed transitions, marked with vertical arrows, are $^3T_{1g}(F) \rightarrow {}^3T_{2g}, {}^3T_{1g}(F) \rightarrow {}^3A_{2g}$, and $^3T_{1g}(F) \rightarrow {}^3T_{1g}(P)$. The two different $^3T_{1g}$ levels are distinguished by indicating, in parentheses, the free-ion states from which they are derived. Inasmuch as the $^3A_{2g}$ level crosses the $^3T_{1g}(P)$ level, the highest-energy transition of a d^2 octahedral complex can be either $^3T_{1g}(F) \rightarrow {}^3T_{1g}(P)$ or $^3T_{1g}(F) \rightarrow {}^3A_{2g}$, depending on the magnitude of Δ_o.

Although the detailed methods of construction of the correlation diagram in Fig. 17.2 are beyond the scope of this book,[5] it is important to have an understanding of the physical basis of the various splittings shown in the diagram. Let us consider the triplet states, which are those of main concern in the spectra of octahedral d^2 complexes. The ground state, based on the $t_{2g}{}^2$ configuration, is threefold degenerate (actually a $^3T_{1g}$ state[6]) because there are three ways of placing

---

[5] See F. A. Cotton, "Chemical Applications of Group Theory," 2d ed., pp. 254–260, Wiley-Interscience, New York, 1971.

[6] A and B states are nondegenerate, E states are twofold degenerate, and T states are threefold degenerate.

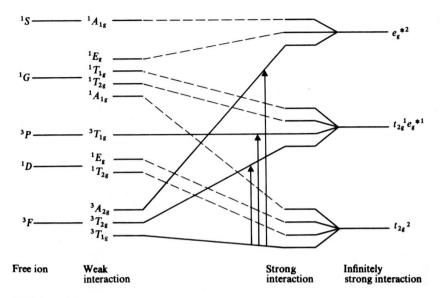

FIGURE 17.2
Correlation diagram for an octahedral d^2 complex.

the electrons in the t_{2g} orbitals (with the restriction that the spins be parallel):

$$(d_{xy})^1(d_{xz})^1$$
$$(d_{xy})^1(d_{yz})^1$$
$$(d_{xz})^1(d_{yz})^1$$

There are two triplet states based on the $t_{2g}{}^1 e_g{}^{*1}$ configuration. The one of lower energy (the triply degenerate $^3T_{2g}$ state) has the electrons in orbitals which are as far apart as possible:

$$(d_{xy})^1(d_{z^2})^1$$
$$(d_{xz})^1(d_{y^2})^1$$
$$(d_{yz})^1(d_{x^2})^1$$

The state of higher energy (the triply degenerate $^3T_{1g}$ state) has the electrons in orbitals which are relatively close together:

$$(d_{xy})^1(d_{x^2-y^2})^1$$
$$(d_{xz})^1(d_{z^2-x^2})^1$$
$$(d_{yz})^1(d_{y^2-z^2})^1$$

The reader will remember from Chap. 1 that the choice of five linear combinations of d functions, corresponding to the five d orbitals, is completely arbitrary. Here we have chosen to use unconventional orbitals such as d_{y^2} and $d_{z^2-x^2}$ when convenient. There is only one triplet state derived from the e_g^{*2} configuration (the nondegenerate $^3A_{2g}$ state), corresponding to the following orbital occupancy:

$$(d_{x^2-y^2})^1(d_{z^2})^1$$

By similar procedures it is possible to deduce the number, the degeneracies, and the relative energies of the states derived from other electron configurations: d^3, d^4, etc.

A correlation diagram of the type shown in Fig. 17.2 can be drawn for each electron configuration in the series d^1 to d^9 for both octahedral and tetrahedral ions. However, when only qualitative information regarding the allowed spectral transitions of high-spin complexes is desired, the Orgel diagrams[7] shown in Figs. 17.3 and 17.4 can be consulted. These diagrams were constructed by taking advantage of the hole formalism, the spherical symmetry of d^5 ions in weak fields, and the fact that the splitting of a given free-ion state is inverted on going from an octahedral to a tetrahedral field. The diagrams give the term symbols of the ground state and all higher states having the same multiplicity. States having the same term symbol are distinguished by parenthetical indication of their free-ion states. The diagrams give the energies of states derived from free-ion ground states, in units of Δ relative to the energies of the free-ion ground states. [These energies are very approximate for $T_1(F)$ states, especially at $\Delta \gtrsim 5000$ cm^{-1}, because these states have their energies depressed by "mixing" with the higher-lying $T_1(P)$ states.] To use the Orgel diagrams, it is only necessary to know that the complex is high-spin (weak-field) and to know the number of d electrons and whether the complex is octahedral (O_h symmetry) or tetrahedral (T_d symmetry). The term symbols with the subscripts g (for "gerade") correspond to octahedral complexes, which have centers of symmetry; the symbols without the g subscripts correspond to tetrahedral complexes, which lack centers of symmetry. The multiplicities of the states should be indicated by left superscripts. As a simple example of the

---

[7] L. E. Orgel, *J. Chem. Phys.*, **23**, 1004 (1955).

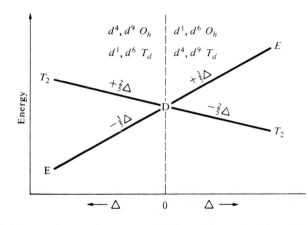

FIGURE 17.3
Qualitative Orgel diagram for high-spin octahedral and tetrahedral complexes with 1, 4, 6, or 9 d electrons.

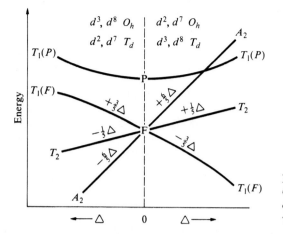

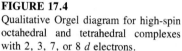

FIGURE 17.4
Qualitative Orgel diagram for high-spin octahedral and tetrahedral complexes with 2, 3, 7, or 8 d electrons.

use of the Orgel diagrams, let us determine the spectroscopic notation for the lowest-energy ligand-field transition of the octahedral $Ni(H_2O)_6^{2+}$ ion. Because the complex is a d^8 system, we obtain the ground-state and first excited-state terms from the left-hand side of Fig. 17.4. The complex has two unpaired electrons, corresponding to a multiplicity of 3, and so the transition is $^3A_{2g} \rightarrow {}^3T_{2g}$. From the diagram, we see that the ground state lies $\frac{6}{5}\Delta$ below the free-ion energy and that the first excited state lies $\frac{1}{5}\Delta$ below the free-ion energy. Hence the transition energy is the difference Δ.

The Ligand-Field Splitting, Δ, and the Interelectronic Repulsion Parameter, B

Before we consider electronic transition energies quantitatively, it is important to have some understanding of the factors which determine the magnitudes of the ligand-field splitting, Δ, and the so-called Racah parameter, B. The following generalizations regarding Δ_o and Δ_t have been drawn from a large body of spectroscopic data.

1. For complexes of the first transition series, Δ_o ranges 7500 to 12,500 cm^{-1} for $+2$ ions and from 14,000 to 25,000 cm^{-1} for $+3$ ions.

2. For metal ions of the same group and with the same charge, Δ_o increases by 30 to 50 percent from the first transition series to the second, and from the second to the third.

3. We have already pointed out that, other things being equal, $\Delta_t \approx \frac{4}{9}\Delta_o$.

4. Ligands may be arranged in a series in the order of their abilities to split d-orbital levels, i.e., in the order of the Δ values for the complexes of a given metal ion. This series is called the "spectrochemical series." For some common ligands, it is $I^- < Br^- < Cl^- < F^- < OH^- < C_2O_4^{2-} \sim H_2O < -NCS^- <$ pyridine $\sim NH_3 <$ en $<$ bipyridyl $< o$-phenanthroline $< NO_2^- < CN^-$. Slightly different orders are found for different metal ions, and this series should be used with caution.

The spectrochemical series constitutes experimental evidence for the superiority of the MO treatment of bonding in complexes over the electrostatic ligand-field treatment. Although the order of the halide ions, $I^- < Br^- < Cl^- < F^-$, is quite logical, the electrostatic theory alone cannot explain why F^- does not give the strongest field of all ligands, as would be expected because of its small size. Undoubtedly $p\pi$-$d\pi$ repulsions are responsible for the relatively low Δ values of halide complexes. Ammonia and the amines have no $p\pi$ lone-pair orbitals and therefore give relatively high Δ values. The water molecule has only one $p\pi$ lone-pair orbital in contrast to two for the OH^- ion; consequently water gives greater Δ values than OH^-. The very high Δ values of CN^- and NO_2^- complexes are due, of course, to $d\pi \rightarrow p\pi^*$ back bonding. The valence bond representations of the bonding of these ligands to metal ions containing d electrons are resonance hybrids of the following type:

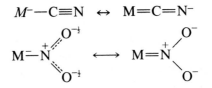

Jørgensen has shown that Δ_o can be roughly estimated from the empirical relation $\Delta_o = fg$, where f is a parameter characteristic of the ligand and g is a parameter characteristic of the metal.[3] The relation can sometimes be helpful when experimental spectral data are not available. Some f and g values are given in Table 17.2.

The energy separation between states having the maximum possible multiplicity (e.g., between the ^3F and ^3P states of a d^2 ion) is proportional to a theoretical parameter, B, called the Racah parameter:

$$E(^3P) - E(^3F) = 15B$$

Values of B for some common ions in the free state are given in Table 17.3. The value of B for a complex is always less than that for the free ion. This reduction of B upon complexation corresponds to a reduction of interelectronic repulsion, due to delocalization of the metal electrons onto the ligands, and is known as the "nephelauxetic" (cloud-expanding) effect. The factor, β, by which B is reduced from the free-ion value has been shown to be related to empirical parameters h for the ligand and k for the metal by the equation

$$\beta = \frac{B_{complex}}{B_{free\ ion}} = 1 - hk$$

Some h and k values are given in Table 17.2.

Moderate success has been attained in calculation of the energies of the electronic states of octahedral transition-metal complexes using semiempirical methods. The results of Tanabe and Sugano are presented in the form of graphs in Fig. 17.5. Instead of using absolute units for the electronic energies and Δ_o, which

TABLE 17.2
Values of f and g for estimating Δ_o and of h and k for estimating β †

Ligand	f	h		Metal ion	$g \times 10^{-3}$, cm^{-1}	k
Br$^-$	0.72	2.3		Mn^{2+}	8.0	0.07
—SCN$^-$	0.75			Ni^{2+}	8.7	0.12
Cl$^-$	0.78	2.0		Co^{2+}	9	0.09
OPCl$_3$	0.82			V^{2+}	12.0	0.08
N$_3^-$	0.83			Fe^{3+}	14.0	0.24
F$^-$	0.9	0.8		Cu^{3+}	15.7	
OSMe$_2$	0.91			Cr^{3+}	17.4	0.21
OCMe$_2$	0.92	1.2		Co^{3+}	18.2	0.35
EtOH	0.97			Ru^{2+}	20	
Me$_2$NCHO	0.98			Ag^{3+}	20.4	
C$_2$O$_4^{2-}$	0.99	1.5		Ni^{4+}	22	
H$_2$O	1.00	1.0		Mn^{4+}	24	0.5
SC(NH$_2$)$_2$	1.01			Mo^{3+}	24.6	0.15
—NCS$^-$	1.02			Rh^{3+}	27	0.30
—NCSe$^-$	1.03			Pd^{4+}	29	
—NH$_2$CH$_2$CO$_2$—$^-$	1.18			Tc^{4+}	31	
CH$_3$CN	1.22			Ir^{3+}	32	0.3
C$_5$H$_5$N	1.23			Pt^{4+}	36	0.5
NH$_3$	1.25	1.4				
en	1.28	1.5				
—SO$_3^{2-}$	1.2					
diars	1.33					
dipy	1.33					
—NO$_2^-$	1.4					
CN$^-$	1.7	2.0				

† C. K. Jørgensen, "Modern Aspects of Ligand Field Theory," pp. 347–348, American Elsevier, New York, 1971; C. K. Jørgensen, "Absorption Spectra and Chemical Bonding in Complexes," p. 113, Pergamon, Oxford, 1962.

TABLE 17.3
Free-ion values of B for transition-metal ions (in cm^{-1})

Metal	M^{2+}	M^{3+}
Ti	695	
V	755	861
Cr	810	918
Mn	860	965
Fe	917	1015
Co	971	1065
Ni	1030	1115

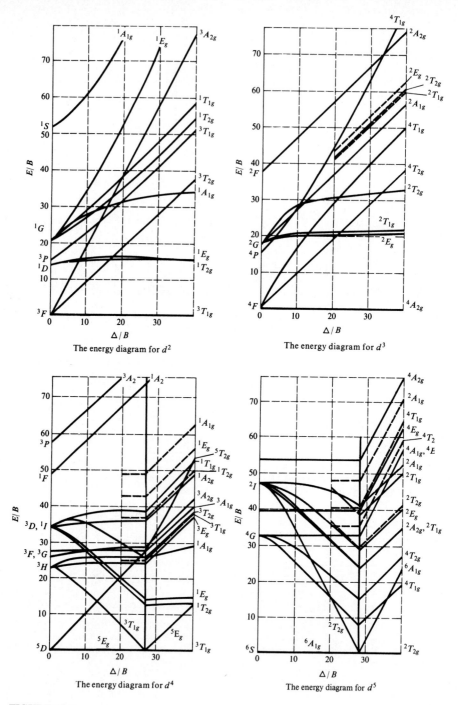

FIGURE 17.5
Tanabe-Sugano energy level diagrams for octahedral complexes with d^2-d^8 electron configurations. [*After Y. Tanabe and S. Sugano, J. Phys. Soc. Japan,* **9**, *753, 766 (1954), reproduced with permission from B. N. Figgis, "Introduction to Ligand Fields," Interscience, New York, 1966.*]

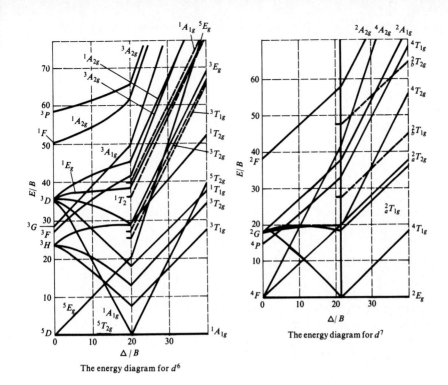

The energy diagram for d^6

The energy diagram for d^7

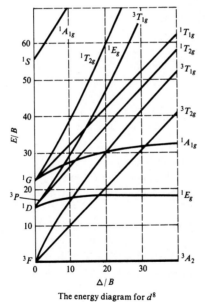

The energy diagram for d^8

would restrict each diagram to just the one case in which the separations of the free-ion terms matched those in the diagram, the state energies and Δ_o are expressed in units of the Racah parameter, B. The diagrams are drawn so that the ground state has zero energy for all values of Δ_o/B. Thus sharp changes in the slopes of the lines occur where the ground state changes. Of course, these sharp changes are merely artifacts of the method of plotting.

Spectral Calculations[8]

Figure 17.6 shows the absorption spectra of the $Ni(H_2O)_6^{2+}$ and $Ni(en)_3^{2+}$ complex ions. The spectrum of $Ni(H_2O)_6^{2+}$ has two main bands in the visible region, with a "window" in the green region ($\sim$ 5500 Å); hence this complex is green. The spectrum of $Ni(en)_3^{2+}$ has one band in the visible region, with windows in the blue and red regions; hence this complex is purple. Each spectrum has three bands, as expected for d^8 octahedral complexes. The $Ni(en)_3^{2+}$ bands are at higher frequencies than the corresponding $Ni(H_2O)_6^{2+}$ bands, in accord with the spectrochemical series. On the basis of the Orgel and Tanabe-Sugano diagrams, one would assign the lowest-frequency band in each spectrum to a $^3A_{2g} \rightarrow {}^3T_{2g}$ transition, the intermediate-frequency band to a $^3A_{2g} \rightarrow {}^3T_{1g}(F)$ transition, and the highest-frequency band to a $^3A_{2g} \rightarrow {}^3T_{1g}(P)$ transition. The plausibility of these assignments can be verified by using the d^8 Tanabe-Sugano diagram. Consider, for example, $Ni(H_2O)_6^{2+}$, which has transitions at 9000, 14,000, and 25,000 cm^{-1}. Clearly the lowest transition energy, 9000 cm^{-1}, corresponds to Δ_o. (This value is in fair agreement with the value estimated from f and g parameters, 8700 cm^{-1}.) From the appropriate data in Tables 17.2 and 17.3, one calculates $\beta = 0.88$ and $B = 906$ cm^{-1} for $Ni(H_2O)_6^{2+}$. By reading up from the point corresponding to $\Delta/B = 9000/906 = 9.9$ on the abscissa of the d^8 Tanabe-Sugano diagram, one finds E/B values of 16.5 and 29, or transition energies of 15,000 and 26,000 cm^{-1}, respectively, for the $^3A_{2g} \rightarrow {}^3T_{1g}(F)$ and $^3A_{2g} \rightarrow {}^3T_{1g}(P)$ transitions. The agreement with the experimental values is quite good.

The $Co(H_2O)_6^{2+}$ ion has absorption bands at 8350 and 19,000 cm^{-1} which have been assigned to the transitions $^4T_{1g}(F) \rightarrow {}^4T_{2g}$ and $^4T_{1g}(F) \rightarrow {}^4T_{1g}(P)$, respectively.[9] Using parameters from Tables 17.2 and 17.3, we estimate $B = 884$ cm^{-1} and $\Delta_o = 9000$ cm^{-1} for $Co(H_2O)_6^{2+}$. Using these values and the d^7 Tanabe-Sugano diagram, we obtain 8600 and 20,000 cm^{-1} for the $^4T_{1g}(F) \rightarrow {}^4T_{2g}$ and $^4T_{1g}(F) \rightarrow {}^4T_{1g}(P)$ transitions, respectively, in fair agreement with the observed bands. The $^4T_{1g}(F) \rightarrow {}^4A_{2g}$ band is estimated to be at 16,000 cm^{-1}. The latter band is expected to be relatively weak (because it corresponds

---

[8] For more detail, see R. S. Drago, "Physical Methods in Chemistry," pp. 359–410, Saunders, Philadelphia, 1977.

[9] C. J. Ballhausen, "Introduction to Ligand Field Theory," McGraw-Hill, New York, 1962.

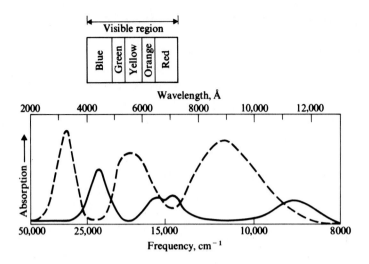

FIGURE 17.6
Absorption spectra of $Ni(H_2O)_6^{2+}$ (solid line) and $Ni(en)_3^{2+}$ (dashed line). The colors of the visible region of the spectrum are indicated above. [*Reproduced with permission from F. A. Cotton, J. Chem. Educ., **41**, 466 (1964).*]

to the two-electron excitation $t_{2g}^5 e^{*2} \to t_{2g}^3 e^{*4}$) and is probably obscured by the relatively strong $^4T_{1g}(F) \to {}^4T_{1g}(P)$ band.

Let us now consider the tetrahedral complex, $CoCl_4^{2-}$. Although from the Orgel diagram we predict three transitions, only two have been observed,[9] at 6300 and 15,000 cm^{-1}. We shall assign these transitions by comparing the observed frequencies with those estimated for the three allowed transitions. We have already estimated $\Delta_o = 9000$ cm^{-1} for $Co(H_2O)_6^{2+}$. Because chloride gives about 0.78 as strong a ligand field as water (see Table 17.2) and because $\Delta_t \approx \frac{4}{9}\Delta_o$, we predict Δ_t for $CoCl_4^{2-}$ to be $0.78(\frac{4}{9})9000 = 3100$ cm^{-1}. From Fig. 17.4, we see that the energy of the $^4A_2 \to {}^4T_2$ transition is Δ_t and that the energy of the $^4A_2 \to {}^4T_1(F)$ transition should be about $\frac{9}{5}\Delta_t$. From Fig. 17.4 we also see that the energy of the $^4A_2 \to {}^4T_1(P)$ transition should be $\frac{6}{5}\Delta_t$ plus the energy of the $^4T_1(P)$ state. Now a P state is not split by a tetrahedral field, and as a rough approximation we may assume that the $^4T_1(P)$ energy is the same as that of the ^4P state of the free Co^{2+} ion, or $15B = 14,600$ cm^{-1}. Thus the energy of the $^4A_2 \to {}^4T_1(P)$ transition should be approximately $\frac{6}{5}\Delta_t + 14,600$. Using our estimated value of Δ_t, we calculate 3100, 5600, and 18,000 cm^{-1} for the transitions $^4A_2 \to {}^4T_2$, $^4A_2 \to {}^4T_1(F)$, and $^4A_2 \to {}^4T_1(P)$, respectively. It seems reasonable to assign the observed bands to the latter two transitions and to assume that the band predicted to be at 3100 cm^{-1} (in the infrared region) has not been observed because of the inconvenient spectral region.

Almost all cobalt(III) complexes of the type CoL_6 are strong-field (low-spin) complexes, corresponding to the right-hand side of the d^6 Tanabe-Sugano diagram. The $^1A_{1g}$ ground state, based on the t_{2g}^6 configuration, is represented by the electron configuration

$$(d_{xy})^2(d_{xz})^2(d_{yz})^2$$

There are two singlet states based on the $t_{2g}^5 e_g^{*1}$ configuration. The $^1T_{1g}$ state corresponds to the orbital occupancies

$$(d_{xy})^1(d_{xz})^2(d_{yz})^2(d_{x^2-y^2})^1$$
$$(d_{xy})^2(d_{xz})^1(d_{yz})^2(d_{z^2-x^2})^1$$
$$(d_{xy})^2(d_{xz})^2(d_{yz})^1(d_{y^2-z^2})^1$$

and the $^1T_{2g}$ state corresponds to the orbital occupancies

$$(d_{xy})^1(d_{xz})^2(d_{yz})^2(d_{z^2})^1$$
$$(d_{xy})^2(d_{xz})^1(d_{yz})^2(d_{y^2})^1$$
$$(d_{xy})^2(d_{xz})^2(d_{yz})^1(d_{x^2})^1$$

The $^1T_{1g}$ state is of lower energy because in that state the t_{2g} hole is as close as possible to the e_g^* electron. Two transitions, from the $^1A_{1g}$ ground state to the $^1T_{1g}$ and $^1T_{2g}$ states, are generally observed for cobalt(III) complexes. Transitions to higher singlet states, based on the $t_{2g}^4 e_g^{*2}$ and $t_{2g}^3 e_g^{*3}$ configurations, are of extremely high energy and are usually not observed.

In a cobalt(III) complex of type CoL_4X_2, the symmetry is lowered and the $^1T_{1g}$ and $^1T_{2g}$ levels are split. The splitting of the $^1T_{2g}$ level is smaller than the width of the $^1A_{1g} \rightarrow {}^1T_{2g}$ absorption band, and therefore there is no significant change in that absorption band. However, the splitting of the $^1T_{1g}$ level is relatively large and causes the first absorption band to split into two components. We shall discuss the relative magnitudes of this splitting of the first absorption band in *trans*- and *cis*-CoL_4X_2. Let us suppose that the change in energy of the $d_{x^2-y^2}$ orbital when an L ligand is replaced by an X ligand is zero when X is on the z axis, and δ when X is on the x or y axis. Let us also suppose that the t_{2g} orbitals are unaffected by the ligands (i.e., that π interactions are unimportant). Then the energy levels of the components of the $^1T_{1g}$ state will be changed on going from CoL_6 to *trans*-CoL_4X_2 and *cis*-CoL_4X_2, as shown in Table 17.4.[10] It can be seen that the splitting of the trans complex is twice as great as that of the cis complex and that one component of the trans complex has the same energy as that of the original $^1T_{1g}$ state. These results are in agreement with the experimental spectra. The splitting of the first band of a trans complex is about twice that of a cis complex, and a subband unshifted from the first band of the CoL_6 complex is found only for the trans isomer of CoL_4X_2. Because a cis complex lacks a center of symmetry, its bands are more intense than those of the corresponding trans

---

10 H. Yamatera, *Bull. Chem. Soc. Jpn.*, **31**, 95 (1958).

TABLE 17.4
Predicted energy shifts of the components of the
$^1T_{1g}$ **state on going from CoL$_6$ to CoL$_4$X$_2$**

Electron configuration	trans-CoL$_4$X$_2$ (X at +z, −z)		cis-CoL$_4$X$_2$ (X at +x, +y)	
$(d_{xy})^1(d_{xz})^2(d_{yz})^2(d_{x^2-y^2})^1$	0	$^1A_{2g}$	2δ	$^1B_1$
$(d_{xy})^2(d_{xz})^1(d_{yz})^2(d_{z^2-x^2})^1$	$2\delta\}$	^1E_g	$\delta\}$	$^1A_2, ^1B_2$
$(d_{xy})^2(d_{xz})^2(d_{yz})^1(d_{y^2-z^2})^1$	2δ		δ	

complex. The spectra of *cis*- and *trans*-Co(en)$_2$F$_2$$^+$, shown in Fig. 17.7, illustrate these effects.[11]

CHARGE-TRANSFER SPECTRA

Ligand-to-Metal Charge Transfer

An electronic transition between orbitals that are centered on different atoms is called a charge-transfer transition, and the absorption band is usually very strong. Such bands are often prominent in the spectra of complexes in which there are electrons in π orbitals of the ligands. An energy level diagram for a complex of this type is shown in Fig. 15.11. The spectra of RuCl$_6$$^{2-}$ and IrBr$_6$$^{2-}$ (d^4 and d^5 complexes, respectively) show two sets of bands that have been assigned to transitions from the weakly bonding π orbitals on the ligands to the antibonding $t_{2g}{}^*$ and $e_g{}^*$ orbitals of the metal atom. In IrBr$_6$$^{3-}$ (a d^6 complex), the $t_{2g}{}^*$ orbitals are filled, and only the transitions to the $e_g{}^*$ orbitals can be observed.

---

[11] F. Basolo, C. J. Ballhausen, and J. Bjerrum, *Acta Chem. Scand.*, **9**, 810 (1955).

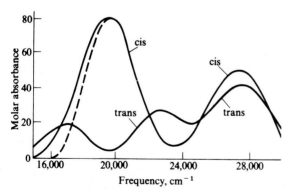

FIGURE 17.7
Absorption spectra of *cis*- and *trans*-Co(en)$_2$F$_2$$^+$. The splitting of the $^1A_{1g} \rightarrow {}^1T_{1g}$ band is very apparent in the spectrum of the trans complex and is evidenced only by a band asymmetry in the spectrum of the cis complex. The dashed curve shows the probable shape of the high-frequency component of the split band for the cis isomer. [*Reproduced with permission from F. Basolo et al., Acta Chem. Scand., 9, 810 (1955).*]

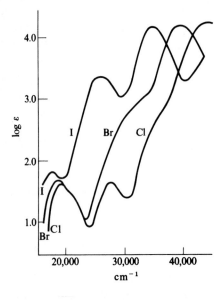

FIGURE 17.8
The spectra of the $Co(NH_3)_5X^{2+}$ ions, where X is a halogen. *(Reproduced with permission from W. L. Jolly, "The Synthesis and Characterization of Inorganic Compounds," Prentice-Hall, Englewood Cliffs, N.J., 1970.)*

   In the halogenopentaammine complexes of the type $Co(NH_3)_5X^{2+}$, strong charge-transfer bands are observed in the ultraviolet region, as shown in Fig. 17.8. These bands appear at progressively lower frequencies on going from the chloro to the bromo to the iodo complex, as one might expect from the trend in reduction potentials for these halogens. Indeed, in $Co(NH_3)_5I^{2+}$, the charge-transfer bands largely obscure the weaker *d-d* transitions.

Metal-to-Ligand Charge Transfer

The visible absorption spectra of iron(II) complexes with ligands containing the α-diimine unit

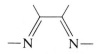

have intense charge-transfer bands associated with the transfer of charge from metal t_{2g} orbitals to the antibonding orbitals of the α-diimine group.[12] In the case of the 1,10-phenanthroline complex $Fe(phen)_3^{2+}$, the transition occurs at 19,600 cm^{-1}. A series of 6-coordinate iron(II) complexes containing the tetraimine macro-cyclic ligand (TIM) shown below, have been prepared with various monoden-

---

[12] E. Konig, *Coord. Chem. Rev.*, **3**, 471 (1969); P. Krumholz, *Struct. Bonding (Berlin)*, **9**, 139 (1971).

TABLE 17.5
Charge-transfer band maxima of Fe(II)-TIM complexes†

Complex	Band maximum, cm^{-1}
$Fe(TIM)CH_3CN(CO)^{2+}$	23,200
$Fe(TIM)(P(OEt)_3)_2{}^{2+}$	19,500
$Fe(TIM)(CH_3CN)_2{}^{2+}$	18,200
$Fe(TIM)(imidazole)_2{}^{2+}$	15,200
$Fe(TIM)(NH_3)_2{}^{2+}$	14,300

† M. J. Incorvia and J. I. Zink, *Inorg. Chem.*, **16**, 3161 (1977).

tate ligands occupying the two axial sites.[13] The band maxima of these complexes, corresponding to metal-to-TIM charge transfer, are listed in Table 17.5. It can be seen that the transition energy is a function of the π-acceptor ability of the axial ligands. As the π-acceptor ability increases, the energy of the d_{xz} and d_{yz} orbitals decreases relative to that of the $\pi^*(TIM)$ orbital, causing the $t_{2g} \rightarrow \pi^*(TIM)$ charge-transfer band to move to higher energies.

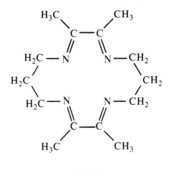

TIM

SPECTRA OF COMPOUNDS WITH METAL-METAL BONDS[14]

Compounds which contain metal-metal bonds are intensely colored. For example, $Mn_2(CO)_{10}$ is bright yellow, $Re_2Cl_8{}^{2-}$ is deep blue, $Co_2(CO)_8$ is purple-black, $Mo_2Cl_8{}^{4-}$ is cherry red, and $Fe_2(CO)_9$ is gold. The MO energy level diagrams

---

[13]   M. J. Incorvia and J. I. Zink, *Inorg. Chem.*, **16**, 3161 (1977).

[14]   W. C. Trogler and H. B. Gray, *Acc. Chem. Res.*, **11**, 232 (1978); W. C. Trogler, *J. Chem. Educ.*, **57**, 424 (1980).

shown in Fig. 15.13 are appropriate for these compounds; the observed transitions are of the types $\sigma \rightarrow \sigma^*$, $\pi \rightarrow \pi^*$, and $\delta \rightarrow \delta^*$ and are fully allowed. In the case of $Mn_2(CO)_{10}$, the $\sigma \rightarrow \sigma^*$ band peaks at 29,400 cm^{-1}. One expects that this excitation should be accompanied by weakening or dissociation of the Mn—Mn bond. Indeed, ultraviolet irradiation of $Mn_2(CO)_{10}$ in carbon tetrachloride solution leads to the intermediate formation of $Mn(CO)_5$ radicals, which abstract chlorine atoms from the solvent to form the chlorine-substituted mononuclear complex:

$$Mn_2(CO)_{10} \xrightarrow[\text{CCl}_4]{hv} 2Mn(CO)_5Cl$$

Photochemical cleavage of metal-metal bonds is fairly common in metal cluster compounds. However, it appears that excitation to the relatively weakly antibonding π^* and δ^* states is ineffective for such cleavage.

PHOTOELECTRON SPECTRA

The various electronic energy levels of a complex can be directly identified by photoelectron spectroscopy. Figure 17.9 shows the ultraviolet photoelectron spec-

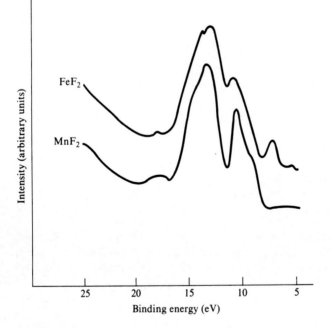

FIGURE 17.9
Ultraviolet photoelectron spectra of MnF_2 and FeF_2. [*Adapted from R. T. Poole, J. D. Riley, J. G. Jenkin, J. Liesegang, and R. C. G. Leckey, Phys. Rev. B, 13, 2620 (1976).*]

tra of crystalline MnF_2 and FeF_2, which have the rutile structure (octahedral coordination of the metal ions and trigonal coordination of the fluoride ions).[15] The MnF_2 spectrum has two main bands. The band at lower energy corresponds to photoionization from the half-filled $3d$ shell. the asymmetry of this band is probably due to the splitting of the d shell by the octahedral field into the $t_{2g}{}^3$ and $e_g{}^{*2}$ subshells; the low-energy shoulder can reasonably be assigned to the antibonding $e_g{}^{*2}$ electrons. The intense band at higher energy is due to the fluoride $2p$ electrons. The asymmetry of this band is probably due to spin-orbit splitting into $2p_{3/2}$ and $2p_{1/2}$ components.

The spectrum of FeF_2 is very similar to that of MnF_2, except for an extra small band at the low-energy end of the spectrum. This new feature is part of a final-state multiplet. Photoemission from Mn^{2+} ($3d^5$), with a half-filled d shell, can only lead to quintet final states. In Fe^{2+} ($3d^6$), the extra d electron is antiparallel to the other five d electrons, and photoemission can give both quartet and sextet final states. The small band is due to the $^6A_{1g}$ final state; the larger band is due to quartet states.

PROBLEMS

17.1 The $Mn(H_2O)_6{}^{2+}$ ion has an extremely pale pink color, attributable to transitions which are formally forbidden. Explain why there are no completely allowed d-d transitions.

17.2 Two d-d bands can be seen, at 17,000 and 26,000 cm^{-1}, in the absorption spectrum of $V(H_2O)_6{}^{3+}$. Assign the bands.

17.3 Predict the colors of the following complexes, which have absorption bands at the indicated frequencies: $CrF_6{}^{3-}$ (14,900, 22,700 cm^{-1}), $Cu(NH_3)_4{}^{2+}$ (14,000 cm^{-1}), $FeO_4{}^{2-}$ (12,700, 19,600 cm^{-1}), $Co(en)_3{}^{2+}$ (21,400, 29,600 cm^{-1}).

17.4 Predict the d-d absorption spectra (frequencies and spectroscopic assignments for all bands) of $CoF_6{}^{3-}$ (high-spin), $NiCl_4{}^{2-}$, and $Fe(H_2O)_6{}^{2+}$ (high-spin). Estimate parameters when necessary.

17.5 Without referring to any data, indicate how many spin-allowed ligand-field electron transitions you would expect to observe for the following complexes: VCl_4, $CrO_4{}^{2-}$, $FeO_4{}^{2-}$, $Fe(H_2O)_6{}^{2+}$, $Cr(H_2O)_6{}^{3+}$, $Ni(NH_3)_6{}^{2+}$, $Fe(CN)_6{}^{4-}$, $Co(H_2O)_6{}^{2+}$, $Mn(H_2O)_6{}^{2+}$, $MnO_4{}^{2-}$, $Zn(OH)_4{}^{2-}$.

17.6 The $CrO_4{}^{2-}$ ion is a d^0 complex and yet is colored. (*a*) Explain. (*b*) Would you expect the energy of the transition to be higher or lower than than for $MnO_4{}^-$? (Hint: Consider the appropriate redox potentials.)

*17.7 Deduce the number, degeneracies, and relative energies of the quartet states for an octahedral d^3 complex.

---

[15] R. T. Poole, J. D. Riley, J. G. Jenkin, J. Liesegang, and R. C. G. Leckey, *Phys. Rev. B*, **13**, 2620 (1976).

18

OTHER PROPERTIES OF COORDINATION COMPOUNDS

We continue our study of the properties of simple transition-metal compounds. In this chapter we cover static properties that are affected by the metal valence d electrons: magnetic properties, structural effects of the Jahn-Teller theorem, and both structural and thermodynamic consequences of ligand-field stabilization energy and metal-ligand π bonding. Reactions of transition-metal compounds will be covered in later chapters.

PARAMAGNETISM

Magnetic Moment

When a substance is subjected to a magnetic field, H, a magnetization, I, is induced. The ratio I/H is called the "volume susceptibility," κ, and can be measured by a variety of techniques, including the Gouy balance method, the Faraday method, and an nmr method.[1] The volume susceptibility is simply related to the

---

[1] W. L. Jolly, "The Synthesis and Characterization of Inorganic Compounds," pp. 369–384, Prentice-Hall, Englewood Cliffs, N.J., 1970.

"gram susceptibility," χ, and the "molar susceptibility," χ_M,

$$\chi = \frac{\kappa}{d} \qquad \chi_M = \frac{\kappa M}{d}$$

where d and M are the density and molecular weight of the substance, respectively. For most substances, κ, χ, and χ_M have negative values; such substances are weakly repelled by a magnetic field and are called "diamagnetic." For substances having unpaired electrons that do not strongly interact with one another, κ, χ, and χ_M have relatively large positive values; these substances are attracted into a magnetic field and are called "paramagnetic." A list of miscellaneous substances and their molar susceptibilities is given in Table 18.1.

When a paramagnetic substance is placed in a magnetic field, the moments of the paramagnetic molecules or ions tend to align with the field; however, thermal agitation tends to randomize the orientations of the individual moments. Theoretical analysis of the situation leads to the relation

$$\chi_M^{\text{corr}} = \frac{N\mu^2}{3kT}$$

where χ_M^{corr} is the molar susceptibility which has been corrected both for the diamagnetic contribution to the susceptibility (due to the nonparamagnetic atoms in the sample) and for any small temperature-independent paramagnetism arising

TABLE 18.1
Some representative molar magnetic susceptibilities†

Compound	State	$\chi_M \times 10^6$	Compound	State	$\chi_M \times 10^6$
H_2	g	-4.0	CCl_4	l	-66.8
He	g	-1.9	CBr_4	l	-93.7
N_2	g	-12.0	C_6H_6	l	-54.85
NO	g	1,472 (292 K)	C_6H_{12}	l	-66.1
	g	2,324 (147 K)	$[Cr(H_2O)_6]Cl_3$	s	5,950
O_2	g	3,450	$MnCl_2 \cdot 4H_2O$	s	14,600
H_2O	g	-13.1 (373 K)	$FeCl_3 \cdot 6H_2O$	s	15,250
	l	-12.97	$K_3Fe(CN)_6$	s	2,290
	l	-12.93 (273 K)	$FeCl_2 \cdot 4H_2O$	s	12,900
	s	-12.65 (273 K)	$K_4Fe(CN)_6$	s	-130
KCl	s	-39.0	$CuSO_4 \cdot 5H_2O$	s	1,460
	soln	-39.4	$ZnSO_4 \cdot 7H_2O$	s	-143
KBr	s	-49.1	$Ce(NO_3)_3 \cdot 5H_2O$	s	2,310
KI	s	-63.8	$Ce(SO_4)_2 \cdot 4H_2O$	s	-97
Hg	l	-33.4	$Gd(C_2H_5SO_4)_3 \cdot 9H_2O$	s	24,100
Na	s	$+16.1$	$U(C_2O_4)_2 \cdot 5H_2O$	s	3,760

† Values are for 290 to 295 K unless otherwise indicated. They are taken from G. Foëx, Tables of Constants and Numerical Data (U.I.C.P.A.), No. 7, "Constantes Sélectionnées Diamagnétisme et Paramagnétisme," Masson, Paris, 1957. They are in units of 10^{-6} cm^3 mol^{-1}.

from paramagnetic excited states of the system,[2] N is Avogadro's number, k is the Boltzmann constant, μ is the "magnetic moment" of the molecule, and T is the absolute temperature. By substituting numerical values for N and k, we obtain

$$\chi_M^{\text{corr}} = \frac{0.125 \mu^2}{T} \qquad \text{or} \qquad \mu = 2.83 \sqrt{\chi_M^{\text{corr}} T} \qquad (18.1)$$

The importance of the magnetic moment, μ, in transition-metal chemistry lies in the fact that, for most compounds, it is both experimentally measurable and theoretically calculable. Often the magnitude of the calculated moment is markedly changed by a change in the assumed structure or a change in the assumed type of bonding. In such cases comparison of experimental μ values with calculated μ values can be of considerable value in characterization of the compounds.

The magnetic moment of a transition-metal complex is a combination of spin and orbital moments. However, in many complexes the orbital contribution is almost completely quenched, and the magnetic moment can be calculated by the following "spin-only" formula,

$$\mu = 2 \sqrt{S(S + 1)}$$

where S is the total spin of the complex. Because in the ground state S is one-half the number of unpaired electrons, n, we may write

$$\mu = \sqrt{n(n + 2)}$$

In Table 18.2, the experimental μ values for various octahedral complexes of the first transition series are listed with the μ values calculated by the spin-only formula. The data permit us to distinguish readily between strong- and weak-field complexes. For example, $[Cr(H_2O)_6]SO_4$ is a weak-field d^4 complex with $\mu = 4.8$, whereas $[Cr(dipy)_3]Br_2 \cdot 4H_2O$ is a strong-field d^4 complex with $\mu = 3.3$. Similar pairs of weak- and strong-field complexes, respectively, are the d^5 complexes $K_2[Mn(H_2O)_6](SO_4)_2(\mu = 5.9)$ and $K_4[Mn(CN)_6] \cdot 3H_2O(\mu = 2.2)$ and the d^7 complexes $(NH_4)_2[Co(H_2O)_6](SO_4)_2(\mu = 5.1)$ and $K_2Pb[Co(NO_2)_6](\mu = 1.8)$. The d^6 weak-field complex $(NH_4)_2[Fe(H_2O)_6](SO_4)_2$ $(\mu = 5.5)$ may be compared with the strong-field $K_4[Fe(CN)_6]$, which is not listed in the table because it is diamagnetic.

Consider the effect on a d^8 octahedral complex of a tetragonal elongation. Two possible electron configurations are shown in Fig. 18.1. In the case of a regular octahedral complex, there are two unpaired electrons. However, in the

---

[2]   If the uncorrected value of χ is used, the corresponding value of μ is usually called the "effective" magnetic moment, μ_{eff}. For a discussion of diamagnetic contributions and temperature-independent paramagnetism, see F. A. Cotton and G. Wilkinson, "Advanced Inorganic Chemistry," 4th ed., pp. 1359–1365, Wiley-Interscience, New York, 1980; and Landolt-Börnstein, "Magnetic Properties of Coordination and Organometallic Transition Metal Compounds," E. König, vol. II/2, 1966; E. König and G. König, vol. II/8, 1976, Springer-Verlag, New York.

TABLE 18.2
Experimental and calculated values of the magnetic moment

No. of unpaired electrons, n	No. of d electrons	Compound	Ground state	Experimental μ	$[n(n+2)]^{1/2}$
		O_h complexes			
1	1	$CsTi(SO_4)_2 \cdot 12H_2O$	$^2T_{2g}$	1.8	1.73
	5	$K_4Mn(CN)_6 \cdot 3H_2O$†	$^2T_{2g}$	2.2	1.73
	7	$K_2PbCo(NO_2)_6$†	^2E_g	1.8	1.73
	9	$(NH_4)_2Cu(SO_4)_2 \cdot 6H_2O$	^2E_g	1.9	1.73
2	2	$(NH_4)V(SO_4)_2 \cdot 12H_2O$	$^3T_{1g}$	2.7	2.83
	4	$Cr(dipy)_3Br_2 \cdot 4H_2O$†	$^3T_{1g}$	3.3	2.83
	8	$(NH_4)_2Ni(SO_4)_2 \cdot 6H_2O$	$^3A_{2g}$	3.2	2.83
3	3	$KCr(SO_4)_2 \cdot 12H_2O$	$^4A_{2g}$	3.8	3.87
	7	$(NH_4)_2Co(SO_4)_2 \cdot 6H_2O$	$^4T_{1g}$	5.1	3.87
4	4	$CrSO_4 \cdot 6H_2O$	^5E_g	4.8	4.90
	6	$(NH_4)_2Fe(SO_4)_2 \cdot 6H_2O$	$^5T_{2g}$	5.5	4.90
5	5	$K_2Mn(SO_4)_2 \cdot 6H_2O$	$^6A_{1g}$	5.9	5.92
		T_d complexes			
1	1	$V(Nor)_4$‡	^2E	1.82	1.73
	5	$Co(Nor)_4$ † ‡	$^2T_2$	2.00	1.73
	9	$Co[P(O\text{-}i\text{-}C_3H_7)_3]_4$	$^2T_2$	2.0	1.73
2	2	FeO_4^{2-}	$^3A_2$	2.8	2.83
	8	NiX_4^{2-}	$^3T_1$	3.2–4.1	2.83
3	3	$Mn(Nor)_4$‡	$^4T_1$	3.78	3.87
	7	CoX_4^{2-}	$^4A_2$	4.4–4.8	3.87
4	6	FeX_4^{2-}	^5E	5.0–5.2	4.90
5	5	$FeCl_4^-$	$^6A_1$	5.9	5.92

† Low-spin complexes.

‡ Nor = 1-norbornyl group.

case of strong distortion or a square planar complex, a zero-spin state results. Indeed, all square planar d^8 complexes are diamagnetic unless the ligands have unpaired electrons. In tetrahedral complexes, high-spin and low-spin states are possible for the configurations d^3, d^4, d^5, and d^6, but because of the relatively small magnitude of Δ_t, there are few low-spin tetrahedral complexes.

High-Spin–Low-Spin Crossover

Octahedral complexes with 4, 5, 6, or 7 d electrons can be either high-spin or low-spin, depending on the magnitude of the ligand-field splitting, Δ_o. When the ligand-field splitting has an intermediate value such that the two states of the complex have similar energies, the two states can coexist in measurable amounts

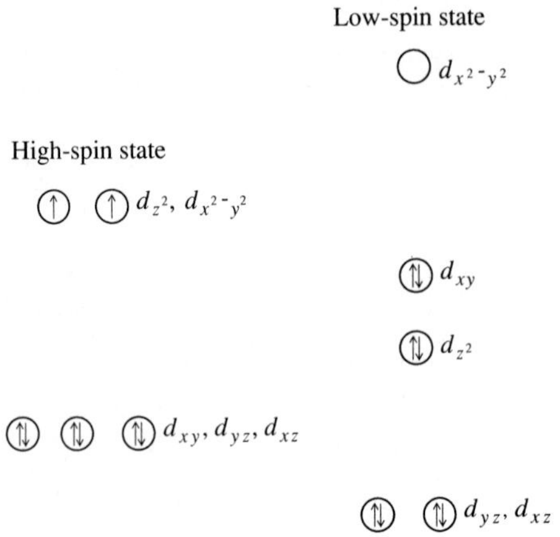

Low-spin state

$\bigcirc d_{x^2-y^2}$

High-spin state

$\textcircled{\uparrow} \quad \textcircled{\uparrow} \; d_{z^2}, d_{x^2-y^2}$

$\textcircled{\uparrow\downarrow} \; d_{xy}$

$\textcircled{\uparrow\downarrow} \; d_{z^2}$

$\textcircled{\uparrow\downarrow} \quad \textcircled{\uparrow\downarrow} \quad \textcircled{\uparrow\downarrow} \; d_{xy}, d_{yz}, d_{xz}$

$\textcircled{\uparrow\downarrow} \quad \textcircled{\uparrow\downarrow} \; d_{yz}, d_{xz}$

FIGURE 18.1
Energy level diagrams for regular and strongly tetragonally distorted octahedral d^8 complexes, showing the electron occupancies.

at equilibrium. Many "crossover" systems of this type have been studied.[3] In solutions, these systems are fairly straightforward; the change in magnetic susceptibility with temperature can be interpreted in terms of the heat of conversion of one isomer to another. For example, susceptibility-versus-temperature data for CH_2Cl_2 solutions of the iron(II) d^6 complex $Fe[(pz)_3BH]_2$ [where the $(pz)_3BH$ ligand has the structure shown on the next page] are shown in Fig. 18.2.[4]

---

[3]   R. L. Martin and A. H. White, *Transition Met. Chem.*, **4,** 113 (1968); H. A. Goodwin, *Coord. Chem. Rev.*, **18,** 293 (1976).

[4]   J. P. Jesson, S. Trofimenko, and D. R. Eaton, *J. Am. Chem. Soc.*, **89,** 3158 (1967).

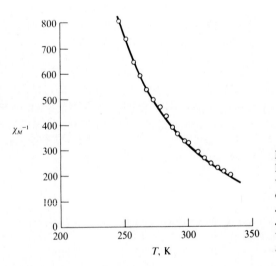

FIGURE 18.2
Plot of reciprocal molar susceptibility versus temperature for CH_2Cl_2 solution of $Fe[(pz)_3BH]_2$. [*Adapted from J. P. Jesson, S. Trofimenko, and D. R. Eaton, J. Am. Chem. Soc., 84, 3158 (1967), with permission. Copyright 1967 American Chemical Society.*]

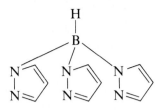

Clearly the data do not conform to Eq. 18.1. However, treatment of the system as an equilibrium between two spins yields $\Delta H = 3.85$ kcal mol^{-1} and $\Delta S = 11.4$ eu for the high-spin $\rightarrow$ low-spin conversion. On the other hand, spin crossover in solids is a complex phenomenon because of cooperative structural changes and changes in the energy separation of the high-spin and low-spin states with temperature.[5] Thus the magnetism of $Fe(phen)_2(NCS)_2$ changes sharply at 174 K, as shown in Fig. 18.3.[6]

Orbital Angular Momentum[7]

About half of the experimental μ values listed in Table 18.2 are close to the theoretical μ values calculated from the spin-only formula. However, the remaining experimental μ values differ significantly from the simple theoretical values. The deviations are generally due to contributions of the orbital magnetic moment. The question is: When can we expect to find such orbital contribution to the magnetic moment, and when can we expect the orbital contribution to be completely quenched?

An electron must be able to circulate about an axis if it is to have orbital angular momentum. Therefore an orbital must be available, in addition to the orbital containing the electron, which has the following properties. It must have the same shape and energy as the orbital containing the electron. It must be superimposable on the orbital containing the electron by rotation about an axis. And, finally, it cannot contain an electron having the same spin as the first electron. These conditions are fulfilled in the case of d electrons in an octahedral or tetrahedral field whenever any two t_2 or t_{2g} orbitals (d_{xy}, d_{xz}, d_{yz}) contain one or three electrons. Thus unquenched orbital angular momentum will remain for a metal complex with 1, 2, 4, or 5 electrons in the t_2 or t_{2g} orbitals, corresponding to T_1, T_2, T_{1g}, or T_{2g} ground states. In Table 18.2 practically all the complexes for which μ deviates markedly from the spin-only μ value have T ground states.

[5]   E. König and S. Kremer, *Theoret. Chim. Acta,* **20,** 143 (1971).

[6]   E. König and K. Madeja, *Inorg. Chem.,* **6,** 48 (1967). For similar systems, see M. G. Burnett, V. McKee, and S. M. Nelson, *J. Chem. Soc. Dalton Trans.,* 1492 (1981).

[7]   B. N. Figgis and J. Lewis, *Prog. Inorg. Chem.,* **6,** 37 (1964).

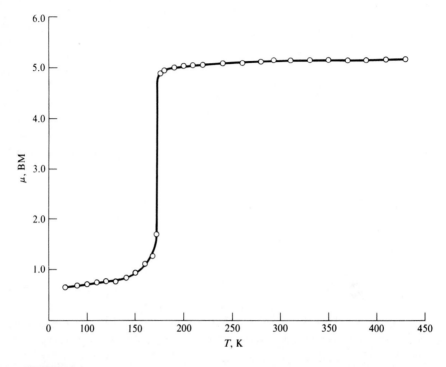

FIGURE 18.3
The magnetic moment of $Fe(phen)_2(NCS)_2$ as a function of temperature. [*Adapted from E. König and K. Madeja, Inorg. Chem.,* **6,** *48 (1967). Copyright 1967 American Chemical Society.*]

For the purpose of estimating μ values for complexes with unquenched orbital angular momentum, the values in Table 18.3 can be used. These are theoretical values,[8] calculated from free-ion parameters, and one should not expect agreement with experimental values for complexes to better than ± 0.2.

Smaller but significant discrepancies between experimental moments and calculated spin-only moments are found for complexes with A_2, A_{2g}, E, and E_g ground states. These discrepancies are caused by angular momentum introduced to the ground state by a mixing of higher T states. Theory shows that the magnetic moment of such a complex can be calculated from the relation

$$\mu = \left(1 - \frac{\lambda'}{\Delta}\right)[n(n+2)]^{1/2} \tag{18.2}$$

---

[8] B. N. Figgis, "Introduction to Ligand Fields," Interscience, New York, 1966.

TABLE 18.3
Calculated values of μ at 300 K for octahedral and tetrahedral stereochemistries, assuming free-ion values of spin-orbit coupling†

$^2T_{2(g)}$ states		$^3T_{1(g)}$ states		$^4T_{1(g)}$ states		$^5T_{2(g)}$ states	
Ti^{3+}	1.9	V^{3+}	2.7	Cr^{3+}	3.4	Cr^{2+}	4.7
Mo^{5+}	1.0	Cr^{2+}	3.5	Co^{2+}	5.1	Mn^{3+}	4.5
Mn^{2+}	2.55	Mo^{4+}	1.9			Fe^{2+}	5.65
Fe^{3+}	2.45	W^{4+}	1.5			Co^{3+}	5.75
Ru^{3+}	2.1	Mn^{3+}	3.65				
Os^{3+}	1.85	Re^{3+}	2.4				
Ir^{4+}	1.8	Fe^{4+}	3.55				
Cu^{2+}	2.2	Ru^{4+}	3.2				
		Os^{4+}	1.9				
		Ni^{2+}	4.0				

† B. N. Figgis, "Introduction to Ligand Fields," Interscience, New York, 1966.

Appropriate values of the spin-orbit coupling parameter, λ', for metal ions in octahedral and tetrahedral environments are given in Table 18.4.

Magnetic data may be used to determine the type of bonding in cobalt(II) complexes. Tetrahedral Co(II) complexes have room-temperature magnetic moments that range from 4.40 to 4.88. For example, $\mu = 4.59$ for $CoCl_4^{2-}$. We can show that this is a reasonable value by use of Eq. 18.2:

$$\mu = \left(1 + \frac{688}{3100}\right)(3.87) = 4.73$$

TABLE 18.4
Values of λ' for several metal ions†

Metal ion	Symmetry	Ground state	λ', cm^{-1}
V^{4+}	T_d	^2E	500
V^{2+}	O_h	$^4A_{2g}$	228
Cr^{3+}	O_h	$^4A_{2g}$	368
Cr^{2+}	O_h	^5E_g	116
Mn^{4+}	O_h	$^4A_{2g}$	552
Mn^{3+}	O_h	^5E_g	178
Fe^{2+}	T_d	^5E	-200
Co^{2+}	T_d	$^4A_2$	-688
Co^{2+}	O_h	^2E_g	-1030
Ni^{2+}	O_h	$^3A_{2g}$	-1260
Cu^{2+}	O_h	^2E_g	-1660

† B. N. Figgis, "Introduction to Ligand Fields," Interscience, New York, 1966.

Octahedral weak-field Co(II) complexes such as $(NH_4)_2[Co(H_2O)_6](SO_4)_2$ have room-temperature moments of 5.0 to 5.1, in remarkably close agreement with the value of 5.1 from Table 18.3. Octahedral strong-field Co(II) complexes appear to be very rare. However, $K_2Pb[Co(NO_2)_6]$ has been reported to have $\mu = 1.8$, in fair agreement with the value estimated by using Eq. 18.2:

$$\mu = \left(1 + \frac{1030}{12,600}\right)(1.73) = 1.87$$

FERROMAGNETISM, FERRIMAGNETISM, AND ANTIFERROMAGNETISM

If there is coupling between the individual magnetic moments of a paramagnetic sample, spontaneous ordering of the moments will occur below a critical temperature. If this ordering involves alignment of all the moments in the same direction to produce a permanent magnetic moment, the substance is said to be "ferromagnetic" and the critical temperature is called the "Curie temperature." If some of the moments are systematically aligned opposed to the others, but the relative numbers or magnitudes of the moments are such as to give a finite resultant magnetic moment, the substance is said to be "ferrimagnetic." The phenomenological properties of ferrimagnetism are essentially the same as those of ferromagnetism. If half the moments are aligned opposed to the other half, giving zero net moment, the substance is said to be "antiferromagnetic" and the critical temperature is called the Néel temperature. Schematic representations of the various types of magnetism are shown in Fig. 18.4, and rough plots of χ versus temperature for a paramagnet, a ferromagnet (or ferrimagnet), and an antiferromagnet are shown in Fig. 18.5.

We have already discussed ferromagnetic metals in Chap. 12. However, many compounds also exhibit ferromagnetism. For example CrO_2, which is used in recording tapes, is ferromagnetic, with a room-temperature magnetization of 515 gauss and a Curie temperature of 386 K. A familiar example of a ferrimagnetic

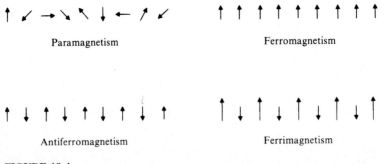

Paramagnetism

Ferromagnetism

Antiferromagnetism

Ferrimagnetism

FIGURE 18.4
A schematic representation of the various types of magnetism.

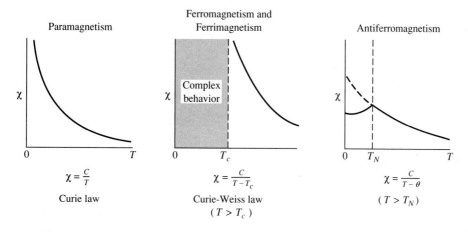

FIGURE 18.5
Schematic plots of magnetic susceptibility versus temperature for the various types of magnetism. T_C = Curie temperature; T_N = Néel temperature.

material is magnetite, Fe_3O_4. Here the Fe(II) moments are aligned parallel, but the Fe(III) moments are antiparallel, and the observed net moment arises only from the Fe(II) moments. The classic example of antiferromagnetism is MnO, which has the NaCl structure and a Néel temperature of 116 K. The three-dimensional distribution of spins in the antiferromagnetic form of this compound has been elucidated by neutron diffraction,[9] which can distinguish between sets of atoms having opposed moments. The neutron diffraction patterns at 80 and 293 K are shown in Fig. 18.6. The 80-K pattern has more lines than the 293-K pattern, and the two lines in the 293-K pattern are shifted from their corresponding positions in the 80-K pattern. The reflections at 80 K correspond to a unit cell of lattice constant 8.85 Å; those at 293 K correspond to a cell of lattice constant 4.43 Å. However, the lattice constant determined by x-ray diffraction (which cannot distinguish spins) is 4.43 Å at both temperatures. These results are most plausibly interpreted in terms of the low-temperature arrangement of Mn^{2+} spins shown in Fig. 18.7.[10] It can be seen that the repeat distance for spins aligned in the same direction is twice as great as that for manganese atoms.

The coupling between the individual moments of most ferromagnetic, ferrimagnetic, and antiferromagnetic substances is stronger than can be accounted for by simple dipole-dipole interactions. It is believed that the dipoles are coupled through the electrons of intervening nonmetal atoms, by a process called "superex-

[9] C. G. Shull, W. A. Strauser, and E. O. Wollan, *Phys. Rev.*, **83**, 333 (1951).
[10] Y. Y. Li, *Phys. Rev.*, **100**, 627 (1955).

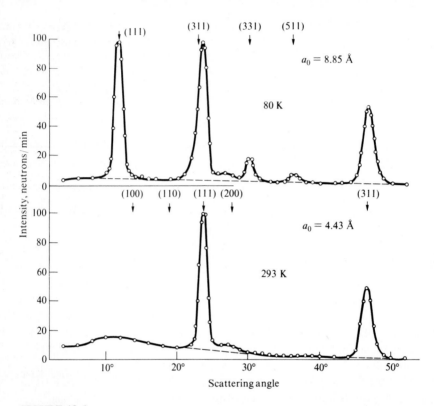

FIGURE 18.6
Neutron diffraction patterns for MnO below and above the Néel temperature of 116 K. [*Reproduced with permission from C. G. Shull, W. A. Strauser, and E. O. Wollan, Phys. Rev., 83, 333 (1951).*]

change."[11] For example, in MnO, one superexchange mechanism is believed to involve σ interactions between the half-filled e_g orbitals of Mn^{2+} ions and the filled $p\sigma$ orbitals of oxide ions, as follows.

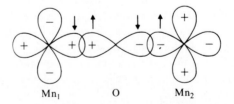

At any instant, the oxide electron which interacts with Mn_1 must have its spin opposed to that of the Mn_1 electron, leaving an electron of opposite spin to interact with Mn_2. Thus the manganese electrons are antiferromagnetically coupled.

[11]   J. B. Goodenough, "Magnetism and the Chemical Bond," Interscience, New York, 1963.

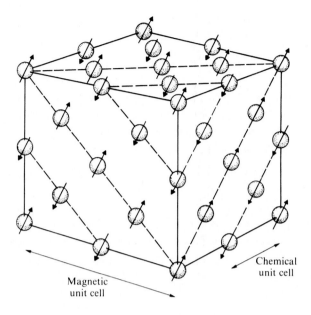

FIGURE 18.7
Arrangement of Mn^{2+} spins in the antiferromagnetic form of MnO. Above 116 K the Mn^{2+} ions are still magnetic, but they are no longer ordered. Note that the oxygen atoms are not shown.

Chemical unit cell

Magnetic unit cell

JAHN-TELLER DISTORTIONS

A regular octahedral environment is the most stable one for a spherically symmetric metal ion surrounded by six ligand atoms. For metal ions with certain d electron configurations which are not spherically symmetric, the regular octahedral configuration is not the most stable. The situation can be expressed in a general and rigorous way by the Jahn-Teller theorem:[12] Any nonlinear molecule in a degenerate electronic state will undergo distortion to remove the degeneracy and to lower the energy. For example, consider a high-spin d^4 complex. In a regular octahedral environment the $e_g{}^*$ electron is doubly degenerate; it can occupy either the d_{z^2} or the $d_{x^2-y^2}$ orbital. However, if the complex undergoes a tetragonal distortion, the $e_g{}^*$ levels are split and the electron can then occupy the lower of the two orbitals (the d_{z^2} orbital in the case of tetragonal elongation; the $d_{x^2-y^2}$ orbital in the case of tetragonal contraction). The same sort of distortion is expected in the case of d^9 and low-spin d^7 complexes. Examples of Jahn-Teller distortions of octahedral coordination are found in CrF_2 and MnF_3 (d^4 configurations), in $NaNiO_2$ (low-spin d^7 configuration), and in many copper(II) compounds, such as $CuCl_2$ and CuF_2 (d^9 configurations). In each of these cases the metal ion is surrounded by six anions at the vertices of an elongated octahedron. Complexes with d^8 configurations are a special case. In general, the complexes are either regular octahedral (with two unpaired electrons) or square planar and diamagnetic. The reason for the lack of d^8 octahedral complexes with small or intermediate

---

[12] H. A. Jahn and E. Teller, *Proc. R. Soc. London Ser. A,* **161,** 220 (1937).

tetragonal elongation can be seen by reference to Fig. 18.8. The energy of the singlet state either will never be below that of the undistorted triplet state (as in Fig. 18.8a) or will become lower than that of the undistorted triplet state in a highly distorted complex (as in Fig. 18.8b). In either case, the triplet state is much more stable than the singlet state in the undistorted octahedral complex. In Fig. 18.8a no degree of distortion will make the complex more stable than the undistorted triplet-state complex. In Fig. 18.8b, a large tetragonal distortion is required to stabilize the singlet state relative to the undistorted triplet state. In this latter situation, it is obvious that the distortion causes the energy gap between the d_{z^2} and $d_{x^2-y^2}$ orbitals to exceed the spin pairing energy.

Degeneracies arising from partially filled t_{2g} levels should also produce Jahn-Teller distortions, but the effects are relatively small because the t_{2g} orbitals are relatively unaffected by the ligands. Evidence for Jahn-Teller distortions and the consequent splitting of energy levels is found in the spectra of many complexes. For example, consider the absorption spectrum of $Ti(H_2O)_6^{3+}$, shown in Fig. 18.9. This band is obviously not a simple symmetric band such as one might expect for a $^2T_{2g} \rightarrow {}^2E_g$ transition. The breadth and asymmetry of the band are evidence of transitions to two excited states caused by distortion of the complex.

Relatively weak distortions are expected for tetrahedral complexes with d^3, d^4, d^8, and d^9 configurations.[13] In the d^3 and d^8 complexes, one of the

---

[13]   L. E. Orgel, "An Introduction to Transition-Metal Chemistry. Ligand Field Theory," 2d ed., Chapman and Hall, London, 1966.

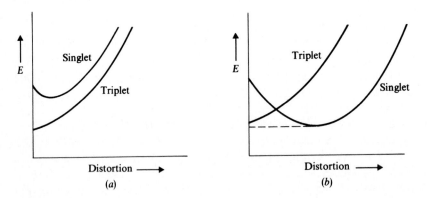

FIGURE 18.8
Potential energy curves illustrating stability of the high-spin regular octahedral d^8 complex (a) and stability of the highly distorted (essentially square planar) low-spin d^8 complex (b). *(Adapted with permission from L. E. Orgel, "An Introduction to Transition-Metal Chemistry. Ligand Field Theory," 2d ed., Chapman and Hall, London, 1966.)*

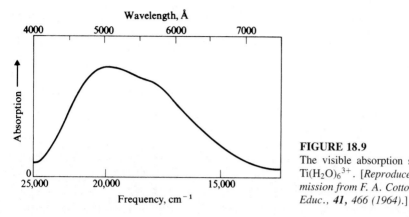

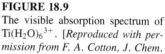

FIGURE 18.9
The visible absorption spectrum of $Ti(H_2O)_6^{3+}$. [*Reproduced with permission from F. A. Cotton, J. Chem. Educ., **41**, 466 (1964).*]

t_2 orbitals has one more electron than the other two. This situation should cause an elongation of the tetrahedron, as shown in Fig. 18.10a. In the d^4 and d^9 complexes, one of the t_2 orbitals has one fewer electron than the other two (i.e., an electron hole). This situation should cause a flattening of the tetrahedron, as shown in Fig. 18.10b. In agreement with these predictions, it has been found that the $CuCl_4^{2-}$ complex has a flattened tetrahedral configuration and that the oxide environment around the Ni^{2+} in $NiCr_2O_4$ is that of an elongated tetrahedron.

There are apparent exceptions to the Jahn-Teller theorem, in which certain physical properties correspond to symmetric structures when distorted structures are expected. All such apparent exceptions to the theorem are probably examples of the so-called *dynamic Jahn-Teller effect,* in which the direction of distortion of a complex (say an elongation of one axis of an octahedron) randomly moves among the available symmetry axes of the complex more rapidly than the physical measurement can "follow." The physical measurement thus "sees" an average structure, with no apparent distortion. For example, the room-temperature epr spectrum of a dilute solution of $Cu(H_2O)_6^{2+}$ in a host crystal of $[Zn(H_2O)_6]SiF_6$ shows only one transition, corresponding to a regular octahedral geometry for

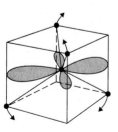

(a)

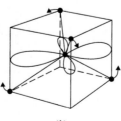

(b)

FIGURE 18.10
Distortions of tetrahedral coordination. (a) Elongation caused by t_2^1 and t_2^4 configurations; (b) flattening caused by t_2^2 and t_2^5 configurations. *(Adapted with permission from L. E. Orgel, "An Introduction to Transition-Metal Chemistry. Ligand Field Theory," 2d ed., Chapman and Hall, London, 1966.)*

the $Cu(H_2O)_6{}^{2+}$ ions.[14] However, it is believed that, in the crystal, each cation is tetragonally distorted along one of three mutually perpendicular axes. At room temperature only one transition is observed because, on the epr time scale, each cation appears to be symmetric. However, when the crystal is cooled to 20 K, the interchange of distortion axes slows down, and the spectrum becomes that expected for three magnetic sites.

LIGAND-FIELD STABILIZATION ENERGY

When a transition-metal ion is placed in an octahedral field, three of the d orbitals are stabilized by $\frac{2}{5}\Delta_o$ and two of the d orbitals are destabilized by $\frac{3}{5}\Delta_o$. In the case of an octahedral complex with the configuration $t_{2g}{}^{p}e_g{}^{*q}$, the net stabilization energy is $(\frac{2}{5}p - \frac{3}{5}q)\Delta_o$. For example, a d^3 complex such as $Cr(H_2O)_6{}^{3+}$ is stabilized by $\frac{6}{5}\Delta_o$, that is (using $\Delta_o = 17,400$ cm^{-1}), by 20,900 cm^{-1} or 59.7 kcal mol^{-1}, relative to a hypothetical $Cr(H_2O)_6{}^{3+}$ ion in which the three d electrons have a spherically symmetric disposition. Obviously these stabilization energies are of great importance in determining the relative thermodynamic properties of transition-metal complexes.

Consider the hydration energies of the $+2$ ions of the first transition series:

$$M^{2+}(g) \quad \rightarrow \quad M^{2+}(aq) \qquad \Delta H^\circ = -\Delta H_{hyd}$$

These energies, obtained from thermochemical data, are plotted as a function of atomic number in Fig. 18.11. If one draws a smooth curve through the three points corresponding to the spherically symmetric metal ions (Ca^{2+}, Mn^{2+}, and Zn^{2+}), for which one would expect no net ligand-field stabilization, it is obvious

---

[14]   B. Bleaney and D. J. E. Ingram, *Proc. Phys. Soc. London Sect. A,* **63,** 408 (1950); A. Abragam and M. H. L. Pryce, *Proc. R. Soc. London Ser. A,* **206,** 164 (1951).

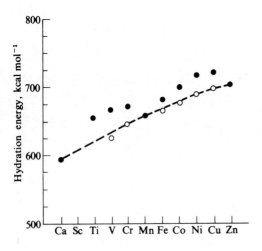

FIGURE 18.11
Hydration energies of the $+2$ ions of the first transition series plotted versus atomic number. Solid circles are experimental points; open circles correspond to values from which spectrally evaluated ligand-field stabilization energies have been subtracted. [*Reproduced with permission from F. A. Cotton, J. Chem. Educ., 41, 466 (1964).*]

that all the other points lie above the curve. It is reasonable to assume that the deviations from the curve are due to the extra ligand-field stabilization energies of the $M(H_2O)_6^{2+}$ ions. If we subtract the appropriate values of $(\frac{2}{5}p - \frac{3}{5}q)\Delta_o$ (calculated from spectral Δ_o data) from the hydration energies, we obtain the points indicated by open circles, which fall on the smooth curve. This result gives strong support to the concept of the splitting of d orbitals in octahedral fields.

In Fig. 18.12 the lattice energies of the dichlorides of the first-transition-series metals are plotted as a function of atomic number. The plot shows that the lattice energies of the salts other than $CaCl_2$, $MnCl_2$, and $ZnCl_2$ are exceptionally large, undoubtedly because of ligand-field stabilization energy. Similar plots are obtained for the lattice energies of other MX_2 salts and for MX_3 salts.

In the absence of ligand-field stabilization effects the ionic radii of the transition-metal ions would be expected to decrease steadily with increasing atomic number. However, from the festoons which appear in the plot of the radii of the $+2$ ions versus atomic number, shown in Fig. 18.13, it is clear that the stabilization energies correlate with the distances between atoms in ionic solids. The greater the ligand-field stabilization energy, the smaller the ionic radius, relative to the hypothetical spherically symmetric ion. Octahedral ions with high ligand-field stabilization energies have their d electrons mainly or entirely in the t_{2g} orbitals, directed between the ligands. Thus ligands can approach the ions closely, with minimal repulsion by the d electrons. On the other hand, if the d electrons are uniformly distributed among the t_{2g} and antibonding e_g orbitals, there is no ligand-field stabilization energy. Ligands are repelled by the e_g electrons, and therefore the effective ionic radius is greater.

In the case of an ion which can exist in either a low-spin or high-spin form, there are two corresponding ionic radii. The low-spin ion, which has fewer antibonding e_g electrons, has the smaller ionic radius. Various examples of low-spin and high-spin ionic radii can be found in Appendix F.

A ligand-field stabilization energy can be calculated for an ion of any geometry. The relative energies, in units of Δ_o, for the various d orbitals in several

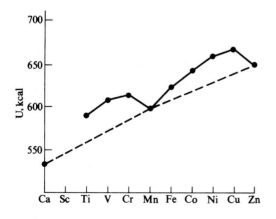

FIGURE 18.12
Lattice energies of the dichlorides of the metals of the first transition series.

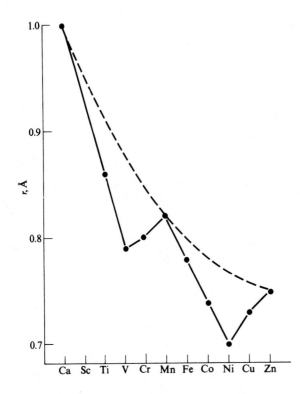

FIGURE 18.13
Ionic radii of the $+2$ ions of the first transition series, as in high-spin compounds.

coordination geometries are given in Table 18.5. From these data one can readily calculate that a 4-coordinate d^8 complex has a ligand-field stabilization energy of $2.456\Delta_o$ in a square planar geometry, and $0.356\Delta_o$ in a tetrahedral geometry. On this basis, one would expect all 4-coordinate d^8 complexes of a d^8 ion such as Ni^{2+} to be square planar. Indeed, most 4-coordinate Ni(II) complexes are square planar, but a few are tetrahedral. The tetrahedral complexes are mainly of the following types: NiX_4^{2-}, NiX_3L^-, NiX_2L_2, and $Ni(AA)_2$, where X is a halogen or SPh, and L and AA are bulky monodentate and bidendate ligands, respectively. Obviously steric repulsions between ligands favor tetrahedral structures over square planar structures. Thus we can readily rationalize the formation of tetrahedral complexes with bulky ligands. In the case of halide ligands, the ligand-field strengths are so low that ligand-field stabilization energies are less important than steric repulsion energies. It is significant that Pd(II) and Pt(II) complexes, which have relatively large ligand-field splittings (and therefore relatively large ligand-field stabilization energies), are almost always square planar.

Ligand-field stabilization effects are clearly seen in the structures of various spinels. The spinel structure, named after the mineral $MgAl_2O_4$, is adopted by a large number of compounds of the type $M^{II}M^{III}_2O_4$. The structure consists of a close-packed lattice of oxide ions with one-third of the metal ions in tetrahedral holes and two-thirds of the metal ions in octahedral holes. In a "normal"

TABLE 18.5
The energy levels of d orbitals in various coordination geometries (in units of Δ_o)

Coordination no.	Structure	d_{z^2}	$d_{x^2-y^2}$	d_{xy}	d_{xz}	d_{yz}
2	Linear†	1.028	−0.628	−0.628	0.114	0.114
3	Trigonal‡	−0.321	0.546	0.546	−0.386	−0.386
4	Tetrahedral	−0.267	−0.267	0.178	0.178	0.178
4	Square planar‡	−0.428	1.228	0.228	−0.514	−0.514
5	Trigonal bipyramid*	0.707	−0.082	−0.082	−0.272	−0.272
5	Square pyramid*	0.086	0.914	−0.086	−0.457	−0.457
6	Octahedron	0.600	0.600	−0.400	−0.400	−0.400
6	Trigonal prism	0.096	−0.584	−0.584	0.536	0.536
7	Pentagonal bipyramid	0.493	0.282	0.282	−0.528	−0.528

† Ligands lie on z axis.

‡ Ligands lie in xy plane.

* Pyramid base in xy plane.

spinel, the M(II) ions occupy tetrahedral holes and the M(III) ions occupy octahedral holes, thus: $[M^{II}]_{tet}[M^{III}_2]_{oct}O_4$. In an "inverse" spinel, half of the M(III) ions exchange positions with the M(II) ions, thus: $[M^{III}]_{tet}[M^{II}M^{III}]_{oct}O_4$. The oxides Mn_3O_4 and Co_3O_4 are normal spinels, and Fe_3O_4 and $NiAl_2O_4$ are inverse spinels. These differences can be explained in terms of ligand-field stabilization energies.[15] Because $\Delta_o > \Delta_t$, octahedral site stabilization energies are usually much greater than tetrahedral site stabilization energies. In Mn_3O_4, the Mn^{3+} ions, but not the Mn^{2+} ions, are ligand-field-stabilized, and therefore the Mn^{3+} ions occupy the sites which provide the maximum stabilization. In Co_3O_4, the Co^{3+} ions occupy octahedral sites as low-spin d^6 ions; if the structure were to invert, half of the Co^{3+} ions would be transferred to tetrahedral sites where they would probably become high-spin and lose a tremendous amount of ligand-field stabilization energy. In Fe_3O_4, the Fe^{2+} ions, but not the Fe^{3+} ions, are ligand-field-stabilized; hence the Fe^{2+} ions occupy octahedral sites. Similarly, in $NiAl_2O_4$, only the Ni^{2+} ions are ligand-field-stabilized, and therefore they occupy octahedral sites.

Spinels containing ions with magnetic moments have interesting magnetic properties. In the spinel structure, these ions are close enough to interact with one another, and one often observes ferromagnetism and antiferromagnetism. For example, in the black mineral magnetite, Fe_3O_4, all the ions in the octahedral holes have their magnetic spins aligned, and the Fe^{3+} ions in the tetrahedral holes have their spins aligned in the opposite direction, thus:

---

[15] For a different explanation, see J. K. Burdett, G. D. Price, and S. L. Price, *J. Am. Chem. Soc.*, **104,** 92 (1982).

$$\overset{\downarrow}{[Fe^{III}]_{tet}} \overset{\uparrow \ \uparrow}{[Fe^{III}Fe^{II}]_{oct}} O_4$$

Hence the Fe^{3+} ions are antiferromagnetically coupled; their mutual interaction yields no magnetic moment. The net ferrimagnetism of the compound is entirely attributable to the Fe^{2+} ions. If the Fe^{2+} are replaced with diamagnetic ions, as in $ZnFe_2O_4$, the resulting spinel is normal and completely antiferromagnetic, with no net magnetic moment.

STRUCTURAL EFFECTS OF LIGAND-METAL PI BONDING

π-Donor Bonding

Most of the transition-metal dioxo complexes of the type MO_2X_4 have d^0 or d^2 electron configurations. All the d^0 complexes have the cis structure, and the d^2 complexes have the trans structure.[16] These structures are logical consequences of the maximization of $p\pi \rightarrow d\pi$ bonding between the strongly π-donating oxygen atoms and the metal atoms. In the d^0 cases, the cis configuration is clearly preferred. If we assume the oxygen atoms are on the x and y axes, then they have exclusive use of one $d\pi$ orbital each (d_{xz} for O_x; d_{yz} for O_y) and share a third (d_{xy}). In the trans form (oxygen atoms on the $+x$ and $-x$ axes) the oxygen atoms would have to share two $d\pi$ orbitals (d_{xy} and d_{xz}) and leave one unused (d_{yz}). This latter orbital accommodates the electron pair in the d^2 trans complexes. Examples of d^0 cis complexes are $MoO_2Cl_4^{2-}$ and $VO_2F_4^{3-}$. Examples of d^2 trans complexes are $OsO_2(CN)_4^{2-}$ and $ReO_2py_4^{+}$.

π-Acceptor Bonding

After the discovery in 1890 that nickel carbonyl can be prepared by the reaction of nickel metal with carbon monoxide,[17] it was suspected that the $Ni(CO)_4$ molecule consists of a nickel atom tetrahedrally bonded to four carbon monoxide molecules, essentially as indicated in the following valence bond structure.

16 W. P. Griffith and T. D. Wickins, *J. Chem. Soc. (A)*, 400 (1968).

17 L. Mond, C. Langer, and F. Quincke, *J. Chem. Soc.*, **57,** 749 (1890).

This structure seemed plausible because, in it, the nickel atom achieves the stable electron configuration of the rare gas krypton. Of course incidentally the nickel atom has acquired an extremely negative formal charge, a situation which we now recognize as electronically unstable. When, in 1935, Brockway and Cross[18] determined the structure by electron diffraction, they verified the tetrahedral configuration, but they found that the Ni—C bond length is surprisingly small, only 1.82 ± 0.03 Å. This is about 0.34 Å less than the sum of the estimated single-bond covalent radii of carbon and nickel, and Brockway and Cross proposed that the Ni—C bonds have multiple-bond character, corresponding to significant contributions from resonance structures such as the following.

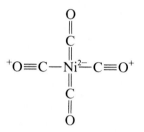

Thus they rationalized the short Ni—C bond and eliminated the extremely negative formal charge associated with the first structure. This proposal constituted the first clear-cut application of the concept of "back bonding," so called because it involves the shift of electron density from otherwise nonbonding π orbitals on the metal atom back to antibonding π^* orbitals on the ligands.

Back bonding is favored by an increase in electron density on the metal atom and therefore is most prominent in low-oxidation-state metal complexes. The tendency for a transition metal to back-bond also increases as one descends a periodic-table family, i.e., on going from the first row to the second row, and then to the third row. For example, although Fe^{2+} and Ru^{3+} show relatively weak back bonding, Ru^{2+} is a strong back-bonder. Striking evidence for the strong back-bonding ability of Ru^{2+} can be seen in the structural data for the *cis*-tetraamminebis(isonicotinamide)ruthenium(II) and -(III) ions,[19]

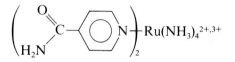

In the Ru(III) complex, the Ru—N(isn) bond distances (2.099 Å average) are only slightly less than the Ru—NH_3 bond distances (2.125 Å average). Presumably

---

[18] L. O. Brockway and P. C. Cross, *J. Chem. Phys.*, **3**, 828 (1935).

[19] D. E. Richardson, D. D. Walker, J. E. Sutton, K. O. Hodgson, and H. Taube, *Inorg. Chem.*, **18**, 2216 (1979).

both the isonicotinamide and ammonia molecules are principally σ donors in this complex where back bonding is unimportant. On the other hand, in the Ru(II) complex, the Ru—N(isn) bond distances (2.060 Å average) are significantly shorter than the Ru—NH$_3$ bond distances (2.143 Å average). The difference is explained by assuming $d\pi \rightarrow \pi^*$ interactions in the Ru^{II}—N(isn) bonds. Isonicotinamide, like many other aromatic molecules, has empty π^* MOs which can accept metal $d\pi$ electron density. The fact that the Ru^{II}—N(isn) bond is shorter than the Ru^{III}—N(isn) bond is particularly striking; ordinarily, when π bonding is not involved, metal-ligand distances decrease with increasing oxidation state, as in the case of the Ru—NH$_3$ bonds.

It has been noted that, in d^8 trigonal bipyramidal complexes with ligands of different π-acceptor strength, the stronger π acceptors have a distinct preference for the equatorial sites, at least in the case of ligands with cylindrically symmetric acceptor orbitals, such as CO, CN$^-$, and PR$_3$. Thus in Mn(CO)$_4$NO and Co(PPh$_2$Me)$_2$Cl$_2$NO, the strongly π-accepting nitrosyl groups occupy equatorial positions, and in Fe(CO)$_4$CN$^-$ and Fe(CO)$_4$py, the relatively weakly π-accepting cyanide and pyridine ligands occupy axial positions.[20] This structural preference can be rationalized in terms of the relative number of $d\pi$ orbitals available to ligands at the axial and equatorial sites.[21] The $d_{x^2-y^2}$ and d_{xy} orbitals are available only to the three equatorial ligands, corresponding to $\frac{2}{3}$ orbital per equatorial ligand. The d_{xz} and d_{yz} orbitals are available to all five ligands. If we apportion these orbitals to the axial and equatorial ligands in proportion to the number of symmetry-allowed π bonds that can be formed, each axial ligand gets only $\frac{4}{7}$ orbital whereas each equatorial ligand gets a total of ($\frac{2}{3} + \frac{2}{7}$) or $\frac{20}{21}$ orbital.

It is interesting to consider the configurations of metal-olefin complexes. The structure of the anion of Zeise's salt, K[PtCl$_3$(C$_2$H$_4$)] · H$_2$O, the first metal-olefin complex prepared, is shown in Fig. 18.14.[22] Several features of the structure are worthy of comment. (1) The C—C bond length, 1.375 Å, is significantly greater than that of free ethylene, 1.337 Å. This bond lengthening reflects the shift of electron density into the antibonding π^* orbital of the coordinated ethylene molecule. (2) The hydrogen atoms are bent back from the metal atom, as expected for an approach to sp^3 hybridization of the carbon atoms. (3) The trans chlorine atom is slightly farther from the metal atom than the cis chlorine atoms. This slight

[20]   We assign relative π-acceptor abilities to ligands as consistently as possible with both theoretical expectations and experimental data. The structural data cited for trigonal bipyramidal complexes are merely a small part of a large body of data which are consistent with NO$^+$ being a better π acceptor than CN$^-$. This result is of course expected. Differences in the properties of these isoelectronic species can be logically ascribed to the difference in their net charge. Because of the positive charge of NO$^+$, one expects it to be a better π acceptor than CN$^-$.

[21]   For a general MO study of the favored sites for π acceptors in d^0-d^{10} trigonal bipyramidal complexes, see A. R. Rossi and R. Hoffmann, *Inorg. Chem.*, **14**, 365 (1975).

[22]   R. A. Love, T. F. Koetzle, G. J. B. Williams, L. C. Andrews, and R. Bau, *Inorg. Chem.*, **14**, 2653 (1975).

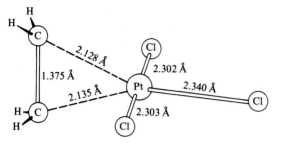

FIGURE 18.14

Configuration of the anion of Zeise's salt, $K[PtCl_3(C_2H_4)] \cdot H_2O$.

weakening of the trans Pt—Cl bond is probably a consequence of the relatively strong Pt—C_2H_4 σ bond. (The synergism of the $d\pi \rightarrow \pi^*$ back bonding and the $C_2H_4 \rightarrow Pt$ σ bond makes the latter bond much stronger than it would be otherwise.) The d,s,p hybridization of the metal orbital involved in σ bonding to the ethylene is such as to give good overlap with the π orbital of C_2H_4; consequently, the hybridization of the orbital involved in σ bonding to the trans chlorine is such as to give relatively poor overlap. (4) The C—C bond is perpendicular to the plane of the complex. This configuration is not easily rationalized in terms of ease of back bonding; it appears that back bonding from the d_{xy} orbital would be just as effective as back bonding from the d_{xz} orbital. Indeed, solution nmr studies indicate that the coordinated ethylene molecule undergoes rotation about the platinum-olefin axis, thus indicating that the configuration in which the olefin lies in the plane of the $PtCl_3$ group has a stability comparable to that in which the olefin is perpendicular to the plane.[23] The structure observed for the salt is probably a consequence of the minimization of steric repulsions between the ethylene ligand and the cis chlorine atoms combined with subtle crystal packing forces.

We generally find stronger evidence for back bonding when the metal atom has a low oxidation state. Consider, the example, tris(ethylene)nickel(0), which is believed to have the structure shown in Fig. 18.15a, in which the six carbon atoms and the nickel atom are coplanar. It is instructive to consider the reasons for the stability of this geometry as opposed to the "upright" structure of Fig. 18.15b, in which the C—C bonds are perpendicular to the coordination plane.[24] Presumably the σ bonding is equally strong in each case, and therefore the difference in stability must be due to differences in the $d\pi \rightarrow \pi^*$ back bonding. Let us first consider the effects of hypothetical σ interactions in the absence of π interactions. In both structures, σ interaction of the metal d orbitals with the bonding π orbitals of the C_2H_4 molecules yields three sets of MOs:[25] an antibonding e' set

[23] S. Maricic, C. R. Redpatch, and J. A. S. Smith, *J. Chem. Soc.*, 4905 (1963).

[24] N. Rösch and R. Hoffmann, *Inorg. Chem.*, **13**, 2656 (1974); R. M. Pitzer and H. F. Schaefer, *J. Am. Chem. Soc.*, **101**, 7176 (1979).

[25] The z axis is perpendicular to the coordination plane. We shall ignore the d_{z^2} orbital.

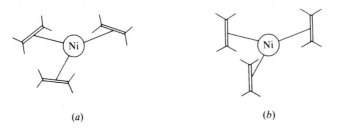

(a)   (b)

FIGURE 18.15
Two conceivable geometries for Ni(C₂H₄)₃. (a) The "planar" geometry; (b) the "upright" geometry.

consisting essentially of the destabilized d_{xy} and $d_{x^2-y^2}$ orbitals, a nonbonding e'' set consisting of the d_{xz} and d_{yz} orbitals, and a bonding e' set consisting essentially of stabilized π orbitals of the C_2H_4 ligands. This set of MOs is shown for both structures in Fig. 18.16. If now we allow interactions with the empty π^* orbitals of the C_2H_4 molecules, we find that the two structures give different results. In the planar structure, the upper e' MO engages in back bonding and is strongly stabilized, and the e'' MO, which has no interaction partner, stays at the same energy. In the upright structure, the otherwise nonbonding e'' MO engages in weak back bonding and is slightly stabilized, and the upper e' MO stays at the same energy. The stabilization of the e'' MO in the upright structure is relatively slight because the e'' MO is farther away in energy from the C_2H_4 π^* than the e' MO which is destabilized by the σ interaction. In the complex, the MOs of Fig. 18.15 are occupied by 12 electrons (8 from the nickel atom and 4 from the

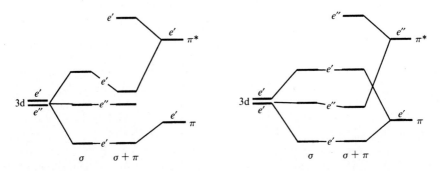

FIGURE 18.16
Schematic MO energy level diagrams of π bonding in Ni(C₂H₄)₃. The left-hand diagram is for the planar structure; the right-hand diagram for the upright structure. The orbital interactions are introduced stepwise. First, interaction with the $C_2H_4\pi$ levels is shown in the columns marked σ. Second, the effects of the $C_2H_4\pi*$ levels are shown, in the columns marked $\sigma + \pi$. [*Reproduced with permission from N. Rösch and R. Hoffmann, Inorg. Chem.,* **13**, *2656 (1974). Copyright 1974 American Chemical Society.*]

TABLE 18.6
Comparison of ethylene and tetracyanoethylene complexes†

Compound	r_{c-c}(complex) $-$ r_{c-c}(free olefin), Å
$Rh(C_2H_4)(C_2F_4)(C_5H_5)$	0.021
$K[PtCl_3(C_2H_4)] \cdot H_2O$	0.038
$Nb(C_2H_4)(C_2H_5)(C_5H_5)_2$	0.069
$Ni(t\text{-}BuNC)_2C_2(CN)_4$	0.132
$Ir(CO)(Br)(PPh_3)_2C_2(CN)_4$	0.162
$Ir(CO)(C_6N_4H)(PPh_3)_2C_2(CN)_4$	0.182

† Data from R. A. Love, T. F. Koetzle, G. J. B. Williams, L. C. Andrews, and R. Bau, *Inorg. Chem.*, **14**, 2653 (1975).

ethylene ligands); consequently the energy of the planar structure should be less than that of the upright structure.

Olefins with electronegative or π-accepting substituents, such as tetracyanoethylene, function as much better π acceptors than ethylene. Table 18.6 gives the lengthening of the olefin C—C bond on going from the free olefin to the coordinated olefin for several complexes of ethylene and tetracyanoethylene. The lengthening is much greater for tetracyanoethylene, consistent with stronger back bonding in the tetracyanoethylene complexes.

OTHER EFFECTS OF BACK BONDING

There is controversy regarding the question as to whether the trifluoromethyl group, CF_3, can act as a π-acceptor ligand when bonded to a transition metal. Vibrational frequencies for CF_3I and $CF_3Mn(CO)_5$ have yielded the C—F stretching force constants 5.9 and 4.6 mdyn/Å, respectively.[26] The marked weakening of the C—F bond in $CF_3Mn(CO)_5$ has been attributed to overlap of filled metal $d\pi$ orbitals with C—F σ^* antibonding orbitals. A similar rationale has been used to explain the fact that the Fe—C distance in $(CHF_2CF_2)_2Fe(CO)_4$ is about 0.05 Å shorter than in analogous unfluorinated compounds.[27] However, theoretical calculations have failed to show any good evidence for such $d\pi \rightarrow \sigma^*$ back bonding.[28] Perhaps the strengthening of the metal-carbon bond and weakening of the C—F bond on going from a metal alkyl to a metal fluoroalkyl is merely a consequence of rehybridization of the carbon orbitals. Because of the high electronegativity of fluorine, the carbon orbitals involved in C—F bonds should have more p character than those involved in C—H bonds. Hence the carbon orbital

---

[26]   F. A. Cotton and R. M. Wing, *J. Organomet. Chem.*, **9**, 511 (1967).

[27]   M. R. Churchill, *Inorg. Chem.*, **6**, 185 (1967).

[28]   M. B. Hall and R. F. Fenske, *Inorg. Chem.*, **11**, 768 (1972).

involved in the metal-CF_3 bond should have more s character than that involved in the metal-CH_3 bond. Because of the generally observed increase in bond strength with increase in orbital s character,[29] the changes that occur on going from a metal alkyl to a metal fluoroalkyl are rationalized.

Back bonding from a transition metal to a carbon monoxide ligand involves a shift of electron density from the metal atom to the antibonding π^* orbital of the CO. Thus one expects increased back bonding to be accompanied by decreased C—O bond strength and increased negative charge on the carbon and oxygen atoms. We can follow changes in C—O bond strength with the C—O stretching force constant, calculated from vibrational spectroscopic data. We can follow changes in the carbon and oxygen atomic charges with the corresponding core-electron binding energies, obtained from x-ray photoelectron spectroscopy. Figure 18.17 shows a plot of carbon $1s$ binding energy versus C—O force constant for carbonyl compounds, covering a wide variety of transition metals and

---

[29] See pp. 123–124.

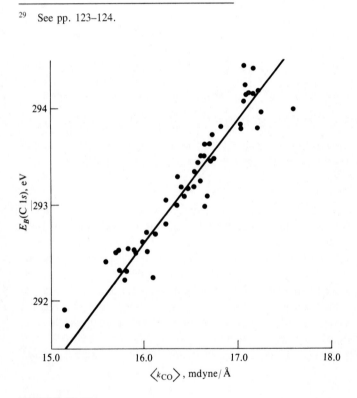

FIGURE 18.17

Plot of carbon $1s$ binding energy for gas-phase metal carbonyl compounds versus C—O stretching force constant. [*Reproduced with permission from S. C. Avanzino, A. A. Bakke, H. -W. Chen, C. J. Donahue, W. L. Jolly, T. H. Lee, and A. J. Ricco, Inorg. Chem., **19**, 1931 (1980). Copyright 1980 American Chemical Society.*]

other ligands.[30] An increase in force constant corresponds to an increase in bond strength, and an increase in core binding energy corresponds to a decrease in negative charge. The observed linear correlation between these physical quantities is strong support for the idea that these quantities are good measures of back bonding. Of the compounds represented in Fig. 18.17, the one with the lowest C—O force constant is η^7-$C_7H_7V(CO)_3$. Vanadium is on the left side of the transition series, with a relatively low nuclear charge, and therefore does not hold its valence electrons as tightly as metals to its right in the periodic table. This fact, coupled with the facts that the C_7H_7 group is not a strong π acceptor and that only three carbonyl groups are competing for $d\pi$ electron density, rationalizes the strong metal-CO back bonding in this molecule. The compound with the highest C—O force constant is $Mn(NO)_3CO$. The relatively weak metal-CO back bonding in this molecule is undoubtedly due to strong electron withdrawal by the three NO groups, which are exceedingly strong π acceptors.

Chemical reactivity can be affected by back bonding.[31] The pyrazinium ion,

has a pK of 0.6. When the free nitrogen atom of this ion is coordinated to the cationic species $Ru(NH_3)_5{}^{3+}$, the acidity of the coordinated pyrazinium ion increases, presumably because of the inductive effect of the metal cation.

$$\left[(NH_3)_5Ru-N\bigcirc NH\right]^{4+} \qquad pK \approx -0.8$$

For the same reason one might expect $Ru(NH_3)_5{}^{2+}$ to increase the acidity of the pyrazinium ion, only not as much as $Ru(NH_3)_5{}^{3+}$. However, the acidity is actually decreased.

$$\left[(NH_3)_5Ru-N\bigcirc NH\right]^{3+} \qquad pK = 2.5$$

From this we conclude that back bonding from the Ru(II) to the pyrazine ring more than compensates for the inductive effect.

[30]   S. C. Avanzino, A. A. Bakke, H. -W. Chen, C. J. Donahue, W. L. Jolly, T. H. Lee, and A. J. Ricco, *Inorg. Chem.,* **19,** 1931 (1980).

[31]   P. Ford, D. F. P. Rudd, R. Gaunder, and H. Taube, *J. Am. Chem. Soc.,* **90,** 1197 (1968); H. Taube, *Surv. Prog. Chem.,* **6,** 1 (1973).

PROBLEMS

*18.1 The following compounds have effective magnetic moments of 5.85 to 5.95:

$$Fe(NO_3)_3 \cdot 9H_2O \qquad\qquad FeCl_3 \cdot 6H_2O$$
$$K_3Fe(C_2O_4)_3 \cdot 3H_2O \qquad\qquad Fe(urea)_6Cl_3 \cdot 6H_2O$$
$$Na_3FeF_6 \qquad\qquad [Fe(dipy)_2Cl_2][FeCl_4]$$
$$[Fe(phen)_2Cl_2][FeCl_4] \qquad\qquad Fe(acac)_3$$

The following have μ_{eff} values of 2.3 to 2.4:

$$[Fe(phen)_2(CN)_2]NO_3 \cdot 4H_2O \qquad [Fe(dipy)_2(CN)_2]NO_3$$
$$H_3O[Fe(phen)(CN)_4] \cdot H_2O \qquad K_3Fe(CN)_6$$
$$Fe(dipy)_3(ClO_4)_3 \cdot 3H_2O \qquad Fe(phen)_3(ClO_4)_3 \cdot 3H_2O$$

Using this and other information, estimate the ligand-field strength corresponding to crossover in Fe^{3+} O_h complexes.

*18.2 Which would you expect to have the greater magnetic moment, $CoCl_4{}^{2-}$ or $CoI_4{}^{2-}$? Why?

18.3 The complex $Cr_2(OAc)_4 \cdot 2H_2O$ (see structure, page 376) is diamagnetic. What can you say regarding the nature of the bonding in this compound?

*18.4 The following compounds have the indicated effective magnetic moments. Describe the structure and bonding of these compounds on the basis of the μ_{eff} values. K_2NiF_6, 0.0; $Ni(NH_3)_2Cl_2$, 3.3; $Ni(PEt_3)_2Cl_2$, 0.0; $Ni(Ph_3AsO)_2Cl_2$, 3.95.

18.5 $Ni(C_5H_5)_2$ has the same type of structure as that of ferrocene. Would you expect $Ni(C_5H_5)_2$ to be paramagnetic or diamagnetic? Why?

18.6 Although a d^8 tetrahedral complex is somewhat stabilized by flattening of the tetrahedron (show this by consideration of the energy level diagram), more stabilization is obtained by elongation of the tetrahedron. Similar, but opposite, statements apply to d^9 complexes. Explain.

*18.7 Equilibrium constants for the stepwise formation of mono-, bis-, and tris(ethylenediamine) complexes of Co^{2+}, Ni^{2+}, and Cu^{2+} in aqueous solution are as follows:

Ion	log K_1	log K_2	log K_3
Co^{2+}	5.89	4.83	3.10
Ni^{2+}	7.52	6.28	4.26
Cu^{2+}	10.55	9.05	-1.0

$$M(H_2O)_6{}^{2+} + en \rightleftharpoons M(H_2O)_4(en)^{2+} + 2H_2O \qquad K_1$$

$$M(H_2O)_4(en)^{2+} + en \rightleftharpoons M(H_2O)_2(en)_2{}^{2+} + 2H_2O \qquad K_2$$

$$M(H_2O)_2(en)_2{}^{2+} + en \rightleftharpoons M(en)_3{}^{2+} + 2H_2O \qquad K_3$$

Explain the anomalously low value of K_3 for Cu^{2+}.

18.8 Prepare a table in which you indicate, for each of the following complexes, the number of unpaired electrons, the estimated magnetic moment, whether or not one would expect an appreciable Jahn-Teller distortion, the number of d-d transitions, and whether or not one would expect ligand-field stabilization: $Co(CO)_4^-$, $Cr(CN)_6^{4-}$, $Fe(H_2O)_6^{3+}$, VCl_4, $Co(NO_2)_6^{4-}$, $Co(NH_3)_6^{3+}$, $Ir(CO)Cl(PPh_3)_2$, $CuCl_2^-$, MnO_4^{2-}, $Cu(H_2O)_6^{2+}$.

*18.9 Suggest a reason for the fact that a number of tetrahedral Co(II) complexes are stable, whereas the corresponding Ni(II) complexes are not.

18.10 Using ligand-field stabilization energy as the criterion, indicate whether you would expect the following spinels to be normal or inverse: $CuFe_2O_4$, $CuCr_2O_4$, $NiFe_2O_4$, $MnCr_2O_4$.

18.11 Calculate the difference in ligand-field stabilization energy (in units of Δ_o) between octahedral and tetrahedral coordination for high-spin configurations from d^1 to d^9. Assume that $\Delta_t = \frac{4}{9}\Delta_o$.

*18.12 Can you explain why it might be possible to observe the effects of Jahn-Teller distortion in the electronic spectrum of a crystal even though x-ray diffraction data conform to a symmetric structure?

18.13 Explain the fact that the high-spin d^6 complex CoF_6^{3-} shows two absorption bands in the visible spectral region.

18.14 List three different experimental methods which could be used for measuring back bonding in a transition-metal carbonyl.

18.15 Without referring to any data, describe the structures of the following complexes: $Fe(CO)_2(NO)_2$, $Fe(CO)_4H_2$, $Mn(CO)_4NO$, $Fe(CO)_4CN^-$, $Fe(CO)_4C_2H_4$. For the latter complex, indicate precisely the orientation of the C_2H_4 group.

18.16 Can you explain why the Néel temperature of MnS is higher than that of MnO? (Consider the factors influencing superexchange.) Similarly, can you explain the following trend in Néel temperature? MnO, 116 K; FeO, 198 K; CoO, 291 K; NiO, 525 K.

*18.17 Why would you not expect TiO or VO to be antiferromagnetic?

18.18 Using data from Tables 18.3 and 18.4 (and estimated Δ values when necessary), calculate values of μ for the first 11 compounds of Table 18.2. Show that the average deviation from the experimental values is much less than in the case of the $[n(n + 2)]^{1/2}$ values.

18.19 A compound, once formulated as $K_2OsO_4 \cdot 2H_2O$, has been shown to be diamagnetic. Suggest a better formulation, more indicative of the structure, and explain the diamagnetism.

CHAPTER
19

KINETICS AND MECHANISMS OF REACTIONS OF TRANSITION-METAL COMPLEXES

In the remaining chapters of this book we will be concerned with the reactions of transition-metal compounds. In this chapter we discuss the kinetics and mechanisms of ligand substitution and redox reactions for Werner-type complexes (i.e., complexes like the classical octahedral and square planar complexes studied in the early part of this century by Werner). Reactions of organometallic and bioinorganic systems will be covered in the following chapters.

SUBSTITUTION REACTIONS OF OCTAHEDRAL COMPLEXES[1]

Lability

The rates of replacement of the ligands of hexacoordinate complexes by other ligands vary over an extremely wide range—from very high rates ($k \approx 10^9 M^{-1} s^{-1}$ at room temperature) to very low rates ($k \approx 10^{-9} M^{-1} s^{-1}$ at room temperature).[2]

---

[1] F. Basolo and R. G. Pearson, "Mechanisms of Inorganic Reactions," 2d ed., Wiley, New York, 1967; R. G. Wilkins, "The Study of Kinetics and Mechanism of Reactions of Transition Metal Complexes," Allyn and Bacon, Boston, 1974.

[2] A rate constant of $10^9 M^{-1} s^{-1}$ corresponds to a half-life of 10^{-9} with 1 M reagents, and $k = 10^{-9} M^{-1} s^{-1}$ corresponds to a half-life of about 30 years under the same conditions.

482

The complex $Co(NH_3)_6{}^{3+}$ is an example of a relatively inert, or slow-reacting, complex. Although the thermodynamic driving force of the following reaction is enormous,

$$Co(NH_3)_6{}^{3+} + H^+ + Cl^- \longrightarrow Co(NH_3)_5Cl^{2+} + NH_4{}^+$$

the complex must be heated in 6 M hydrochloric acid for many hours to obtain any perceptible amount of the chloro complex. On the other hand, $Cu(NH_3)_4(H_2O)_2{}^{2+}$ is a labile, or fast-reacting, complex. When a solution containing this blue species is added to concentrated hydrochloric acid, the solution turns green essentially as rapidly as the solutions are mixed.

$$Cu(NH_3)_4(H_2O)_2{}^{2+} + 4H^+ + 4Cl^- \longrightarrow CuCl_4{}^{2-} + 4NH_4{}^+ + 2H_2O$$

A chemist who wishes to synthesize an octahedral complex must have some idea of the lability of the complex in order to choose appropriate experimental conditions for the synthesis. When a particular geometric or optical isomer is to be prepared, it is necessary that the rate of isomerization or racemization be slow enough to permit the isolation of a pure isomer. Obviously any generalizations or rules that can be used to predict the relative reactivities of complexes are important.

The size of the central atom affects the ease with which ligands can be replaced. The smaller the central atom (other factors being equal), the more tightly the ligands are held and the more inert the complex is. Thus, the rate constants for the exchange of solvent and coordinated water molecules in the aqueous ions $Mg(H_2O)_6{}^{2+}$, $Ca(H_2O)_6{}^{2+}$, and $Sr(H_2O)_6{}^{2+}$ are $\sim 10^5$, $\sim 2 \times 10^8$, and $\sim 4 \times 10^8$ s^{-1}, respectively, at 25°C. The charge of the central atom is also important; the higher the charge, the more inert the complex. This effect can be seen in the water exchange rate constants for the following aqueous ions:

$$\begin{cases} Na(H_2O)_6{}^+ (k \approx 8 \times 10^9 \, s^{-1}; \, r_{Na^+} = 0.95 \, \text{Å}) \\ Ca(H_2O)_6{}^{2+} (k \approx 2 \times 10^8 \, s^{-1}; \, r_{Ca^{2+}} = 0.99 \, \text{Å}) \end{cases}$$

and

$$\begin{cases} Mg(H_2O)_6{}^{2+} (k \approx 10^5 \, s^{-1}; \, r_{Mg^{2+}} = 0.65 \, \text{Å}) \\ Ga(H_2O)_6{}^{3+} (k \approx 10^3 \, s^{-1}; \, r_{Ga^{3+}} = 0.62 \, \text{Å}) \end{cases}$$

The net effect of changes in both charge and size can be seen in the marked decrease in reactivity in the series of complexes $AlF_6{}^{3-}$, $SiF_6{}^{2-}$, $PF_6{}^-$, and SF_6. The addition of base to a solution of $AlF_6{}^{3-}$ causes the instant precipitation of $Al(OH)_3$, whereas SF_6 undergoes no detectable reaction with hot concentrated base solutions over long periods of time.

The degree of lability or inertness of a transition-metal complex can be correlated with the d electron configuration of the metal ion. Let us consider a simple qualitative approach first used by Taube.[3] If a complex contains electrons in the antibonding $e_g{}^*$ orbitals, the ligands are expected to be relatively weakly

---

[3] H. Taube, *Chem. Rev.*, **50**, 69 (1952).

TABLE 19.1
Kinetic classification of octahedral complexes

Electron configuration	Complex
	Labile
d^0	$CaEDTA^{2-}$, $Sc(H_2O)_5OH^{2+}$, $TiCl_6^{2-}$
d^1	$Ti(H_2O)_6^{3+}$, $VO(H_2O)_5^{2+}$, $MoOCl_5^{2-}$
d^2	$V(phen)_3^{3+}$, $ReOCl_5^{2-}$
d^4 (high-spin)	$Cr(H_2O)_6^{2+}$
d^5 (high-spin)	$Mn(H_2O)_6^{2+}$, $Fe(H_2O)_4Cl_2^{+}$
d^6 (high-spin)	$Fe(H_2O)_6^{2+}$
d^7	$Co(NH_3)_6^{2+}$
d^8	$Ni(en)_3^{2+}$
d^9	$Cu(NH_3)_4(H_2O)_2^{2+}$
d^{10}	$Ga(C_2O_4)_3^{3-}$
	Inert
d^3	$V(H_2O)_6^{2+}$, $Cr(en)_2Cl_2^{+}$
d^4 (low-spin)	$Cr(CN)_6^{4-}$, $Mn(CN)_6^{3-}$
d^5 (low-spin)	$Mn(CN)_6^{4-}$, $Fe(CN)_6^{3-}$
d^6 (low-spin)	$Fe(CN)_6^{4-}$, $Co(en)_2(H_2O)_2^{3+}$

bound and to be easily displaced. If a complex contains an empty t_{2g} orbital, the four lobes of that orbital correspond to directions from which an incoming ligand (which is to displace one of the bound ligands) can approach the complex with relatively little electrostatic repulsion. Therefore one concludes that a complex with one or more $e_g{}^*$ electrons or with fewer than three d electrons should be relatively labile, and that a complex with any other electron configuration should be relatively inert. In Table 19.1, examples of octahedral complexes with all possible electron configurations are listed and categorized as labile or inert, depending on whether ligand substitution takes place in less than or more than 1 min at room temperature with 0.1 M reactants. All the data are consistent with Taube's simple rule.

Electrostatic ligand-field theory can be used to predict the relative labilities of octahedral transition-metal complexes if the geometric configurations of the activated complexes of the substitution reactions are known or assumed. The change in ligand-field stabilization energy on going from the reactant complex to the activated complex corresponds to the ligand-field contribution to the activation energy of the reaction. If the reaction is assumed to proceed by a mechanism in which the rate-determining step involves considerable dissociation of the original complex, a 5-coordinate square pyramid activated complex is a reasonable assumption. The ligand-field contributions to the activation energy, based on that assumption, are given in Table 19.2 for complexes with various electron configurations. If the reaction is assumed to proceed by a mechanism in which the rate-determining

TABLE 19.2
Ligand-field stabilization energies and contributions to activation energies (in units of Δ_o) for dissociative mechanisms (octahedron $\rightarrow$ square pyramid)†

Electron configuration	Octahedron	Square pyramid	Contribution to E_a
d^0	0	0	0
d^1	0.400	0.457	−0.057
d^2	0.800	0.914	−0.114
d^3	1.200	1.000	0.200
d^4 (high-spin)	0.600	0.914	−0.314
d^4 (low-spin)	1.600	1.457	0.143
d^5 (high-spin)	0	0	0
d^5 (low-spin)	2.000	1.914	0.086
d^6 (high-spin)	0.400	0.457	−0.057
d^6 (low-spin)	2.400	2.000	0.400
d^7 (high-spin)	0.800	0.914	−0.114
d^7 (low-spin)	1.800	1.914	−0.114
d^8	1.200	1.000	0.200
d^9	0.600	0.914	−0.314
d^{10}	0	0	0

† F. Basolo and R. G. Pearson, "Mechanisms of Inorganic Reactions," 2d ed., Wiley, New York, 1967.

step involves association of the incoming ligand with the original complex, a 7-coordinate pentagonal bipyramid activated complex is a reasonable assumption. The corresponding ligand-field contributions to the activation energy are given in Table 19.3.

It is remarkable that, in spite of the different assumptions involved, the data in Tables 19.1 to 19.3 are to a large extent in qualitative agreement. Both sets of ligand-field data indicate that complexes with d^3, d^8, and low-spin d^6 configurations should be inert. The main disagreement between the ligand-field results and the qualitative predictions of Taube is in the case of d^8 complexes, which are predicted to be labile by the Taube rule. However, examination of quantitative rate data shows that, in a sense, both predictions are correct. Consider the rate constant data for +2 ions given in Table 19.4. Although the rate constants for $Ni(H_2O)_6{}^{2+}$ are not as low as the corresponding constants for the $d^3 V(H_2O)_6{}^{2+}$ ion, they are definitely lower than those for all the other +2 ions listed.

Substitution reactions of $Cr(CO)_6$ are very slow, as expected for a low-spin d^6 complex. However, the reactivity of the isoelectronic molecule $V(CO)_5NO$ is greater by many orders of magnitude, for both dissociative (first-order) and associative (second-order) processes. This increased reactivity can be explained for both types of processes in terms of the exceptionally high π-acceptor character of the nitrosyl ligand.[4] The back bonding to the CO group trans to the NO group

---

4   Q. Shi, T. G. Richmond, W. C. Trogler, and F. Basolo, *Inorg. Chem.*, **23**, 957 (1984).

TABLE 19.3
Ligand-field stabilization energies and contributions to activation energies (in units of Δ_o) for associative mechanisms (octahedron $\rightarrow$ pentagonal bipyramid)†

Electron configuration	Octahedron	Pentagonal bipyramid	Contribution to E_a
d^0	0	0	0
d^1	0.400	0.528	−0.128
d^2	0.800	1.056	−0.256
d^3	1.200	0.774	0.426
d^4 (high-spin)	0.600	0.493	0.107
d^4 (low-spin)	1.600	1.302	0.298
d^5 (high-spin)	0	0	0
d^5 (low-spin)	2.000	1.830	0.170
d^6 (high-spin)	0.400	0.528	−0.128
d^6 (low-spin)	2.400	1.548	0.852
d^7 (high-spin)	0.800	1.056	−0.256
d^7 (low-spin)	1.800	1.266	0.534
d^8	1.200	0.774	0.426
d^9	0.600	0.493	0.107
d^{10}	0	0	0

† F. Basolo and R. G. Pearson, "Mechanisms of Inorganic Reactions," 2d ed., Wiley, New York, 1967.

in V(CO)$_5$NO is weakened by the strong withdrawal of t_{2g} electron density into the π^* orbitals of the NO group.

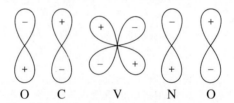

$$O \quad C \quad V \quad N \quad O$$

Thus this CO group undergoes *dissociation* much more readily than the other four CO groups in the molecule. The *associative* substitution process for Cr(CO)$_6$ is slow because it involves an unstable 7-coordinate activated complex with 20 valence electrons on the metal atom. In the analogous process for V(CO)$_5$NO, the 20-electron activated complex can be avoided by the transfer of an electron pair to the NO ligand:

$$^-V = \overset{+}{N} = O \longrightarrow {}^+V = N\overset{\displaystyle O^-}{\diagup}$$

Mechanisms

The data in Table 19.4 and a large number of similar kinetic data have the following general features:

TABLE 19.4
Log k for substitution reactions of aqueous
+2 ions at 25°C†

Metal ion	Entering ligand			
	$H_2O\ddagger$	NH_3*	HF*	phen*
V^{2+}	2.0	....	....	0.5
Cr^{2+}	8.5	....	....	~8.0
Mn^{2+}	7.5	....	6.3	~5.4
Fe^{2+}	6.5	....	6.0	5.9
Co^{2+}	6.0	5.1	5.7	5.3
Ni^{2+}	4.3	3.7	3.5	3.4
Cu^{2+}	8.5	....	7.5	7.9
Zn^{2+}	7.5	....	....	6.8

† R. G. Wilkins, *Acc. Chem. Res.*, **3**, 408 (1970).
‡ Bulk water-coordinated water exchange rate constants (s^{-1}).
* Second-order rate constants ($M^{-1}\,s^{-1}$).

1. For a given aqueous ion, the rate of replacement of a coordinated water molecule is not strongly affected by the nature of the entering ligand.
2. The water exchange reaction is always faster than substitution reactions involving other ligands.

These results suggest that the reactions proceed in two steps, the first of which is the formation of an aquo ion-ligand "outer-sphere" complex, and the second of which is the exchange of an "inner-sphere" coordinated water molecule with the outer-sphere coordinated ligand.

$$M(H_2O)_6{}^{n+} + L \; \overset{K_1}{\rightleftharpoons} \; M(H_2O)_6 \cdot L^{n+} \tag{19.1}$$

$$M(H_2O)_6 \cdot L^{n+} \; \overset{k_2}{\longrightarrow} \; M(H_2O)_5L^{n+} + H_2O \tag{19.2}$$

When the overall rate constants (K_1k_2) are divided by estimated outer-sphere complex formation constants (K_1), the resulting values of k_2 are very close to the corresponding water exchange rate constants. This result shows that the formation of the activated complex in the second, rate-determining step consists mainly of the breaking of the $M—OH_2$ bond and that the entering ligand influences k_2 only very slightly. However, the volumes of activation, ΔV_a, for solvent exchange reactions in methanol[5] have the following values in cubic centimeters per mole: -5.0 for Mn^{2+}, $+0.4$ for Fe^{2+}, $+8.9$ for Co^{2+}, and $+11.4$ for Ni^{2+}. Presumably a similar trend would be found in water. Because a negative volume of activation corresponds to an associative-type process (in which the rate is affected by the entering ligand) and a positive volume of activation corresponds to a dissociative-

---

5 F. K. Meyer, K. E. Newman, and A. E. Merbach, *J. Am. Chem. Soc.*, **101**, 5588 (1979).

type process (in which the rate is unaffected by the entering ligand),[6] the data suggest that reaction step 19.2 undergoes a changeover from associative to dissociative on going from left to right in the transition series. Further work is needed to clarify the mechanisms of these substitution reactions.[7]

In some substitution reactions, the entering ligand is not at all involved in the dissociation step, and a 5-coordinate intermediate is formed. The reactions of the $Co(CN)_5H_2O^{2-}$ ion with various ligands are examples of such reactions:[8]

$$Co(CN)_5H_2O^{2-} \underset{k_2}{\overset{k_1}{\rightleftharpoons}} Co(CN)_5^{2-} + H_2O$$

$$Co(CN)_5^{2-} + X^- \underset{K_4}{\overset{k_3}{\rightleftharpoons}} Co(CN)_5X^{3-}$$

Kinetic data for $X^- = Br^-, I^-, SCN^-$, and N_3^- give the same value, $1.6 \times 10^{-3} \, s^{-1}$, for k_1. In fact, essentially the same value has been obtained for the rate constant of the water exchange reaction.

The base hydrolyses of pentaamminecobalt(III) complexes have been extensively studied.

$$Co(NH_3)_5X^{2+} + OH^- \rightarrow Co(NH_3)_5OH^{2+} + X^-$$

The reactions are first order in the complex and first order in hydroxide ion, and one might reasonably suspect a mechanism involving an ion pair, analogous to reactions 19.1 and 19.2, or a simple bimolecular displacement in which the hydroxide ion enters the coordination sphere as the X^- ion leaves. However, an extensive set of experimental data indicates that these base hydrolyses proceed by the following mechanism:[1,9]

$$Co(NH_3)_5X^{2+} + OH^- \overset{fast}{\rightleftharpoons} Co(NH_3)_4(NH_2)X^+ + H_2O$$

$$Co(NH_3)_4(NH_2)X^+ \overset{slow}{\longrightarrow} Co(NH_3)_4(NH_2)^{2+} + X^-$$

$$Co(NH_3)_4(NH_2)^{2+} + H_2O \overset{fast}{\longrightarrow} Co(NH_3)_5OH^{2+}$$

In this mechanism the rate-determining and dissociative step takes place after the removal of a proton from one of the coordinated ammonia molecules. A deprotonation of this type is not unreasonable in view of the finite acidities of metal ammine complexes.[10] [For example, $pK = 8.4$ for $Pt(NH_3)_5Cl^{3+}$, and $pK \approx 6.5$ for $Au(en)_2^{3+}$.] Strong evidence for the mechanism is obtained from studies[11] of the hydrolysis in the presence of various anions (Y^-). These studies show that $Co(NH_3)_5Y^{2+}$ ions are formed in addition to $Co(NH_3)_5OH^{2+}$ ions.

[6]   G. A. Lawrance and D. R. Stranks, *Acc. Chem. Res.*, **12**, 403 (1979).

[7]   K. E. Newman and K. M. Adamson-Sharpe, *Inorg. Chem.*, **23**, 3818 (1984).

[8]   A. Haim and W. K. Wilmarth, *Inorg. Chem.*, **1**, 573 (1962); A. Haim, R. J. Grassie, and W. K. Wilmarth. *Adv. Chem. Ser.*, **49**, 31 (1965).

[9]   M. L. Tobe, *Acc. Chem. Res.*, **3**, 377 (1970).

[10]   Basolo and Pearson, op. cit., p. 33.

[11]   D. A. Buckingham, I. I. Olsen, and A. M. Sargeson, *J. Am. Chem. Soc.*, **88**, 5443 (1966).

The product ratio $Co(NH_3)_5Y^{2+}/Co(NH_3)_5OH^{2+}$, shows little dependence on the leaving group X^-, is independent of the OH^- concentration, but depends on the Y^- concentration and varies from one Y^- to another. The results clearly indicate a common reactive intermediate in all the reactions.

The following arguments have been offered to rationalize the mechanism. The X^- ion can dissociate from the deprotonated complex more easily than from the original complex because of the lower positive charge of the deprotonated complex. Loss of X^- is probably somewhat aided by the deprotonation because the NH_2^- ligand can donate some of its π lone-pair electron density to the electron-deficient cobalt atom of the 5-coordinate intermediate.

SUBSTITUTION REACTIONS OF SQUARE PLANAR COMPLEXES[1]

It is believed that substitution reactions of square planar complexes almost always proceed by an associative mechanism involving an intermediate which contains both the leaving ligand and the entering ligand. It is assumed that the entering ligand approaches the complex from one side of the plane, over the ligand to be displaced. The leaving ligand presumably moves down as the entering ligand approaches, so that the intermediate has a trigonal bipyramidal configuration.

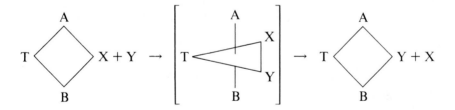

A 5-coordinate intermediate, or at least a 5-coordinate activated complex, is quite reasonable in view of the fact that there are many stable 5-coordinate d^8 complexes.

For reactions of Pt(II) complexes in aqueous solution, the rate law often has the form

$$\text{Rate} = k_1[\text{complex}] + k_2[\text{complex}][Y^-] \qquad (19.3)$$

where Y^- is the entering ligand. The second term corresponds to the bimolecular displacement of the leaving ligand by the entering Y^- ligand, and the first term corresponds to a two-step path in which first the leaving ligand is displaced by a water molecule in a rate-determining step and then the coordinated water molecule is relatively rapidly displaced by Y^-. Kinetic data on a wide variety of Pt(II) complexes show that the rate of a ligand substitution reaction depends on the natures of the entering ligand, the leaving ligand, and the ligand trans to the leaving ligand. The relation between the reaction rate and the ligand trans to the leaving ligand is called the "trans effect," and we shall discuss this effect before considering the effects of the entering and leaving ligands.

The Kinetic Trans Effect

Ligands can be put in order according to their abilities to labilize (make susceptible to substitution) trans leaving groups. In the following list, the ligands are given in the order of increasing trans effect (i.e., increasing trans labilizing ability): $H_2O < OH^- < F^- \approx RNH_2 \approx py \approx NH_3 < Cl^- < Br^- < SCN^- \approx I^- \approx NO_2^- \approx C_6H_5^- < SC(NH_2)_2 \approx CH_3^- < NO \approx H^- \approx PR_3 < C_2H_4 \approx CN^- \approx CO$.

The effect is illustrated by the reaction of the $PtCl_4^{2-}$ ion with two molecules of ammonia. After one chloride ion has been displaced, there are two kinds of chloride ions remaining in the complex: the chloride ion which is trans to an ammonia molecule and the chloride ions which are trans to each other. Because chloride is more trans-directing than ammonia, the chloride ions which are trans to each other are more labile than the chloride ion trans to the ammonia molecule. Hence the second molecule of ammonia displaces one of the more labile chloride ions and yields cis-$Pt(NH_3)_2Cl_2$.

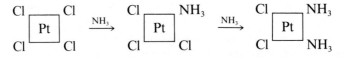

$Trans$-$Pt(NH_3)_2Cl_2$ can be prepared by the reaction of the $Pt(NH_3)_4^{2+}$ ion with two chloride ions. The first step yields $Pt(NH_3)_3Cl^+$, which contains two ammonia molecules trans to each other and an ammonia molecule trans to the chloride ion. The latter ammonia molecule is labilized by the chloride ion and is displaced next:

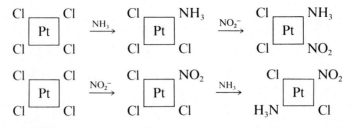

The trans effect can be used as the basis for the synthesis of many other isomeric Pt(II) complexes. For example, cis- and $trans$-$PtNH_3NO_2Cl_2^-$ are prepared from $PtCl_4^{2-}$ by the following routes.

Notice that a different isomer is obtained by simply reversing the order of introduction of the groups.

We shall now show that it is possible to rationalize the trans effect on the basis of a trigonal bipyramidal intermediate. In this intermediate, the leaving ligand, the entering ligand, and the trans ligand occupy the three equatorial positions of the trigonal bipyramid. We have already pointed out in Chap. 18 that, in a trigonal bipyramidal d^8 complex, there is a greater interaction of the filled $d\pi$ orbitals with the equatorial ligands than with the axial ligands. Therefore $d\pi \rightarrow \pi^*$ back bonding and $d\pi$-$p\pi$ nonbonding repulsions affect the bonding of the equatorial ligands more than that of the axial ligands. Sigma-bonding ligands presumably can interact with the metal s orbital about equally well at equatorial and axial sites. However, interaction with metal p orbitals is favored at equatorial sites because, on the average, two-thirds of a p orbital is available at each equatorial site, whereas only one-half of a p orbital is available at each axial site. Thus covalent σ-bonding is favored at equatorial sites, and strongly basic ligands with relatively electropositive donor atoms tend to favor the equatorial sites over the axial sites (other factors being equal).

In the following reactions,

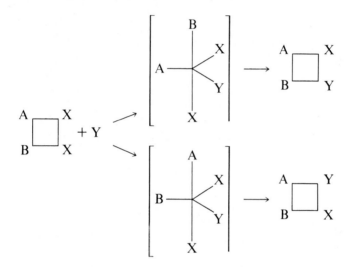

the geometry of the principal product is determined by the relative trans-directing abilities of A and B. The faster reaction corresponds to the more stable trigonal bipyramidal intermediate. Obviously the more strongly trans-directing ligand is the one which gives the more stable intermediate when it is in an equatorial, rather than an axial, position. Hence strong trans-directing ligands are those which engage in strong back bonding (CN^-, CO, olefins, etc.) or are very strong σ donors (CH_3^-, H^-, PR_3, etc.). Ligands which are strong π donors (H_2O, OH^-, NH_2^-, etc.) are weakly trans-directing.

Effect of Entering and Leaving Ligands[12]

One of the fascinating problems associated with nucleophilic displacement reactions is that of predicting reactivity in terms of the properties of the entering and

---

[12] Basolo and Pearson, ref. 1.

leaving ligands. Consider a series of reactions of the type

$$trans\text{-}PtL_2Cl_2 + Y^- \quad \rightarrow \quad trans\text{-}PtL_2ClY + Cl^-$$

in which the ligand L is kept constant but the ligand Y is varied. On the basis of the general rate law for Pt(II) substitution reactions (Eq. 19.3), we may take the bimolecular rate constant k_2 as a measure of the nucleophilicity of the entering ligand Y. It has been found that the nucleophilicity order of the Y ligands is remarkably constant on going from one Pt(II) complex to another, i.e., on changing L. In most cases, the reaction rate increases in the following order of entering ligands: $ROH < OR^- < Cl^- \approx py \approx NO_2^- < N_3^- < Br^- < I^- < SCN^- < SO_3^{2-} < CN^- < C_6H_5S^- < PR_3$. In fact the k_2 values of a wide variety of Pt(II) substitution reactions may be correlated using the following linear free-energy relation.

$$\log k_2 = s\eta_{Pt}^\circ + \text{constant}$$

The parameter s depends on the complex and is called the "nucleophilic discrimination factor"; the parameter η_{Pt}° depends on the entering ligand and is called the "nucleophilic reactivity constant." A large value of s means that $\log k_2$ is very sensitive to changes in the nucleophilic character of Y, and a small value of s means that $\log k_2$ is relatively insensitive to changes in Y. The effects of changes in Y on the activation energy of the reaction can be understood by reference to Fig. 19.1. If the entering ligand forms a stronger bond to the platinum atom than the leaving ligand (as shown in Fig. 19.1a), then changes in the entering ligand

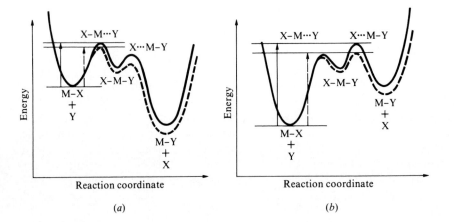

(a)   (b)

FIGURE 19.1
Plots of energy versus reaction coordinate for displacement reactions of Pt(II) complexes, showing the effect on the activation energy of changing the bonding energy of the entering ligand, Y. In example (a), the entering ligand is more strongly bound than the leaving ligand and the energy of the activated complex is only slightly affected by the nature of Y. In example (b), the leaving ligand is more strongly bound than the entering ligand and the energy of the activated complex is strongly affected by the nature of Y.

cause relatively small changes in activation energy (i.e., s is small). On the other hand, if the entering ligand forms a weaker bond to the platinum atom than the leaving atom (as shown in Fig. 19.1b), then changes in the entering ligand cause relatively large changes in activation energy (i.e., s is large). Incidentally, the fact that the reaction of Fig. 19.1b is endothermic (uphill energetically) should be no cause for alarm. Reactions for which $\Delta H^\circ > 0$ and $\Delta G^\circ > 0$ can proceed spontaneously if the reaction quotient (the product of the concentrations of the product species divided by the product of the concentrations of the reactant species, each concentration being raised to a power equal to the coefficient of that species in the net reaction) is small enough to make ΔG less than zero. In other words, mass action can overcome an unfavorable equilibrium constant.

The effect of the leaving ligand on the rate of a displacement from Pt(II) can be seen by reference to Fig. 19.2. If the entering ligand forms a stronger bond than the leaving ligand (as shown in Fig. 19.2a), changes in the leaving ligand cause relatively small changes in activation energy; the rate is essentially independent of the nature of the leaving ligand. On the other hand, if the leaving ligand forms a stronger bond than the entering ligand (as shown in Fig. 19.2b), changes in the leaving ligand cause relatively large changes in activation energy; the rate is a sensitive function of the nature of the leaving ligand. Both situations are found in reactions of thiourea with complexes of the type Pt(dien)X^+:

$$\text{Pt(dien)}X^+ + \text{tu} \quad \rightarrow \quad \text{Pt(dien)tu}^{2+} + X^-$$

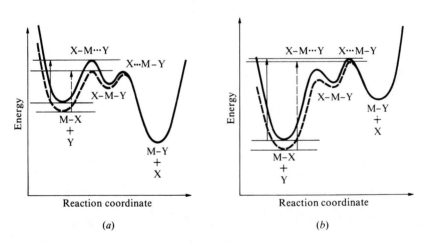

(a)

(b)

FIGURE 19.2
Plots of energy versus reaction coordinate for displacement reactions of Pt(II) complexes, showing the effect on the activation energy of changing the bonding energy of the leaving ligand, X. In example (a), the entering ligand is more strongly bound than the leaving ligand and the energy of the activated complex is strongly affected by the nature of X. However, note that ΔE_a changes very little. In example (b), the leaving ligand is more strongly bound than the entering ligand and the energy of the activated complex is only slightly affected by the nature of X. However, ΔE_a changes markedly.

The rate is essentially the same for the relatively weakly bound leaving ligands Cl^-, Br^-, and I^- but shows a definite trend with the more strongly bound leaving ligands: $Cl^- \approx Br^- \approx I^- \gg N_3^- \approx NO_2^- > SCN^- > CN^-$.

Pyridine is a relatively weakly bound ligand, and in reactions of the type

$$Pt(dien)X^+ + py \quad \rightarrow \quad Pt(dien)py^{2+} + X^-$$

the reaction rate shows a marked trend even among weakly bound leaving groups: $NO_3^- > H_2O > Cl^- > Br^- > I^- > N_3^- > SCN^- > NO_2^- > CN^-$.

The entering- and leaving-ligand series have a close similarity to the trans-effect series, but distinct differences can be seen. Some of these differences are probably due to the fact that the entering- and leaving-group series are affected by differences in the solvation of ligands, whereas the trans-effect series is independent of such solvation effects.

OXIDATION-REDUCTION REACTIONS[12-14]

Outer-Sphere Electron Transfer[15]

Many complexes undergo oxidation-reduction reactions more rapidly than they undergo substitution reactions. When two such complexes engage in a redox reaction, it is clear that electron transfer must occur through the intact coordination shells of the metal ions. These reactions are called "outer-sphere" electron-transfer reactions. The following is an example of such a reaction.

$$Fe(CN)_6^{4-} + IrCl_6^{2-} \quad \rightarrow \quad Fe(CN)_6^{3-} + IrCl_6^{3-}$$

Even when one of the complexes is labile, the reaction will generally proceed by an outer-sphere mechanism if the inert complex does not possess a donor atom which can be used to form a bridge to the labile complex. For example, this is the situation in the reduction of $Co(NH_3)_6^{3+}$ by $Cr(H_2O)_6^{2+}$.

The simplest type of outer-sphere redox reactions are those in which no net reaction occurs, i.e., electron-exchange reactions such as the following.

$$Fe(H_2O)_6^{2+} + Fe(H_2O)_6^{3+} \quad \rightarrow \quad Fe(H_2O)_6^{3+} + Fe(H_2O)_6^{2+}$$

The rates of such reactions can be followed by the use of isotopic tracers. An energy-versus-reaction coordinate diagram for symmetric outer-sphere electron exchange is shown in Fig. 19.3. The curve represents the energy of the system along a particular configurational coordinate. In this case the coordinate could well be, for example, the average metal-ligand distance for the complex on the

[13] H. Taube, "Electron Transfer Reactions of Complex Ions in Solution," Academic, New York, 1970.

[14] A. G. Sykes, "Kinetics of Inorganic Reactions," Pergamon, Oxford, 1966.

[15] See T. J. Meyer, *Acc. Chem. Res.*, **11**, 94 (1978).

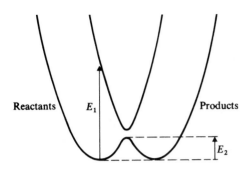

Reactants E_1 Products

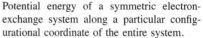

FIGURE 19.3
Potential energy of a symmetric electron-exchange system along a particular configurational coordinate of the entire system.

right minus the average metal ligand distance for the complex on the left, while maintaining a constant metal-metal distance. There are two possible ground states for the system, in either of which the reduced complex has a greater metal-ligand distance than the oxidized complex. That is, the electron in question can be on the left complex (corresponding to the left minimum in the diagram) or on the right complex (corresponding to the right minimum). Orbital overlap between the two sites leads to the level splitting at the midpoint of the diagram, where the two complexes have the same size. In principle, electron transfer from left to right could suddenly occur by absorption of a photon with the energy indicated by E_1 in Fig. 19.3. The immediately formed product species would be highly excited, inasmuch as the oxidized species would have a large metal-ligand distance and the reduced species would have a small metal-ligand distance. As we shall soon see, optical transitions of this type have been observed in mixed-valence systems, in which the two metal atoms are permanently linked together by a bridging ligand. In ion-pair systems such as we are now discussing, electron exchange proceeds by thermal activation. By absorption of a relatively small amount of thermal energy, the system can be brought to the midpoint of the diagram. This involves shortening the metal-ligand distance of the larger complex and lengthening the metal-ligand distance of the smaller complex to make the system structurally symmetric. At this point, the electron is delocalized, and transfer can occur in either direction.

The barrier height, indicated by E_2 in the figure, corresponds to the structural reorganizational contribution to the activation energy of the electron exchange. Thus the more the two complexes differ initially in size, the higher will be the activation energy and the slower will be the electron exchange. Relatively large changes in size can accompany the removal or addition of an $e_g{}^*$ electron, and electron exchange reactions which involve changes in the number of $e_g{}^*$ electrons are generally slower than those involving only changes in the number of t_{2g} electrons. This electronic effect can be seen in the rate-constant data of Table 19.5. The $Cr(H_2O)_6{}^{2+}-Cr(H_2O)_6{}^{3+}$ exchange and several of the Co(II)–Co(III) exchange reactions are very slow, whereas all the other exchange reactions are relatively fast. However, the Co(II)–Co(III) exchange reactions require special comment.

TABLE 19.5
Rate constants of electron-exchange reactions†

Reaction	k at 25°C, M^{-1} s^{-1}
$Cr(H_2O)_6{}^{2+} + Cr(H_2O)_6{}^{3+}$	$\leqslant 2 \times 10^{-5}$
$Co(NH_3)_6{}^{2+} + Co(NH_3)_6{}^{3+}$	$< 10^{-9}$
$Co(en)_3{}^{2+} + Co(en)_3{}^{3+}$	1.4×10^{-4}
$CoEDTA^{2-} + CoEDTA^{-}$	1.4×10^{-4}
$Co(H_2O)_6{}^{2+} + Co(H_2O)_6{}^{3+}$	~ 5
$Co(o\text{-phen})_3{}^{2+} + Co(o\text{-phen})_3{}^{3+}$	1.1
$V(H_2O)_6{}^{2+} + V(H_2O)_6{}^{3+}$	1.0×10^{-2}
$Fe(H_2O)_6{}^{2+} + Fe(H_2O)_6{}^{3+}$	4
$Fe(CN)_6{}^{4-} + Fe(CN)_6{}^{3-}$	740
$Fe(o\text{-phen})_3{}^{2+} + Fe(o\text{-phen})_3{}^{3+}$	$> 10^5$
$Ru(NH_3)_6{}^{2+} + Ru(NH_3)_6{}^{3+}$	8×10^2
$Os(bipy)_3{}^{2+} + Os(bipy)_3{}^{3+}$	5×10^4
$IrCl_6{}^{3-} + IrCl_6{}^{2-}$	10^3

† F. Basolo and R. G. Pearson, "Mechanisms of Inorganic Reactions," 2d ed., Wiley, New York, 1967; R. G. Wilkins, "The Study of Kinetics and Mechanism of Reactions of Transition Metal Complexes," Allyn and Bacon, Boston, 1974; H. Taube, "Electron Transfer Reactions of Complex Ions in Solution," Academic. New York, 1970; J. Halpern, *Quart. Rev.*, **15**, 207 (1961).

The simple transfer of one electron from Co(II) to Co(III) leads to an excited state of each product ion and involves an asymmetric activated complex:

$$Co^{II}(t_{2g}{}^5 e_g{}^{*2}) + Co^{III}(t_{2g}{}^6) \quad \rightarrow \quad Co^{III}(t_{2g}{}^5 e_g{}^{*1}) + Co^{II}(t_{2g}{}^6 e_g{}^{*1})$$

Inasmuch as the symmetry of the reaction requires the steps leading to and from the activated complex to be symmetric, this process is unacceptable. It is believed that the reaction actually involves an excited state of one reactant. If the Co(III) is excited to the $t_{2g}{}^4 e_g{}^{*2}$ configuration, the following symmetric reaction can occur:

$$Co^{II}(t_{2g}{}^5 e_g{}^{*2}) + Co^{III}(t_{2g}{}^4 e_g{}^{*2}) \quad \rightarrow \quad Co^{III}(t_{2g}{}^4 e_g{}^{*2}) + Co^{II}(t_{2g}{}^5 e_g{}^{*2})$$

If the Co(II) is excited to the $t_{2g}{}^6 e_g{}^{*1}$ configuration, the exchange is also symmetric:

$$Co^{II}(t_{2g}{}^6 e_g{}^{*1}) + Co^{III}(t_{2g}{}^6) \quad \rightarrow \quad Co^{III}(t_{2g}{}^6) + Co^{II}(t_{2g}{}^6 e_g{}^{*1})$$

In either case, the excitation energy contributes to the activation energy, and therefore most Co(II)–Co(III) exchange reactions are slow. However, in the case of $Co(H_2O)_6{}^{3+}$, the ligand-field strength is low and excitation to the high-spin state is fairly easy. Hence the $Co(H_2O)_6{}^{2+}$–$Co(H_2O)_6{}^{3+}$ exchange is rapid. In the case of $Co(o\text{-phen})_3{}^{2+}$, the ligand-field strength is very high and excitation to the low-spin state is easy. Hence the $Co(o\text{-phen})_3{}^{2+}$–$Co(o\text{-phen})_3{}^{3+}$ exchange is also fast.

Part of the enhanced rate of the $Co(o\text{-phen})_3{}^{2+}$–$Co(o\text{-phen})_3{}^{3+}$ exchange may be due to another effect, which also causes the $Fe(o\text{-phen})_3{}^{2+}$–$Fe(o\text{-phen})_3{}^{3+}$

exchange to be faster than other Fe(II)–Fe(III) exchanges. The latter enhancement is probably due to the facts that (1) o-phenanthroline is an aromatic system which can be reduced relatively easily to a radical anion and (2) electron delocalization in the o-phen complexes allows the transferring electron to penetrate the coordination shells easily.

Spontaneous outer-sphere redox reactions in which there is a net chemical change are usually faster than the corresponding electron-exchange reactions. Some data are given in Table 19.6. Usually the greater the thermodynamic driving force for a redox reaction, the greater the rate. A rough estimate of the rate constant k_{12} can be obtained from the theoretically derived Marcus equation,[16]

$$k_{12} \approx (k_{11}k_{22}K_{12})^{1/2}$$

where k_{11} and k_{22} are the rate constants of the electron-exchange reactions and K_{12} is the equilibrium constant.

Inner-Sphere Redox Reactions[13]

In an "inner-sphere" redox reaction, the two metal ions are connected in the activated complex through a bridging ligand common to both coordination shells. The first definitive proof of such a mechanism was obtained by Taube and his

---

[16]   R. A. Marcus, *Ann. Rev. Phys. Chem.,* **15**, 155 (1964).

TABLE 19.6
A comparison of redox and electron-exchange rate constants at 25°C†

Reaction	$k, M^{-1} s^{-1}$
$Fe(CN)_6^{4-} + IrCl_6^{2-}$	3.8×10^5
$Fe(CN)_6^{4-} + Fe(CN)_6^{3-}$	7.4×10^2
$IrCl_6^{3-} + IrCl_6^{2-}$	10^3
$Cr(H_2O)_6^{2+} + Fe(H_2O)_6^{3+}$	$\sim 2 \times 10^3$
$Cr(H_2O)_6^{2+} + Cr(H_2O)_6^{3+}$	$\leqslant 2 \times 10^{-5}$
$Fe(H_2O)_6^{2+} + Fe(H_2O)_6^{3+}$	4
$Fe(o\text{-phen})_3^{2+} + Co(H_2O)_6^{3+}$	1.4×10^4
$Fe(o\text{-phen})_3^{2+} + Fe(o\text{-phen})_3^{3+}$	10^5
$Co(H_2O)_6^{2+} + Co(H_2O)_6^{3+}$	~ 5
$Fe(H_2O)_6^{2+} + Co(H_2O)_6^{3+}$	$10‡$
$Fe(H_2O)_6^{2+} + Fe(H_2O)_6^{3+}$	$0.9‡$
$Co(H_2O)_6^{2+} + Co(H_2O)_6^{3+}$	$\sim 1‡$

† F. Basolo and R. G. Pearson, "Mechanisms of Inorganic Reactions," 2d ed., Wiley, New York, 1967; R. G. Wilkins, "The Study of Kinetics and Mechanism of Reactions of Transition Metal Complexes," Allyn and Bacon, Boston, 1974; A. G. Sykes, "Kinetics of Inorganic Reactions," Pergamon, New York, 1966.
‡ At 0°C.

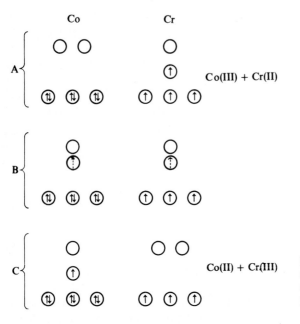

FIGURE 19.4
Changes in the *d* orbitals of Co and Cr as X^- moves from Co to Cr. Electron transfer occurs at stage B.

coworkers,[17] who showed that aqueous chromium(II) is oxidized by a pentaam-minecobalt(III) complex, $Co(NH_3)_5X^{2+}$, to give the species $Cr(H_2O)_5X^{2+}$. The result proves that, in the redox reaction, the X^- ion is transferred directly from Co(III) to Cr(II) and that in the activated complex the two metal atoms are bridged by the X^- ion, as shown in the following structure:

$$(NH_3)_5CoXCr(H_2O)_5{}^{4+}$$

This particular reaction system was chosen because the Co(III) complex retains its integrity in solution for long periods of time, both Cr(II) and Co(II) complexes are labile, and Cr(III) complexes are inert. Each of these chemical characteristics was essential to the success of the experiment.

It can be shown that the transfer of X^- from Co(III) to Cr(II) favors the transfer of the electron from Cr(II) to Co(III). The *d* electron structures of the two metals are shown in Fig. 19.4, stage A before electron transfer, stage B in the activated complex, and stage C immediately after electron transfer. In stage A, before X^- has shifted toward the chromium atom, the electron in the $d\sigma$ orbital has an energy below that of the $d\sigma$ orbitals of the cobalt atom. When X^- moves toward the chromium atom, the cobalt $d\sigma$-orbital energy levels split as their center of gravity is lowered and the chromium $d\sigma$-orbital energy levels move together as their center of gravity is raised. Electron transfer occurs when the energy of the lower cobalt $d\sigma$ orbital is approximately the same as that of the

---

[17] H. Taube, H. Myers, and R. L. Rich, *J. Am. Chem. Soc.*, **75**, 4118 (1953); H. Taube and H. Myers. *J. Am. Chem. Soc.*, **76**, 2103 (1954).

lower chromium $d\sigma$ orbital. Further shift of X^- traps the electron on the cobalt atom and completes the transfer of X^- to the chromium atom.

Electron Transfer Through Extended Bridges

The reduction of $Co(NH_3)_5py^{3+}$ by $Cr(H_2O)_6^{2+}$ ($k = 4.0 \times 10^{-3}\,M^{-1}s^{-1}$) yields $Cr(H_2O)_6^{3+}$ as the only Cr-containing product. The reaction undoubtedly proceeds by an outer-sphere mechanism. On the other hand, the complex

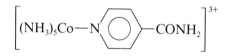

reacts with $Cr(H_2O)_6^{2+}$ much more rapidly ($k = 17.4\,M^{-1}s^{-1}$) to give the following Cr(III) complex.

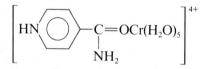

Separate experiments have verified that the latter species is the immediate, primary product; hence it is certain that the Cr(II) attacks the original complex at the remote keto group and that the electron is transferred to the Co(III) through the conjugated bonds of the isonicotinamide.[18]

Isied and Taube[19] prepared the complex

$$\left[(NH_3)_5CoO_2C-\left\langle\bigcirc\right\rangle N-Ru(NH_3)_4SO_4\right]^{2+}$$

in the Co(III)—L—Ru(II) form and, by spectrophotometric measurements, determined the rate ($k \sim 1 \times 10^2\,s^{-1}$) of the *intra*molecular electron-transfer reaction:

$$Co(III)—L—Ru(II) \quad \rightarrow \quad Co(II)—L—Ru(III)$$

They found that the following complex, with a CH_2 group between the carboxylate group and the aromatic ring, underwent the analogous electron transfer much more slowly ($k \sim 2 \times 10^{-2}\,s^{-1}$).

$$\left[(NH_3)_5CoO_2CCH_2-\left\langle\bigcirc\right\rangle N-Ru(NH_3)_4H_2O\right]^{4+}$$

18   H. Taube and E. S. Gould, *Acc. Chem. Res.*, **2**, 321 (1969).
19   S. S. Isied and H. Taube, *J. Am. Chem. Soc.*, **95**, 8198 (1973).

The difference is undoubtedly mainly due to blockage of the π bonding between the metal atoms. In fact, electron transfer in the latter complex may proceed by a different sort of route. Molecular models of the complex show that the carbonyl group is close to the π cloud of the ring, and thus the CH_2 linkage may be completely bypassed in the electron transfer.

The reactions of aqueous chromium(II) with the complexes (I, II, and III) shown here illustrate outer- and inner-sphere redox processes and give information regarding the ability of the terephthalic acid ring to function as a bridge for

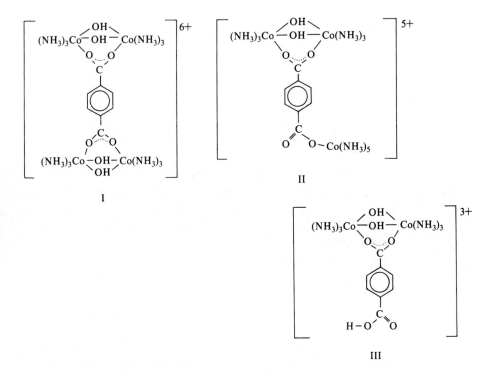

I

II

III

electron transfer.[20] In all three reactions, the Co(III) atoms are reduced stepwise. In the reactions of complexes I and III, reduction of the first Co(III) is relatively slow and rate-determining. These processes are believed to be outer-sphere processes, and their slowness is attributable to the fact that the numbers of $e_g{}^*$ electrons on both the chromium and cobalt atoms change in the reaction. In the reaction of complex II, the first Co(III) is reduced rapidly, and the remaining Co(III) atoms are reduced slowly. It should be noted that complex II has one uncomplexed carboxylic oxygen atom adjacent to a $Co(NH_3)_5$ moiety. The first step of the reaction presumably involves the rapid inner-sphere reduction of this Co(III) center, with attack of the Cr^{2+} at the uncomplexed carbonyl oxygen; a mixed $[Co(III)]_2$–Cr(III)

[20] M. Hery and K. Wieghardt, *Inorg. Chem.*, **15**, 2315 (1976).

complex is formed which then is further reduced relatively slowly by outer-sphere processes. Although complex III has an uncomplexed carboxylic oxygen atom, its reduction proceeds by a slow outer-sphere mechanism because the carboxylate group is not directly attached to a Co(III) atom. It is clear that terephthalic does not significantly facilitate the transfer of electrons from one carboxylate group to the other in such reactions.

Mixed-Valence Compounds[21–23]

Symmetric mixed-valence complexes are of considerable theoretical interest because of their close relation to electron-transport phenomena. Some of these are valence-localized species, in which two metal ions of different oxidation state are trapped at specific sites having different dimensions. Such systems can be represented by an energy-versus-configuration diagram such as Fig. 19.3. Others are valence-delocalized species, in which the two metal ions have the same, average oxidation state and the system is structurally symmetric. Such systems are represented by the energy-versus-configuration diagram of Fig. 19.5. We shall first describe a valence-localized complex, shown below.[24]

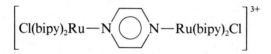

Each ruthenium atom of this mixed-valence complex is coordinated to two bipyridine ligands, a chloride ion, and a bridging pyrazine (pyz) ligand. The complex has a weak absorption band in the near infrared (1300 nm), whereas the corresponding +2 and +4 complexes show no absorption in this region. The band has been interpreted as an "intervalence transition" corresponding to E_1 in Fig. 19.3. As expected from Franck-Condon considerations, the absorption band is broad.

[21]   H. Taube, *Ann. N.Y. Acad. Sci.,* **313,** 481 (1978).

[22]   T. J. Meyer, *Ann. N.Y. Acad. Sci.,* **313,** 496 (1978).

[23]   T. J. Meyer, *Acc. Chem. Res.,* **11,** 94 (1978).

[24]   R. W. Callahan, F. R. Keene, T. J. Meyer, and D. J. Salmon, *J. Am. Chem. Soc.,* **99,** 1064 (1977).

FIGURE 19.5
Energy-versus-configuration diagram for a mixed-valence complex with complete delocalization. The energy levels correspond to the bonding and antibonding combinations of the metal orbitals which hold the electron.

This interpretation of the spectrum implies that the oxidation states of the metal atoms are $+2$ and $+3$, rather than $+2.5$ and $+2.5$. Confirmation is provided by the appearance of a symmetric pyrazine stretch band in the infrared. The latter band does not appear in the infrared spectra of the $+2$ and $+4$ complexes, nor in that of free pyrazine, but it does appear in the spectrum of the monomeric complex $(bipy)_2Ru(pyz)Cl^+$. The band is forbidden in a symmetric bridging pyrazine and in the free ligand but is allowed when the twofold symmetry of the ligand is broken by asymmetric complexation.

The following cyanogen complex is a well-characterized example of a valence-delocalized complex.[21]

$$(NH_3)_5Ru—N\equiv C—C\equiv N—Ru(NH_3)_5{}^{5+}$$

In the $+6$ complex, the C—N stretching frequency is $2330\ cm^{-1}$. In the $+4$ complex, this frequency is reduced (by $d\pi \rightarrow \pi^*$ delocalization) to $1960\ cm^{-1}$. It is significant that in the $+5$ (mixed-valence) complex, a single C—N stretching frequency is observed at $2210\ cm^{-1}$, intermediate between those of the $+4$ and $+6$ complexes. Thus one can say that the lifetime of any state of the complex in which one ruthenium is $+2$ and the other $+3$ must be less than the period of a molecular vibration, $\sim 10^{-13}$ s. Further evidence for delocalization is found in the observation of a narrow band in the infrared corresponding to the electron transition indicated in Fig. 19.5. The fact that the band is narrow indicates that it cannot correspond to an intervalence transition such as that indicated in Fig. 19.3.

Another example of a valence-delocalized complex is the following osmium complex.[21]

$$(NH_3)_5Os—N\equiv N—Os(NH_3)_5{}^{5+}$$

Because of the symmetry of the complex, the N—N stretch is inactive in the infrared. It has been determined that the equilibrium constant for the reaction of the $+4$ complex with the $+6$ complex,

$$[2,2]^{4+} + [3,3]^{6+} \rightleftharpoons 2[2.5,2.5]^{5+}$$

is greater than 10^{20}, corresponding to a free-energy change of more than $27\ kcal\ mol^{-1}$. Clearly there is a marked stabilization of the mixed-valence complex by electron delocalization. The related complex,

$$Cl(NH_3)_4Os—N\equiv N—Os(NH_3)_4Cl^{3+}$$

has a pK of 12, but when oxidized to the $[3,3]^{4+}$ state, the pK drops to 6.6. These data prove that the mixed-valence complex is delocalized, because if it were valence-trapped [Os(II) on one side and Os(III) on the other], the pK would be near 6.6, as in the case of the $[3,3]^{4+}$ complex.[25]

---

[25] H. Taube, private communication.

The complex

$$\left[(NH_3)_5Ru-N\bigcirc N-Ru(NH_3)_5\right]^{5+}$$

appears to be delocalized according to some criteria, but localized according to others. Probably it represents an example of intermediate coupling between the metal atoms. The x-ray photoelectron spectrum of the complex shows two ruthenium core level peaks with binding energies close to those observed for the $[2,2]^{4+}$ and $[3,3]^{6+}$ complexes and with relative intensities of 1:1. This result is obviously consistent with a valence-trapped complex.[26] However, Hush has pointed out that, even in cases where the ground state is delocalized, it is possible to have two core-ionized states, leading to two photopeaks.[27] Thus the x-ray photoelectron spectrum cannot be interpreted unambiguously. It is significant that this complex at least shows intermediate coupling between the metal centers, whereas the analogous complex, in which the five ammonias on each ruthenium are replaced by two bipyridines and a chloride ion, is definitely valence-trapped. It is not clear whether the valence trapping is a consequence of four π-acceptor atoms on Ru(II) competing with the bridging ligand for $d\pi$ electron density (this would diminish coupling between the centers) or whether it is a result of having a chloride ion in the coordination sphere. The Ru(II)–Cl$^-$ distance is expected to be much larger than the Ru(III)–Cl$^-$ distance, and this difference may affect valence trapping.

Various physical methods have shown that the iron atoms in the following mixed-valence biferrocenylene complex are equivalent, and a $[2.5, 2.5]^+$ description seems appropriate.[28]

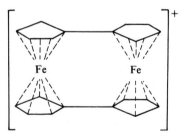

However, high-quality calculations for this system predict valence trapping.[29] Clearly this system deserves further theoretical study.

[26] P. H. Citrin, *J. Am. Chem. Soc.*, **95**, 6472 (1973); P. H. Citrin and A. P. Ginsberg, *J. Am. Chem. Soc.*, **103**, 3673 (1981).

[27] N. S. Hush, *Chem. Phys.*, **10**, 361 (1975).

[28] D. O. Cowan, D. LeVanda, J. Park, and F. Kaufman, *Acc. Chem. Res.*, **6**, 1 (1973).

[29] M. C. Böhm, R. Gleiter, F. Delgado-Pena, and D. O. Cowan, *Inorg. Chem.*, **19**, 1081 (1980).

Unstable Intermediate Oxidation States[14]

Redox reactions involving changes in oxidation state of more than one unit often involve intermediates in unusual oxidation states. For example, the reaction of Tl^{3+} with Fe^{2+}

$$2Fe^{2+} + Tl^{3+} \quad \rightarrow \quad 2Fe^{3+} + Tl^{+}$$

is first order in each of the reactants during the first part of the reaction, suggesting the formation of either Tl(II) or Fe(IV) as an intermediate:

$$Fe^{2+} + Tl^{3+} \quad \rightarrow \quad Fe^{3+} + Tl(II)$$

or

$$Fe^{2+} + Tl^{3+} \quad \rightarrow \quad Fe(IV) + Tl^{+}$$

When the products Fe^{3+} and Tl^{+} are added to the reaction mixture, Fe^{3+}, but not Tl^{+}, decreases the reaction rate. This result is in accord with the following mechanism:

$$Fe^{2+} + Tl^{3+} \quad \rightleftharpoons \quad Fe^{3+} + Tl(II)$$

$$Tl(II) + Fe^{2+} \quad \rightarrow \quad Tl^{+} + Fe^{3+}$$

The reaction of chromium(VI) with iodide and iron(II) in dilute acid solution has some remarkable features. The oxidation of iodide by chromium(VI) is normally slow, but in the presence of iron(II), which is rapidly oxidized by chromium(VI), iodine is rapidly formed. Iron(III) does not oxidize iodide rapidly in dilute acid; therefore one concludes that a reactive intermediate, presumably an intermediate oxidation state of chromium, is responsible for the oxidation of the iodide. When a large excess of iodide is present, two equivalents of iodide are oxidized per mole of Fe^{2+}:

$$Fe^{2+} + 2I^{-} + Cr(VI) \quad \rightarrow \quad Fe^{3+} + I_2 + Cr(III)$$

This result suggests that the first reactive intermediate formed is chromium(V),

$$Fe^{2+} + Cr(VI) \quad \rightarrow \quad Cr(V) + Fe^{3+}$$

and that the iodide reduces the Cr(V) to Cr(III):

$$Cr(V) + 2I^{-} \quad \rightarrow \quad Cr(III) + I_2$$

It is not known whether or not the latter reaction involves the intermediate formation of Cr(IV). This oxidation of iodide in the presence of iron(II) is an example of an *induced* reaction.

PROBLEMS

19.1 Arrange the following complexes in order of increasing lability. $Fe(H_2O)_6{}^{3+}$, $Cr(H_2O)_6{}^{3+}$, $Y(H_2O)_6{}^{3+}$, $Al(H_2O)_6{}^{3+}$, $Sr(H_2O)_6{}^{2+}$, $K(H_2O)_6{}^{+}$.

*19.2 Explain why the following reaction is catalyzed by Co^{2+}:

$$Co(NH_3)_6{}^{3+} + 6CN^{-} \quad \rightarrow \quad Co(CN)_6{}^{3-} + 6NH_3$$

19.3 There are two isomers of $Pt(NH_3)_2Cl_2$, A and B. When A is treated with thiourea (tu), $Pt(tu)_4^{2+}$ is formed. When B is treated with thiourea, $Pt(NH_3)_2(tu)_2^{2+}$ is formed. Identify the isomers and explain the data.

19.4 Using the trans effect, predict the product of the reaction of 2 mol of ethylenediamine with 1 mol of $PtCl_6^{2-}$.

*19.5 The reaction of $CrCl_3$ with liquid ammonia ordinarily gives principally $[Cr(NH_3)_5Cl]Cl_2$, but when a trace of KNH_2 is present, the main product is $[Cr(NH_3)_6]Cl_3$. Explain.

*19.6 Suggest a method for preparing the complex $(NC)_5Fe-CN-Co(CN)_5^{6-}$.

*19.7 The oxidation of Sn(II) by Fe^{3+} is first order in each reagent. How would you try to determine whether Sn(III) is involved in the mechanism?

19.8 Suggest methods for the preparation of the three isomers of $PtNH_3py(NO_2)Br$.

19.9 The rate of the aqueous vanadium(II)–vanadium(III) electron-exchange reaction may be expressed Rate $= k[V^{II}][V^{III}]$, where k shows the following dependence on hydrogen-ion concentration:

$$k = a + \frac{b}{[H^+]}$$

Explain this result in terms of two parallel reaction paths. (Hint: The so-called hydrolysis of an aquo metal ion is really the ionization of a Brönsted acid.)

19.10 Slow substitution reactions are found for octahedral complexes with what electron configurations?

*19.11 The reaction $Ni(CO)_4 + L \rightarrow Ni(CO)_3L + CO$ is first order in $Ni(CO)_4$ and zero order in L, whereas the reaction $Co(CO)_3NO + L \rightarrow Co(CO)_2(NO)L + CO$ is second order (first order in the complex and in L). Explain the difference.

19.12 It has been observed that two of the water molecules in the aqueous $Cu(H_2O)_6^{2+}$ ion undergo exchange with the bulk solvent molecules much more rapidly than the other four. Explain.

19.13 Show that the iron(II)-induced reaction of iodide with chromium(VI) could be explained by assuming an Fe(V) species as the only unusual intermediate.

CHAPTER
20

HOMOGENEOUS CATALYSIS OF ORGANIC REACTIONS

A reaction is said to be catalyzed when its rate is increased by the presence of a substance (the catalyst) which does not appear in the equation for the net, overall reaction. In homogeneous catalysis the catalyst and the reactants are molecularly dispersed in the same phase, either liquid or gaseous. In heterogeneous catalysis the catalyst and reactants are in different phases; usually heterogeneous catalysts are solids, and the reactants are liquids or gases. In this chapter we shall discuss various examples of homogeneous catalysis by transition-metal compounds. Heterogeneous catalysis will be covered in Chap. 21.

A transition-metal catalyst provides a site where the rate-determining step of a reaction can take place more easily than it can in the absence of the catalyst. The activation energy of the catalyzed reaction is less than that of the uncatalyzed reaction, and therefore the catalyzed reaction is more rapid than the uncatalyzed reaction. Generally a catalyzed reaction is the sum of a sequence of reaction steps, only one of which is rate-determining. In the case of reactions catalyzed by transition-metal complexes, the various reaction steps can usually be categorized into three broad types. We shall discuss these three types of reaction before describing the detailed mechanisms of particular catalyzed reactions.

1. *Dissociation and addition.* We have seen that there are many transition-metal complexes with only 16, or even 14, electrons in the valence shell of the metal atom. Sixteen-electron complexes are often found with d^8 metal ions, such as Co(I), Rh(I), Ni(II), and Pd(II). Such complexes are said to be coordinatively unsaturated; they can add ligands to form coordinatively saturated 18-electron compounds, and conversely the latter can undergo dissociation to form coordinatively unsaturated compounds. Compounds of both kinds are commonly

involved in catalytic processes. Typically, a d^8 metal complex, by gain or loss of ligands, reversibly converts between 4-coordination and 5-coordination, as follows:

$$L_4M + :X \rightleftharpoons \underset{d^8}{L_4MX}$$

with L_4M labeled d^8.

Here X is a species such as CO, an olefin, or a halide ion, which can donate an electron pair to the metal atom.

2. *Oxidative addition and reductive elimination.*[1] If, in the addition of a species to a complex, the metal atom formally provides part or all of the electrons used in bonding the metal atom to the added species, the process is called "oxidative addition," and the reverse process is called "reductive elimination." The simplest example of oxidative addition is the addition of a proton to a metal complex. For example,

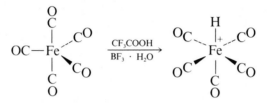

According to the usual methods of calculating oxidation states in such compounds, the hydrogen of the adduct exists as a coordinated hydride ion and the iron is in the +2 oxidation state.

More commonly, oxidative addition involves the addition of a molecule which cleaves into two fragments that bond separately to the metal atom. The oxidation state of the metal effectively increases by two units, and, unless the metal atom loses a ligand in the process, the coordination number also increases by two units.

$$L_4M + Y:Z \rightleftharpoons L_4M\overset{\displaystyle Y}{\underset{\displaystyle Z}{\big\langle}}$$

Note that the metal atom formally must provide two of the four electrons used to bond groups Y and Z. Typically, the added molecule is a small molecule such as H_2, RX, HX, X_2, etc., and the oxidative addition converts a coordinatively unsaturated d^6 complex to a coordinatively saturated d^6 complex. For example,

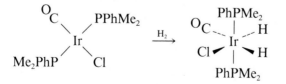

[1] J. P. Collman and L. S. Hegedus, "Principles and Applications of Organotransition Metal Chemistry," pp. 176–258, University Science Books, Mill Valley, Calif., 1980.

However, relatively large molecules can also add oxidatively,

$$C_6H_5-\underset{\underset{H}{|}}{\overset{\overset{F}{|}}{C}}-\underset{\underset{CO_2Et_2}{|}}{\overset{\overset{H}{|}}{C}}-Br \quad \xrightarrow{Ir(PPhMe_2)_2(CO)Cl} \quad C_6H_5-\underset{\underset{H}{|}}{\overset{\overset{F}{|}}{C}}-\underset{\underset{CO_2Et_2}{|}}{\overset{\overset{H}{|}}{C}}-Ir(PPhMe_2)_2(CO)BrCl$$

And there are some examples of oxidative addition to coordinatively saturated complexes, with the loss of a neutral ligand such as CO, PR$_3$, or N$_2$:

$$Os(CO)_5 \quad \xrightarrow[80 \text{ atm}, 120°C]{H_2} \quad H_2Os(CO)_4 + CO$$

Many intramolecular oxidative additions are known. For example, oxidative addition of an ortho C—H bond of one of the phenyl groups in Ir(PPh$_3$)$_3$Cl yields a four-membered metallacycle:[2]

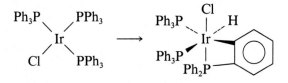

Transition-metal complexes which easily undergo one-electron oxidations (such as complexes of Co^{2+} and Cr^{2+}) can undergo oxidative addition with an overall one-electron change. The net reaction is of the form

$$2L_nM + Y-Z \quad \longrightarrow \quad L_nM-Y + L_nM-Z$$

However, such reactions usually occur in two steps, the first being rate-determining. For example, alkyl halides react with Co(CN)$_5{}^{3-}$ as follows:

$$Co^{II}(CN)_5{}^{3-} + RX \quad \xrightarrow{slow} \quad X-Co^{III}(CN)_5{}^{3-} + R^{\cdot}$$

$$Co^{II}(CN)_5{}^{3-} + R^{\cdot} \quad \xrightarrow{fast} \quad R-Co^{III}(CN)_5{}^{3-}$$

3. *Insertion reactions.* A third important type of reaction involved in catalysis amounts to the insertion of a group between the metal atom and a ligand. Two examples of migratory insertions (i.e., intramolecular insertions) follow:

---

[2] The prefix "metalla" indicates that one of the carbon atoms in the ring has been replaced by a metal atom.

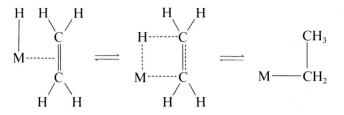

In each of these examples, the coordination number of the metal atom decreases by one.

Carbonyl insertion generally accompanies the substitution reactions of $RMn(CO)_5$:

$$RMn(CO)_5 + L \quad \rightarrow \quad cis\text{-}R(CO)Mn(CO)_4L$$

The kinetics of the reaction, as well as the fact that the substituent goes to the *cis* position (even when $L = {}^{13}CO$), are consistent with the following mechanism, involving a 16-electron intermediate.[3]

$$RMn(CO)_5 \quad \rightleftharpoons \quad R(CO)Mn(CO)_4$$

$$R(CO)Mn(CO)_4 + L \quad \rightarrow \quad cis\text{-}R(CO)Mn(CO)_4L$$

A remaining question is whether a CO group actually inserts into the R-Mn bond or whether the alkyl group migrates to a coordinated CO. A study of the reaction

$$cis\text{-}CH_3(CO)Mn(CO)_4({}^{13}CO) \quad \rightarrow \quad CH_3Mn(CO)_4({}^{13}CO) + CO$$

(in which the ratio of cis to trans isomers in the product was found to be 2:1) showed that probably the methyl group, rather than a CO group, migrates.[4]

*Inter*molecular insertion reactions involve the direct attack of nucleophilic reagents on unsaturated ligands without prior coordination of the nucleophile to the metal. For example, alkyl- and aryllithium reagents react directly with iron pentacarbonyl to form anionic acyl derivatives.

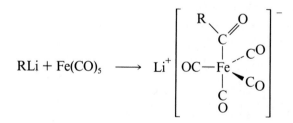

Nucleophilic attack at a terminal carbon atom of an η^3-allyl group leads to the corresponding olefin complex and a formal two-electron reduction of the metal:

---

[3]   I. S. Butler, F. Basolo, and R. G. Pearson, *Inorg. Chem.*, **6**, 2074 (1967).

[4]   K. Noack and F. Calderazzo, *J. Organomet. Chem.*, **10**, 101 (1967).

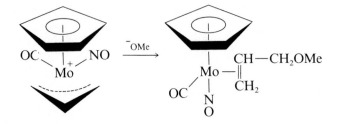

It is significant that, in the above example, the η^3-allyl group is attacked preferentially over three other unsaturated ligands in the complex.

OLEFIN HYDROGENATION

Most catalytic processes for the hydrogenation of double bonds require the use of high temperatures and high hydrogen pressures. The first practical system for the homogeneous reduction of olefins and other unsaturated molecules at 25°C and 1 atm pressure involved the complex $RhCl(PPh_3)_3$, often known as Wilkinson's catalyst.[5] In solution this complex dissociates to a small extent at 25°C:

$$RhCl(PPh_3)_3 \rightleftharpoons RhCl(PPh_3)_2 + PPh_3 \qquad K = 1.4 \times 10^{-4}$$

A simplified mechanism for the catalytic hydrogenation of olefins or acetylenes, involving species derived from Wilkinson's catalyst, is shown by the following cyclic scheme:

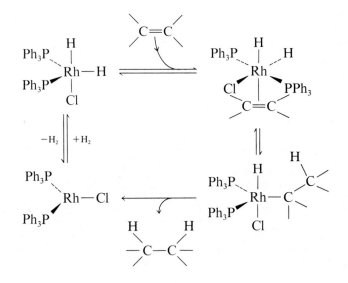

The 14-electron species $RhCl(PPh_3)_2$ adds a molecule of hydrogen oxidatively to

---

[5] F. A. Cotton and G. Wilkinson, "Advanced Inorganic Chemistry," 5th ed., pp. 1245–1247, Wiley-Interscience, New York, 1988.

form the 5-coordinate, 16-electron dihydro complex. This in turn adds a molecule of olefin to form a 6-coordinate, 18-electron complex. Transfer of a hydrogen atom to the β-carbon atom yields a 5-coordinate alkyl intermediate which then undergoes another rearrangement to give the hydrogenated product and to reform the $RhCl(PPh_3)_2$ species. All of the three broad types of reaction steps (addition, oxidative addition, and insertion) are represented in this mechanism.

It has been shown that the hydrogenation of olefins with chiral (optically active) catalysts can lead to the formation of optically active saturated compounds.[6] Rhodium(I) complexes of the type $Rh(BINAP)S_2^+$, where S is a solvent molecule and BINAP is a bidentate phosphine with the following structure,

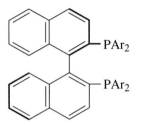

can be prepared in chiral form, in which the dissymmetry is caused by the nonparallel naphthyl groups. Such complexes can be used to catalyze the hydrogenation of many compounds to give the corresponding optically active saturated compounds, with optical yields higher than 90%. For example,

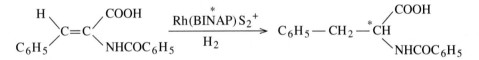

OLEFIN DIMERIZATION AND METATHESIS

It has been shown that metallacyclic intermediates participate in homogeneous catalytic reactions such as olefin dimerization and olefin metathesis (disproportionation).[7] An example is the nickel-catalyzed cyclodimerization of ethylene to give cyclobutane,[8] a compound that is otherwise difficult to obtain:

$$2C_2H_4 \longrightarrow \begin{matrix} H_2C-CH_2 \\ | \quad\quad | \\ H_2C-CH_2 \end{matrix}$$

It is believed that an important step in the process is the conversion of a bis(ethylene)nickel(0) complex to a nickelacyclopentane:

[6] R. Noyori, *Science*, **248**, 1194 (1990).

[7] G. Wilke, *Pure Appl. Chem.*, **50**, 677 (1978).

[8] R. H. Grubbs and A. Miyashita, *J. Am. Chem. Soc.*, **100**, 7416 (1978).

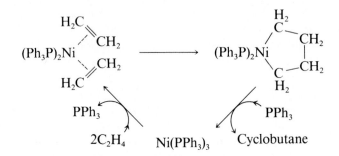

Phosphine-promoted reductive elimination of the two Ni—C bonds forms cyclobutane and Ni(PPh₃)₃, which can react with more ethylene to regenerate the bis(ethylene) complex.

The olefin metathesis (or disproportionation) reaction may be represented by the following equation.

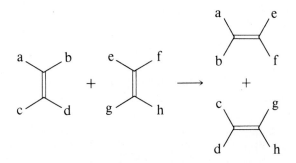

The reaction is catalyzed by a wide variety of catalysts, generally prepared by the reaction of halides of molybdenum, tungsten, or rhenium with alkylating agents such as $EtAlCl_2$, AlR_3, and ZnR_2. It was once believed that olefin metatheses proceed by a concerted process involving a cyclobutane-like intermediate:

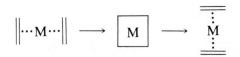

Although such a mechanism might be thought to be symmetry-forbidden according to the orbital symmetry considerations discussed in Chap. 6, it has been shown that, because of the overlap of the metal d orbitals with the olefin MOs, the process is symmetry-allowed.[9] Nevertheless, it is now believed that olefin metatheses are chain reactions involving transition-metal carbenes ($L_nM{=}CR_2$) and metallacyclobutane intermediates.[10] The chain-propagating steps are believed to be of

---

[9] F. D. Mango and J. H. Schachtschneider, *J. Am. Chem. Soc.*, **93**, 1123 (1971).

[10] T. J. Katz and J. McGinnis, *J. Am. Chem. Soc.*, **97**, 1592 (1975); M. T. Mocella, M. A. Busch, and E. L. Muetterties, *J. Am. Chem. Soc.*, **98**, 1283 (1976); R. H. Grubbs, *Prog. Inorg. Chem.*, **24**, 1 (1978).

the following type:

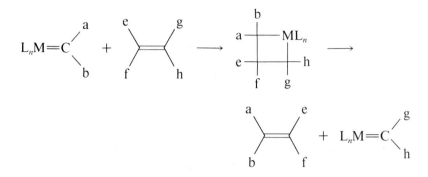

There are a variety of kinetic data that tend to support such a mechanism and to rule out mechanism 20.1.[11] Consider, for example, an experiment in which cyclooctene, 2-butene, and 4-octene are allowed to react simultaneously and the relative amounts of the hydrocarbons formed are determined by analysis near the very beginning of the reaction:

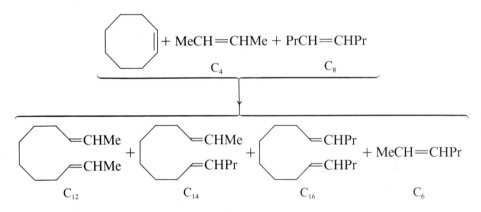

It is obvious that, whatever the reaction mechanism, the concentration of C_6 is much less than that of C_4 or C_8 early in the reaction. Therefore if the reaction proceeded according to mechanism 20.1, at first the ratios C_{14}/C_{12} and C_{14}/C_{16} would be near zero, because C_{14} would not form from C_6 as quickly as C_{12} and C_{16} form from C_4 and C_8. On the other hand, if the reaction proceeded according to mechanism 20.2, C_{14} would initially predominate over C_{12} and C_{16}. (From purely statistical considerations, the initial rate of formation of C_{14} should be twice that of C_{12} the reaction C_{14} is formed in a greater amount than either C_{12} or C_{16}.

Further support for mechanism 20.2 is found in the actual isolation of catalytically active carbene complexes, such as $(CO)_5WCR_2$ and $Cl_2(PEt_3)_2$-

---

[11] T. J. Katz, *Adv. Organomet. Chem.*, **16**, 283 (1977).

$W(O)CHCMe_3$, and metallacyclobutanes such as $(\eta^5-C_5H_5)_2Ti$ 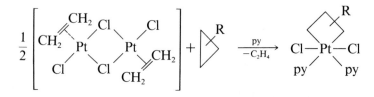 [1,12]

Both $(CO)_5WCR_2$ and $Cl_2(PEt_3)_2W(O)CHCMe_3$ have been shown to react with olefins, as in general reaction 20.2, and to catalyze the metathesis of olefins.

Stable platinacyclobutanes have been formed by the oxidative addition of cyclopropanes to Pt(II) complexes,[13]

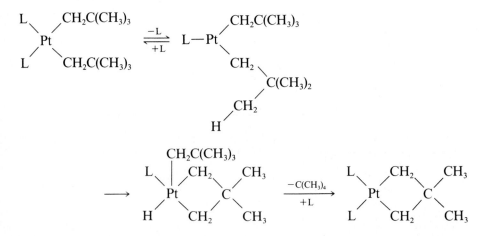

and by the cyclometallation of bis(neopentyl)platinum complexes,[14]

These platinacyclobutanes apparently are too stable to engage in olefin metathesis reactions. At high temperatures, however, they decompose to form cyclopropane and propylene (or their substituted derivatives) by competitive reductive elimination and β-hydride elimination.

OLEFIN ISOMERIZATION

Many transition-metal compounds catalyze the migration of double bonds in alkenes. The mechanism involves the reversible transfer of a hydrogen atom from the transition metal to the coordinated olefin to give a σ-bonded alkyl group. For

---

[12] R. H. Grubbs et al., *J. Am Chem Soc.*, **102**, 6876 (1980); **103**, 7358 (1981).

[13] J. Rajaram and J. A. Ibers, *J. Am Chem. Soc.*, **100**, 829 (1978).

[14] P. Foley and G. M. Whitesides, *J. Am. Chem. Soc.*, **101**, 2732 (1979).

example, the HCo(CO)₄-catalyzed isomerization of allyl alcohol to propionaldehyde is believed to proceed by the following mechanism:

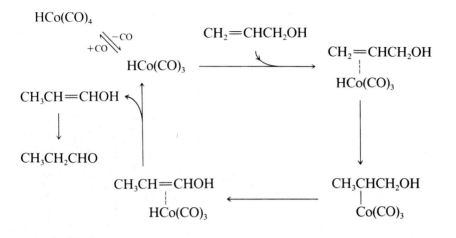

HYDROFORMYLATION

The hydroformylation reaction is the addition of H_2 and CO to an olefin to form an aldehyde:

$$RCH{=}CH_2 + H_2 + CO \longrightarrow RCH_2CH_2CHO$$

The classical hydroformylation process involves the use of $Co_2(CO)_8$ as a catalyst, temperatures in the range 150 to 180°C, and pressures of 200 to 400 atm. Under these reaction conditions, the $Co_2(CO)_8$ is converted to $HCo(CO)_4$,

$$H_2 + Co_2(CO)_8 \longrightarrow 2HCo(CO)_4$$

and the mechanism is as follows:

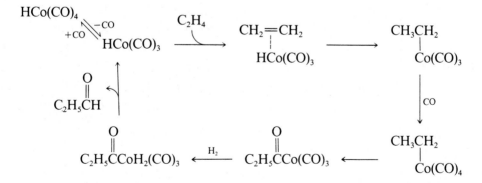

Disadvantages of this process are that some 15 percent of the olefin is hydrogenated to the saturated hydrocarbon, that ketones are formed as by-products, and that some of the aldehyde is reduced to the alcohol. A more effective process, which is carried out industrially at 100°C and a few atmospheres pressure, involves rhodium triphenylphosphine complexes.[15] The mechanism is essentially the same as that of the cobalt-catalyzed process, except that the $(Ph_3P)_2RhCO$ group replaces the $Co(CO)_3$ group.

WATER GAS SHIFT REACTION

The water gas shift reaction,

$$H_2O + CO \rightleftharpoons CO_2 + H_2$$

has received renewed attention because of the need to produce hydrogen from nonpetroleum sources (e.g., coal). Commercially, the reaction is best carried out over solid metal oxide catalysts at elevated temperatures. However, the reaction can proceed homogeneously using alkaline solutions of various metal carbonyls.[16] A postulated general cycle for such homogeneous catalysis is the following:

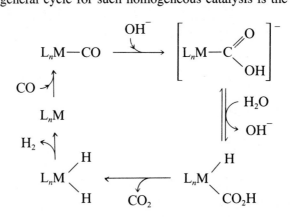

In this mechanism, ML_n might represent, for example, the $Fe(CO)_4$ group. The attack of hydroxide on L_nMCO presumably forms an anionic carboxylic acid intermediate which in turn abstracts a proton from the solvent to form $L_nMH(COOH)$. This species undergoes decarboxylation to form the dihydride, which reductively eliminates H_2 and adds CO to re-form L_nMCO, thus completing the cycle.

ACETIC ACID FROM ETHYLENE[17]

The first step of the overall oxidation of ethylene to acetic acid is the oxidation of ethylene to acetaldehyde, a reaction which is carried out industrially by the Wacker process. It has been known since 1894 that ethylene is readily oxidized by aqueous $PdCl_2$ solutions to acetaldehyde. Chemists at Wacker-Chemie combined

---

[15] Cotton and Wilkinson, op. cit., pp. 1235–1237.

[16] P. C. Ford, *Acc. Chem. Res.*, **14**, 31 (1981). Also see D. J. Darensbourg, A. Rokicki, and M. Y. Darensbourg, *J. Am. Chem. Soc.*, **103**, 3223 (1981).

[17] G. W. Parshall, *Science*, **208**, 1221 (1980).

this reaction with a copper-catalyzed reoxidation of palladium by air to achieve the catalytic air oxidation of ethylene:

$$C_2H_4 + \tfrac{1}{2}O_2 \longrightarrow CH_3CHO$$

The major mechanistic features of the process are shown in the following schematic catalytic cycle:[18]

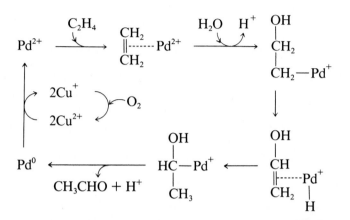

The second step in the production of acetic acid from ethylene is the oxidation of acetaldehyde. The process involves heating acetaldehyde with air at 65°C in the presence of a Co^{2+} catalyst. Acetaldehyde is initially oxidized by a noncatalytic free-radical process to give peracetic acid:

$$\underset{\text{CH}_3\overset{\displaystyle O}{\overset{\|}{C}}H}{} + O_2 \longrightarrow \underset{\text{CH}_3\overset{\displaystyle O}{\overset{\|}{C}}OOH}{}$$

The Co^{2+} catalyzes the reaction between peracetic acid and acetaldehyde to form two molecules of acetic acid:

$$\underset{\text{CH}_3\overset{\displaystyle O}{\overset{\|}{C}}OOH}{} + CH_3CHO \xrightarrow{\;Co^{2+}\;} 2CH_3COOH$$

Although the details of this step are not known, it is believed that both Co(II) and Co(III) are involved, and that the acetoxy radical, $CH_3\overset{\displaystyle O}{\overset{\|}{C}}O\cdot$, is an intermediate.[19]

Another way of making acetic acid, methanol carbonylation, appears to be very promising for future technology. The process consists of three parts:

---

[18] J. E. Bäckvall, B. Akermark, and S. O. Ljunggren, *J. Am. Chem. Soc.*, **101**, 2411 (1979).

[19] G. C. Allen and A. Aguilo, *Adv. Chem. Ser.*, **76**, 363 (1968).

$$CH_3OH + HI \longrightarrow CH_3I + H_2O$$
$$CH_3I + CO \longrightarrow CH_3COI$$
$$\underline{CH_3COI + H_2O \longrightarrow CH_3COOH + HI}$$

$$CH_3OH + CO \longrightarrow CH_3COOH$$

The second part, the carbonylation of methyl iodide, is catalyzed by rhodium carbonyl iodide complexes, as shown by the following scheme:[20]

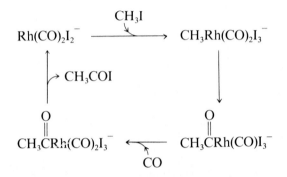

Oxidative addition of methyl iodide to $Rh(CO)_2I_2^-$ forms a 6-coordinate Rh(III) complex which undergoes rearrangement to give a 5-coordinate acetyl complex. After the addition of CO, reductive elimination of the product, CH_3COI, regenerates the $Rh(CO)_2I_2^-$.

TEMPLATE SYNTHESES[21]

The term "template synthesis" has been suggested for reactions in which a central metal ion serves as a "template" on which to coordinate ligands which then react to form chelate rings. In some cases this process can surround the metal ion with a fused cyclic ring system, i.e., a "macrocyclic" ligand. A common procedure involves the cyclic condensation of an α-diketone with a 1,2- or 1,3-diamine. Thus the following planar nickel complex has been prepared by treating 2 mol of 1,3-diaminopropane monohydrochloride with 2 mol of biacetyl, followed by 1 mol of nickel acetate.

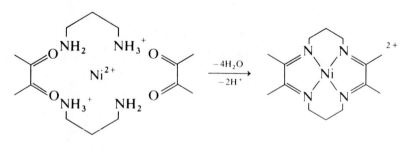

[20]   D. Forster, *Adv. Organomet. Chem.*, **17**, 255 (1979).

[21]   L. F. Lindoy and D. H. Busch, *Prep. Inorg. React.*, **6**, 1 (1971).

Mercaptoethylamine reacts with biacetyl in the presence of nickel ion to form a Schiff-base complex which in turn can be condensed with α,α'-dibromo-o-xylene to form a macrocyclic ligand:

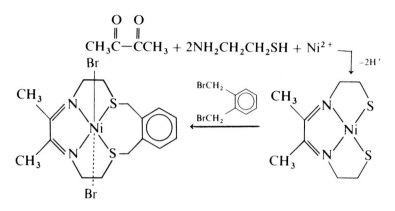

PROBLEMS

20.1 Treatment of $Cr(CO)_6$ with $LiCH_3$, followed by $[(CH_3)_3O]BF_4$, yields the carbene complex $(CO)_5CrC\overset{\diagup CH_3}{\diagdown OCH_3}$. Propose a mechanism for this synthesis.

20.2 When $[(C_6H_5O)_3P]_4RhH$ is heated with a large excess of D_2 at 100°C, the ortho-hydrogen atoms undergo exchange to form $[(C_6H_3D_2O)_3P]_4RdD$. Explain.

20.3 The hydrosilylation reaction is

$$RCH{=}CH_2 + HSiR_3 \quad \rightarrow \quad RCH_2CH_2SiR_3$$

The reaction is catalyzed by $Pd(PPh_3)_4$. Suggest a mechanism.

20.4 Although H_2 has a high dissociation energy, it readily adds oxidatively to many transition-metal complexes. Explain.

20.5 Give a plausible mechanism for the hydroformylation reaction using $RhH(CO)\cdot(PPh_3)_2$ as the catalyst.

20.6 $Ni[P(OEt)_3]_4$ catalyzes the dimerization of ethylene in acid solution to give 1-butene. Propose separate mechanisms for this reaction which (*a*) involve insertion of alkenes into Ni—C bonds, coupled with β elimination, and (*b*) involve carbene intermediates.

20.7 Many low-oxidation-state transition-metal complexes form adducts with Lewis acids such as BF_3:

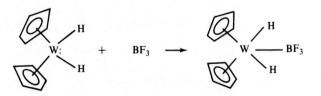

Do you think that such reactions should be classed as oxidative additions?

*20.8 Predict the product of the reaction of

(a) $(\eta^5\text{-}C_5H_5)Fe(CO)_2{}^-$ with CH_3COCl

(b) $Ir(PPh_3)_2(CO)Cl$ with CH_2N_2 (CH_2N_2 readily yields the methylene group and molecular nitrogen)

(c) $Co_2(CO)_8$ with $RC{\equiv}CR$

20.9 Formate ion is suspected as a by-product in the metal-carbonyl-catalyzed water gas shift reaction. Explain how it might form.

20.10 Assuming that the Wacker process proceeds by the mechanism described in this chapter, do you expect that some of the hydrogen of the acetaldehyde originates in the solvent? How would you ascertain this experimentally?

*20.11 Suppose that cyclooctene and 2-hexene were allowed to react in the presence of an olefin metathesis catalyst. What would you expect for the relative yields of C_{12}, C_{14}, and C_{16} hydrocarbons near the beginning of the reaction (a) if mechanism 20.1 were followed and (b) if mechanism 20.2 were followed?

chapter 21

HETEROGENEOUS
CATALYSIS

In this chapter we restrict the scope of heterogeneous catalysis to systems with solid catalysts and liquid or gaseous reactants. The catalyst provides a surface where effective interaction of adsorbed reactant molecules can take place. Hence the most effective catalysts are usually those with high surface areas per unit volume, e.g., powders or porous solids. For many years research and development in the field of heterogeneous catalysis was mainly empirical, because there were few physicochemical tools available for the characterization of surfaces and adsorbed species. Now, however, there are at least a dozen techniques for studying surfaces, some of which are listed and briefly described in Table 21.1.[1] Unfortunately, none of these techniques gives as much chemical information as can be obtained from the nmr spectrum of a coordination compound in solution or from an x-ray diffraction study of a crystalline coordination compound. Generally several of the techniques in Table 21.1 must be applied to a given surface problem in order to characterize the system even approximately.

EFFECTS OF SURFACE SITE ON ADSORPTION

The faces of a crystal and the internal layers of atoms are identified by a set of three numbers known as the "Miller indices." These are the smallest integers which are proportional to the reciprocals of the intercepts of the plane on the a, b, and c axes of the crystal. Several examples of crystal planes in cubic crystals are

---

[1]   E. L. Muetterties, *Angew. Chem. Int. Ed.*, **17,** 545 (1978); G. A. Somorjai, *Acc. Chem. Res.,* **9,** 248 (1976); G. A. Somorjai, "Chemistry in Two Dimensions: Surfaces," Cornell University Press, Ithaca, N.Y., 1981.

TABLE 21.1
Physical methods for studying surfaces

Technique	Principle	Application	Sensitivity (monolayer)	Penetration depth
Auger electron spectroscopy (AES)	Stimulated (typically electron-stimulated) emission from valence or inner shell creates an excited ion. Auger electron is emitted in the deexcitation of the excited ion	Analysis of surface composition and electronic features	$10^{-2}-10^{-3}$	1–7 layers
Ion scattering spectroscopy (ISS)	Energy loss for rare-gas ion scattering from a surface reveals the mass of the surface atom involved in the scattering process	Analysis of surface composition	$\sim 10^{-3}-10^{-4}$	1 layer
Secondary ion mass spectroscopy (SIMS)	Ion (typically rare-gas ion)-stimulated secondary ion emission	Analysis of surface composition	$\sim 10^{-6}$	1 layer
Ion neutralization spectroscopy (INS)	Incident low-energy ions are neutralized by electron transfer from surface atoms which then stimulate an Auger electron emission	Analysis of surface composition and valence band structure	10^{-2}	1 or > 1 layer
X-ray photoelectron spectroscopy (XPS)	X-ray stimulated electron emission from valence and inner shells of surface atoms	Analysis of surface composition and of surface atom oxidation states	$\sim 10^{-1}-10^{-2}$	1–7 layers
Ultraviolet photoelectron spectroscopy (UPS)	Ultraviolet radiation-stimulated electron emission from valence shell of surface atoms	Electronic structure and orientation of chemisorbed molecules	$10^{-1}-10^{-2}$	1–3 layers; ~1 layer in synchrotron radiation mode

Technique	Principle	Surface information	Approximate sensitivity	Sampling depth
Low-energy electron diffraction (LEED)	Elastic scattering of low-energy electrons from the surface and the near-surface atoms	Symmetry features of surface atom environments, unit cell dimensions, and structural features of the surface	$\sim 10^{-1}\text{–}10^{-2}$	1–7 layers
High-resolution electron loss spectroscopy (HRELS)	High-resolution analysis of energy loss in inelastically scattered electrons from a surface	Analysis of vibrational states	10^{-2}	1–7 layers
Appearance potential spectroscopy (APS)	Programmed escalation of incident electron energy with scanning of the resultant x-rays from the surface	Analysis of surface composition and electronic features	$10^{-2}\text{–}10^{-3}$	1–7 layers
Infrared reflection-absorption	Spectral analysis of infrared radiation reflected from a surface	Analysis of vibrational states	$\sim 10^{-2}$	1 layer
Thermal desorption spectroscopy	Mass-spectral detection of desorbed species during programmed surface warm-up	Characterization of chemisorbed species	$\sim 10^{-3}$	
Molecular beam surface scattering	Measurement of angular distribution of scattered beam	Determination of residence times on surface and activation energies of adsorption and reaction		
Scanning tunneling microscopy (STM)	The electron tunneling current from a scanning point probe to a specimen is maintained constant by adjusting the probe-surface distance, thus revealing the surface topology	Atomic-scale imaging of surfaces and of atoms adsorbed on the surfaces		1 layer

illustrated in Fig. 21.1. In Fig. 21.1*a*, one of the set of parallel planes intercepts the *a* axis at 2 unit cell lengths and is parallel to the *b* and *c* axes (i.e., it intercepts the *b* and *c* axes at ∞). Thus the Miller indices for these planes are the smallest set of integers proportional to $(\frac{1}{2}, \frac{1}{\infty}, \frac{1}{\infty})$, or (100). In Fig. 21.1*d*, one of the set of planes intercepts the axes at 1, 1, and $\frac{1}{2}$ unit cell lengths; hence these planes are designated the (112) planes.

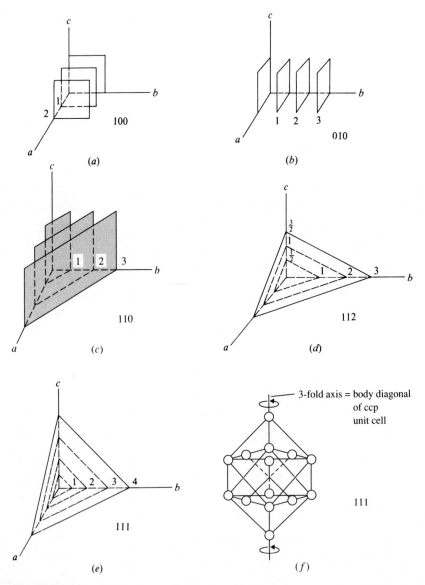

FIGURE 21.1
Examples of crystal planes. Unit cell spacings and Miller indices are indicated in (*a*)–(*e*).

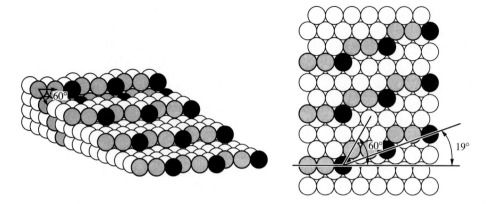

FIGURE 21.2
A representation of a stepped surface of a ccp metal on which there are (111) terraces (unshaded circles), steps (light shaded circles), and kinks (dark shaded circles). *(Drawing courtesy of C. E. Smith and G. A. Somorjai.)*

Even a highly polished single-crystal surface is usually quite rough on a microscopic or atomic level. For example, consider the surface of a cubic close-packed metal. If viewed microscopically, the crystal would show scratches and defects. On the atomic level, "steps" and "kinks", illustrated in Fig. 21.2, are generally present.[2] Thus the structural environment of a surface atom is not only determined by the particular crystal plane involved but also by whether the atom occupies a terrace, step, or kink site. On the terrace or flat section of a (111) face of a cubic close-packed metal, the coordination number is 9. At a step between terraces, the metal atom at the top of the riser has a coordination number of 7 or 8. If the step is so irregular as to have a kink, the kink atom has a coordination number of 6. The (110) face of a cubic close-packed metal (Fig. 21.3) is a continuously stepped surface, like corduroy fabric, and the protruding atoms have a coordination number of only 7.

---

[2] Muetterties, op. cit.

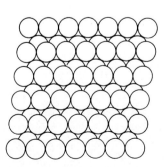

FIGURE 21.3
A perspective view of the (110) face of a ccp metal. The coordination numbers of the exposed atoms and those in the troughs are 7 and 11, respectively.

One might expect the reactivity between adsorbed molecules and surface metal atoms to increase with a decrease in metal-metal coordination number. Indeed, data on the reactivity of molecules with various metal faces agree with this prediction. For example, the (110) surface of platinum reversibly chemisorbs cyanogen (C_2N_2) relatively strongly, whereas the (111) surface of platinum interacts with cyanogen only slightly.[2] Similar results are found in the adsorption of acetonitrile (CH_3CN) on nickel metal. Acetonitrile is weakly and reversibly bound by nickel (111) surfaces; 1 or 2 percent irreversible decomposition, probably due to a few step or kink sites, is observed when the acetonitrile is desorbed from these surfaces by heating in high vacuum to approximately 90°C.[3] The decomposition is characterized by the evolution of gaseous hydrogen and the retention of carbon and nitrogen on the remaining surface. On the other hand, thermal desorption of acetonitrile from the highly stepped nickel (110) surface requires a slightly higher temperature (about 110°C), and about 90 percent decomposition occurs. Figure 21.4 shows the bonding configurations for acetonitrile (at a terrace site, at a step site, and immediately below a step site) which have been proposed to explain these results. At a terrace site, the C—N bond is believed to be perpendicular to the surface, and the hydrogen atoms do not get close enough to the surface to permit C—N bond cleavage. However, the hydrogen atoms of molecules at or near step sites can interact with the surface and thus lead to irreversible decomposition.

It is interesting, with respect to the study of acetonitrile on nickel surfaces, that methyl isocyanide (CH_3NC) is strongly and irreversibly bound to *all* clean nickel surfaces. It cannot be thermally desorbed except with decomposition and evolution of N_2 and H_2 gas. This result is perhaps not surprising from studies of molecular complexes. Acetonitrile is known to be a fairly weak σ donor and a weak π acceptor, whereas methyl isocyanide is a strong σ donor and a strong π acceptor. An interesting, but unexplained, fact is that methyl isocyanide

---

[3] C. M. Friend, J. Stein, and E. L. Muetterties, *J. Am. Chem. Soc.*, **103**, 767 (1981).

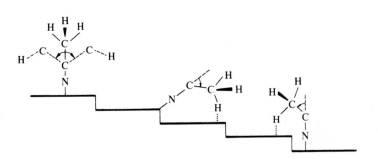

FIGURE 21.4
Possible bonding configurations for acetonitrile at a terrace site, a step site, and immediately below a step site. Reasonable distortion of the molecule can bring the hydrogen atoms in contact with the surface in the case of molecules at a step or immediately below a step. [*Reproduced with permission from C. M. Friend, J. Stein, and E. L. Muetterties, J. Am. Chem. Soc., **103**, 767 (1981). Copyright 1981 American Chemical Society.*]

chemisorbed on carbon-containing nickel (111) surfaces rearranges on heating and can be thermally desorbed *in the form of acetonitrile*. Indeed, ordinary Raney nickel (prepared by dissolving the aluminum out of an Al–Ni alloy with aqueous alkali) is an effective catalyst for the $CH_3NC \rightarrow CH_3CN$ isomerization.[3]

EFFECTS OF SURFACE COVERAGE ON ADSORPTION OF CARBON MONOXIDE

The concentration of a species adsorbed in a single layer on a surface is generally expressed in terms of the fractional coverage, θ. In principle, the coverage can range from zero (no adsorbed species) to one, corresponding to one adsorbed species per surface atom. Figure 21.5 shows the change in the infrared absorption

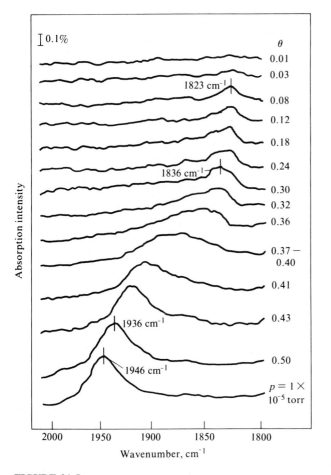

FIGURE 21.5

Infrared absorption band in the C—O stretch region for CO on a Pd(111) surface at 300 K as a function of increasing coverage. [*Reproduced with permission from A. M. Bradshaw and F. M. Hoffmann, Surf. Sci., 72, 513 (1978).*]

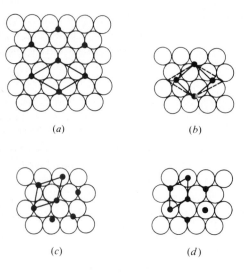

(a)   (b)

(c)   (d)

FIGURE 21.6
Structural models, based on LEED studies, of CO adsorbed on a Pd(111) surface. (a) $\theta = 0.33$; (b) $\theta = 0.5$; (c) $\theta = 0.63$; (d) $\theta = 0.66$. Although the spacing and pattern of the CO molecules were determined by LEED, the CO positions relative to the metal atoms were not determined. [*Reproduced with permission from T. Engel and G. Ertl, Adv. Catal., 28, 1 (1979).*]

band for the C—O stretch as a function of increasing coverage of carbon monoxide on a palladium (111) surface at 300 K.[4] At low coverage, the C—O stretch frequency is 1823 cm^{-1} and remains constant until about $\theta = 0.18$. On further increase in coverage, a very broad band forms whose maximum shifts to higher frequencies. This shift is particularly rapid at $0.3 < \theta < 0.4$. Beyond $\theta = 0.41$, a relatively narrow band emerges around 1920 cm^{-1} and continues to shift to a higher frequency. The 1823 cm^{-1} stretch frequency observed at low coverage is believed to correspond to coordination of the CO molecules at threefold coordination sites. This type of coordination is consistent with the results of low-energy electron diffraction (LEED) studies, which show that, at $\theta = \frac{1}{3}$, the CO groups are distributed on the surface in a hexagonal pattern, as shown in Fig. 21.6a. An increase in coverage beyond $\theta = \frac{1}{3}$ causes compression of the unit cell of the adsorbed CO molecules until, at $\theta = 0.5$, a more compact pattern of CO molecules is obtained. This compression is connected with an enormous shift in the infrared band, corresponding to the shift of the CO molecules to twofold coordination sites, as shown in Fig. 21.6b. Further increase in coverage is believed to lead to the structure of Fig. 21.6c at $\theta = 0.63$ and to that of Fig. 21.6d at $\theta = 0.66$.

Different results have been obtained in a study of the adsorption of CO on a rhodium (111) surface.[5] The vibrational spectrum of CO as a function of coverage at 300 K is shown in the electron energy loss spectrum of Fig. 21.7. At very low exposures only one C—O stretching peak is observed, at 1990 cm^{-1}. This peak is believed to be due to CO groups attached to single metal atoms. The slight gradual shift of this peak to higher frequencies with increasing

[4]   A. M. Bradshaw and F. M. Hoffmann, *Surf. Sci.,* **72,** 513 (1978).
[5]   L. H. Dubois and G. A. Somorjai, *Surf. Sci.,* **91,** 514 (1980).

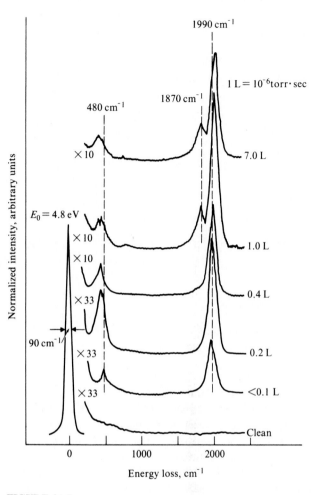

FIGURE 21.7

Energy loss spectrum showing vibrational bands of CO adsorbed on an Rh(111) surface at 300 K as a function of exposure. Note that a Langmuir (L) corresponds to an exposure of the surface to 10^{-6} torr for a period of 1 s. [*Reproduced with permission from L. H. Dubois and G. A. Somorjai, Surf. Sci.,* **91**, *514 (1980).*]

coverage may be due to a decrease in $d\pi \to \pi^*$ back bonding because of the increased competition for electron density among the CO groups. The effect is consistent with the slight shift of the metal-carbon frequency at 480 cm^{-1} to lower frequencies with increasing coverage. Such competitive effects have been seen in molecular carbonyl complexes. At higher coverage, a shoulder near 1870 cm^{-1} appears. This peak presumably corresponds to bridging CO groups. The vibration spectral data are consistent with LEED studies of the system, which show that, at $\theta \approx \frac{1}{3}$, the CO molecules are situated on top of the metal atoms in the hexagonal pattern shown in Fig. 21.8a. The more compact structure seen at $\theta \approx 0.75$, in which the CO molecules are in both bridging and nonbridging sites (Fig. 21.8b),

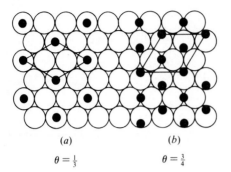

FIGURE 21.8
Structural representations of CO adsorbed on an Rh(111) surface. (a) $\theta = 0.33$; (b) $\theta = 0.75$. [*Reproduced with permission from L. H. Dubois and G. A. Somorjai, Surf. Sci.,* **91**, *514 (1980).*]

(*a*)   (*b*)

$\theta = \frac{1}{3}$ $\theta = \frac{3}{4}$

is consistent with the vibrational spectra obtained at higher coverages. Note that in this compact structure the ratio of nonbridging to bridging CO groups is 2:1, in reasonable agreement with the intensity ratio in the high-coverage vibrational spectra.

THE FISCHER-TROPSCH PROCESS[6]

In this process, hydrocarbons are synthesized by passing a mixture of hydrogen and carbon monoxide over certain transition-metal surfaces.

$$nCO + (2n + 1)H_2 \quad \rightarrow \quad C_nH_{2n+2} + nH_2O$$

A previously widely accepted mechanism for this reaction involves the generation of a methyl group on the surface, followed by a series of steps in which, in effect, methylene groups are successively inserted between the metal and the alkyl group, thus building up a linear alkyl group. The alkyl group is severed from the metal in a hydrogenation step, thus releasing a product hydrocarbon molecule. The methyl group is presumably formed by the reaction of adsorbed hydrogen atoms (formed by the surface dissociation of H_2) with adsorbed CO in a sequence such as the following:

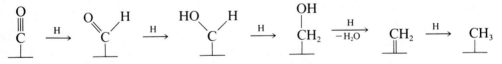

The alkyl-chain growth steps in this mechanism are of the following type,

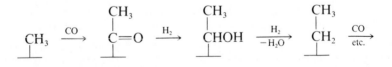

---

[6] For a general discussion of the mechanistic features, see E. L. Muetterties and J. Stein, *Chem. Rev.,* **79**, 479 (1979), and W. A. Herrmann, *Angew. Chem. Int. Ed.,* **21**, 117 (1982).

and the hydrocarbon production step is presumably as follows:

$$R \xrightarrow{\text{H}_2} H + RH$$

This mechanism derived some credibility from the fact that the various proposed steps have known analogs in homogeneous organometallic reactions.

However, both x-ray and ultraviolet photoelectron spectroscopy have shown that carbon monoxide is dissociated to surface carbide and oxide atoms on some metal surfaces. For example, consider the oxygen $1s$ spectra of CO on tungsten shown in Fig. 21.9.[7] The data indicate that CO is adsorbed molecularly at 100 K, but that, on warming, the oxygen atoms become directly bonded to the metal surface and acquire a more negative charge than they have in the undissociated CO groups. The relative ease with which various metal surfaces break the C—O bond of carbon monoxide can be related to the stability of the corresponding metal carbides. Thus tungsten and molybdenum dissociate CO below 170 K, whereas metals with thermally unstable carbides, such as iron and nickel, dissociate CO between 300 and 420 K, and adsorption on the platinum metals (most of which do not form carbides) is mainly nondissociative. On this basis it has been proposed that CO dissociation is involved in Fischer-Tropsch synthesis.[8] It is proposed that adsorbed hydrogen atoms convert the oxygen atoms to water, which is desorbed, and convert the carbon atoms to CH or CH_2 groups, which undergo polymerization, perhaps as follows.[9]

$$H \quad CH_2 \quad CH_2 \longrightarrow CH_3 \quad CH_2 \longrightarrow CH_3{-}CH_2$$

This mechanism provides an explanation of the relative activities of metals in Fischer-Tropsch synthesis. On iron, the CO is expected to be completely dissociated at synthesis temperature. The concentration of CH_2 groups is therefore high and, in agreement with observations, formation of higher hydrocarbons is facilitated. On nickel, CO dissociation is more difficult. Thus the concentration of CH_2 is expected to be lower, and in agreement with observation, hydrogenation, with formation of methane, predominates. On tungsten and molybdenum, which are poor Fischer-Tropsch catalysts, the carbide species are probably too stable to be easily hydrogenated.

It has been observed that diazomethane, CH_2N_2, reacts on metal surfaces to give mainly ethylene, together with nitrogen. Presumably the ethylene forms by the simple dimerization of CH_2 groups:

$$CH_2 \quad CH_2 \longrightarrow CH_2{-}CH_2 \longrightarrow CH_2{=}CH_2$$

---

[7]   J. T. Yates, Jr., T. E. Madey, and N. E. Erickson, *Surf. Sci.*, **43**, 257 (1974).

[8]   R. W. Joyner, *J. Catal.*, **50**, 176 (1977).

[9]   R. C. Brady and R. Pettit, *J. Am. Chem. Soc.*, **102**, 6181 (1980).

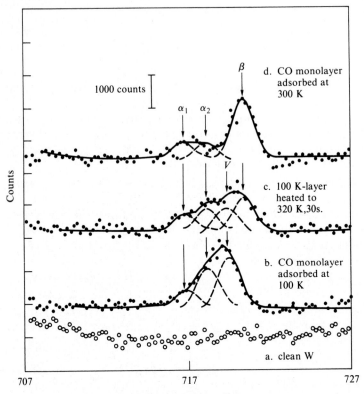

FIGURE 21.9

X-ray photoelectron spectrum of CO on tungsten, showing oxygen $1s$ photolines. Binding energy increases from right to left; lower binding energies correspond to a more negative charge. (*a*) Clean tungsten; (*b*) CO adsorbed at 100 K; (*c*) same as (*b*) after warming to 320 K for 30 s; (*d*) CO adsorbed at 300 K. Peaks α_1 and α_2 correspond to nonbridging CO molecules, V to bridging CO molecules and β to oxide ions on the surface (i.e., dissociated CO). A peak with the same binding energy as β-CO is obtained for oxygen adsorbed on W. [*Reproduced with permission from J. T. Yates, Jr., T. E. Madey, and N. E. Erickson, Surf. Sci., **43**, 257 (1974).*]

However, when a mixture of hydrogen and diazomethane is passed over Co, Fe, and Ru, a linear hydrocarbon mixture, similar to that produced in the Fischer-Tropsch synthesis, is obtained. These data give strong support to a mechanism involving the dissociation of CO and the formation of adsorbed CH_2 groups. It is interesting that this mechanism is essentially the same as that proposed by Fischer and Tropsch in 1926.[10]

Oil crises of recent years has caused chemists to seek chemical raw materials from sources other than petroleum. The Fischer-Tropsch process could compete

---

[10] F. Fischer and H. Tropsch, *Brennst.-Chem.*, **7**, 97 (1926).

with petroleum if (1) the process selectivity were improved (i.e., if high relative yields of particular hydrocarbons, rather than a broad spectrum of hydrocarbons, could be obtained), (2) crude oil were to become even scarcer, and (3) coal were to become relatively inexpensive.[11] Considerable effort is being expended to improve the efficiency and selectivity of the process.

ZIEGLER-NATTA OLEFIN POLYMERIZATION[12]

The Ziegler-Natta process for the polymerization of olefins uses a catalyst formed from $TiCl_4$ and $AlEt_3$. The $AlEt_3$ reduces the $TiCl_4$ to a fibrous form of $TiCl_3$ and replaces some of the surface chlorine atoms with ethyl groups. The rate of polymerization is proportional to the total amount of $TiCl_3$ and the pressure of the olefin and is independent of the $AlEt_3$ concentration. A critical step in the process is believed to be the interposition of an olefin molecule between the titanium atom and an alkyl group. This step requires a titanium-alkyl bond and the possibility of coordinating the olefin. The simplest configuration meeting these requirements is an essentially octahedrally coordinated titanium atom at a surface step site, coordinated to only four chlorine atoms. A fifth octahedral site is occupied by an alkyl group, and the sixth site is available for the olefin. The olefin forms a σ bond to the metal using its filled π orbital. In a concerted step, the alkyl group breaks its bond to the metal and forms a bond to one of the olefin carbon atoms, as the other olefin carbon atom becomes attached to the metal. The process is shown in the following scheme:

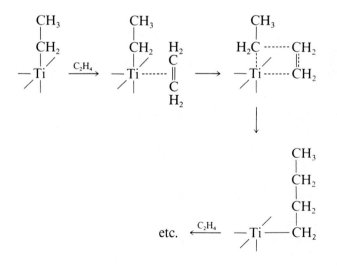

---

[11]   The latter two conditions were met in Germany during World War II and are currently met in South Africa and Australia.

[12]   P. Cossee, *J. Catal.*, **3**, 80 (1964); E. J. Arlman and P. Cossee, *J. Catal.*, **3**, 99 (1964).

When propylene undergoes Ziegler-Natta polymerization, dense, highly stereo-regular polymers are obtained, in contrast to the low-density, branched polymers obtained by free-radical processes. It is assumed that, because of the nonequivalence of the coordination sites of olefin and alkyl group at the active titanium center, the growing alkyl group moves back to its original position after each insertion of a new monomer. Thus the polymerization consists of a sequence of sterically identical steps which lead to stereoregularity in the product.

AMMONIA SYNTHESIS[13]

The synthesis of ammonia from nitrogen and hydrogen is one of the most important processes of chemical industry; tens of millions of tons of ammonia are produced annually throughout the world. Practically all ammonia is made by the Haber process, in which the synthesis is carried out on "promoted" iron catalysts (i.e., iron containing a few percent of oxides such as Al_2O_3, K_2O, and CaO) at pressures around 300 atm and temperatures near 500°C.

In addition to surface reaction steps in which N—H bonds are formed, the following surface reaction steps must be considered in the Haber process[13]:

$$H_2 \;\rightleftharpoons\; 2H_{(ad)}$$

$$N_2 \;\rightleftharpoons\; N_{2(ad)}$$

$$N_2 \;\rightleftharpoons\; 2N_{(ad)}$$

$$NH_{3(ad)} \;\rightleftharpoons\; NH_3$$

The (ad) subscripts refer to metal-adsorbed species. There has been considerable controversy as to whether the initial N—H bond formation involves the reaction of molecular nitrogen or the reaction of atomic nitrogen. If $N_{2(ad)}$ were the reacting species, then the surface would become essentially saturated with $N_{(ad)}$ under steady-state reaction conditions (at least under conditions far from $2NH_3 \rightleftharpoons N_2 + 3H_2$ equilibrium, at low ammonia pressures). Photoelectron spectroscopic studies of the catalyst surface under such conditions show that the $N_{(ad)}$ concentration is far below the saturation value. Hence it is believed that NH_3 is formed on the surface by a sequence starting with nitrogen atoms:

$$N_{(ad)} \;\xrightarrow{\;H_{(ad)}\;}\; NH_{(ad)} \;\xrightarrow{\;H_{(ad)}\;}\; NH_{2(ad)} \;\xrightarrow{\;H_{(ad)}\;}\; NH_{3(ad)}$$

When the hydrogen pressure is not high, the reaction rate is independent of the hydrogen pressure and is first order in nitrogen. That is, the dissociative adsorption of nitrogen is the rate-determining step. At higher hydrogen pressures, the reaction

---

[13] G. Ertl, *Catal. Rev.*, **21**, 201 (1980).

rate is of a fractional order in hydrogen, corresponding to a partial shifting of the rate-determining step to the $N_{(ad)} + H_{(ad)} \rightarrow NH_{(ad)}$ step. Hence the critical steps of the reaction mechanism are

$$N_2 \rightleftharpoons N_{2(ad)} \rightleftharpoons 2N_{(ad)}$$

and

$$N_{(ad)} + H_{(ad)} \rightleftharpoons NH_{(ad)}$$

The overall potential energy diagram for the reaction on an iron surface is schematically shown in Fig. 21.10. Obviously if the reaction proceeded through gas-phase species such as N and H, the activation energy for the reaction would be enormous. The catalyst provides a surface which greatly stabilizes the atomic and radical species and allows the reaction to proceed with no large energy barrier.

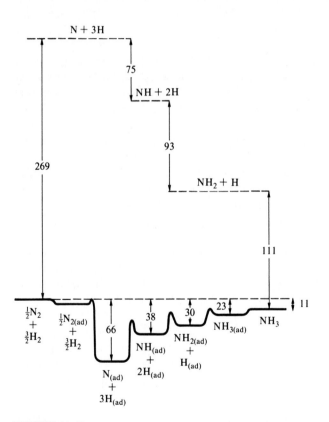

FIGURE 21.10

Potential energy diagram for ammonia synthesis on an iron surface (lower curve) and via gas-phase radicals (upper dashed lines). Energy values in kilocalories per mole. [*Adapted from G. Ertl, Catal. Rev., **21**, 201 (1980).*] 1 kcal = 4.1840 kJ.

THE PHOTOGRAPHIC PROCESS[14]

Practical photographic processes were first devised in the 1830s by Daguerre in France and by Talbot in England.[15] In Daguerre's method, a silver iodide–coated silver plate was exposed to light in a camera, whereby the light-exposed silver iodide was decomposed to metallic silver and iodine. A clear image was produced by treating the plate with mercury vapor (which amalgamated the silver) and by rinsing in a strong salt solution to remove the remaining silver iodide. Talbot's procedure consisted of washing paper in successive baths of salt water and silver nitrate solution, thus depositing silver chloride in the fibers of the paper. The still-wet paper was then exposed in a camera until a silver image appeared, and the remaining silver chloride was removed by washing with a concentrated salt solution or a sodium thiosulfate solution. By waxing or oiling the negative sheet, Talbot made the paper transparent, and by then making a second exposure through the negative onto another sensitized sheet, he produced a positive image.

In 1840 Talbot modified his process by adding a step that may have been the first practical application of heterogeneous catalysis. He found that a very short camera exposure (about 1/60 of that required to yield a visible image) left an invisible "latent" image on the sensitized paper. The latent image was then "developed" into a visible image by treatment with a solution of gallic acid and silver nitrate. This modification, together with the negative/positive feature, made Talbot's process so superior that it has survived, in its general form, to the present day. The main difference between Talbot's process and modern photographic practice is that now the silver halide (usually silver bromide or a mixture of silver halides), in the form of approximately micron-sized crystals, or "grains," is suspended in a gelatin matrix. In this section we shall explain how exposure of a silver bromide grain to a few photons can cause the formation on the grain of a tiny silver speck, and how that latent image speck subsequently catalyzes the reduction of the entire grain to metallic silver by a suitable reducing agent (developer).

Silver bromide in the dark has only a very low conductivity at normal temperatures, and a negligible conductivity at temperatures below $-200°C$. This low dark conductivity is due to the migration of interstitial silver ions. When a crystal of silver bromide is illuminated, there is an immediate increase in conductivity, which occurs even at low temperatures. This photoconductivity is due to the ejection of valence electrons from bromide ions in the AgBr lattice:

$$Br^-(lattice) + h\nu \longrightarrow Br(lattice) + e^-(conduction\ band)$$

Some of the conduction electrons back-react with bromine atoms to re-form bromide ions. However, a surprisingly large fraction (as much as one-half or more) of the conduction electrons are at least temporarily trapped at defect sites in

---

[14]   T. H. James, ed., "The Theory of the Photographic Process," 4th ed., Macmillan, New York, 1977; B. H. Carroll, G. C. Higgins, and T. H. James, "Introduction to Photographic Theory," Wiley, New York, 1980.

[15]   J. L. Ennis, *Chem. & Eng. News*, **67**(51), 26 (1989); B. Newhall, "The History of Photography from 1839 to the Present Day," rev. ed., Museum of Modern Art, New York, 1964.

the crystal at an energy level slightly lower than the bottom of the conduction band. Typical defect sites are surface kinks (similar to those in Figure 21.2), crystal dislocations, and interstitial silver ions (Frenkel defects). If the defect has a residual positive charge, it can trap electrons. Specks of silver sulfide on the silver bromide grain surfaces, formed during digestion of the AgBr-gelatin suspension, are also believed to be efficient electron traps.

It is believed that latent image formation involves both the migration of interstitial silver ions and the production and trapping of photoelectrons. According to the Gurney-Mott[16] mechanism, an electron is captured at a defect site, where it is temporarily localized for a millisecond or so. A silver ion can migrate to this negative site and combine with the trapped electron to form an atom. The atom is not stable; i.e., it can decompose into a silver ion and a free electron. However, the atom is even a better electron trap than the original defect site, and during its lifetime it can trap a second electron, if one becomes available. If this second electron remains trapped until the arrival of a second interstitial silver ion, a two-atom cluster forms. This buildup of a silver cluster can continue as long as photoelectrons are available. The minimum-sized cluster corresponding to a stable latent image speck is believed to consist of three or four silver atoms. Specks of silver of this size or greater, on the silver halide crystal surface, can catalyze the subsequent action of a developer. The Gurney-Mott mechanism is represented by the following equations, in which e^- stands for a conduction electron and Ag^+ stands for an interstitial silver ion.

$$\text{trap} + e^- \rightleftharpoons e^-/\text{trap}$$

$$e^-/\text{trap} + Ag^+ \rightleftharpoons Ag/\text{trap}$$

$$Ag/\text{trap} + e^- \rightleftharpoons Ag^-/\text{trap}$$

$$Ag^-/\text{trap} + Ag^+ \rightleftharpoons Ag_2/\text{trap}$$

$$\text{etc.}$$

The bromine atom formed in the initial photoionization has a much lower mobility in the crystal than the photoelectron—a fact which is partially responsible for the high quantum efficiency of the silver cluster formation. However, elemental bromine attacks silver metal; therefore, the quantum yield would drop off because of back-reaction if it were not for the presence of halogen scavengers, both naturally occurring and added, in the gelatin. Acetone semicarbazone, $(CH_3)_2C{=}N{-}NHCONH_2$, is often added as a scavenger to photographic emulsions.

The latent image is submicroscopic; the ratio of silver ions to Ag_n latent-image silver atoms in a developable AgBr grain can be as great as 10^8. Development with a reducing agent is required to completely convert the developable grains to metallic silver. The following procedure is believed to occur in a typi-

---

16   R. W. Gurney and N. F. Mott, *Proc. R. Soc. London Ser. A,* **164,** 151 (1938).

cal development process. The developing agent is adsorbed on the surface of the silver cluster, to which it releases an electron. The negative charge on the cluster attracts a silver ion, which attaches itself to the cluster, thus increasing the number of silver atoms in the cluster by one. (The silver ion can either be a mobile interstitial silver ion or it can be produced by the dissociation of a dissolved complex silver ion formed by reaction of some of the AgBr with complexing agents such as sulfite or bromide in the developer solution.) A bromide ion from the crystal goes into solution, thus maintaining the crystal at electrical neutrality, and the oxidized developing agent is desorbed and replaced by unoxidized developing agent. The overall process can continue until the entire crystal has been converted to silver.

Developer solutions commonly contain a mixture of two developing agents: metol (N-methyl paraaminophenol) and hydroquinone, with the structures shown below.

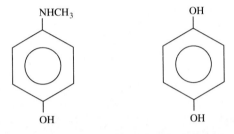

The *combination* of these developing agents has a much greater reactivity than the weighted average of the individual reactivities of the developing agents. This "superadditivity" of the developer combination is believed to be due to a development mechanism in which the anionic deprotonated metol is more or less permanently adsorbed on the growing silver cluster and serves as a conduit for electrons from hydroquinone molecules to the silver cluster. The development is relatively fast because of the elimination of the steps corresponding to the adsorption of the developing agent and the desorption of its oxidation product. This mechanism is represented schematically in the following diagram.

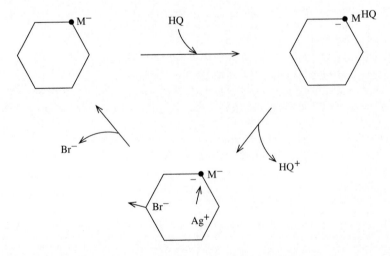

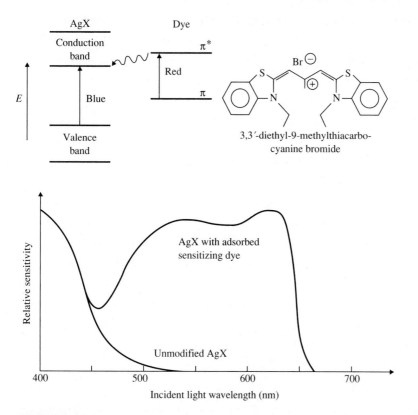

FIGURE 21.11

Spectral sensitization by an adsorbed dye in a black-and-white photographic emulsion. *(Reproduced with permission from W. W. Porterfield, "Inorganic Chemistry: A Unified Approach," Addison-Wesley, Reading, Mass., 1984.)*

The energy gap between the valence and conduction bands in silver bromide is about 2.6 eV, corresponding to a maximum wavelength of about 480 nm for the photons that generate latent-image specks. Thus unmodified AgBr is sensitive only to blue light, because light of longer wavelength lacks the energy to excite electrons to the conduction band. Panchromatic films are made by adding sensitizing dyes that adsorb on the AgBr grains. A relatively low-energy $\pi \rightarrow \pi^*$ transition in the dye molecule populates the conduction band. The energy relationships are shown in Fig. 21.11. It should be noted that the π^* level of the dye must lie higher than the bottom of the conduction band for this sensitization process to be effective. The structure of a typical sensitizing dye and the sensitization spectrum are also shown in the figure.[17]

---

[17] W. W. Porterfield, "Inorganic Chemistry: A Unified Approach," pp. 659–663, Addison-Wesley, Reading, Mass., 1984.

ELECTRODE SURFACE MODIFICATION

The kinetic properties of electrodes are directly related to the chemical and structural features of their surfaces. Thus electrode properties can be controlled by suitable chemical modification of their surfaces.

In Chap. 13 we discussed how light can be used to drive redox reactions at the junction between a semiconductor and a solution. However, semiconducting photoanode materials are susceptible to oxidative decomposition when photogenerated holes rise to the top of the valence band. That is, the semiconductor itself may be oxidized. Thus ordinary *n*-type silicon photoanodes develop an insulating coat of oxide on their surface when exposed to light and are useless for any practical photooxidation. However, it has been found that the silicon surface can be stabilized by covering the surface with ferrocene groups.[18] This is accomplished by the reaction of a compound such as trichlorosilylferrocene with the surface hydroxyl groups that are present on the normal air-oxidized surface of silicon:

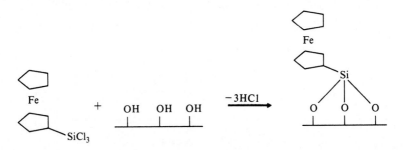

This ferrocene derivatization causes a dramatic improvement in the photoanode durability. The derivatized silicon can photooxidize anything oxidizable by the ferricenium ion. For example, the photooxidation of ferrocyanide can proceed continuously by the following cycle:

$$\text{Surface ferrocene} \underset{Fe(CN)_6{}^{4-}}{\overset{light}{\rightleftharpoons}} \text{Surface ferricenium}^+$$

It is hoped that further study of derivatized photoelectrodes may lead to an improved photoelectrolytic system for deriving energy from sunlight.

The development of an electrode at which the reduction of molecular oxygen to water proceeds at the reversible potential (1.23 V versus the standard hydrogen electrode) would constitute a major advance in fuel cell technology. Present acid electrolyte oxygen cathodes made of platinum are expensive and produce useful current densities only at overpotentials of several hundred millivolts. Some encouraging results toward the development of an efficient oxygen cathode have been obtained by adsorbing on a graphite surface dimeric cobalt porphyrin molecules in which the two porphyrin rings are constrained to lie parallel to one another.[19] The

[18]   A. B. Bocarsly, E. G. Walton, and M. S. Wrighton, *J. Am. Chem. Soc.,* **102,** 3390 (1980); M. S. Wrighton, *Chem. Eng. News,* 29–47 (Sept. 3, 1979).

[19]   J. P. Collman, P. Denisevich, Y. Konai, M. Marrocco, C. Koval, and F. C. Anson, *J. Am. Chem. Soc.,* **102,** 6027 (1980).

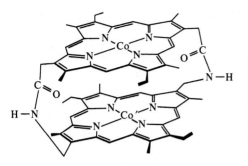

FIGURE 21.12
Structure of the dimeric cobalt-porphyrin complex which, when adsorbed on graphite, acts as an effective catalyst for the reduction of O_2 to H_2O. [*Adapted from J. P. Collman, P. Denisevich, Y. Konai, M. Marrocco, C. Koval, and F. C. Anson, J. Am. Chem. Soc., 102, 6027 (1980).*]

complex shown in Fig. 21.12 has been found to catalyze the reduction of oxygen to water very effectively. There is little doubt that this complex functions by binding the O_2 molecule jointly to the two cobalt atoms during the four-electron reduction process. A plausible mechanism is shown in Fig. 21.13.

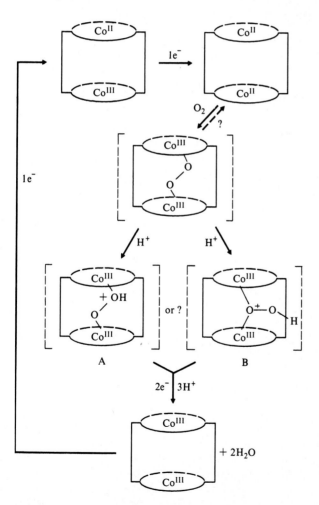

FIGURE 21.13
Proposed mechanism of the electrode surface-catalyzed reduction of O_2 to H_2O. [*Reproduced with permission from J. P. Collman et al., J. Am. Chem. Soc., 102, 6027 (1980). Copyright 1980 American Chemical Society.*]

CATALYSIS BY METAL CLUSTERS

Metal cluster compounds are obviously intermediate in character between mononuclear molecular complexes and metals. A question of considerable interest is: Can cluster compounds be used as valid models of metal surfaces with chemisorbed ligands?[20] It has been shown that some ligands can interact with clusters in ways that are unknown with monomeric complexes. For example, acetonitrile reacts with either $Fe_2(CO)_8^{2-}$ or $HFe_3(CO)_{11}^-$ to form a mixture of $Fe_3(CH_3C=NH)(CO)_9^-$ and $Fe_3(N=CHCH_3)(CO)_9^-$, which undergo the interconversion reactions shown in the following scheme:[21]

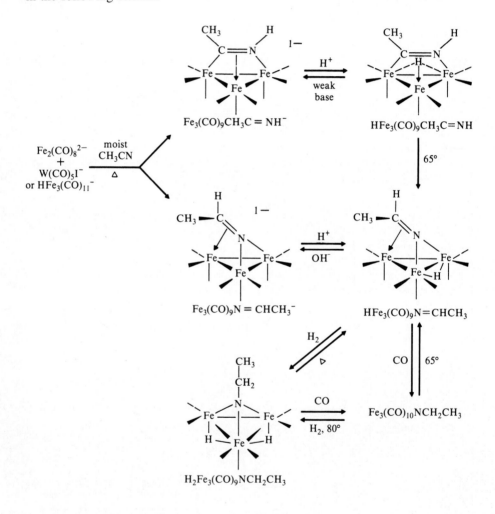

[20] E. L. Muetterties, T. N. Rhodin, E. Band, C. F. Brucker, and W. R. Pretzer, *Chem. Rev.,* **79,** 91 (1979); J. F. Hamilton and R. C. Baetzold, *Science,* **205,** 1213 (1979).

[21] M. A. Andrews and H. D. Kaesz, *J. Am. Chem. Soc.,* **101,** 7238 (1979).

Of particular interest is the fact that $HFe_3(CO)_9N\!\!=\!\!CHCH_3$ readily reacts with either H_2 or CO at room temperature to form complexes containing the ethylimido group, CH_3CH_2N. Unfortunately attempts to displace ethylamine from these complexes leads to decomposition of the cluster to iron metal. However, the reactions do show that several metal atoms acting in concert can lead to the reduction of nitriles. Thus the system probably serves as a rough model for what happens at the surface of the iron catalysts that actually catalyze the hydrogenation of nitriles.

Treatment of the square pyramidal complex $Fe_5C(CO)_{14}{}^{2-}$ with hydrogen chloride leads to the formation of $HFe_4(CH)(CO)_{12}$, a butterfly Fe_4 cluster with an η^2-CH group spanning the terminal Fe atoms.[22]

$$Fe_5C(CO)_{14}{}^{2-} + 4HCl \quad \rightarrow \quad HFe_4(CH)(CO)_{12} + 2CO + FeCl_2 + 2Cl^- + H_2$$

The latter complex can be reversibly deprotonated to give $Fe_4C(CO)_{12}{}^{2-}$. Oxidation of $Fe_4C(CO)_{12}{}^{2-}$ with $AgBF_4$ in the presence of H_2 re-forms $HFe_4(CH)(CO)_{12}$ by a process that may involve oxidative addition of hydrogen to an $Fe_4C(CO)_{12}$ intermediate. This conversion of a carbide atom in a metal cluster to a CH group is a formal analog of a proposed step in Fischer-Tropsch reactions. The reactions are summarized in the following scheme.

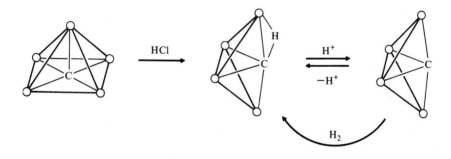

In another study of the Fe_nC cluster system,[23] it has been shown that the octahedral complex $Fe_6C(CO)_{16}{}^{2-}$ is oxidized by tropylium bromide in methanol to give the complex $Fe_4(CO)_{12}CCO_2CH_3$, in which the $C-\overset{\overset{\displaystyle O}{\|}}{C}-OCH_3$ group bridges

[22] M. Tachikawa and E. L. Muetterties, *J. Am. Chem. Soc.*, **102**, 4541 (1980); M. A. Beno, J. M. Williams, M. Tachikawa, and E. L. Muetterties, *J. Am. Chem. Soc.*, **102**, 4542 (1980).

[23] J. S. Bradley, G. B. Ansell, and E. W. Hill, *J. Am. Chem. Soc.*, **101**, 7417 (1979); also see *Chem. Eng. News,* 27 (April 28, 1980); J. S. Bradley, G. B. Ansell, M. E. Leonowicz, and E. W. Hill, *J. Am. Chem. Soc.*, **103**, 4968 (1981).

the terminal iron atoms of an Fe_4 butterfly cluster. This complex, in turn, can be hydrogenated under mild conditions to give methyl acetate and the original $Fe_6C(CO)_{16}^{2-}$ complex. A proposed mechanism for the process is shown in the following scheme.

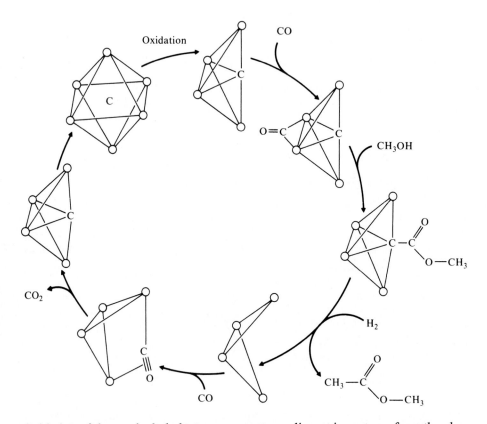

Oxidation of the octahedral cluster removes two adjacent iron atoms from the cluster, exposing the central carbon atom. Carbon monoxide, formed in the oxidation reaction, presumably reacts with the exposed carbon atom, and then reaction with the methanol solvent gives the methyl ester fragment. Regenerating the cluster in its active form requires addition of an oxidizing agent, so the cycle cannot be considered catalytic. However, the process essentially involves the synthesis of an organic molecule from CO and H_2 and thus is analogous to the Fischer-Tropsch process.

We have already pointed out that the steps

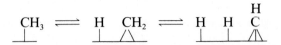

are probably important in metal-catalyzed processes such as the Fischer-Tropsch synthesis. Steps if this type have actually been observed[24] in a reversible equilibrium between $Os_3(\mu\text{-H})(\mu\text{-CH}_3)(CO)_{10}$ and $Os_3(\mu\text{-H})_2(\mu\text{-CH}_2)(CO)_{10}$ and the conversion of these compounds to $Os_3(\mu\text{-H})_3(\mu_3\text{-CH})(CO)_9$:

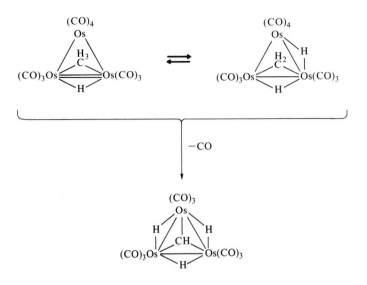

Thus we have a model of the process by which C—H bonds break at metal surfaces.

The study of cluster compounds suggests that the main advantage of clusters as catalysts will be their ability to provide novel molecular transformations because of the unique coordination environments provided by the metal atoms. In any case, it is clear that clusters do provide analogs for structures and processes that have been postulated in surface catalysis.

PROBLEMS

21.1 Carbon dioxide adsorbed on an Rh (111) surface gives essentially the same electron energy loss spectrum as carbon monoxide adsorbed on the same surface. Both systems show absorption in the C—O stretching region. What can you say with respect to the nature of the adsorption of CO_2?

*21.2 Copper metal has an interatomic distance of 2.55 Å. Calculate the surface concentration of atoms (in atoms per square centimeter) on the (111), (100), and (110) crystal faces.

21.3 List several reasons why discrete metal cluster complexes are (a) good analogs of catalytic metal surfaces and (b) poor analogs of such surfaces.

21.4 Why would you expect platinum to be a poor catalyst for the Haber process?

---

24   R. B. Calvert and J. R. Shapley, *J. Am. Chem. Soc.*, **99**, 5225 (1977).

CHAPTER

22

SOME
BIOLOGICAL
SYSTEMS

Metallobiomolecules are natural products which contain one or more metallic elements. A classification scheme for some of the more extensively studied metallobiomolecules is given in Fig. 22.1.[1] These molecules are complex coordination compounds whose metal-containing sites (the "active sites") are usually involved in electron transfer, the binding of exogenous molecules, and catalysis. The comparative study of these sites and analogously structured synthetic coordination compounds is the main part of the interdisciplinary field of bioinorganic chemistry. In this chapter we shall describe just a few of the many bioinorganic systems that have been studied in recent years.

METALLOPORPHYRINS AND RELATED SYSTEMS

The porphyrin ligands have the general structure shown in Fig. 22.2; they are derivatives of porphine (the compound in which all the R groups are hydrogen atoms). The most commonly used synthetic porphyrins are octaethylporphyrin ($R_p = C_2H_5$, $R_m = H$) and tetraphenylporphyrin ($R_m = C_6H_5$, $R_p = H$). Many metalloporphyrins are known, in which the two protons bonded to the pyrrole nitrogen atoms have been removed and in which a metal ion is coordinated to the four nitrogen atoms. These compounds are important because many natural products (e.g., chlorophylls, hemes, cytochromes, and vitamin B_{12}) contain similar groups.

---

[1] J. A. Ibers and R. H. Holm, *Science*, **209**, 223 (1980).

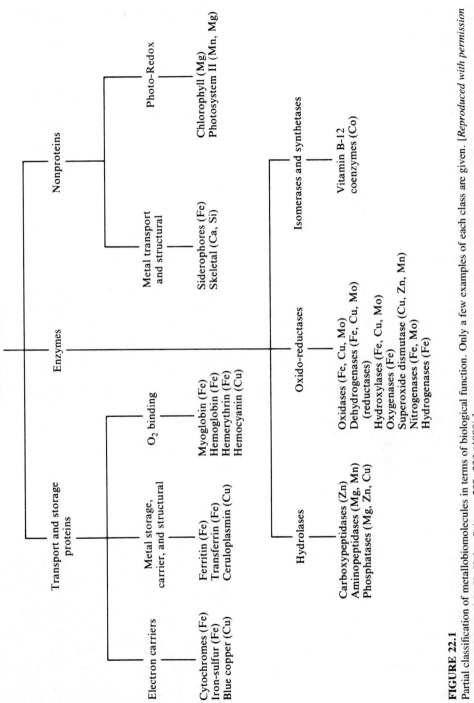

FIGURE 22.1
Partial classification of metallobiomolecules in terms of biological function. Only a few examples of each class are given. [*Reproduced with permission from J. A. Ibers and R. H. Holm, Science, **209**, 223 (1980).*]

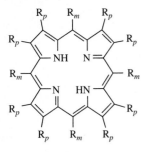

FIGURE 22.2
The porphyrin skeleton. Note that there are two main types of substituents—those on the pyrrole rings (R_p) and those in the intermediate, meso, positions (R_m). The parent nucleus, porphine, corresponds to $R_p = R_m = H$.

Hemoglobin and Myoglobin[2]

The proteins hemoglobin and myoglobin serve to transport and to store oxygen, respectively. These are life functions essential to all vertebrates. The active site in both proteins is the planar heme group (Fig. 22.3) embedded in a convoluted protein chain (globin), with a coordinate bond between the iron atom and the nitrogen atom of the imidazole side chain of a histidine residue (the "proximal" histidine). Myoglobin, shown in Fig. 22.4, consists of one globin and one heme group. Hemoglobin consists of four myoglobinlike subunits, with two of the globins somewhat different from the other two. In each subunit, an oxygen molecule can bond to the iron atom on the side of the porphyrin opposite the proximal histidine, thus forming a hexacoordinate iron complex.

Hemoglobin carries oxygen via the circulatory system from the lungs to tissues. There the oxygen can be transferred to myoglobin, where it is stored until it is required in a metabolic process. Probably the most intriguing property of hemoglobin is the cooperative (or "autocatalytic") nature of its O_2 binding. The effect of this cooperativity can be seen in the oxygen-binding curves of Fig. 22.5.

---

[2]   M. F. Perutz, *Sci. Am.*, **239**, 92 (December 1978); N. M. Senozan and R. L. Hunt, *J. Chem. Educ.*, **59**, 173 (1982).

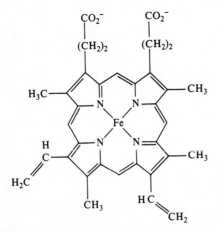

FIGURE 22.3
The heme group.

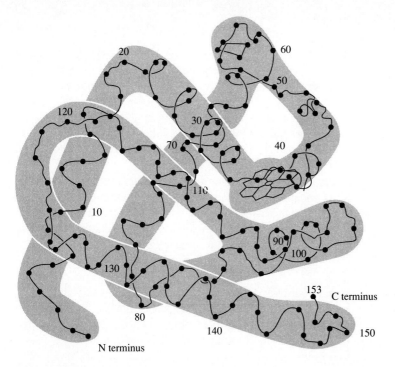

FIGURE 22.4
A schematic representation of the myoglobin molecule. Note the heme unit attached to the chain near residue 40.

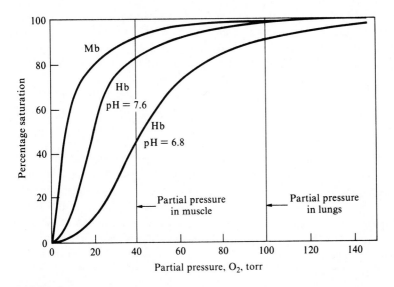

FIGURE 22.5
The oxygen-binding curves for myoglobin (Mb) and hemoglobin (Hb), showing the pH dependence for the latter. *(Reproduced with permission from F. A. Cotton and G. Wilkinson, "Advanced Inorganic Chemistry," 4th ed., Wiley-Interscience, New York, 1980.)*

At high oxygen concentrations, hemoglobin and myoglobin bind O_2 approximately equally strongly. However, at low oxygen concentrations, such as in muscle tissue during or immediately after muscular activity, hemoglobin is a poorer oxygen binder and therefore can pass its oxygen on to myoglobin. The difference in O_2-binding ability between myoglobin and hemoglobin is accentuated at low pH values; hence the transfer from hemoglobin to myoglobin has a greater driving force where it is most needed, viz., in tissues where oxygen has been converted to CO_2.

The oxygen-binding curve of hemoglobin indicates that the four subunits of hemoglobin do not act independently: they somehow communicate with each other. When one heme group reacts with an O_2 molecule, the other three show an increased ability to react with O_2 molecules. Thus hemoglobin behaves somewhat as if it could combine either with four oxygen molecules or none at all:

$$Hb + 4O_2 \rightleftharpoons Hb(O_2)_4$$

It is generally believed that coordination of an oxygen molecule to the 5-coordinate iron(II) in a deoxy heme causes conformational changes within the protein which affect the ability of the other heme groups to combine with oxygen.[3] This is the basis of the "allosteric" theory, which explains heme-heme interaction without postulating any direct communication between the heme groups. The hemoglobin molecule is believed to have two alternative structures, designated T for "tense" and R for "relaxed," which favor the deoxygenated and oxygenated states, respectively. These structures are very similar but differ significantly in the bonding both between subunits and within subunits. These structural changes are closely integrated and occur in a concerted action.

In both structures, the four subunits are held together in essentially a tetrahedral arrangement by hydrogen bonds and "salt bridges." (A salt bridge is an electrostatic bond between a quaternary nitrogen atom, with a positive formal charge, and an oxygen atom with a negative formal charge.) In the transition from the T structure to the R structure, one pair of subunits rotates by $15°$ relative to the other pair of subunits, and various groups within each subunit shift position so as to favor the coordination of molecular oxygen. The rearrangement of the subunits involves changes in hydrogen bonding and salt bridging. There are more salt bridges in the T structure than in the R structure. Thus the salt bridges explain the influence of protons (and CO_2) on the oxygen-binding curve. The addition of protons favors the formation of the T state and lowers the oxygen affinity of hemoglobin by increasing the number of quaternary, positively charged, nitrogen atoms which can form salt bridges. Conversely, the transition from the R structure to the T structure brings negatively charged oxygen atoms into proximity with nitrogen atoms and thus favors the protonation of these nitrogen atoms. A schematic representation of the transition from the T structure to the R structure

---

[3] M. F. Perutz, op. cit.; D. K. White, J. B. Cannon, and T. G. Traylor, *J. Am. Chem. Soc.*, **101**, 2443 (1979); F. Basolo, B. M. Hoffman, and J. A. Ibers, *Acc. Chem. Res.*, **8**, 384 (1975).

T structure

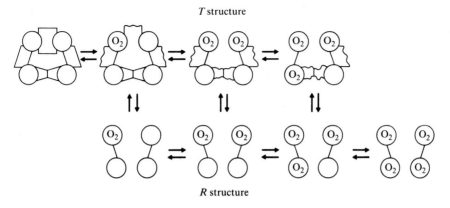

R structure

FIGURE 22.6
A schematic representation of the equilibrium between the *T* and *R* structures of hemoglobin. The straight lines joining subunits represent salt bridges; the wavy lines represent weakened salt bridges. [*Adapted from M. F. Perutz, Sci. Am.,* **239**, *92 (December 1978).*]

is shown in Fig. 22.6. The salt bridges linking the subunits in the *T* structure break progressively as oxygen is coordinated, and even salt bridges that have not yet broken are weakened, a situation represented in Fig. 22.6 by wavy lines. The *T* ⟶ *R* transition does not take place after a fixed number of oxygen molecules have been bound, but it becomes more probable with each additional oxygen molecule bound.

Perutz has postulated that a change in configuration of the heme group upon coordination of an oxygen molecule triggers interconversion of the *T* and *R* structures. In the deoxy form, the high-spin iron(II) atom is 5-coordinate and is situated between the plane of the porphyrin and the coordinated nitrogen atom. Upon oxygenation the iron atom, which is variously thought to be Fe(II) or Fe(III), becomes low-spin (and of smaller size), and it is believed to move into the plane of the porphyrin. This stereochemical change is illustrated in Fig. 22.7. Perutz proposes that the iron atom is held in a tense 5-coordinate state by the protein and that this tension is released upon oxygenation. The change in coordination triggers relaxation throughout the subunit, and the movement is transmitted to

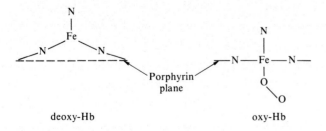

FIGURE 22.7
Illustration of the movement of the proximal histamine upon coordination of oxygen to the iron atom of the heme group in hemoglobin.

salt bridges. Recent studies indicate that it may be more appropriate to look upon the tension of the T state as a set of small strains in many bonds.[4] Indeed, other workers[5] propose that the T structure should be characterized by restraint rather than tension. The restraint may consist of steric repulsion which prevents the proximal histidine from approaching the iron atom with an orientation favorable for bonding. Transition to the R structure presumably involves rearrangements which allow the iron atom to move toward the porphyrin and to bind oxygen. Nevertheless, in spite of uncertainties about the mechanism, it is certain that the change in coordination geometry at the iron atom plays an important part in the cooperativity phenomenon of hemoglobin.

It is interesting to note that the reactions of hemoglobin with protons when it liberates oxygen explain how hemoglobin aids the transport of CO_2 to the lungs. About two protons are taken up for every four molecules of oxygen released, and two protons are released again when four molecules of oxygen are absorbed. The CO_2 released in tissues is too insoluble to be transported entirely as such, but it can be solubilized by conversion to bicarbonate. Thus the reaction occurring during the release of oxygen by hemoglobin can be roughly represented as follows.

$$Hb(O_2)_4 + 2CO_2 + 2H_2O \ \rightleftharpoons \ H_2Hb^{2+} + 2HCO_3^- + 4O_2$$

Considerable effort has been expended in attempts to synthesize model compounds which mimic the oxygen-carrying properties of the heme group in natural proteins. The main problem in such work has been the fact that oxygen reacts irreversibly with simple iron-porphyrin systems to form binuclear iron(III) oxides of the following type.

One strategy has been to inhibit this irreversible oxidation by preparing porphyrins which have a protective enclosure for binding oxygen on one side of the porphyrin ring, with the other side protected by a bulky axially coordinated nitrogen base. Two examples of this type which have been intensively studied are the "picket-fence"[6] and "capped"[7] porphyrins, shown in Figs. 22.8 and 22.9, respectively.

---

[4] J. M. Hopfield, *J. Mol. Biol.*, **77**, 207 (1973).

[5] J. P. Collman, *Acc. Chem. Res.*, **10**, 265 (1977); M. F. Perutz, op. cit.

[6] J. P. Collman, *Acc. Chem. Res.*, **10**, 265 (1977); J. P. Collman et al., *J. Am. Chem. Soc.*, **102**, 4182 (1980).

[7] J. R. Budge, P. E. Ellis, R. D. Jones, J. E. Linard, F. Basolo, J. E. Baldwin, and R. L. Dyer, *J. Am. Chem. Soc.*, **101**, 4760 (1979); R. D. Jones, D. A. Summerville, and F. Basolo, *Chem. Rev.*, **79**, 139 (1979).

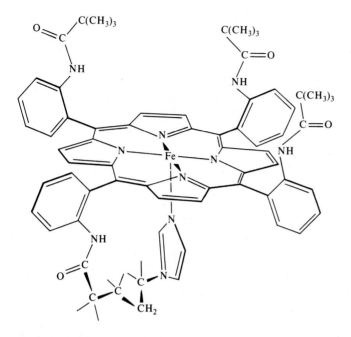

FIGURE 22.8
A picket-fence porphyrin with an appended imidazole base. [*Reproduced with permission from J. P. Collman, Acc. Chem. Res., **10**, 265 (1977). Copyright 1977 American Chemical Society.*]

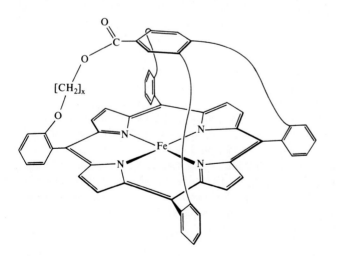

FIGURE 22.9
Schematic representation of the capped porphyrin ($x = 2$ or 3). [*Reproduced with permission from Budge et al., J. Am. Chem. Soc. **101**, 4762 (1979). Copyright 1979 American Chemical Society.*]

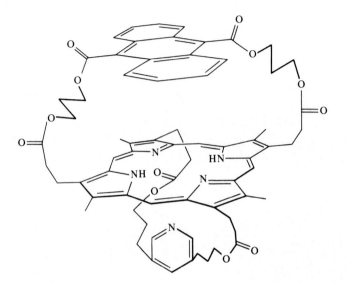

FIGURE 22.10
A porphyrin with an anthracene bridge across one face and a pyridine bridge across the other. The corresponding Fe(II) metalloporphyrin has the hemoglobinlike properties described in the text. [*Reproduced with permission from A. R. Battersby and A. D. Hamilton, J. Chem. Soc. Chem. Commun., 117 (1980).*]

Another example, the "doubly bridged"[8] porphyrin, is shown in Fig. 22.10. The latter compound has not received the attention it probably deserves. It is a fairly good analog of a natural heme group because it possesses the following features: (1) a hydrophobic pocket for the O_2 molecule, (2) no substituents at the four meso positions of the porphyrin, (3) a basic fifth ligand covalently bound to the porphyrin, (4) the ability to bind oxygen readily and reversibly, and (5) the ability to bind carbon monoxide reversibly. Incidentally, the poisonous character of carbon monoxide is due to its ability to displace oxygen from its site in the heme group. In the blood of heavy smokers, as much as 20 percent of the oxygen sites can be blocked by carbon monoxide.

The irreversible oxidation of simple iron porphyrin systems, although fast, is slower than the oxygenation step. Thus a second strategy for the study of model heme compounds is the use of fast reaction and fast spectroscopy methods.[9] In a typical flash photolysis study, a mixture of O_2 and CO is equilibrated with a solution of a heme model compound. Under these conditions, the compound is almost entirely in the form of the heme–CO complex. A short laser pulse causes complete dissociation of this complex, and the deoxy heme then reacts with O_2 at a high, but measurable, rate. Subsequently, the heme–O_2 complex dissociates

---

8 A. R. Battersby and A. D. Hamilton, *J. Chem. Soc. Chem. Commun.*, 117 (1980).

9 T. G. Traylor, *Acc. Chem. Res.*, **14**, 102 (1981).

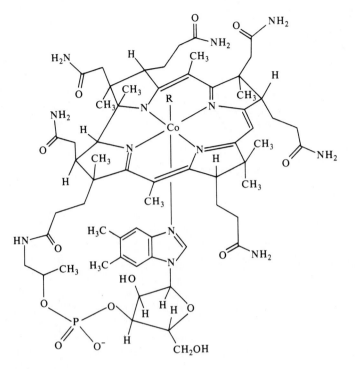

FIGURE 22.11
General structure of vitamin B_{12} and its derivatives.

and is reconverted to the heme–CO complex. These reactions can be followed by fast ultraviolet-visible spectroscopy.

Vitamin B_{12}[10]

This compound and its derivatives are cobalt complexes with the general structure shown in Fig. 22.11. In the isolated vitamin, R is a cyanide ion. However, in vivo, the cyanide ion is not present, and R is probably a loosely bound water molecule. In both the cyano and aquo forms of the vitamin, the cobalt is present in the +3 oxidation state. The cobalt(III) can be reduced to cobalt(II) (vitamin B_{12r}) by catalytic hydrogenation, by treatment with chromium(II) acetate at pH 5, or by controlled potential electrolysis. Further reduction to cobalt(I) (vitamin B_{12s}) is effected by treatment with chromium(II) acetate at pH 9.5, with zinc dust in aqueous ammonium chloride, or with sodium borohydride. In vitamin B_{12s}, the R site is vacant, and the 5-coordinate cobalt(I) atom is extremely reactive. Some of the reactions of vitamin B_{12s} are summarized in the following scheme:

---

[10] B. M. Babior, *Acc. Chem. Res.*, **8**, 376 (1975); R. H. Abeles and D. Dolphin, *Acc. Chem. Res.*, **9**, 114 (1976).

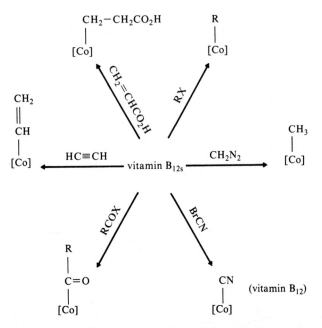

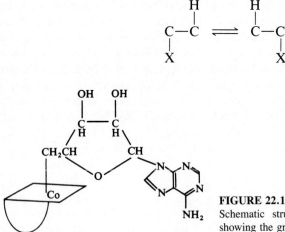

Vitamin B_{12s} may be intimately involved in biological functions, where it can act as a methyl-transfer agent (e.g., in the biosynthesis of methionine) or as a reducing agent (e.g., in the reduction of ribose).

The reaction of adenosine triphosphate with vitamin B_{12s} forms a cobalt-carbon bond, as shown in Fig. 22.12. The resulting molecule is known as vitamin B_{12} coenzyme and was the first organometallic compound discovered in living systems. This coenzyme is required in various biological processes, but the detailed mechanisms of these processes are unknown. Many of these processes are 1,2 rearrangements in which a group X and an H atom exchange positions:

$$
\begin{array}{ccc}
& \text{H} & \qquad \text{H} \\
& | & \qquad | \\
\text{C} & -\text{C} & \rightleftharpoons \quad \text{C} - \text{C} \\
| & & \qquad\qquad | \\
\text{X} & & \qquad\qquad \text{X}
\end{array}
$$

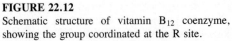

FIGURE 22.12
Schematic structure of vitamin B_{12} coenzyme, showing the group coordinated at the R site.

The nature of the group X is highly variable; it may be a complex alkyl [e.g., $CH(NH_2)COOH$] or acyl (COR) group, or a relatively simple nucleophilic group (e.g., OH or NH_2). For example, in conjunction with glutamate mutase, the following reaction is catalyzed:

$$
\begin{array}{ccc}
CO_2^- & & CO_2^- \\
| & & | \\
CHNH_3^+ & & CHNH_3^+ \\
| & \rightleftharpoons & | \\
CH_2 & & CH-CH_3 \\
| & & | \\
CH_2 & & CO_2^- \\
| & & \\
CO_2^- & &
\end{array}
$$

One possible mechanism of the hydrogen transfer in such reactions is shown in the following scheme.[11]

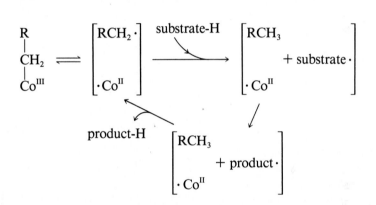

Here RCH_2 represents the adenosyl group of the coenzyme. An enzyme-catalyzed cleavage of the carbon-cobalt bond gives cobalt(II) and the adenosyl radical. The radical then abstracts the hydrogen atom from the enzyme-bound substrate molecule to form adenosine and the substrate radical. After migration of the group X on the substrate radical, the rearranged radical retrieves a hydrogen atom from the adenosine to give the product molecule and thus completes the catalytic cycle. Little is known about the mechanism of the migration of group X; current research focuses on elucidating this step.

Certain relatively simple cobalt complexes, such as bis(dimethylglyoximato)-cobalt complexes,

---

[11]   Babior, op. cit.; for another mechanism, see R. Hamilton, T. R. B. Mitchell, E. A. McIlgorm, J. J. Rooney, and M. A. McKervey, *J. Chem. Soc. Commun.*, 686 (1981).

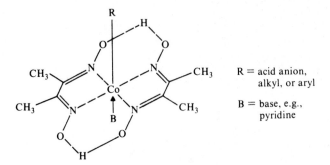

R = acid anion,
alkyl, or aryl

B = base, e.g.,
pyridine

mimic many of the reactions of vitamin B_{12}.[12] The cobalt atom exhibits four oxidation states, Co(I), Co(II), Co(III), and Co(IV). The monovalent species, in analogy to vitamin B_{12s}, is a strong nucleophile that undergoes Michael additions and nucleophilic displacements to give complexes with alkyl groups directly bonded to cobalt(III). It is hoped that the study of such model complexes will, by the use of analogy, clarify the reactions of vitamin B_{12}.

IRON-CONTAINING PROTEINS

Iron-Sulfur Proteins[13]

The iron-sulfur proteins have relatively low molecular weights and contain iron atoms tetrahedrally coordinated to sulfur atoms. Representative structures of the iron-sulfur centers in these proteins are pictured in Fig. 22.13. The proteins containing single iron atoms coordinated tetrahedrally by four mercapto sulfur atoms are called "rubredoxins." The proteins containing two-iron and four-iron clusters

[12] G. N. Schrauzer, *Angew. Chem. Int. Ed.*, **15**, 417 (1976); **16**, 223 (1977); R. M. Magnuson, J. Halpern, I. Y. Levitin, and M. G. Vol'pin, *Chem. Commun.*, 44 (1978); J. Topich and J. Halpern, *Inorg. Chem.*, **18**, 1339 (1979).

[13] W. Lovenberg, ed., "Iron-Sulfur Proteins," vols. 1–3, Academic, New York, 1973, 1977.

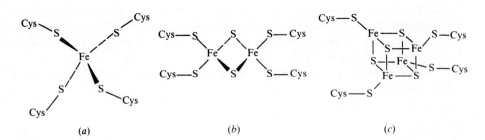

(a) (b) (c)

FIGURE 22.13
The mononuclear unit (a) found in rubredoxin, and the binuclear complex (b) and tetranuclear cluster (c) of ferredoxins. The S-Cys groups are cysteine amino acid residues of the protein chain.

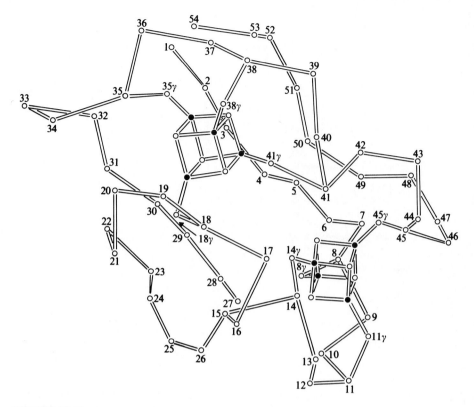

FIGURE 22.14

The skeleton of *M. aerogenes* ferredoxin. *(Reproduced with permission from K. T. Yasunobu and M. Tanaka, in W. Lovenberg, ed., "Iron-Sulfur Proteins," vol. 2, chap. 2, Academic, New York, 1973.)*

are called "ferredoxins." Both rubredoxins and ferredoxins can be oxidized and reduced; they function as electron carriers in a wide variety of biological systems. The ferredoxin from *Micrococcus aerogenes* contains two cubic Fe_4S_4 groups connected to a protein chain with a very complicated conformation,[14] as shown in Fig. 22.14. The four tetrahedrally arranged iron atoms in each Fe_4S_4 cluster are connected to the protein by four more sulfur atoms which are in cysteine groups.[15]

The complex $Fe_4S_4(SCH_2Ph)_4{}^{2-}$, which has the structure shown in Fig. 22.15, is a synthetic ferredoxin analog.[16] This complex is synthesized by the reaction of iron(III) chloride, sodium methoxide, sodium hydrosulfide, and benzyl

---

[14]   K. T. Yasunobu and M. Tanaka, in "Iron-Sulfur Proteins," W. Lovenberg, ed., vol. 2, chap. 2, Academic, New York, 1973.

[15]   L. C. Sieker, E. Adman, and L. H. Jensen, *Nature*, **235**, 40 (1972).

[16]   T. Herskovitz, B. A. Averill, R. H. Holm, J. A. Ibers, W. D. Phillips, and J. F. Weiher, *Proc. Natl. Acad. Sci. USA*, **69**, 2437 (1972). Also see R. H. Holm, *Acc. Chem. Res.*, **10**, 427 (1977).

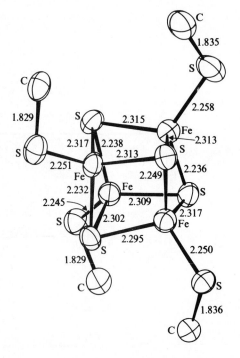

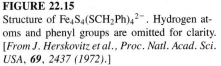

FIGURE 22.15
Structure of $Fe_4S_4(SCH_2Ph)_4^{2-}$. Hydrogen atoms and phenyl groups are omitted for clarity. [*From J. Herskovitz et al., Proc. Natl. Acad. Sci. USA*, **69**, 2437 (1972).]

mercaptan in methanol solution. Many of the properties of the complex (magnetic susceptibility, ^{57}Fe Mössbauer spectrum, electronic spectrum, and redox properties) are analogous to those of ferredoxins which contain Fe_4S_4 clusters. Although the compound might be formulated as a complex of two Fe(II) atoms and two Fe(III) atoms, all the physical methods, including x-ray photoelectron spectroscopy, indicate that the iron atoms are equivalent. Hence a delocalized electronic description, in terms of identical iron atoms in the "+2.5" oxidation state, is probably appropriate.

Nitrogen Fixation

The enzyme "nitrogenase," present in various bacteria, catalyzes the reduction of molecular nitrogen to ammonia. Nitrogenase is composed of two proteins: an iron protein and a molybdenum-iron protein. The Fe protein obtained from the bacterium *Clostridium pasteurianum* is a dimer of a subunit which has an approximate molecular weight of 28,000, two iron atoms, two acid-labile sulfide groups, and six "free" cysteine residues. The Mo–Fe protein from the same bacterium has a molecular weight of approximately 230,000 and contains two molybdenum atoms and approximately 30 atoms each of iron and acid-labile sulfide groups. About half the iron content of the Mo–Fe protein is in Fe_4S_4 centers which probably serve as intramolecular electron-transfer sites. Most of the remaining iron is

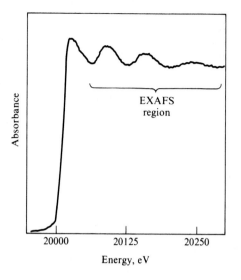

Absorbance

EXAFS
region

20000   20125   20250

Energy, eV

FIGURE 22.16
The x-ray absorption spectrum of lyophilized, semireduced *C. pasteurianum* Mo–Fe protein, showing the EXAFS region associated with the molybdenum K absorption edge. [*Adapted with permission from Cramer et al., J. Am. Chem. Soc.*, **100**, *3814 (1978). Copyright 1978 American Chemical Society.*]

associated with molybdenum in the "Mo–Fe cofactor," which presumably is the catalytic site of the enzyme. In the cofactor, the Fe/Mo ratio is 7 ± 1.

The Mo–Fe protein has been studied by x-ray absorption spectroscopy.[17] In the region of the molybdenum and iron K absorption edges, one observes a series of peaks (due to transitions of the $1s$ electron to unoccupied molecular orbitals having mainly transition-metal character) superimposed on a steeply rising absorption due to transitions of the $1s$ electron into continuum levels. The spectrum at the molybdenum absorption edge is shown in Fig. 22.16. Extended absorption fine structure of this type is generally referred to as EXAFS. The shape of the edge and the positions of the edge features give information about the complex in the vicinity of the metal atom. By Fourier transform analysis and curve fitting of the molybdenum and iron EXAFS spectra, the information summarized in Table 22.1 has been obtained. This information includes the approximate atomic weight (and hence the identity) of the atoms which are neighbors of the molybdenum and iron atoms, the approximate number of these neighbor atoms, and their approximate distances from the molybdenum or iron atom. The uncertainties in the structural data are so great that we have only a rough idea of the nature of the Mo–Fe site of nitrogenase, but they do serve as the basis for reasonable conjecture. Thus, it has been proposed[18] that the molybdenum site has a structure like that shown

---

[17] S. P. Cramer, K. O. Hodgson, W. O. Gillum, and L. E. Mortenson, *J. Am. Chem. Soc.*, **100**, 3398 (1978); M. R. Antonio, B.-K. Teo, W. H. Orme-Johnson, M. J. Nelson, S. E. Groh, P. A. Lindahl, S. M. Kauzlarich, and B. A. Averill, *J. Am. Chem. Soc.*, **104**, 4703 (1982). For a discussion of the application of the EXAFS method to molybdenum complexes, see S. P. Cramer et al., *J. Am. Chem. Soc.*, **100**, 2748 (1978).

[18] B.-K. Teo and B. A. Averill, Abstracts of the 178th National A. C. S. Meeting, September 1979, INOR 238.

TABLE 22.1
Structural features of the Mo-Fe cofactor of nitrogenase obtained from molybdenum and iron EXAFS spectra

Atom	Neighboring atoms	No. of neighboring atoms	Distance to neighboring atoms, Å
Mo	Fe	2–3	~ 2.7
Mo	S	3–4	2.32–2.38
Mo	S	1–2	~ 2.5
Fe	Fe	2.3 ± 0.9	2.66 ± .03
Fe	Mo	0.4 ± 0.1	2.76 ± .03
Fe	S	3.4 ± 1.6	2.25 ± .02
Fe	O or N	1.2 ± 1.0	1.81 ± .07

in Fig. 22.17, in which the molybdenum atom is linked to two Fe_4S_4 cubes by sulfide groups and Mo—Fe bonds. In this proposed structure, the iron atoms are of two structural types (two Fe^A and six Fe^B). Antiferromagnetic coupling of the six Fe^B atoms via a diamagnetic $Fe^AS_2MoS_2Fe^A$ unit could conceivably result in a net spin of $\frac{3}{2}$ for the entire site, and would be consistent with the experimental epr spectrum, which indicates that the protein is a spin $\frac{3}{2}$ system.[19] Molybdenum-iron complexes having some of the structural features of nitrogenase have been synthesized.[20-23] The structures of some of these complexes are shown in Fig. 22.18. The molybdenum EXAFS spectrum of the complex $Mo_2Fe_6S_9(SEt)_8{}^{3-}$ (Fig. 22.18a) is sufficiently similar to that of *C. pasteurianum* nitrogenase to indicate that the complex is a fair approximation to the molybdenum site of nitrogenase. Thus the cubane-like $MoFe_3S_4$ cluster of this species is looked upon

[19]   G. Palmer et al., *Arch. Biochem. Biophys.*, **153**, 325 (1972).

[20]   T. E. Wolff, J. M. Berg, K. O. Hodgson, R. B. Frankel, and R. H. Holm, *J. Am. Chem. Soc.*, **101**, 4140 (1979).

[21]   T. E. Wolff, J. M. Berg, P. P. Power, K. O. Hodgson, and R. H. Holm, *Inorg. Chem.*, **19**, 430 (1980).

[22]   D. Coucouvanis et al., *J. Am. Chem. Soc.*, **102**, 1732 (1980). D. Coucouvanis, *Acc. Chem. Res.*, **14**, 201 (1981).

[23]   G. Christou and C. D. Garner, *J. Chem. Soc. Dalton Trans.*, 2354 (1980).

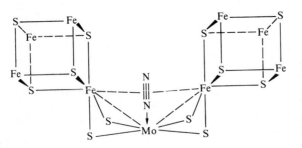

FIGURE 22.17
One of many proposed models for the molybdenum site of the Mo–Fe protein of nitrogenase. [*From B.-K. Teo and B. A. Averill, Abstracts of the 178th National A.C.S. Meeting, September 1979, INOR 238.*]

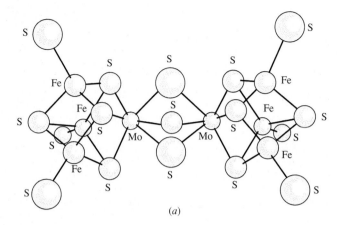

(a)

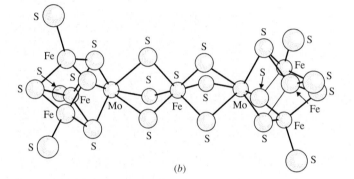

(b)

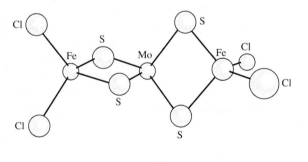

(c)

FIGURE 22.18
Synthetic complexes having some of the structural features of nitrogenase. (a) $Mo_2Fe_6S_9(SEt)_8^{3-}$; the two ethyl groups of the bridge and the six ethyl groups of the terminal thiolate ligands are omitted. [*Reproduced with permission from Wolff et al., J. Am. Chem. Soc.,* **101**, *4140 (1979).*] (b) $Mo_2Fe_7S_8(SEt)_{12}^{3-}$; the ethyl groups of the bridging and terminal thiolate ligands are omitted. [*Reproduced with permission of Wolff et al., J. Am. Chem. Soc.,* **101**, *4140 (1979).*] (c) $MoFe_2S_4Cl_4^{2-}$. [*Reproduced with permission from Coucouvanis et al., J. Am. Chem. Soc.,* **102**, *1732 (1980).*] *(Copyright 1979 and 1980 American Chemical Society.)*

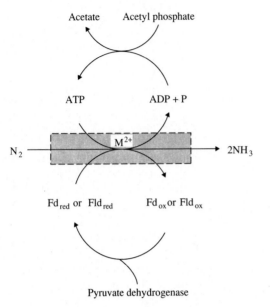

Acetate   Acetyl phosphate

ATP   ADP + P

M^{2+}

N_2 ⟶ $2NH_3$

Fd_{red} or Fld_{red} Fd_{ox} or Fld_{ox}

Pyruvate dehydrogenase

FIGURE 22.19
Schematic representation of the nitrogen-ase-catalyzed reduction of nitrogen. Fd = ferredoxin, Fld = flavodoxin.

as a preliminary model of the coordination unit of molybdenum in the enzyme.[24] However, much more synthetic effort will be required in order to obtain model systems in which the metal composition approaches that of the Mo–Fe cofactor of the enzyme (6 to 8 Fe/Mo).

It is traditionally believed that the reduction of molecular nitrogen occurs at the molybdenum site of the enzyme. The evidence is all circumstantial, and it is not even known whether the initial coordination of nitrogen occurs at an iron atom or a molybdenum atom, although most opinions favor the molybdenum atom. The overall mechanism of the reduction process is schematically represented by the diagram in Fig. 22.19. The electrons required in the reduction are transferred to nitrogenase by the reduced forms of ferredoxins and flavodoxins. The source of these electrons is the oxidation of pyruvate. The electrons are first transferred to the smaller protein (Fe protein). The reduced Fe protein forms a complex with the Mo–Fe protein and the monomagnesium salt of ATP and then transfers its reducing electron to the Mo–Fe protein and thus to the dinitrogen bound at the active site. A series of such electron-transfer steps, with concomitant proton transfer from water to the dinitrogen, results in the production of the magnesium salt of ADP, free phosphate (P), and ammonia, with regeneration of the proteins in their original states. Energy for the process is provided by the hydrolysis of ATP to ADP plus P. For an example of one conceivable mechanism, consider the model of Fig. 22.17.[18] Bonding of N_2 to the molybdenum in a linear, Mo—N≡N, fashion would perturb the Fe^A—Mo—Fe^A bonding and result in a complex in

---

[24]   R. H. Holm, *Chem. Soc. Rev.*, **10**, 455 (1981).

which the activated dinitrogen is σ-bonded to the molybdenum and π-bonded to the two Fe^A atoms. Stepwise injection of electrons via the Fe_4S_4 cubes and successive protonation of the dinitrogen might give a sequence of intermediates such as the following:

$$\text{Mo—N} \equiv \text{N} \xrightarrow[2H^+]{2e^-} \text{Mo—N} = \text{NH}_2 \xrightarrow[2H^+]{2e^-} \text{Mo—NH—NH}_3 \xrightarrow[2H^+]{2e^-} \text{Mo} + 2NH_3$$

It is significant that H_2 inhibits the fixation of nitrogen by nitrogenase.[25] In the presence of N_2, H_2, and D_2O or N_2, D_2, and H_2O, the enzyme produces HD. This latter reaction and the H_2 inhibition of N_2 reduction have been interpreted as evidence for the formation of the intermediate coordinated diimide (N_2H_2), which reacts with H_2 to form $2H_2$ and N_2 (or with D_2 to form 2HD and N_2).

Ferritin[26]

Ferritin is widely distributed among mammals, especially in liver, spleen, and bone marrow. It functions as a reversible iron storage site from which iron can be released in usable form in various cells of the animal. X-ray crystallography on horse spleen ferritin reveals a structure consisting of 24 protein subunits surrounding a core that contains up to 4500 iron atoms in the form of a microcrystalline material of approximate composition $(FeOOH)_8(FeOH_2PO_4)$. (See Fig. 22.20.) The core structure is based on a close-packed array of O^{2-} and OH^- ions, with iron atoms occupying octahedral interstices. A perplexing and poorly understood feature of ferritin is the way in which iron enters and leaves the unit. The ability to solubilize Fe(III) at physiological pH (ca 7) is remarkable; under these conditions Fe(III) is ordinarily almost completely hydrolyzed to insoluble oxide-hydroxide polymers.

---

[25] See E. I. Stiefel, *Prog. Inorg. Chem.*, **22**, 1 (1977).

[26] D. W. Rice, G. C. Ford, J. L. White, J. M. A. Smith, and P. M. Harrison, *Adv. Inorg. Biochem.*, **5**, 39 (1983); and S. J. Lippard et al., *J. Am. Chem. Soc.*, **109**, 3337 (1987).

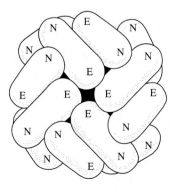

FIGURE 22.20
A sketch of the packing of the 24 subunits that surround the core of horse spleen ferritin. Each subunit is represented as a "sausage" with structurally distinguishable ends, E and N. [*Reproduced with permission from D. W. Rice, G. C. Ford, J. L. White, J. M. A. Smith, and P. M. Harrison, Adv. Inorg. Biochem., 5, 39 (1983).*]

cis-DICHLORODIAMMINEPLATINUM(II) ANTICANCER ACTIVITY[27]

In an attempt to determine the effect of oscillating electric fields on bacterial growth, Rosenberg and VanCamp suspended *Escherichia coli* bacteria in an ammonium chloride solution between two platinum electrodes connected to an audio amplifier. This treatment inhibited cell division (the bacterial population decreased), but cell growth continued, leading to long filaments. After considerable further study, it was discovered that this peculiar effect on the bacteria was due to the electrolytic formation of *cis*-Pt(NH$_3$)$_2$Cl$_4$ and that the corresponding trans complex has essentially no effect on the bacteria. It was then shown that *cis*-Pt(NH$_3$)$_2$Cl$_2$ has the same sort of effect on *E. coli* as *cis*-Pt(NH$_3$)$_2$Cl$_4$ and that *trans*-Pt(NH$_3$)$_2$Cl$_2$, like the corresponding trans Pt(IV) complex, has no effect. Tests with various animals showed that *cis*-Pt(NH$_3$)$_2$Cl$_2$ is a very effective antitumor agent, causing regression of both slow- and fast-growing cancers. This compound, under the name cisplatin, has been approved for use in humans. The mechanism of the anticancer activity is not completely known, but certain features are clear.[28] In the high chloride concentration of extracellular fluids, replacement of the chloride ions of the complex by water molecules is suppressed. However, after passing into a cell, where the chloride concentration is about $\frac{1}{30}$ of that outside the cell, the following reactions can take place:

$$Pt(NH_3)_2Cl_2 + H_2O \rightleftharpoons Pt(NH_3)_2(H_2O)Cl^+ + Cl^-$$

$$Pt(NH_3)_2(H_2O)Cl^+ + H_2O \rightleftharpoons Pt(NH_3)_2(H_2O)_2^{2+} + Cl^-$$

It is believed that the diaquo species reacts to form *cis*-Pt(NH$_3$)$_2$ bridges between the nitrogen atoms of DNA bases, producing mainly intrastrand DNA cross-links rather than DNA-protein cross-links.

A related chemotherapeutic agent, diammine(1,1-cyclobutane-dicarboxylato) platinum(II) (known as carboplatin), has the following structural formula:

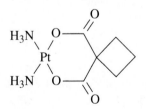

This compound undergoes aquation more slowly, and produces less severe side-effects, than cisplatin.

---

[27] B. Rosenberg, *Interdiscip. Sci. Rev.*, **3**, 134 (1978); B. Rosenberg, L. VanCamp, J. E. Trosko, and V. H. Mansour, *Nature*, **222**, 385 (1969).

[28] S. J. Lippard, *Science*, **218**, 1075 (1982); M. J. Egorin et al., *Cancer Res.*, **44**, 5432 (1984); C. F. J. Barnard, *Platinum Metals Rev.*, **33**, 162 (1989); Bristol-Myers pamphlet on Paraplatin, Feb. 1989.

CARBONIC ANHYDRASE[29]

In Chap. 8 it was pointed out that the hydration of CO_2 to give carbonic acid is a slow process. This property of CO_2 is not consistent with the fact that many physiological processes require rapid equilibration of CO_2 with HCO_3^- and H_2CO_3 at about pH 7. It is the function of the zinc-containing enzyme called carbonic anhydrase to catalyze this equilibration. The active site of the enzyme contains a zinc ion bonded to three histidine imidazole nitrogen atoms and either a water molecule or a hydroxide ion, as follows:

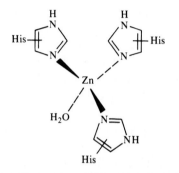

A water molecule coordinated to a zinc ion has a pK near 7, and thus the role of the zinc ion is to reduce the basicity of the hydroxide ion such that a high effective concentration of hydroxide can exist in an approximately neutral medium. The probable mechanism of the hydration process is shown in the following cyclic scheme.

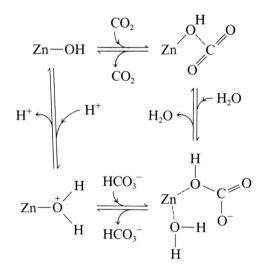

---

[29] F. A. Cotton and G. Wilkinson, "Advanced Inorganic Chemistry," 5th ed., pp. 1361–1362, Wiley-Interscience, New York, 1988.

In effect, the enzyme allows the relatively fast reaction, $CO_2 + OH^- \rightarrow HCO_3^-$, to be the major pathway for hydration at pH 7, where the hydroxide concentration is so low that the slow reaction of CO_2 with H_2O would ordinarily predominate.

SUPEROXIDE DISMUTASES[30]

Because oxygen is vital to many biological respiration processes, its redox chemistry has been the subject of many investigations.[31] It is well established that the initial step in the electron-transfer reduction of O_2 produces the superoxide ion, O_2^-. It has been postulated that the superoxide ion, or some product derived from it (such as the hydroxyl radical, $\cdot OH$) is toxic to cellular systems. Although there are cogent arguments against this postulate,[32] it is nevertheless true that all organisms that use oxygen, and many that have to survive in an oxygen environment, contain at least one variety of the enzyme (superoxide dismutase) that very efficiently catalyzes the dismutation (disproportionation) of superoxide into hydrogen peroxide and oxygen:

$$2O_2^- + 2H^+ \longrightarrow H_2O_2 + O_2$$

The superoxide dismutase isolated from bovine liver is very effective in suppressing swelling and inflammation of tissue and has been found to be valuable in the treatment of inflammatory osteoarthritis.

There are three varieties of superoxide dismutase: (1) a copper-zinc protein found in cells that have nuclei, (2) a manganese protein found in bacteria, and (3) an iron protein found in bacteria. A crystal structure determination of the copper-zinc protein has shown that the Cu^{2+} and Zn^{2+} ions are coordinated to the same imidazole ring of a histidine residue, as follows:

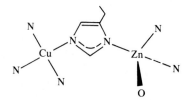

Biochemical studies indicate that an arginine residue in the protein, positioned close to the copper atom, plays an important role in the catalytic activity

---

[30] W. H. Bannister and J. V. Bannister, eds., "Biological and Clinical Aspects of Superoxide Dismutase." Proceedings of the International Symposium on Superoxide and Superoxide Dismutases, Malta, 1979; J. V. Bannister and H. A. O. Hill, eds., "Chemical and Biochemical Aspects of Superoxide and Superoxide Dismutase," Proceedings of the International Symposium on Superoxide and Superoxide Dismutases, 2d, Malta, 1979.

[31] J. Wilshire and D. T. Sawyer, *Acc. Chem. Res.*, **12**, 105 (1979).

[32] J. A. Fee, in Bannister and Bannister, op. cit., p. 41.

of the enzyme.[33] The following mechanism has been proposed for the catalyzed dismutation of the superoxide ion:[34]

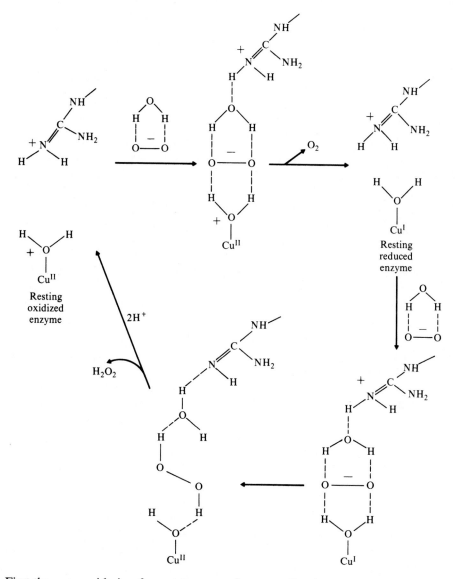

First the superoxide ion forms an outer-sphere coordination complex with the hydrated Cu^{2+} ion, and a hydrogen-bonded water molecule connects the superoxide

---

[33] D. P. Malinowski and I. Fridovich, in Bannister and Hill, op. cit., p. 299.
[34] A. E. G. Cass and H. A. O. Hill, in Bannister and Hill, op. cit., p. 290.

ion to the guanidinium group of the arginine. An outer-sphere electron transfer then produces Cu^+ and molecular oxygen. Another superoxide ion is then bound between the hydrated Cu^+ and the arginine. Electron transfer from the copper then produces Cu^{2+} and peroxide ion, while simultaneously the copper-bound water molecule and the guanidinium group each donate a proton to the developing peroxide dianion. The hydrogen peroxide then leaves, and both the Cu^{2+}-bound hydroxide ion and the guanidine group are reprotonated by nearby water molecules. The function of the zinc ion is completely unknown.

PROBLEMS

22.1 Draw the structure of porphine. Indicate how the heme structure is related to porphine and how the macrocyclic ligand system in vitamin B_{12} differs from porphine.

22.2 Describe the functions of hemoglobin and myoglobin and describe the changes in the heme groups of hemoglobin on going from the oxy to the deoxy form.

22.3 Describe a possible mechanism for the vitamin B_{12}-catalyzed dehydration of a diol:

$$RCHOHCH_2OH \quad \rightarrow \quad RCH_2CHO + H_2O$$

How could you test the mechanism, using tritium as a hydrogen tracer?

22.4 Outline the functions of nitrogenase. What is known about the structure of the enzyme in the vicinity of a molybdenum atom?

22.5 (a) Show that the hyperbolic curve for myoglobin in Fig. 22.5 is consistent with the equilibrium

$$Mb(aq) + O_2(g) \quad \rightleftharpoons \quad MbO_2(aq)$$

(b) Show that the sigmoidal curve for hemoglobin in Fig. 22.5 can be explained by an equilibrium of the following type, where $1 < n < 4$:

$$Hb(aq) + nO_2(g) \quad \rightleftharpoons \quad Hb(O_2)_n(aq)$$

APPENDIX
A

UNITS AND CONVERSION FACTORS

The tables in this section give values for some important physical constants (Table A.3) and conversion factors for units of energy (Table A.4). In recent years there has been a trend among scientists toward the use of SI units[1] (i.e., units based on the *Systeme Internationale d'Unites*). Although there are some advantages to the use of SI units, undue rigidity should be avoided. Indeed, the uncritical adoption of SI units is usually undesirable.[2] Most modern chemistry journals retain the use of traditional units such as angstroms, atmospheres, molarity (moles per liter), electronvolts, and kilocalories, although there is a marked trend toward the use of joules instead of calories. This text employs the traditional units; thus it conforms with the practice of most working chemists. When necessary, the tables of this appendix can be used to convert from one unit to another.

---

[1] *Nat. Bur. Stand.* (U.S.), *Spec. Publ.* 330 (1981).

[2] A. W. Adamson, *J. Chem. Educ.*, **55,** 634 (1978); D. R. Lide, *Chem. Eng. News*, 4, 54. (May 10, 1982)

TABLE A.1

Physical quantity	Name	Symbol
Length	meter	m
Mass	kilogram	kg
Time	second	s
Electric current	ampere	A
Temperature	kelvin	K
Luminous intensity	candela	cd

TABLE A.2

Physical quantity	Name	Symbol
Force	newton	N (kg m s^{-2})
Energy	joule	J (N m)
Power	watt	W(J s^{-1})
Electric charge	coulomb	C (A s)
Electric potential	volt	V (W A^{-1})
Electric capacitance	farad	F (A s V^{-1})
Electric resistance	ohm	Ω(V A^{-1})
Frequency	hertz	Hz (s^{-1})
Magnetic flux	weber	Wb (V s)
Magnetic flux density	tesla	T (Wb m^{-2})
Inductance	henry	H (V s A^{-1})

TABLE A.3
Some useful constants†

Constant and symbol	Value
Gas constant, R	1.98722 cal K^{-1} mol^{-1}
	82.0578 mL atm K^{-1} mol^{-1}
	62,363.9 mL torr K^{-1} mol^{-1}
	8.31451 J K^{-1} mol^{-1}
Ideal gas volume, std. cond.	22,414.1 mL mol^{-1}
Avagadro's number, N	6.022137 × 10^{23} mol^{-1}
Faraday constant, F	96,485.3 C mol^{-1}
	23,060.5 cal volt^{-1} mol^{-1}
Boltzmann's constant, k	1.38066 × 10^{-23} J K^{-1}
Electron charge, e	1.602177 × 10^{-19} C
	4.80321 × 10^{-10} abs. esu
Electron mass, m	9.109390 × 10^{-28} g
Proton mass, M_p	1.672623 × 10^{-24} g
Planck's constant, h	6.626076 × 10^{-34} Js
Newtonian constant of gravity, G	6.6726 × 10^{-11} m^3 kg^{-1} s^{-2}
Bohr radius, a_0	0.529177 Å
Ice point (0°C)	273.150 K
Speed of light, c	2.99792458 × 10^{10} cm s^{-1}

† Mainly from E. R. Cohen and B. N. Taylor, *Codata Bull.*, no. 63, Nov. 1986, "The 1986 Adjustment of the Fundamental Physical Constants."

TABLE A.4
Units of molecular energy

	=	erg molecule^{-1}	J molecule^{-1}	cal mol^{-1}	eV molecule^{-1}	Wave number (cm^{-1})	J mol^{-1}
1 erg molecule^{-1}	=	1	10^{-7}	1.4393×10^{16}	6.2415×10^{11}	5.0340×10^{15}	6.0221×10^{16}
1 J molecule^{-1}	=	10^7	1	1.4393×10^{23}	6.2415×10^{18}	5.0340×10^{22}	6.0221×10^{23}
1 cal mol^{-1}	=	6.9477×10^{-17}	6.9477×10^{-24}	1	4.3363×10^{-5}	0.34975	4.1840
1 eV molecule^{-1}	=	1.6022×10^{-12}	1.6022×10^{-19}	23,061	1	8065.5	96,485
1 wave number (cm^{-1})	=	1.9865×10^{-16}	1.9865×10^{-23}	2.8592	1.2398×10^{-4}	1	11.963
1 J mol^{-1}	=	1.6605×10^{-17}	1.6605×10^{-24}	0.23901	1.0364×10^{-5}	8.3594×10^{-2}	1

APPENDIX
B

TERM
SYMBOLS
FOR
FREE ATOMS
AND IONS

The statement that a free atom or ion has a particular valence electron configuration, say np^2, is an incomplete description of the electronic state of the species. In order to completely describe the state, one must specify the values of the total *orbital* angular momentum quantum number, L, the total *spin* angular momentum quantum number, S, and the total angular momentum quantum number, J.

The components, in a reference direction, of L and S can have the following quantized values:

$$M_L = L, L - 1, \ldots, -(L - 1), -L$$

$$M_S = S, S - 1, \ldots, -(S - 1), -S$$

Here M_L and M_S are the vector sums of the m_l and m_s values for the individual electrons:

$$M_L = m_{l_1} + m_{l_2} + m_{l_3} + \cdots$$

$$M_S = m_{s_1} + m_{s_2} + m_{s_3} + \cdots$$

The set of individual quantum states corresponding to a particular value of L and a particular value of S constitutes a "term" and is represented by a term symbol, $^{2S+1}L_J$, where the value of L is indicated by one of the capital letters S, P, D, F, G, H, ... (corresponding to 0, 1, 2, 3, ...). The quantum number J is the vector sum of L and S and may have the values

$$J = L + S, L + S - 1, L + S - 2, \ldots, |L - S|$$

J can only have positive values or be zero. The components, in a reference direction, of J can have values

$$M_J = J, J-1, \ldots, -(J-1), -J$$

For example, one of the terms of an atom with an np^2 configuration is the $^1D_2$ (pronounced singlet D two) term. This term can be represented by the following set of individual quantum states, in which the m_s values are indicated by arrows pointing up or down.

	$m_l = +1$	0	−1	M_L	M_S	M_J
$^1D_2$	(↑↓)	○	○	2	0	2
	(↑)	(↓)	○	1	0	1
	○	(↑↓)	○	0	0	0
	○	(↑)	(↓)	−1	0	−1
	○	○	(↑↓)	−2	0	−2

The np^2 configuration also yields the terms $^3P_2$, $^3P_1$, and $^3P_0$, for which we can write the following individual quantum states:

	$m_l = +1$	0	−1	M_L	M_S	M_J
$^3P_2$	(↑)	(↑)	○	1	1	2
	(↑)	○	(↑)	0	1	1
	○	(↑)	(↑)	−1	1	0
	(↓)	○	(↓)	0	−1	−1
	○	(↓)	(↓)	−1	−1	−2
$^3P_1$	(↓)	(↑)	○	1	0	1
	(↓)	(↓)	○	1	−1	0
	○	(↓)	(↑)	−1	0	−1
$^3P_0$	(↑)	○	(↓)	0	0	0

The only other term of the np^2 configuration is the $^1S_0$ term:

	$m_l = +1$	0	−1	M_L	M_S	M_J
$^1S_0$	(↓)	○	(↑)	0	0	0

Note that we have now written all the possible combinations of m_l and m_s. It is important to recognize that most of the individual quantum state assignments are not unique. For example, the configurations (↓) ○ (↑) and ○ (↑↓) ○ could just as well be interchanged, or suitable linear combinations assigned to particular terms.

It should be obvious that, because of limitations imposed by the Pauli principle, not all conceivable combinations of L and S are possible for a given many-electron system. A list of the allowed states for equivalent s, p, and d electrons is

TABLE B.1
Allowed states for equivalent s, p, and d electrons

Configuration	Terms
s	^2S
s^2	^1S
p or p^5	^2P
p^2 or p^4	^1S, ^1D, ^3P
p^3	^2P, ^2D, ^4S
p^6	^1S
d or d^9	^2D
d^2 or d^8	$^1(SDG)$, $^3(PF)$
d^3 or d^7	^2D, $^2(PDFGH)$, $^4(PF)$
d^4 or d^6	$^1(SDG)$, $^3(PF)$, $^1(SDFGI)$, $^3(PDFGH)$, ^5D
d^5	^2D, $^2(PDFGH)$, $^4(PF)$, $^4(SDFGI)$, $^4(DG)$, ^6S

given in Table B.1. Methods for determining the terms corresponding to particular configurations are described in various books.[1] In order to determine which term for a given configuration represents the ground state, we apply Hund's rules:

1. The most stable state is the one with the largest value of S.
2. Of a group of states with the same value of S, that with the largest value of L lies lowest.
3. Of a group of states with the same values of L and S, corresponding to a shell of electrons less than half full, states with lower J lie lower. When the shell is more than half full, the reverse applies.

In the x-ray photoelectron spectrum of argon (Fig. 1.11), the $2p$ peak is split into two components corresponding to the $^2P_{3/2}$ and $^2P_{1/2}$ ions which are formed. These components have relative intensities of 2:1, corresponding to the ratio of the number of individual quantum states for the two states (i.e., the ratio of the $2J + 1$ values).

---

[1] H. B. Gray, "Electrons and Chemical Bonding," pp. 22–27, W. A. Benjamin, New York, 1964; C. J. Ballhausen, "Introduction to Ligand Field Theory," pp. 8–10, McGraw-Hill, New York, 1962; F. A. Cotton and G. Wilkinson, "Advanced Inorganic Chemistry," 3d ed., pp. 80–85, Wiley-Interscience, New York, 1972; J. E. Huheey, "Inorganic Chemistry," pp. 35–38. Harper & Row, New York, 1972; L. Pauling, "The Nature of the Chemical Bond," 3d ed., pp. 580–588. Cornell University Press, Ithaca, N.Y., 1960.

APPENDIX

C

ELECTRON AFFINITIES

TABLE C.1
Electron affinities of atoms†

Element	eV	kcal mol⁻¹‡	Element	eV	kcal mol⁻¹‡
Aluminum	0.46	10.6	Neon	<0	<0
Antimony	1.05	24.2	Nickel	1.15	26.5
Argon	<0	<0	Niobium	1.0	23
Arsenic	0.80	18.4	Nitrogen	−0.07	−1.6
Barium	<0	<0	Osmium	1.1	25
Beryllium	<0	<0	Oxygen	1.462	33.72
Bismuth	1.1	25	Palladium	0.6	14
Boron	0.28	6.5	Phosphorus	0.743	17.13
Bromine	3.364	77.58	Platinum	2.128	49.07
Cadmium	<0	<0	Polonium	1.9	44
Calcium	<0	<0	Potassium	0.5012	11.56
Carbon	1.268	29.24	Radon	<0	<0
Cesium	0.4715	10.87	Rhenium	0.15	3.5
Chlorine	3.615	83.37	Rhodium	1.2	28
Chromium	0.66	15.2	Rubidium	0.4860	11.21
Cobalt	0.7	16	Ruthenium	1.1	25
Copper	1.226	28.27	Scandium	<0	<0
Fluorine	3.399	78.38	Selenium	2.021	46.61
Gallium	0.3	7	Silicon	1.385	31.94
Germanium	1.2	28	Silver	1.303	30.05
Gold	2.309	53.25	Sodium	0.546	12.59
Hafnium	<0	<0	Strontium	<0	<0
Helium	<0	<0	Sulfur	2.077	47.90
Hydrogen	0.7542	17.39	Tantalum	0.6	14
Indium	0.3	7	Technetium	0.7	16
Iodine	3.061	70.59	Tellurium	1.971	45.45
Iridium	1.6	37	Thallium	0.3	7
Iron	0.25	5.8	Tin	1.25	28.8
Krypton	<0	<0	Titanium	0.2	5
Lanthanum	0.5	12	Tungsten	0.6	14
Lead	1.1	25	Vanadium	0.5	12
Lithium	0.620	14.30	Xenon	<0	<0
Magnesium	<0	<0	Yttrium	~0	~0
Manganese	<0	<0	Zinc	<0	<0
Mercury	<0	<0	Zirconium	0.5	12
Molybdenum	1.0	23			

† H. Hotop and W. C. Lineberger, *J. Phys. Chem. Ref. Data*, **4**, 539 (1975).

‡ 1 kcal = 4.1840 kJ.

TABLE C.2
Electron affinities of molecules and radicals[a]

Molecule	eV	kcal mol^{-1}
Diatomics		
Br$_2$	2.53	58.3
CN	3.82	88.1
CS[b]	0.20	4.7
Cl$_2$	2.35	54.2
F$_2$	3.1	71
HS	2.32	53.5
IBr	2.6	60
I$_2$	2.55	58.8
LiCl	0.61	14.1
NH	0.38	8.8
NO	0.02	0.5
NS[b]	1.19	27.5
OH	1.825	42.09
O$_2$	0.44	10.1
PH	1.03	23.8
SO	1.13	26.1
S$_2$	1.66	38.3
SeH	2.21	51.0
SiH	1.28	29.5
Triatomics		
AsH$_2$	1.27	29.3
CH$_2$	0.21	4.8
NH$_2$	0.74	17.1
N$_3$[c]	2.7	62
NO$_2$	2.3	53
O$_2$H[d]	1.19	27.4
O$_3$[e]	2.10	48.4
PH$_2$	1.25	28.8
SO$_2$	1.1	25
S$_3$	2.0	46
SiH$_2$	1.12	25.8
Tetraatomics		
CH$_3$	0.08	1.8
CO$_3$	2.7	62
GeH$_3$	≤ 1.74	≤ 40.1
SO$_3$	≥ 1.7	≥ 39
SiH$_3$	≤ 1.44	≤ 33.2

TABLE C.2 (*continued*)

Molecule	eV	kcal mol^{-1}
	Larger polyatomics	
CH_3NO_2	0.4	9
CH_3O	1.57	36.2
CH_3S	1.88	43.4
C_3H_5(allyl)	0.55	12.7
C_5H_5(cyclopentadienyl)	1.79	41.3
C_6H_6	-1.14	-26.3
HNO_3	0.6	14
$POCl_3$	1.4	32
SF_6	0.5	12
SeF_6	2.9	67
TeF_6	3.3	76
WF_6[f]	3.5	81
ReF_6[f]	5	115
OsF_6[f]	6.5	150
IrF_6[f]	8	184
PtF_6[g]	8.0	184
AuF_6[h]	8.9	205

[a] B. K. Janousek and J. I. Brauman, in "Gas Phase Ion Chemistry," M. T. Bowers, ed., vol. 2, chap. 10, p. 53, Academic, New York, 1979.

[b] S. M. Burnett, C. S. Feigerle, and W. C. Lineberger, National A. C. S. Meeting, Las Vegas, April 1982, PHYS 170.

[c] R. L. Jackson, M. J. Pellerite, and J. I. Brauman, *J. Am. Chem. Soc.*, **103**, 1803 (1981).

[d] V. M. Bierbaum, R.J. Schmitt, C. H. DePuy, R. D. Mead, P. A. Schultz, and W. C. Lineberger, *J. Am. Chem. Soc., 103*, 6262 (1981).

[e] W. C. Lineberger, *Gov. Rep. Announce. Index (U.S.),* **79**, 92 (1979).

[f] N. Bartlett, E. M. McCarron, B. W. McQuillan, and T. E. Thompson, *Synth. Metals,* **1**, 221 (1980).

[g] M. I. Nikitin et al., *Int. J. Mass Spectrom. Ion Phys.* **37**, 13 (1981).

[h] G. Frenking et al., *J. Am. Chem. Soc.* **111**, 31 (1989).

APPENDIX
D

IONIZATION ENERGIES

TABLE D.1
Ionization energies of some molecules and radicals†

Molecule or radical	IE, eV	Molecule or radical	IE, eV	Molecule or radical	IE, eV
H_2	15.43	CO	14.01	HS	10.41
BH	9.77	CO_2	13.77	H_2S	10.47
BH_2	~9.8	NO	9.26	CS¶	~11.33
BH_3	~12.3	N_2O	12.89	NS	~9.85
B_2H_6	11.39	NO_2	9.75	SO	10.34
CH	10.64	CH_3OH	10.84	SO_2	12.33
CH_2	10.40	F_2	15.69	SO_3	~11.0
CH_3	9.84	HF	16.01	S_2O	~10.3
CH_4	≤ 12.62	BF_3	15.56	SF_4	12.28
C_2H_2	11.41	CF_3	9.17	SF_6	15.35
C_2H_4	10.51	NF_3	13.00	Cl_2	11.49
C_2H_6	11.52	SiH_4	11.66	HCl	12.74
C_6H_6	9.25	Si_2H_6	~10.2	BCl_3	11.62
N_2	15.58	P_4‡	9.2	GeH_4	11.31
NH_3	10.16	PH_3	9.98	Ge_2H_6	~12.5
CN	14.1	PF_3§	11.66	AsH_3	10.04
HCN	13.59	OPF_3	12.75	H_2Se	9.88
CH_3NH_2	8.97	CH_3PH_2§	9.12	Br_2	10.51
$(CH_3)_2NH$	8.24	$(CH_3)_2PH$§	8.47	HBr	11.67
$(CH_3)_3N$	7.81	$(CH_3)_3P$§	8.11	SbH_3	9.58
C_5H_5N	9.27	$P(OCH_3)_3$§	8.50	H_2Te	9.14
O_2	12.07	S_2	9.36	I_2	9.39
H_2O	12.61	S_8	9.04	HI	10.39

† Except as noted, data from H. M. Rosenstock, K. Drexl, B. W. Steiner, and J. T. Herron, *J. Phys. Chem. Ref. Data*, **6**, suppl. 1 (1977). See Table 1.7 for atomic ionization potentials.

‡ C. R. Brundle, N. A. Kuebler, M. B. Robin, and H. Basch, *Inorg. Chem.*, **11**, 20 (1972).

§ R. V. Hodges, F. A. Houle, J. L. Beauchamp, R. A. Montag, and J. G. Verkade, *J. Am. Chem. Soc.*, **102**, 932 (1980).

¶ N. Jonathan, A. Morris, M. Okuda, K. J. Ross, and D. J. Smith, *Faraday Discuss. Chem. Soc.*, **54**, 48 (1972).

APPENDIX
E

SELECTED VALUES OF THERMODYNAMIC DATA

The following table lists a few data selected from the U.S. National Bureau of Standards Technical Notes 270-3, 270-4, 270-5, and 270-6, and Interim Report NBSIR 76-1034. For much more complete sets of data, the reader should consult the latter publications, the JANAF Thermochemical Tables [NSRDS-NBS 37 (1971)], and the CODATA compilation in *J. Chem. Thermodyn.*, **10,** 903 (1978).

It should be noted that other thermodynamic data are given in various tables in this text, e.g., atomic ionization energies (Table 1.7), dissociation energies (Tables 3.9 to 3.11), ΔH_f° values for gaseous atoms (Table 3.12), ΔH_f° values for hydrides (Tables 7.1 and 7.3), proton affinities (Tables 8.1 and 8.2), pK values of acids (Tables 8.3, 8.4, 8.5, and 8.7), thermodynamic constants for the ionization of aqueous acids (Tables 8.12), gas-phase hydration energies (Table 8.8), electron affinities (Appendix C), and molecular ionization energies (Appendix D).

TABLE E.1
Selected thermodynamic data at 25°C†

Species	State	ΔH_f°, kcal mol^{-1}	ΔG_f°, kcal mol^{-1}	S°, cal deg^{-1} mol^{-1}
Ag	c	0	0	10.17
Ag$^+$	aq	25.23	18.43	17.37
Ag^{2+}	4 M HClO$_4$	64.2	64.3	-21
Ag$_2$O	c	-7.42	-2.68	29.0
AgO	c	-2.9	3.3	14
AgCl	c	-30.37	-26.24	23.0

TABLE E.1 (*continued*)

Species	State	ΔH_f°, kcal mol^{-1}	ΔG_f°, kcal mol^{-1}	S°, cal deg^{-1} mol^{-1}
AgBr	c	−23.99	−23.16	25.6
AgI	c	−14.78	−15.82	27.6
Al	c	0	0	6.77
Al^{3+}	aq	−127	−116	−76.9
Al$_2$O$_3$	c	−400.5	−378.2	12.17
Al(OH)$_3$	amorph	−305		
Al(OH)$_4{}^-$	aq	−356.2	−310.2	28
AlF$_3$	c	−359.5	−340.6	15.88
AlCl$_3$	c	−168.3	−150.3	26.45
As	c	0	0	8.4
As$_4$	g	34.4	22.1	75
As$_4$O$_6$	c	−314.04	−275.46	51.2
AsH$_3$	g	15.88	16.47	53.22
HAsO$_2$	aq	−109.1	−96.25	30.1
H$_3$AsO$_4$	aq	−215.7	−183.1	44
AsO$_2{}^-$	aq	−102.54	−83.66	9.9
AsO$_4{}^{3-}$	aq	−212.27	−155.00	−38.9
AsF$_3$	g	−220.04	−216.46	69.07
AsCl$_3$	g	−61.80	−58.77	78.17
AsBr$_3$	g	−31	−38	86.94
Am	c	0	0	11.33
Am(OH)$_3$	c	−101.5	−75.77	45.3
AmCl	c	−8.3		
AmCl$_3$	c	−28.1		
AmCl$_4{}^-$	aq	−77.0	−56.22	63.8
Am(CN)$_2{}^-$	aq	57.9	68.3	41
B	c	0	0	1.40
B	g	134.5	124.0	36.65
B$_2$O$_3$	c	−304.20	−285.30	12.90
H$_3$BO$_3$	aq	−256.29	−231.56	38.8
B(OH)$_4{}^-$	aq	−321.23	−275.65	24.5
BH$_4{}^-$	aq	11.51	27.31	26.4
B$_2$H$_6$	g	8.5	20.7	55.45
BF$_3$	g	−271.75	−267.77	60.71
B$_2$F$_4$	g	−344.2	−337.1	75.8
BCl$_3$	g	−96.50	−92.91	69.31
BBr$_3$	g	−49.15	−55.56	77.47
BI$_3$	g	17.00	4.96	83.43
Ba	c	0	0	15.0
Ba^{2+}	aq	−128.50	−134.02	2.3
BaO	c	−132.3	−125.5	16.83
BaO$_2$	c	−151.6		
Ba(OH)$_2$	c	−225.8		
BaCl$_2$	c	−205.2	−193.7	29.56
BaCO$_3$	c	−290.7	−271.9	26.8
Be	c	0	0	2.27
Be^{2+}	aq	−91.5	−90.75	−31.0
BeO	c	−145.7	−138.7	3.38

TABLE E.1 (*continued*)

Species	State	ΔH_f°, kcal mol^{-1}	ΔG_f°, kcal mol^{-1}	S°, cal deg^{-1} mol^{-1}
BeO_2^{2-}	aq	-189.0	-153.0	-38
BeF_2	c	-245.4	-234.1	12.75
$BeCl_2$	c	-118.5	-107.3	18.12
Bi	c	0	0	13.56
BiO^+	aq		-35.0	
Bi_2O_3	c	-137.16	-118.0	36.2
$BiCl_3$	g	-63.5	-61.2	85.74
$BiCl_4^-$	aq		-115.1	
$BiOCl$	c	-87.7	-77.0	28.8
Bi_2S_3	c	-34.2	-33.6	47.9
Br_2	l	0	0	36.384
Br_2	g	7.387	0.751	58.641
Br	g	26.741	19.701	41.805
HBr	g	-8.70	-12.77	47.463
Br^-	aq	-29.05	-24.85	19.7
BrO^-	aq	-22.5	-8.0	10
BrO_3^-	aq	-20.0	0.4	39.0
BrF_3	g	-61.09	-54.84	69.89
BrF_5	g	-102.5	-83.8	76.50
C	graphite	0	0	1.372
C	diamond	0.4533	0.6930	0.568
C	g	171.291	160.442	37.760
CO	g	-26.416	-32.780	47.219
CO_2	g	-94.051	-94.254	51.06
CO_2	aq	-98.90	-92.26	28.1
HCO_3^-	aq	-165.39	-140.26	21.8
CO_3^{2-}	aq	-161.84	-126.17	-13.6
CH_4	g	-17.88	-12.13	44.492
C_2H_6	g	-20.24	-7.86	54.85
CF_4	g	-221	-210	62.50
COF_2	g	-151.7	-148.0	61.78
CCl_4	l	-32.37	-15.60	51.72
CCl_4	g	-24.6	-14.49	74.03
$COCl_2$	g	-52.3	-48.9	67.74
CBr_4	g	19	16	85.55
Ca	c	0	0	9.90
Ca^{2+}	aq	-129.74	-132.30	-12.7
CaO	c	-151.79	-144.37	9.50
$Ca(OH)_2$	c	-235.68	-214.76	19.93
$CaCl_2$	c	-190.2	-178.8	25.0
$CaCO_3$	c	-288.46	-269.80	22.2
$CaCN_2$	c	-83.8		
Cd	c	0	0	12.37
Cd^{2+}	aq	-18.14	-18.54	-17.5
CdO	c	-61.7	-54.6	13.1
$Cd(OH)_4^{2-}$	aq		-181.3	
CdF_2	c	-167.4	-154.8	18.5
$CdCl_2$	c	-93.57	-82.21	27.55

TABLE E.1 (*continued*)

Species	State	ΔH_f°, kcal mol^{-1}	ΔG_f°, kcal mol^{-1}	S°, cal deg^{-1} mol^{-1}
$CdCO_3$	c	-179.4	-160.0	22.1
Ce	c	0	0	17.2
Ce^{3+}	aq	-166.4	-160.6	-49
Ce^{4+}	aq	-128.4	-120.4	-72
CeO_2	c	-260.2	-244.9	14.89
Ce_2O_3	c	-429.3	-407.8	36.0
$CeCl_3$	c	-251.8	-233.7	36
Cl_2	g	0	0	53.288
Cl	g	29.082	25.262	39.457
HCl	g	-22.062	-22.777	44.646
Cl^-, HCl	aq	-39.952	-31.372	13.5
Cl_2O	g	19.2	23.4	63.60
$HClO$	g	-22	-18	56.5
$HClO$	aq	-28.9	-19.1	34
ClO^-	aq	-25.6	-8.8	10
ClO_2^-	aq	-15.9	4.1	24.2
ClO_3^-	aq	-23.7	-0.8	38.8
ClO_4^-	aq	-30.91	-2.06	43.5
ClF	g	-13.02	-13.37	52.05
ClF_3	g	-39.0	-29.4	67.28
ClO_2F	g	-8.1		
Co	c	0	0	7.18
Co^{2+}	aq	-13.9	-13.0	-27
Co^{3+}	aq	22	32	-73
CoO	c	-56.87	-51.20	12.66
$CoCl_2$	c	-74.7	-64.5	26.09
Cr	c	0	0	5.68
Cr^{2+}	aq	-34.3		
CrO_3	c	-140.9		
CrO_4^{2-}	aq	-210.60	-173.96	12.00
Cr_2O_3	c	-272.4	-252.9	19.4
$CrCl_2$	c	-94.5	-85.1	27.56
$CrCl_3$	c	-133.0	-116.2	29.4
Cs	c	0	0	19.8
Cs^+	aq	-62.6	-70.8	31.8
Cs_2O	c	-75.9		
CsO_2	c	-62.1		
CsN_3	c	-2.4		
$CsHF_2$	c	-219.5		
CsF	c	-131.7		
CsI	c	-83.7		
Cu	c	0	0	7.92
Cu^+	aq	17.13	11.95	9.7
Cu^{2+}	aq	15.48	15.66	-23.8
CuO	c	-37.6	-31.0	10.19
Cu_2O	c	-40.3	-34.9	22.26
$CuCl$	c	-32.8	-28.65	20.6
$CuCl_2$	c	-52.6	-42.0	25.83

TABLE E.1 (*continued*)

Species	State	ΔH_f°, kcal mol^{-1}	ΔG_f°, kcal mol^{-1}	S°, cal deg^{-1} mol^{-1}
CuS	c	−12.7	−12.8	15.9
CuSO$_4$	c	−184.36	−158.2	26
F$_2$	g	0	0	48.44
F	g	18.88	14.80	37.92
F$^-$	aq	−79.50	−66.64	−3.3
HF	g	−64.8	−65.3	41.51
HF	aq	−76.50	−70.95	21.2
HF$_2^-$	aq	−155.34	−138.18	22.1
F$_2$O	g	−5.2	−1.1	59.1
Fe	c	0	0	6.52
Fe^{2+}	aq	−21.3	−18.85	−32.9
Fe^{3+}	aq	−11.6	−1.1	−75.5
Fe$_{0.947}$O	c	−63.64	−58.59	13.74
Fe$_2$O$_3$	c	−197.0	−177.4	20.89
Fe$_3$O$_4$	c	−267.3	−242.7	35.0
FeCl$_2$	c	−81.69	−72.26	28.19
FeCl$_3$	c	−95.48	−79.84	34.0
Ge	c	0	0	7.43
Ge	g	90.0	80.3	40.10
GeO	brown	−50.7	−56.7	12
GeO$_2$	c	−131.7	−118.8	13.21
H$_2$GeO$_3$	aq	−195.73		
GeH$_4$	g	21.7	27.1	51.87
Ge$_2$H$_6$	g	38.8		
Ge$_3$H$_8$	g	54.2		
GeF$_4$	g	−284.4		
GeCl$_4$	g	−118.5	−109.3	83.08
GeBr$_4$	g	−71.7	−76.0	94.66
GeI$_4$	g	−13.6	−25.4	102.5
Ge$_3$N$_4$	c	−15.1		
H$_2$	g	0	0	31.208
H	g	52.095	48.581	27.391
H$^+$	aq	0	0	0
H$^-$	g	33.39		
H$^-$	aq		52	
Hg	l	0	0	18.17
Hg^{2+}	aq	40.9	39.30	−7.7
Hg$_2^{2+}$	aq	41.2	36.70	20.2
HgO	red	−21.71	−13.995	16.80
HHgO$_2^-$	aq		−45.5	
HgCl$_2$	c	−53.6	−42.7	34.9
Hg$_2$Cl$_2$	c	−63.39	−50.38	46.0
HgI$_2$	c	−25.2	−24.3	43
Hg$_2$I$_2$	c	−29.00	−26.53	55.8
I$_2$	c	0	0	27.757
I$_2$	g	14.923	4.627	62.28
I	g	25.535	16.798	43.184
HI	g	6.33	0.41	49.351

TABLE E.1 (*continued*)

Species	State	ΔH_f°, kcal mol^{-1}	ΔG_f°, kcal mol^{-1}	S°, cal deg^{-1} mol^{-1}
I$^-$	aq	-13.19	-12.33	26.6
IO$^-$	aq	-25.7	-9.2	-1.3
IO$_3^-$	aq	-52.9	-30.6	28.3
IO$_4^-$	aq	-35.2		
IF	g	-22.86	-28.32	56.42
ICl	g	4.25	-1.30	59.14
IBr	g	9.76	0.89	61.82
IF$_5$	g	-196.58	-179.68	78.3
IF$_7$	g	-225.6	-195.6	82.8
K	c	0	0	15.34
K$^+$	aq	-60.32	-67.70	24.5
K$_2$O	c	-86.4		
KOH	c	-101.52		
KF	c	-135.58	-128.53	15.91
KI	c	-78.37		
K$_2$SO$_4$	c	-343.69		
K$_2$CO$_3$	c	-275.0	-254.7	37.4
La	c	0	0	13.6
La^{3+}	aq	-169.0	-163.4	-52.0
La$_2$O$_3$	c	-428.7	-407.7	30.43
LaCl$_3$	c	-256.0		
Li	c	0	0	6.70
Li$^+$	aq	-66.55	-70.22	3.4
Li$_2$O	c	-142.4	-133.8	8.97
Li$_2$O$_2$	c	-151.7		
LiN$_3$	c	2.6		
Li$_2$CO$_3$	c	-290.54	-270.66	21.60
LiF	c	-147.1		
LiI	c	-64.6		
Mg	c	0	0	7.81
Mg^{2+}	aq	-111.58	-108.7	-33.0
MgO	c	-143.81	-136.10	6.44
Mg(OH)$_2$	c	-220.97	-199.23	15.10
MgCl$_2$	c	-153.28	-141.45	21.42
MgSO$_4$	c	-307.1	-279.8	21.9
Mg(NO$_3$)$_2$	c	-188.97	-140.9	39.2
Mn	c	0	0	7.65
Mn^{2+}	aq	-52.76	-54.5	-17.6
MnO	c	-92.07	-86.74	14.27
MnO$_2$	c	-124.29	-111.18	12.68
MnO$_4^-$	aq	-129.4	-106.9	45.7
MnCl$_2$	c	-115.03	-105.29	28.26
N$_2$	g	0	0	45.77
N	g	112.979	108.883	36.622
N$_3^-$	aq	65.76	83.2	25.8
NO	g	21.57	20.69	50.347
NO$_2$	g	7.93	12.26	57.35

TABLE E.1 (*continued*)

Species	State	ΔH_f°, kcal mol^{-1}	ΔG_f°, kcal mol^{-1}	S°, cal deg^{-1} mol^{-1}
NO$_2^-$	aq	-25.0	-8.9	33.5
NO$_3^-$	aq	-49.56	-26.61	35.0
N$_2$O	g	19.61	24.90	52.52
N$_2$O$_3$	g	20.01	33.32	74.61
N$_2$O$_4$	l	-4.66	23.29	50.0
N$_2$O$_4$	g	2.19	23.38	72.70
N$_2$O$_5$	c	-10.3	27.2	42.6
N$_2$O$_5$	g	2.7	27.5	85.0
NH$_3$	g	-11.02	-3.94	45.97
NH$_3$	aq	-19.19	-6.35	26.6
N$_2$H$_4$	l	12.10	35.67	28.97
N$_2$H$_4$	aq	8.20	30.6	33
N$_2$H$_4$	g	22.80	38.07	56.97
NH$_4^+$	aq	-31.67	-18.97	27.1
N$_2$H$_5^+$	aq	-1.8	19.7	36
HN$_3$	aq	62.16	76.9	34.9
HNO$_2$	aq	-28.5	-13.3	36.5
NH$_2$OH	c	-27.3		
NF$_3$	g	-29.8	-20.0	62.29
N$_2$F$_4$	g	-1.7	19.4	71.96
NCl$_3$	l	55		
NOF	g	-15.9	-12.2	59.27
NO$_2$F	g	-26.0	-15.9	62.2
NOCl	g	12.36	15.78	62.52
NH$_4$F	c	-110.89	-83.36	17.20
NH$_4$Cl	c	-75.15	-48.51	22.6
NH$_4$Br	c	-64.73	-41.9	27
NH$_4$I	c	-48.14	-26.9	28
Na	c	0	0	12.24
Na$^+$	aq	-57.39	-62.59	14.1
Na$_2$O	c	-99.7		
Na$_2$O$_2$	c	-122.30		
NaOH	c	-101.72		
NaF	c	-137.11		
NaCl	c	-98.27	-91.84	17.33
NaClO$_3$	c	-85.73		
NaI	c	-68.78		
Na$_2$SO$_4$	c	-331.52	-303.39	35.73
NaNO$_3$	c	-111.54	-87.45	27.8
Na$_2$CO$_3$	c	-270.4	-250.5	32.5
Ni	c	0	0	7.14
Ni^{2+}	aq	-12.9	-10.9	-30.8
NiO	c	-57.3	-50.6	9.08
Ni(OH)$_2$	c	-126.6	-106.9	21
Ni(NH$_3$)$_6^{2+}$	aq	-150.6	-61.2	94.3
Ni(CN)$_4^{2-}$	aq	87.9	122.8	52
O$_2$	g	0	0	49.003
O	g	59.553	55.389	38.467

TABLE E.1 (*continued*)

Species	State	ΔH_f°, kcal mol^{-1}	ΔG_f°, kcal mol^{-1}	S°, cal deg^{-1} mol^{-1}
O_3	g	34.1	39.0	57.08
OH^-	g	−33.67		
OH^-	aq	−54.970	−37.594	−2.57
H_2O	l	−68.315	−56.687	16.71
H_2O	g	−57.796	−54.634	45.10
H_2O_2	l	−44.88	−28.78	26.2
H_2O_2	g	−32.58	−25.24	55.6
P	white	0	0	9.82
P	red	−4.2	−2.9	5.45
P	g	79.8		
P_2	g	34.5	24.8	52.108
P_4	g	14.08	5.85	66.89
PO_4^{3-}	aq	−305.3	−243.5	−53
P_4O_6	c	−392.0		
P_4O_{10}	hexagonal	−713.2	−644.8	54.70
PH_3	g	1.3	3.2	50.22
P_2H_4	g	5.0		
H_3PO_3	c	−230.5		
H_3PO_3	aq	−230.6		
HPO_3^{2-}	aq	−231.6		
HPO_4^{2-}	aq	−308.83	−260.34	−8.0
$H_2PO_2^-$	aq	−146.7		
$H_2PO_3^-$	aq	−231.7		
$H_2PO_4^-$	aq	−309.82	−260.17	21.6
H_3PO_4	aq	−307.92	−273.10	37.8
PF_3	g	−219.6	−214.5	65.28
PCl_3	l	−76.4	−65.1	51.9
PCl_3	g	−68.6	−64.0	74.49
PBr_3	g	−33.3	−38.9	83.17
PF_5	g	−381.4		
PCl_5	g	−89.6	−73.0	87.11
PH_4Cl	c	−34.7		
PH_4Br	c	−30.5	−11.4	26.3
PH_4I	c	−16.7	0.2	29.4
Pb	c	0	0	15.49
Pb^{2+}	aq	−0.4	−5.83	2.5
PbO	red	−52.34	−45.16	15.9
PbO_2	c	−66.3	−51.95	16.4
$HPbO_2^-$	aq		−80.90	
$Pb(OH)_2$	c		−108.1	
$PbCl_2$	c	−85.90	−75.08	32.5
PbI_2	c	−41.94	−41.50	41.79
Rb	c	0	0	18.35
Rb^+	aq	−60.03	−67.87	29.04
RbO_2	c	−66.6		
Rb_2O	c	−81		
Rb_2O_2	c	−112.8		
RbOH	c	−99.95		

TABLE E.1 (*continued*)

Species	State	ΔH_f°, kcal mol^{-1}	ΔG_f°, kcal mol^{-1}	S°, cal deg^{-1} mol^{-1}
RbF	c	−133.3		
RbCl	c	−104.05	−97.47	22.92
RbI	c	−79.77	−78.60	28.30
Rb$_2$SO$_4$	c	−343.12	−314.76	47.19
Rb$_2$CO$_3$	c	−271.5	−251.1	43.34
S	rhombic	0	0	7.60
S	g	66.6	56.9	40.094
S^{2-}	aq	7.9	20.5	−3.5
S$_2$	g	30.68	18.96	54.51
S$_8$	g	24.45	11.87	102.98
SO	g	1.5	−4.7	53.02
SO$_2$	g	−70.944	−71.748	59.30
SO$_3$	β-c	−108.63	−88.19	12.5
SO$_3$	g	−94.58	−88.69	61.34
SO$_2$	aq	−77.194	−71.871	38.7
SO$_3{}^{2-}$	aq	−151.9	−116.3	−7
HSO$_3{}^-$	aq	−149.67	−126.15	33.4
HSO$_4{}^-$	aq	−212.08	−180.69	31.5
SO$_4{}^{2-}$	aq	−217.32	−177.97	4.8
H$_2$S	g	−4.93	−8.02	49.16
H$_2$S$_4$	l	−7.85		
H$_2$S$_6$	l	−8.85		
SCl$_2$	g	−4.7		
SF$_4$	g	−185.2	−174.8	69.77
SF$_6$	g	−289	−264.2	69.72
Sb	c	0	0	10.92
Sb	g	62.7	53.1	43.06
Sb$_4$	g	49.0	33.8	84
SbH$_3$	g	34.681	35.31	55.61
SbO$^+$	aq		−42.33	
SbO$_2{}^-$	aq		−81.32	
HSbO$_2$	aq	−116.6	−97.4	11.1
SbCl$_3$	g	−75.0	−72.0	80.71
SbBr$_3$	g	−46.5	−53.5	89.09
Sc	c	0	0	8.28
Sc^{3+}	aq	−146.8	−140.2	−61
Sc$_2$O$_3$	c	−456.22	−434.85	18.4
ScCl$_3$	c	−221.1		
Se	black	0	0	10.144
Se	g	54.27	44.71	42.22
H$_2$Se	g	7.1	3.8	52.32
SeO$_2$	c	−53.86		
H$_2$SeO$_3$	aq	−121.29	−101.87	49.7
SeO$_4{}^{2-}$	aq	−143.2	−105.5	12.9
SeCl$_2$	g	−7.6		
Se$_2$Cl$_2$	g	4		
Se$_2$Br$_2$	g	7		
Si	c	0	0	4.50

TABLE E.1 (*continued*)

Species	State	ΔH_f°, kcal mol^{-1}	ΔG_f°, kcal mol^{-1}	S°, cal deg^{-1} mol^{-1}
Si	g	108.9	98.3	40.12
SiO	g	−23.8	−30.2	50.55
SiO$_2$	quartz	−217.72	−204.75	10.00
H$_4$SiO$_4$	aq	−351.0	−314.7	43
SiH$_4$	g	8.2	13.6	48.88
Si$_2$H$_6$	g	19.2	30.4	65.14
Si$_3$H$_8$	g	28.9		
SiF$_4$	g	−386.0	−375.9	67.49
SiF$_6^{2-}$	aq	−571.0	−525.7	29.2
SiCl$_4$	g	−157.0	−147.5	79.02
SiBr$_4$	g	−99.3	−103.2	90.29
Si$_3$N$_4$	c	−177.7	−153.6	24.2
SiC	cubic	−15.6	−15.0	3.97
Sn	white	0	0	12.32
Sn	gray	−0.50	0.03	10.55
Sn	g	72.2	63.9	40.24
Sn^{2+}	aq HCl	−2.1	−6.5	−4
SnO	c	−68.3	−61.4	13.5
SnO$_2$	c	−138.8	−124.2	12.5
Sn^{4+}	aq HCl	7.3	0.6	−28
SnH$_4$	g	38.9	45.0	54.39
SnCl$_4$	g	−112.7	−103.3	87.4
SnBr$_4$	g	−75.2	−79.2	98.43
Sr	c	0	0	12.5
Sr^{2+}	aq	−130.45	−133.71	−7.8
SrO	c	−141.5	−134.3	13.0
SrO$_2$	c	−151.4		
Sr(OH)$_2$	c	−229.2		
SrCl$_2$	c	−198.1	−186.7	27.45
SrCO$_3$	c	−291.6	−272.5	23.2
Te	c	0	0	11.88
Te	g	47.02	37.55	43.65
H$_2$Te	g	23.8		
TeO$_2$	c	−77.1	−64.6	19.0
TeO$_3^{2-}$	aq	−142.6		
TeF$_6$	g	−315		
Ti	c	0	0	7.32
TiO^{2+}	aq	−164.9		
TiO	c	−124.2	−118.3	8.31
TiO$_2$	c	−225.8	−212.6	12.03
TiCl$_2$	c	−122.8	−111.0	20.9
TiCl$_3$	c	−172.3	−156.2	33.4
TiCl$_4$	g	−182.4	−173.7	84.8
V	c	0	0	6.91
VO	c	−103.2	−96.6	9.3
VO^{2+}	aq	−116.3	−106.7	−32.0
VO$_2^+$	aq	−155.3	−140.3	−10.1

TABLE E.1 (*continued*)

Species	State	ΔH_f°, kcal mol^{-1}	ΔG_f°, kcal mol^{-1}	S°, cal deg^{-1} mol^{-1}
VO$_4$$^{3-}$	aq		-214.9	
V$_2$O$_7$$^{4-}$	aq		-411	
VCl$_2$	c	-108	-97	23.2
VCl$_3$	c	-138.8	-122.2	31.3
VCl$_4$	g	-125.6	-117.6	86.6
Zn	c	0	0	9.95
Zn^{2+}	aq	-36.78	-35.14	-26.8
ZnO	c	-83.24	-76.08	10.43
Zn(OH)$_4$$^{2-}$	aq		-205.23	
ZnF$_2$	c	-182.7	-107.5	17.61
ZnCl$_2$	c	-99.20	-88.30	26.64
ZnCO$_3$	c	-194.26	-174.85	19.7

† 1 kcal = 4.1840 kJ.

APPENDIX
F

IONIC RADII

TABLE F.1
Ionic radii of the elements†

Ion	Coordination number‡	Radius, Å	Ion	Coordination number	Radius, Å
Ac^{3+}	6	1.12	At^{7+}	6	0.62
Ag^+	2	0.67	Au^+	6	1.37
	4	1.00	Au^{3+}	4 sq	0.68
	4 sq	1.02		6	0.85
	5	1.09	Au^{5+}	6	0.57
	6	1.15	B^{3+}	3	0.01
	7	1.22		4	0.11
	8	1.28		6	0.27
Ag^{2+}	4 sq	0.79	Ba^{2+}	6	1.35
	6	0.94		7	1.38
Ag^{3+}	4 sq	0.67		8	1.42
	6	0.75		9	1.47
Al^{3+}	4	0.39		10	1.52
	5	0.48		11	1.57
	6	0.535		12	1.61
Am^{2+}	7	1.21	Be^{2+}	3	0.16
	8	1.26		4	0.27
	9	1.31		6	0.45
Am^{3+}	6	0.975	Bi^{3+}	5	0.96
	8	1.09		6	1.03
Am^{4+}	6	0.85		8	1.17
	8	0.95	Bi^{5+}	6	0.76
As^{3+}	6	0.58	Bk^{3+}	6	0.96
As^{5+}	4	0.335	Bk^{4+}	6	0.83
	6	0.46		8	0.93

TABLE F.1 (*continued*)

Ion	Coordination number	Radius Å	Ion	Coordination number	Radius Å
Br^-	6	1.96		6 HS	0.53
Br^{3+}	4 sq	0.59	Cr^{2+}	6 LS	0.73
Br^{5+}	3 py	0.31		HS	0.80
Br^{7+}	4	0.25	Cr^{3+}	6	0.615
	6	0.39	Cr^{4+}	4	0.41
C^{4+}	3	−0.08		6	0.55
	4	0.15	Cr^{5+}	4	0.345
	6	0.16		6	0.49
Ca^{2+}	6	1.00		8	0.57
	7	1.06	Cr^{6+}	4	0.26
	8	1.12		6	0.44
	9	1.18	Cs^+	6	1.67
	10	1.23		8	1.74
	12	1.34		9	1.78
Cd^{2+}	4	0.78		10	1.81
	5	0.87		11	1.85
	6	0.95		12	1.88
	7	1.03	Cu^+	2	0.46
	8	1.10		4	0.60
	12	1.31		6	0.77
Ce^{3+}	6	1.01	Cu^{2+}	4	0.57
	7	1.07		4 sq	0.57
	8	1.143		5	0.65
	9	1.196		6	0.73
	10	1.25	Cu^{3+}	6 LS	0.54
	12	1.34	D^+	2	−0.10
Ce^{4+}	6	0.87	Dy^{2+}	6	1.07
	8	0.97		7	1.13
	10	1.07		8	1.19
	12	1.14	Dy^{3+}	6	0.912
Cf^{3+}	6	0.95		7	0.97
Cf^{4+}	6	0.821		8	1.027
	8	0.92		9	1.083
Cl^-	6	1.81	Er^{3+}	6	0.890
Cl^{5+}	3 py	0.12		7	0.945
Cl^{7+}	4	0.08		8	1.004
	6	0.27		9	1.062
Cm^{3+}	6	0.97	Eu^{2+}	6	1.17
Cm^{4+}	6	0.85		7	1.20
	8	0.95		8	1.25
Co^{2+}	4 HS	0.58		9	1.30
	5	0.67		10	1.35
	6 LS	0.65	Eu^{3+}	6	0.947
	HS	0.745		7	1.01
	8	0.90		8	1.066
Co^{3+}	6 LS	0.545		9	1.120
	HS	0.61	F^-	2	1.285
Co^{4+}	4	0.40		3	1.30

TABLE F.1 (*continued*)

Ion	Coordination number	Radius, Å	Ion	Coordination number	Radius, Å
	4	1.31		6	0.800
	6	1.33		8	0.92
F^{7+}	6	0.08	Ir^{3+}	6	0.68
Fe^{2+}	4 HS	0.63	Ir^{4+}	6	0.625
	4 sq HS	0.64	Ir^{5+}	6	0.57
	6 LS	0.61	K^+	4	1.37
	HS	0.780		6	1.38
	8 HS	0.92		7	1.46
Fe^{3+}	4 HS	0.49		8	1.51
	5	0.58		9	1.55
	6 LS	0.55		10	1.59
	HS	0.645		12	1.64
	8 HS	0.78	La^{3+}	6	1.032
Fe^{4+}	6	0.585		7	1.10
Fe^{6+}	4	0.25		8	1.160
Fr^+	6	1.80		9	1.216
Ga^{3+}	4	0.47		10	1.27
	5	0.55		12	1.36
	6	0.620	Li^+	4	0.590
Gd^{3+}	6	0.938		6	0.76
	7	1.00		8	0.92
	8	1.053	Lu^{3+}	6	0.861
	9	1.107		8	0.977
Ge^{2+}	6	0.73		9	1.032
Ge^{4+}	4	0.390	Mg^{2+}	4	0.57
	6	0.530		5	0.66
H^+	1	-0.38		6	0.720
	2	-0.18		8	0.89
Hf^{4+}	4	0.58	Mn^{2+}	4 HS	0.66
	6	0.71		5 HS	0.75
	7	0.76		6 LS	0.67
	8	0.83		HS	0.830
Hg^+	3	0.97		7	0.90
	6	1.19		8	0.96
Hg^{2+}	2	0.69	Mn^{3+}	5	0.58
	4	0.96		6 LS	0.58
	6	1.02		HS	0.645
	8	1.14	Mn^{4+}	4	0.39
Ho^{3+}	6	0.901		6	0.530
	8	1.015	Mn^{5+}	4	0.33
	9	1.072	Mn^{6+}	4	0.255
	10	1.12	Mn^{7+}	4	0.25
I^-	6	2.20		6	0.46
I^{5+}	3 py	0.44	Mo^{3+}	6	0.69
	6	0.95	Mo^{4+}	6	0.650
I^{7+}	4	0.42	Mo^{5+}	4	0.46
	6	0.53		6	0.61
In^{3+}	4	0.62	Mo^{6+}	4	0.41

TABLE F.1 (*continued*)

Ion	Coordination number	Radius, Å	Ion	Coordination number	Radius, Å
	5	0.50		3	1.34
	6	0.59		4	1.35
	7	0.73		6	1.37
N^{3-}	4	1.46	Os^{4+}	6	0.630
N^{3+}	6	0.16	Os^{5+}	6	0.575
N^{5+}	3	−0.104	Os^{6+}	5	0.49
	6	0.13		6	0.545
Na^+	4	0.99	Os^{7+}	6	0.525
	5	1.00	Os^{8+}	4	0.39
	6	1.02	P^{3+}	6	0.44
	7	1.12	P^{5+}	4	0.17
	8	1.18		5	0.29
	9	1.24		6	0.38
	12	1.39	Pa^{3+}	6	1.04
Nb^{3+}	6	0.72	Pa^{4+}	6	0.90
Nb^{4+}	6	0.68		8	1.01
	8	0.79	Pa^{5+}	6	0.78
Nb^{5+}	4	0.48		8	0.91
	6	0.64		9	0.95
	7	0.69	Pb^{2+}	4 py	0.98
	8	0.74		6	1.19
Nd^{2+}	8	1.29		7	1.23
	9	1.35		8	1.29
Nd^{3+}	6	0.983		9	1.35
	8	1.109		10	1.40
	9	1.163		11	1.45
	12	1.27		12	1.49
Ni^{2+}	4	0.55	Pb^{4+}	4	0.65
	4 sq	0.49		5	0.73
	5	0.63		6	0.775
	6	0.690		8	0.94
Ni^{3+}	6 LS	0.56	Pd^+	2	0.59
	HS	0.60	Pd^{2+}	4 sq	0.64
Ni^{4+}	6 LS	0.48		6	0.86
No^{2+}	6	1.1	Pd^{3+}	6	0.76
Np^{2+}	6	1.10	Pd^{4+}	6	0.615
Np^{3+}	6	1.01	Pm^{3+}	6	0.97
Np^{4+}	6	0.87		8	1.093
	8	0.98		9	1.144
Np^{5+}	6	0.75	Po^{4+}	6	0.94
Np^{6+}	6	0.72		8	1.08
Np^{7+}	6	0.71	Po^{6+}	6	0.67
O^{2-}	2	1.35	Pr^{3+}	6	0.99
	3	1.36		8	1.126
	4	1.38		9	1.179
	6	1.40	Pr^{4+}	6	0.85
	8	1.42		8	0.96
OH^-	2	1.32	Pt^{2+}	4 sq	0.60

TABLE F.1 (*continued*)

Ion	Coordination number	Radius, Å	Ion	Coordination number	Radius, Å
	6	0.80		8	1.27
Pt^{4+}	6	0.625		9	1.32
Pt^{5+}	6	0.57	Sm^{3+}	6	0.958
Pu^{3+}	6	1.00		7	1.02
Pu^{4+}	6	0.86		8	1.079
	8	0.46		9	1.132
Pu^{5+}	6	0.74		12	1.24
Pu^{6+}	6	0.71	Sn^{4+}	4	0.55
Ra^{2+}	8	1.48		5	0.62
	10	1.70		6	0.690
Rb^{+}	6	1.52		7	0.75
	7	1.56		8	0.81
	8	1.61	Sr^{2+}	6	1.18
	9	1.63		7	1.21
	10	1.66		8	1.26
	11	1.69		9	1.31
	12	1.72		10	1.36
	14	1.83		12	1.44
Re^{4+}	6	0.63	Ta^{3+}	6	0.72
Re^{5+}	6	0.58	Ta^{4+}	6	0.68
Re^{6+}	6	0.55	Ta^{5+}	6	0.64
Re^{7+}	4	0.38		7	0.69
	6	0.53		8	0.74
Rh^{3+}	6	0.665	Tb^{3+}	6	0.923
Rh^{4+}	6	0.60		7	0.98
Rh^{5+}	6	0.55		8	1.040
Ru^{3+}	6	0.68		9	1.095
Ru^{4+}	6	0.620	Tb^{4+}	6	0.76
Ru^{5+}	6	0.565		8	0.88
Ru^{7+}	4	0.38	Tc^{4+}	6	0.645
Ru^{8+}	4	0.36	Tc^{5+}	6	0.60
S^{2-}	6	1.84	Tc^{7+}	4	0.37
S^{4+}	6	0.37		6	0.56
S^{6+}	4	0.12	Te^{2-}	6	2.21
	6	0.29	Te^{4+}	3	0.52
Sb^{3+}	4 py	0.76		4	0.66
	5	0.80		6	0.97
	6	0.76	Te^{6+}	4	0.43
Sb^{5+}	6	0.60		6	0.56
Sc^{3+}	6	0.745	Th^{4+}	6	0.94
	8	0.870		8	1.04
Se^{2-}	6	1.98		9	1.09
Se^{4+}	6	0.50		10	1.13
Se^{6+}	4	0.28		11	1.18
	6	0.42		12	1.21
Si^{4+}	4	0.26	Ti^{2+}	6	0.86
	6	0.400	Ti^{3+}	6	0.670
Sm^{2+}	7	1.22	Ti^{4+}	4	0.42

TABLE F.1 (*continued*)

Ion	Coordination number	Radius, Å	Ion	Coordination number	Radius, Å
	5	0.51	V^{5+}	4	0.355
	6	0.605		5	0.46
	8	0.74		6	0.54
Tl^+	6	1.50	W^{4+}	6	0.66
	8	1.59	W^{5+}	6	0.62
	12	1.70	W^{6+}	4	0.42
Tl^{3+}	4	0.75		5	0.51
	6	0.885		6	0.60
	8	0.98	Xe^{8+}	4	0.40
Tm^{2+}	6	1.03		6	0.48
	8	1.09	Y^{3+}	6	0.900
Tm^{3+}	6	0.880		7	0.96
	8	0.994		8	1.019
	9	1.052		9	1.075
U^{3+}	6	1.025	Yb^{2+}	6	1.02
U^{4+}	6	0.89		7	1.08
	7	0.95		8	1.14
	8	1.00	Yb^{3+}	6	0.868
	9	1.05		7	0.925
	12	1.17		8	0.985
U^{5+}	6	0.76		9	1.042
	7	0.84	Zn^{2+}	4	0.60
U^{6+}	2	0.45		5	0.68
	4	0.52		6	0.740
	6	0.73		8	0.90
	7	0.81	Zr^{4+}	4	0.59
	8	0.86		5	0.66
V^{2+}	6	0.79		6	0.72
V^{3+}	6	0.640		7	0.78
V^{4+}	5	0.53		8	0.84
	6	0.58		9	0.89
	8	0.72			

† The "traditional" scale of ionic radii, as tabulated by R. D. Shannon, *Acta Crystallogr.*, **A32**, 751 (1976). "Effective ionic radii," which probably correspond more closely to the physical sizes of ions in solids, can be obtained from the traditional values by adding 0.14 to the cation radii and subtracting 0.14 from the anion radii.

‡ sq and py indicate square and pyramidal coordination, and LS and HS indicate low- and high-spin states.

APPENDIX
G

INORGANIC NOMENCLATURE

As the complexity of the compounds studied by inorganic chemists increases, the difficulty of unambiguously describing these compounds with simple names increases. Rules of nomenclature seldom anticipate the difficulties associated with naming novel compounds with unusual structures. Nevertheless, it is important to be aware of the approved conventions for naming inorganic compounds. The following paragraphs have been excerpted and adapted from the rules issued by the International Commission on the Nomenclature of Inorganic Chemistry.[1] Current American usage differs from these rules in some respects, and, in this text, a few liberties have been taken which the author believes will not cause the reader any confusion.

FORMULAS AND NAMES OF COMPOUNDS IN GENERAL

In formulas the "electropositive constituent" (cation) should always be placed first, e.g., KCl, $CaSO_4$. In the case of binary compounds between nonmetals, in accordance with established practice, that constituent should be placed first which appears earlier in the sequence Rn, Xe, Kr, B, Si, C, Sb, As, P, N, H, Te, Se, S, At, I, Br, Cl, O, F.

Examples: XeF_2, NH_3, H_2S, S_2Cl_2, Cl_2O, OF_2

---

[1] "Nomenclature of Inorganic Chemistry," 2d ed., Butterworth, London, 1971; also see *Pure Appl. Chem.*, **28**, 67 (1971); adapted and reproduced with permission.

The name of the electropositive constituent is not modified. If the "electronegative constituent" is monatomic or homopolyatomic, its name is modified to end in "-ide." For binary compounds the name of the element standing later in the sequence given above is modified to end in -ide; sodium plumbide, sodium chloride, lithium nitride, arsenic selenide, boron hydride, hydrogen chloride, hydrogen sulfide, silicon carbide, chlorine dioxide, oxygen difluoride, etc.

If the electronegative constituent is heteropolyatomic, it should be designated by the termination "-ate." In certain exceptional cases the terminations "-ide" and "-ite" are used (see below). It is generally possible in a polyatomic group to indicate a "characteristic atom" (as in ClO^-) or a "central atom" (as in ICl_4^-). Such a polyatomic group is designated a "complex," and the atoms, radicals, or molecules bound to the characteristic or central atom are termed "ligands." In this case the name of a negatively charged complex should be formed from the name of the characteristic or central element modified to end in -ate.

Binary hydrogen compounds may be named by the principles given above. Volatile hydrides, except those of group VII and of oxygen and nitrogen, may also be named by citing the root name of the element followed by the suffix "-ane." If the molecule contains more than one atom of that element, the number is indicated by the appropriate Greek prefix.

Recognized exceptions are water, ammonia, and hydrazine, owing to long usage. Phosphine, arsine, stibine, and bismuthine are also allowed. However, for all molecular hydrides containing more than one atom of the element, -ane names should be used.

B_2H_6	diborane	PbH_4	plumbane
Si_3H_8	trisilane	H_2S_n	polysulfane

NAMES FOR IONS AND RADICALS

Monatomic cations should be named as the corresponding element, without change or suffix:

Cu^+	the copper(I) ion
Cu^{2+}	the copper(II) ion
I^+	the iodine(I) cation

This principle should also apply to polyatomic cations corresponding to radicals for which special names are given; i.e., these names should be used without change or suffix: the nitrosyl cation (NO^+), the nitryl cation (NO_2^+).

Names for polyatomic cations derived by the addition of more protons than required to give a neutral unit to monatomic anions are formed by adding the ending "-onium" to the root of the name of the anion element: phosphonium, arsonium, sulfonium, iodonium. The name "ammonium" for the ion NH_4^+ does not conform to the rule but is retained. Substituted ammonium ions derived from nitrogen bases with names ending in "-amine" receive names formed by changing

-amine to -ammonium, for example, $HONH_3^+$, the hydroxylammonium ion. When the nitrogen base is known by a name ending otherwise than in -amine, the cation name is formed by adding the ending "-ium" to the name of the base (if necessary omitting a final "e" or other vowel): hydrazinium, anilinium, glycinium, pyridinium.

The names for monatomic anions consist of the name (sometimes abbreviated) of the elements with the termination -ide:

H^-	hydride ion	N_3^-	azide ion
F^-	fluoride ion	O^{2-}	oxide ion

Certain polyatomic anions have names ending in -ide:

OH^-	hydroxide ion	N_3^-	azide ion
O_2^{2-}	peroxide ion	NH^{2-}	imide ion
S_2^{2-}	disulfide ion	NH_2^-	amide ion
I_3^-	triiodide ion	CN^-	cyanide ion
HF_2^-	hydrogen difluoride ion	C_2^{2-}	acetylide ion

It is quite practical to treat oxygen in the same manner as other ligands, but it has long been customary to ignore the name of this element altogether in anions and to indicate its presence and proportion by means of a series of prefixes and sometimes also by the suffix -ite in place of -ate. The termination -ite has been used to denote a lower state of oxidation and may be retained in certain trivial names such as the following:

NO_2^-	nitrite	SO_3^{2-}	sulfite	ClO_2^-	chlorite
$N_2O_2^{2-}$	hyponitrite	$S_2O_6^{2-}$	disulfite	ClO^-	hypochlorite
AsO_3^{3-}	arsenite	$S_2O_4^{2-}$	dithionite	IO^-	hypoiodite

A radical is a group of atoms which occurs repeatedly in a number of different compounds. Certain neutral and cationic radicals containing oxygen or other chalcogens have, irrespective of charge, special names ending in "-yl," and provisional retention of the following is approved:

HO	hydroxyl	SO	sulfinyl (thionyl)
CO	carbonyl	SO_2	sulfonyl (sulfuryl)
NO	nitrosyl	S_2O_5	disulfuryl
NO_2	nitryl	SeO	seleninyl
PO	phosphoryl	SeO_2	selenonyl
ClO	chlorosyl	CrO_2	chromyl
ClO_2	chloryl	UO_2	uranyl
ClO_3	perchloryl	NpO_2	neptunyl
(similarly for other halogens)		(similarly for other actinides)	

Radicals analogous to the above containing other chalcogens in place of oxygen are named by adding the prefixes "thio-," "seleno-," etc. Examples are:

$COCl_2$	carbonyl chloride
$PSCl_3$	thiophosphoryl chloride
SO_2NH	sulfonyl (sulfuryl) imide

ACIDS

Acids giving rise to -ide anions are named as binary and pseudobinary compounds of hydrogen, e.g., hydrogen chloride, hydrogen sulfide, hydrogen cyanide.

For the oxo acids the -ous or -ic notation to distinguish between different oxidation states is applied in many cases. The -ous names are restricted to acids corresponding to -ite anions.

The prefix "hypo-" is used to denote a lower oxidation state and may be retained in the following cases:

$H_2N_2O_2$	hyponitrous acid	HOBr	hypobromous acid
$H_4P_2O_6$	hypophosphoric acid	HOI	hypoiodous acid
HOCl	hypochlorous acid		

The prefix "per-" has been used to designate a higher oxidation state and is retained for $HClO_4$, perchloric acid, and corresponding acids of the other elements in group VII. This use of the prefix per- should not be extended to elements of other groups.

The prefixes "ortho-" and "meta-" have been used to distinguish acids differing in the "content of water." The following names are approved:

H_3BO_3	orthoboric acid	$(HBO_2)_n$	metaboric acid
H_4SiO_4	orthosilicic acid	$(H_2SiO_3)_n$	metasilicic acid
H_3PO_4	orthophosphoric acid	$(HPO_3)_n$	metaphosphoric acid
H_5IO_6	orthoperiodic acid		
H_6TeO_6	orthotelluric acid		

The names of other oxo acids are given in the following list.

H_2CO_3	carbonic acid
HOCN	cyanic acid
HNCO	isocyanic acid
HONC	fulminic acid
HNO_3	nitric acid
HNO_2	nitrous acid
$H_4P_2O_7$	diphosphoric or pyrophosphoric acid
$H_5P_3O_{10}$	triphosphoric acid

H_2SO_4	sulfuric acid
$H_2S_2O_7$	disulfuric acid
H_2SO_5	peroxomonosulfuric acid
$H_2S_2O_8$	peroxodisulfuric acid
$H_2S_2O_3$	thiosulfuric acid
$H_2S_2O_6$	dithionic acid
H_2SO_3	sulfurous acid
$H_2S_2O_4$	dithionous acid
$HClO_4$	perchloric acid
$HClO_3$	chloric acid
$HClO_2$	chlorous acid

SALTS CONTAINING ACID HYDROGEN

Names are formed by adding the word "hydrogen," with a numerical prefix where necessary, to denote the replaceable hydrogen in the salt. The word "hydrogen" is placed immediately in front of the anion:

$NaHCO_3$	sodium hydrogencarbonate
LiH_2PO_4	lithium dihydrogenphosphate
KHS	potassium hydrogensulfide

COORDINATION COMPOUNDS

In *formulas* the usual practice is to place the symbol for the central atom(s) *first* (except in formulas which are primarily structural), with the ionic and neutral ligands following and the formula for the whole complex enclosed in square brackets. In *names* the central atom(s) should be placed after the ligands.

The names of coordination entities always have been intended to indicate the charge of the central atom (ion) from which the entity is derived. Since the charge on the coordination entity is the algebraic sum of the charges of the constituents, the necessary information may be supplied by giving either the Stock number (oxidation state of the central ion) or the Ewens-Bassett number (overall charge on the complex ion):

$K_3[Fe(CN)_6]$	potassium hexacyanoferrate (III)
	potassium hexacyanoferrate(3−)
	tripotassium hexacyanoferrate
$K_4[Fe(CN)_6]$	potassium hexacyanoferrate(II)
	potassium hexacyanoferrate(4−)
	tetrapotassium hexacyanoferrate

Structural information may be given in formulas and names by prefixes such as *"cis,"* *"trans,"* *"fac,"* *"mer,"* etc. Anions are given the termination

-ate. Cations and neutral molecules are given no distinguishing termination. The ligands are listed in alphabetical order regardless of the number of each. The name of a ligand is treated as a unit. Thus, "diammine" is listed under "a" and "dimethylamine" under "d."

The names of anionic ligands, whether inorganic or organic, end in "o." In general, if the anionic ligand name ends in -ide, -ite, or -ate, the final "e" is replaced by "o," giving "-ido," "-ito," or "-ato," respectively. Enclosing marks are required for inorganic anionic ligands containing numerical prefixes, as (triphosphato), and for "thio," "seleno," and "telluro" analogs of oxo anions containing more than one atom, as (thiosulfato). Examples of organic anionic ligands which are named in this fashion are:

CH_3COO^-	acetato
$(CH_3)_2N^-$	dimethylamido

The anions listed below do not follow exactly the above rule, and modified forms have become established:

	Ion	Ligand
F^-	fluoride	fluoro
Cl^-	chloride	chloro
Br^-	bromide	bromo
I^-	iodide	iodo
O^{2-}	oxide	oxo
H^-	hydride	hydrido (hydro)[2]
OH^-	hydroxide	hydroxo
O_2^{2-}	peroxide	peroxo
CN^-	cyanide	cyano

The letters in each of the ligand names which are used to determine the alphabetical listing are given in boldface type in the following examples to illustrate the alphabetical arrangement. For many compounds, the oxidation number of the central atom and/or the charge on the ion are so well known that there is no need to use either a Stock number or an Ewens-Bassett number. However, it is not wrong to use such numbers, and they are included here.

$K_2[OsCl_5N]$	potassium pentachloron**i**tridoosmate(2−)
	potassium pentachloron**i**tridoosmate(VI)
$[Co(NH_2)_2(NH_3)_4]OC_2H_5$	di**ami**dotetra**amm**inecobalt(1+) ethoxide
	di**ami**dotetra**amm**inecobalt(III) ethoxide

---

[2] Both "hydrido" and "hydro" are used for coordinated hydrogen, but the latter term usually is restricted to boron compounds.

[CoN₃(NH₃)₅]SO₄   pentaammineazidocobalt(2+) sulfate
pentaammineazidocobalt(III) sulfate

NH₄[Cr(NCS)₄(NH₃)₂]   ammonium diamminetetrakis(isothiocyanato)chromate(1−)
ammonium diamminetetrakis(isothiocyanato)chromate(III)

Ba[BrF₄]₂   barium tetrafluorobromate(1−)
barium tetrafluorobromate(III)

The name of a coordinated molecule is used without change. Neutral ligands are generally set off with enclosing marks.

cis-[PtCl₂(Et₃P)₂]   cis-dichlorobis(triethylphosphine)platinum
cis-dichlorobis(triethylphosphine)platinum(II)

[Pt(py)₄][PtCl₄]   tetrakis(pyridine)platinum(2+) tetrachloroplatinate(2−)
tetrakis(pyridine)platinum(II) tetrachloroplatinate(II)

[Co(en)₃]₂(SO₄)₃   tris(ethylenediamine)cobalt(3+) sulfate
tris(ethylenediamine)cobalt(III) sulfate

K[PtCl₃(C₂H₄)]   potassium trichloro(ethylene)platinate(1−)
potassium trichloro(ethylene)platinate(II) or
potassium trichloromonoethyleneplatinate(II)

Water and ammonia as neutral ligands in coordination complexes are called "aqua" (formerly "aquo") and "ammine," respectively. The groups NO and CO, when linked directly to a metal atom, are called "nitrosyl" and "carbonyl," respectively. In computing the oxidation number these ligands are treated as neutral.

[Cr(H₂O)₆]Cl₃   hexaaquachromium(3+) chloride
hexaaquachromium trichloride

Na₂[Fe(CN)₅NO]   sodium pentacyanonitrosylferrate(2−)
sodium pentacyanonitrosylferrate(III)

K₃[Fe(CN)₅CO]   potassium **carbonyl**pentacyanoferrate(3−)
potassium **carbonyl**pentacyanoferrate(II)

Alternative Modes of Linkage of Some Ligands

The different points of attachment of a ligand may be denoted by adding the italicized symbol(s) for the atom or atoms through which attachment occurs at the end of the name of the ligand. Thus the dithiooxalato anion conceivably may be attached through S or O, and these are distinguished as dithiooxalato-S,S' and dithiooxalato-O,O', respectively.

In some cases different names are already in use for alternative modes of attachment, e.g., thiocyanato (—SCN) and isothiocyanato (—NCS), nitro (—NO$_2$), and nitrito (—ONO).

Use of Abbreviations

In the literature of coordination compounds, abbreviations for ligand names are used extensively, especially in formulas. A list of common abbreviations is given in Table 14.2.

Complexes with Unsaturated Molecules or Groups

The name of the ligand group is given with the prefix η^n, where n is the number of bonding atoms of the ligand. This nomenclature has been described in Chap. 14.

Compounds with Bridging Atoms or Groups

1. A bridging group is indicated by adding the Greek letter μ immediately before its name and separating the name from the rest of the complex by hyphens.
2. Two or more bridging groups of the same kind are indicated by "di-μ-" (or "bis-μ-"), etc.
3. The bridging groups are listed with the other groups in alphabetical order *unless the symmetry of the molecule permits simpler names by the use of multiplicative prefixes.*
4. Where the same ligand is present as a bridging ligand and as a nonbridging ligand, it is cited first as a bridging ligand.

Bridging groups between two centers of coordination are of two types: (1) the two centers are attached to the same atom of the bridging group and (2) the two centers are attached to different atoms of the bridging group. For bridging groups of the first type it is often desirable to indicate the bridging atom. This is done by adding the italicized symbol for the atom at the end of the name of the ligand. For bridging groups of the second type, the symbols of all coordinated atoms are added.

[(NH$_3$)$_5$Cr—OH—Cr(NH$_3$)$_5$]Cl$_5$	μ-hydroxobis[pentaamminechromium(5+)] chloride
	μ-hydroxobis[pentaamminechromium(III)] chloride
[(CO)$_3$Fe(CO)$_3$Fe(CO)$_3$]	tri-μ-carbonyl-bis(tricarbonyliron)
[Br$_2$Pt(SMe$_2$)$_2$PtBr$_2$]	bis(μ-dimethylsulfide)-bis[dibromoplatinum(II)]

Homoatomic Aggregates

There are instances of a finite group of metal atoms with bonds directly between the metal atoms but also with some nonmetal atoms or groups (ligands) intimately associated with the "cluster." The geometrical shape of the cluster is designated by *"triangulo," "quadro," "tetrahedro," "octahedro,"* etc., and the nature of the bonds to the ligands by the conventions for bridging bonds and simple bonds.

$Os_3(CO)_{12}$	dodecacarbonyl-*triangulo*-triosmium
$Cs_3[Re_3Cl_{12}]$	cesium dodecachloro-*triangulo*-trirhenate$(3-)$
	tricesium dodecachloro-*triangulo*-trirhenate
B_4Cl_4	tetrachloro-*tetrahedro*-tetraboron
$[Nb_6Cl_{12}]^{2+}$	dodeca-μ-chloro-*octahedro*-hexaniobium$(2+)$ ion

PREFIXES OR AFFIXES USED IN INORGANIC NOMENCLATURE

Multiplying Affixes

(1) "Mono," "di," "tri," "tetra," "penta," "hexa," "hepta," "octa," "nona" ("ennea"), "deca," "undeca" ("hendeca"), "dodeca," etc., are used by direct joining without hyphens; (2) "bis," "tris," "tetrakis," "pentakis," etc., are used by direct joining without hyphens but usually with enclosing marks around each whole expression to which the prefix applies.

Structural Affixes

These affixes are italicized and separated from the rest of the name by hyphens.

antiprismo	8 atoms bound into a rectangular antiprism
asym	asymmetric
catena	a chain structure; often used to designate linear polymeric substances
cis	2 groups occupying adjacent positions; sometimes used in the sense of *fac*
closo	a cage or closed structure, especially a boron skeleton that is a polyhedron having all triangular faces
cyclo	a ring structure[3]
dodecahedro	8 atoms bound into a dodecahedron with triangular faces
fac	3 groups occupying the corners of the same face of an octahedron

---

[3] Cyclo here is used as a modifier indicating structure and hence is italicized. In organic nomenclature, cyclo is considered to be part of the parent name and therefore is not italicized.

hexahedro	8 atoms bound into a hexahedron (e.g., a cube)
hexaprismo	12 atoms bound into a hexagonal prism
icosahedro	12 atoms bound into a triangular icosahedron
mer	meridional; 3 groups on an octahedron in such a relationship that one is cis to the two others which are themselves trans
nido	a nestlike structure, especially a boron skeleton that is very close to a closed or closo structure
octahedro	6 atoms bound into an octahedron
pentaprismo	10 atoms bound into a pentagonal prism
quadro	4 atoms bound into a quadrangle (e.g., a square)
sym	symmetric
tetrahedro	4 atoms bound into a tetrahedron
trans	2 groups directly across a central atom from each other, i.e., in the polar position on a sphere
triangulo	3 atoms bound into a triangle
triprismo	6 atoms bound into a triangular prism
η	signifies that two or more contiguous atoms of the group are attached to a metal
μ	signifies that the group so designated bridges two centers of coordination
σ	signifies that one atom of the group is attached to a metal

Substitutional Affixes[4]

An affix such as "ferra," "nickela," "sila," "metalla," etc., can be used to indicate the formal substitution of a "heteroatom" in a hydrocarbon. For example, sila-benzene is benzene in which one carbon atom is replaced by a silicon atom, and

$$(OC)_4Fe \begin{array}{c} \diagup CH_2 \\ | \\ \diagdown CH_2 \end{array}$$ is a ferracyclopropane and is an example of a metallacycle.

---

[4] Not approved by the International Commission on the Nomenclature of Inorganic Chemistry.

ANSWERS TO
SELECTED
PROBLEMS

CHAPTER 1

1.1 The probability of finding the electron at a radius r is proportional to the product of the electron density and the area of a sphere of radius r. Inasmuch as the latter area is zero when $r = 0$, the probability of finding the electron at a radius $r = 0$ is zero even though the electron density is greatest at $r = 0$.

1.2 The nodal surfaces of a $3p$ orbital are (1) a plane containing the nucleus and perpendicular to the axis of the orbital and (2) a spherical surface centered on the nucleus.

1.4 Al, 1; S, 2; Sc^{3+}, 0; Cr^{3+}, 3; Ir^{3+}, 4; Dy^{3+}, 5.

1.5 Ru^{2+}

1.11 Because $I(He) > I(Kr)$, the first reaction goes as written. Because $I(Cl) > I(Si)$, the second reaction goes as written. Because $A(Cl) > A(I)$, the third reaction goes in the reverse direction.

1.15 In(I) and In(III); Sn(II) and Sn(IV); Sb(III) and Sb(V); Te(IV) and Te(VI).

1.16 $I = 13.60(9)^2 = 1102$ eV.

1.17
$$I_2(Na) = 13.6 \left\{ 8 \left[\frac{11 - 7(0.35) - 2(0.85)}{2} \right]^2 - 7 \left[\frac{11 - 6(0.35) - 2(0.85)}{2} \right]^2 \right\}$$
$$= 42.5 \text{ eV}$$

$$I_2(Mg) = 13.6 \left[\frac{12 - 8(0.85) - 2(1)}{3} \right]^2 = 15.5 \text{ eV}$$

$$E_B(\text{Ne }1s) = 13.6\left\{2\left[\frac{10-0.3}{1}\right]^2 - \left(\frac{10}{1}\right)^2 + 8\left[\frac{10-2(0.85)-7(0.35)}{2}\right]^2 - 8\left[\frac{10-0.85-7(0.35)}{2}\right]^2\right\}$$

$$= 909 \text{ eV}$$

1.21 Although the first and second ionization energies of lead are very similar to those of tin, the third and fourth are considerably greater than the corresponding IEs of tin. In the series Ge, Sn, Pb, all four IEs have a minimum at Sn. The IEs of Pb are probably high because of incomplete nuclear shielding by the fourteen $4f$ electrons and relativity effects.

1.22 For ions of the same charge, the relative strength of binding to an ion-exchange resin, or of complexation in solution, is principally a function of ionic radius. As a consequence of the lanthanide contraction, the elements in the vicinity of holmium have ionic radii very similar to that of yttrium.

CHAPTER 2

2.1 (*a*) C_s; (*b*) C_1; (*c*) C_{2v}; (*d*) C_{3v}; (*e*) C_{3v}; (*f*) $D_{\infty h}$; (*g*) T_d; (*h*) D_{3h}; (*i*) D_{2d}; (*j*) C_{4v}; (*k*) C_{2v}; (*l*) C_{2h}; (*m*) D_{2h}; (*n*) D_{2d}.

2.3 E_g and T_{2g}.

2.4 (*a*) A_{2u}, B_{1g}, E_{1g}, and E_{2u}; (*b*) A_1 and T_2; (*c*) A_2 and B_1; (*d*) A_{1g}, E_g, T_{1u}.

CHAPTER 3

3.1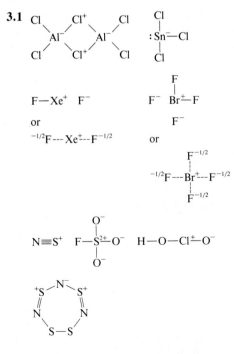

3.2 Hyperconjugated resonance structures such as $O{=}\overset{\displaystyle O^-}{\underset{\displaystyle F}{\overset{|}{\underset{|}{S^{2+}}}}} F^-$ may be important in SO_2F_2,

and therefore the S—O bond order may be comparable to that in SO_2, that is, 1.5. Lone-pair–lone-pair repulsions may somewhat weaken the bonds in SO_2.

3.5 The properties may be estimated from those for the following isoelectronic species: (*a*) NO_2; (*b*) N_2 and CO; (*c*) $H_3C{-}CH_3$; $H_2N{-}NH_2$; and $HO{-}OH$; (*d*) CO_2; (*e*) CH_3; (*f*) SiO_2; (*g*) SO_2; (*h*) N_2O_4.

3.7 Tin is tetravalent (sp^3 hybridization) in tin metal and in SnX_4 molecules. The lower IS values for the latter molecules correspond to higher atomic charges and lower *s*-orbital occupations. In SnX_2 compounds, the tin atoms have lone-pair electrons with relatively high *s*-orbital character. Thus, although the tin atoms in these compounds have positive atomic charges, the IS values are even higher than they are in metallic tin.

3.8 The electronegativity of fluorine is greater than that of oxygen, and the Mn atoms in MnF_2 are more positive than those in MnO_2.

3.11 SO_2, NF_3, SF_4, SF_2, N_2F_4, SiH_3Cl, O_3, BrF_5, S_2Cl_2.

3.13 $\Delta H_f^\circ(N_3H_5) = 58.5$ kcal mol^{-1}; $\Delta H_f^\circ[(GeH_3)_2Se] = 39$ kcal mol^{-1}.

3.15 One estimates an N—F bond length of 1.28 Å. The actual distance in NF_3 is greater because of repulsion between the lone pairs of the N and F atoms.

3.17 Of the various possible resonance structures one can write for $S_4N_3^+$, the one shown as an answer to Prob. 3.1(*h*) is the most satisfactory from the point of view of electrostatic interactions among the formal charges. Hence the S—N bonds indicated as double bonds in that structure are expected to be (and are) the shortest [*Inorg. Chem.*, **19**, 2396 (1980).]

3.19 If 1 mol of diamond is vaporized, $2N$ (not $4N$) C—C bonds are broken. Hence the sublimation energy of diamond is about twice the C—C bond energy.

3.21 $E_B(C\,1s = 309\,eV$; $E_B(C^+1s = 324\,eV$.

3.23 Lone-pair–lone-pair repulsion is much greater for first-row elements than for heavier elements.

CHAPTER 4

4.2 For linear H_3^+, one calculates the following energy levels (in order of increasing energy): $\alpha + \sqrt{2}\beta, \alpha, \alpha - \sqrt{2}\beta$. The two electrons occupy the lowest level, and thus the total energy is $2\alpha + 2\sqrt{2}\beta$. For triangular H_3^+, the energy levels are $\alpha + 2\beta, \alpha - \beta$, and $\alpha - \beta$. With two electrons occupying the lowest level, the total energy is $2\alpha + 4\beta$. Thus we calculate that the triangular form is more stable by $4\beta - 2\sqrt{2}\beta$.

4.4 The bond orders and numbers of unpaired electrons are NeO^+, 1.5, 1; O_2^+, 2.5, 1; CN^+, 2, 0; BN, 2, 0; SiF^+, 3, 0; NO^-, 2, 2; PCl, 2, 2; I_2^+, 1.5, 1; NeH^+, 1, 0.

4.6 The P_4^{2+} ion might be expected to have a square structure, with a total of five bonds (one delocalized P—P π bond and four P—P σ bonds).

4.8 Bi_4^{2-} square planar; Te_4^{2-} zigzag chain; $S_4N_4^{2+}$ planar ring.

4.10 Delocalized π bonding is favored when the atoms involved all have comparable effective electronegativities. In the described compounds, the O and S atoms would have positive formal charges (and effective electronegativities greater than the neutral atom values), and the N and P atoms would have negative formal charges (and effective electronegativities lower than the neutral atom values). In the case of S—N compounds, the effective electronegativities end up with similar magnitudes; in the case of N—O, P—O, and P—S compounds, the effective electronegativity differences are quite large, and alternative structures are favored.

4.12 In the case of N, the $2s$-$2p$ separation is relatively small and therefore the $2s\sigma$ and $2p\sigma$ orbitals of N_2 interact strongly. In the case of F, the increased effective nuclear charge lowers the energy of the $2s$ orbital much more than that of the $2p$ orbital, resulting in a greater $2s$-$2p$ separation and weaker $2s\sigma$-$2p\sigma$ interaction in F_2 than in N_2.

CHAPTER 5

5.1 (a) $2KI + IO_3^- + 6H^+ + 6Cl^- \rightarrow 3ICl_2^- + 6H_2O + 2K^+$

(c) $3Cl_2O + 6OH^- \rightarrow 4Cl^- + 2ClO_3^- + 3H_2O$

(e) $4MnO_4^- + 3H_2PO_2^- + 7H^+ \rightarrow 4MnO_2 + 3H_3PO_4 + 2H_2O$

(g) $S + SO_3^{2-} \rightarrow S_2O_3^{2-}$

(i) $4K_2FeO_4 + 20H^+ \rightarrow 4Fe^{3+} + 3O_2 + 10H_2O + 8K^+$

(m) $2NO_2 + 2OH^- \rightarrow NO_3^- + NO_2^- + H_2O$

(n) $4H_3PO_3 \rightarrow PH_3 + 3H_3PO_4$

(o) $As + 3Ag^+ + 3H_2O \rightarrow 3Ag + H_3AsO_3 + 3H^+$

(t) $Tl(OH)_3 + 4I^- + 3H^+ \rightarrow I_3^- + TlI + 3H_2O$

(u) $3O_3 + Ru^{2+} + H_2O \rightarrow 3O_2 + RuO_4 + 2H^+$

5.4 $E° = -1.73$ V.

5.6 At 25°C, $P_{CO} = 7.1 \times 10^{-13}$ atm and $P_{CO_2} = 1.4 \times 10^{-23}$ atm. At 420°C, $P_{CO} = 0.13$ atm and $P_{CO_2} = 8.9 \times 10^{-6}$ atm.

CHAPTER 6

6.2 Rate $= k[H^+][NO_3^-][HNO_2]$.

6.3 The $\Delta S°$ corresponding to one activated complex is $17 - 40 = -23$ cal deg^{-1} mol^{-1}. This is one-half the $\Delta S°$ value for the reaction as written (with two SO_2 molecules). So only one molecule of SO_2 is involved in the formation of the activated complex.

6.4 $H_2 \overset{K}{\rightleftharpoons} 2H$

$H + p\text{-}H_2 \overset{k}{\longrightarrow} o\text{-}H_2 + H$

6.5 The rate of the reaction $F + H_2 \rightarrow HF + H$ is comparable to that of the reaction $H + F_2 \rightarrow HF + F$ (both reactions are exothermic), and therefore the H-atom concentration is comparable to the F-atom concentration. For this reason one must include the reactions $H + F \rightarrow HF$ and $H + H \rightarrow H_2$ in the mechanism.

6.7 $2NO \rightarrow N_2 + O_2$

$H_2 + N_2 \rightarrow N_2H_2$

$SO_2 + F_2 \rightarrow SO_2F_2$

6.8 The least-motion reaction of singlet CH_2 with H_2 is symmetry-forbidden. However, if the H_2 approaches the CH_2 over the empty p orbital, the reaction is allowed. [See *J. Am. Chem. Soc.*, **99**, 3610 (1977).]

6.11 Both nuclear-fission and branching-chain reactions are exothermic. The neutrons formed in nuclear fission are analogous to the chain-carrying radical species. More neutrons are produced in fission than are consumed, just as is the case with radicals in a branching-chain reaction. The escape of neutrons from the fissioning material is analogous to the chain-terminating steps at the reaction vessel wall. The "critical mass" of nuclear fission is analogous to the first explosion limit.

6.12 Prepare a sample of H_2SO_5 in which the —OOH oxygens are labeled randomly with a certain concentration of ^{18}O. Mix this with an equal amount of ordinary H_2SO_5 and carry out the decomposition. If mass-spectrometric examination of the O_2 shows that one-half of the molecules have the ^{18}O content of the labeled —OOH groups and that one-half of the molecules are isotopically normal, the first mechanism is indicated. If practically all the O_2 molecules contain at least one ^{16}O atom and if the concentration of ^{18}O is one-half that in the labeled —OOH groups, then the second mechanism is indicated.

6.14 1 (*c*); 2 (*b*); 3 (*a*); 4 (*d*).

CHAPTER 7

7.1 (*a*) $CaH_2 + CO_3^{2-} + 2H_2O \rightarrow CaCO_3 + 2H_2 + 2OH^-$

(*c*) $3LiAlH_4 + 4(C_2H_5)_2O \cdot BF_3 \rightarrow 2B_2H_6 + 3LiAlF_4 + 4(C_2H_5)_2O$

(*e*) $AsH_3 + GeH_3^- \rightarrow AsH_2^- + GeH_4$

7.3 $NaNH_2 + GeH_4 \xrightarrow{\text{liq. } NH_3} GeH_3^- + Na^+ + NH_3$

$\quad\quad GeH_3^- + CH_3I \longrightarrow CH_3GeH_3 + I^-$

$\quad CH_3GeH_3 + HBr \longrightarrow CH_3GeH_2Br + H_2$

CHAPTER 8

8.1 If the exchange reaction does not proceed by a mechanism involving the rate-determining step

$$PH_3 + OD^- \xrightarrow{k_1} PH_2^- + HOD$$

then the rate of the actual rate-determining step must be greater than that of the above reaction. Hence $k_1 \leq 0.4\ M^{-1}\ s^{-1}$ and $pK_{PH_3} \geq 27$.

8.3 $NH_3 < CH_3GeH_3 < AsH_3 < C_2H_5OH < H_4SiO_4 < HSeO_4^- < HSO_4^- < H_3O^+ < HSO_3F < HClO_4$.

8.5 Slowest: $NH_3 + OH^- \rightarrow H_2O + NH_2^-$

Fastest: $HSO_4^- + OH^- \rightarrow H_2O + SO_4^{2-}$

8.7 The sublimation of 1 mol of ice involves the cleavage of 2 mol of hydrogen bonds. Therefore the energy per hydrogen bond is 11.9/2 or 6.0 kcal mol^{-1}.

8.9 The stoichiometry of borax corresponds to half-neutralized boric acid, so the solution will consist largely of H_3BO_3 and $B(OH)_4^-$.

8.10 The π bonding between the boron atom and the fluorine atoms of BF_3 is stronger than the analogous π bonding in BCl_3. (The $p\pi$-$p\pi$ overlap for two first-row atoms is greater than that for a first-row and a second-row atom.) Hence it is more difficult to convert the boron atom of BF_3 to sp^3 σ bonding by coordination of a Lewis base because of the greater loss of π-bonding energy.

8.11 (b) $Li_3N + 3NH_4^+ \rightarrow 3Li^+ + 4NH_3$

(d) $H_3BO_3 + 6H_2SO_4 \rightarrow 3H_3O^+ + B(HSO_4)_4^- + 2HSO_4^-$

8.12 $Hphth^- + OH^- \xrightarrow{H_2O} phth^{2-} + H_2O$

$Hphth^- + H^+ \xrightarrow{HOAc} H_2phth$

8.14 The OH^- is strongly hydrogen-bonded to the dissolved water molecules, thus enhancing the solubility of the KOH.

8.16 (b) and (c).

8.18 In the absence of steric effects, $(CH_3)_3N$ would be expected to be the strongest base. However, in $(CH_3)_3BN(CH_3)_3$, the repulsions between methyl groups on adjacent atoms are so great as to make the energy of interaction of $B(CH_3)_3$ with $(CH_3)_3N$ effectively weaker than that of $B(CH_3)_3$ with $(CH_3)_2NH$.

8.19 BH_3, a relatively soft acid, prefers to coordinate to the "soft" phosphorus site. BF_3, a hard acid, prefers the "hard" nitrogen site.

8.20 Cationic species are often good electron acceptors (Lewis acids): in fact they may even be electron-deficient. Hence they can exist only in the absence of good Lewis bases, i.e., in acidic media. The opposite is true for anionic species; they are often good electron donors and can exist only in the absence of good Lewis acids.

CHAPTER 9

9.1 $HI \rightarrow \frac{1}{2}H_2 + \frac{1}{2}I_2$

9.3 Anode: $e^-_{am} \rightarrow e^-$

Cathode: $e^- \rightarrow e^-_{am}$

9.5 (a) $CH_3GeH_3 + e^-_{am} \rightarrow \frac{1}{2}H_2 + CH_3GeH_2^-$

(b) $I_2 + 2e^-_{am} \rightarrow 2I^-$

(c) $(C_2H_5)_2S + 2e^-_{am} + NH_3 \rightarrow C_2H_5S^- + C_2H_6 + NH_2^-$

9.7 The electronic transition of the V band presumably involves ejection of a valence electron from M^- into the solvent. Therefore in the case of a metal with a low electron affinity, such as Cs, the transition should occur at a low energy, and in the case of a metal with a higher electron affinity, such as Na, the transition should occur at a higher energy.

CHAPTER 10

10.1

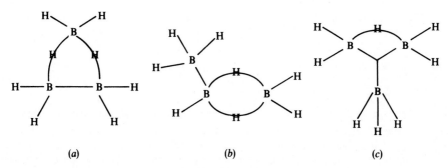

(a)   (b)   (c)

Structure (*a*) is actually observed. In each of the other structures, one boron atom is rather loosely connected to the others by a single bond.

10.3 $b = h = n$, and $q = -2$. Hence $\alpha = n + 3$ and $\beta = n - 2$. If we assume that there are nB—H bonds, then the number of B—B bonds is 3 and the number of B—B—B bonds is $n - 2$.

10.5

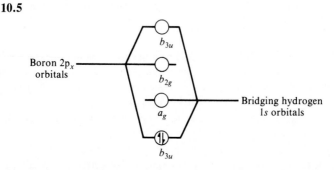

10.8 $B_5H_{12}^-$: 4023, 3114, 2205; B_6H_{10}: 4220, 3311, 2402; B_6H_{12}: 6030, 5121, 4212, 3303.

10.9 Trigonal prism of As atoms.

10.10 The nmr spectrum is explained by a pentagonal pyramidal structure (nido) for $(CCH_3)_6^{2+}$. The $(CH)_5^+$ ion would be expected to have a square pyramidal structure.

CHAPTER 11

11.1

	$\Delta H°$, kcal
$NH_4^+(g) + Cl^-(g) \rightarrow NH_4Cl(s)$	-158.7
$NH_4Cl(s) \rightarrow \frac{1}{2}N_2(g) + 2H_2(g) + \frac{1}{2}Cl_2(g)$	75.2
$\frac{1}{2}Cl_2(g) \rightarrow Cl(g)$	29.1
$Cl(g) + e^-(g) \rightarrow Cl^-(g)$	-83.4
$N(g) + 3H(g) \rightarrow NH_3(g)$	-280.3
$\frac{1}{2}N_2(g) \rightarrow N(g)$	113
$2H_2(g) \rightarrow 4H(g)$	208.4
$H(g) \rightarrow H^+(g) + e^-(g)$	313.6
$NH_4^+(g) \rightarrow NH_3(g) + H^+(g)$	217

11.4 Possible coordination numbers of M and X are 3 and 1, 6 and 2, 9 and 3, etc., respectively.

11.5 The minimum value of r_M/r_X is 0.5275.

11.7 The greater lattice energy of $Mg^{2+}O^{2-}$, compared with that of Mg^+O^-, more than compensates for the higher ionization potential and lower electron affinity. Mg^+O^- would be expected to be paramagnetic, whereas MgO is diamagnetic.

11.9 The S^{2-} and Se^{2-} ions are so much larger than the metal ions that anion-anion contact occurs in the crystals, and the metal ions "rattle" in the anion lattices. For $S^{2-}, r = \frac{1}{2}\sqrt{2}(2.60) = 1.84$ Å.

11.11 The layers stack so as to maximize the electrostatic attraction between the oppositely charged boron and nitrogen atoms.

11.13 Fractional space occupancy 0.3401 for diamond, 0.5236 for simple cubic. The diamond structure is ideal for covalently bonded compounds in which the average number of valence electrons per atom is four. The simple cubic structure (CN 6) is unstable with respect to a close-packed structure (CN 12).

11.15 The Be atoms are so small that they can fit into the tetrahedral holes of a close-packed oxide lattice, thus producing a shorter Be—O distance than would exist if the octahedral holes were occupied.

CHAPTER 12

12.1 Increase in the pressure of Zr favors the bcc structure (d^3s^1 configuration) over the hcp structure ($d^2s^1p^1$ configuration). Increase in the pressure of Fe favors the ccp structure ($d^5s^1p^2$ configuration) over the bcc structure (d^7s^1 configuration).

12.3 The electrical conductivity of a solid-solution alloy passes through a minimum on varying the composition from one pure metal to the other. This result is probably caused by the relative disorder in the lattice in the alloy.

12.5 Ag, ccp; Ag_3Al, bcc; Ag_9Al_4, γ-brass structure; Ag_5Al_3, hcp; Al, ccp.

12.7 (*a*) Yes; (*b*) no, Re melts too high; (*c*) no, Pt (a *d*-electron-rich metal) and Ta (a *d*-electron-poor metal) would react to form a very stable alloy.

12.9 Eu and Yb have cores like those of Gd and Lu, respectively, Hence Eu and Yb crystals involve two bonding electrons per atom, whereas the other lanthanide metals involve three bonding electrons per atom. Thus the "abnormal" densities of Eu and Yb are related to the extraordinary stability of the half- and completely filled $4f$ electronic shell.

12.10 GeF_5^- and/or GeF_6^{2-} ions.

12.11 4.92 Å.

12.13 Three-quarters.

CHAPTER 13

13.1 11,300 Å.

13.3 (*a*) *p*; (*b*) *p*; (*c*) *n*; (*d*) *p*; (*e*) *n*.

13.5 (*c*) $\ll$ (*b*) $<$ (*a*) $<$ (*d*).

13.7 Increase in pressure causes a decrease in internuclear distance and a decrease in band gap. (See Fig. 12.11.) Hence the transition energy decreases and the wavelength increases with increasing pressure. Gallium phosphide, being more ionic (less covalent) than GaAs, has a greater band gap and therefore emits at a shorter wavelength.

CHAPTER 14

14.2 The 2-coordinate silver atoms are Ag(I), and the 4-coordinate silver atoms are Ag(III). These types of coordination are characteristic of d^{10} and d^8 ions, respectively.

14.4 (*a*) A total of three isomers, including a pair of optical isomers. (*b*) A total of five isomers, including a pair of optical isomers. (*c*) Three isomers.

14.5 $Co(en)_3^{3+}$, trigonal distortion, *trans*-$Co(NH_3)_4Cl_2^+$, tetragonal distortion.

14.7 (*a*) For MX_4Y_2 and MX_3Y_3, four isomers each, including a pair of optical isomers.

(*b*) Six isomers, including two pairs of optical isomers. (*c*) Ten isomers, including three pairs of optical isomers.

14.8 Low-temperature ^{13}C or proton nmr spectra would be expected to show either four or five resonances, depending on the structure. The relative intensities of the peaks, corresponding to the atom abundance ratios, would also be diagnostic. Infrared spectroscopy would help identify, in the C—H stretch region, η^5-C_5H_5 and η^5-C_7H_7 groups.

14.10 In the tbp structure, the bidentate ligands can both connect an apical with an equatorial site (optical isomers), or one can connect an apical with an equatorial site while the other connects two equatorial sites. In the *sp* structure, the bidentate ligands can both connect two basal sites, or one can connect two basal sites while the other connects the apical site with a basal site (optical isomers).

CHAPTER 15

15.2 In spite of the negative charge of Cl^-, Δ_0 for Cl^- is less than Δ_0 for NH_3 because of the repulsive interaction between the $p\pi$ electrons of Cl^- and the $d\pi$ electrons of Cr^{3+} (Fig. 15.11). Δ_0 for CN^- is very high because of back bonding, i.e., the shift of electron density from the $d\pi$ orbitals of Cr^{3+} to the antibonding π^* orbitals of CN^-. The latter interaction stabilizes the $d\pi$ orbitals and increases Δ_0 (Fig. 15.12).

15.4 Fe—C bond order 1.5; C—N bond order 2.5. Formal charges of Fe, C and N: -1, 0, and -0.5, respectively.

15.6 The low-spin configuration, corresponding to less electron density directly between the metal and the ligands, would be favored at high pressures.

15.8 Complexes of d^5 iridium (a third-row transition element) are low-spin because of the high Δ_0 in such complexes. The t_{2g} orbitals (which contain the odd electron) overlap with $3p\pi$ orbitals of the chlorine atoms, thus forming a delocalized set of MOs containing the electron spin density.

15.11 The energy level diagram of Fig. 15.13 corresponds to positively charged metal atoms with negative ligands, not to neutral unligated metal atoms. For the latter, the valence *s* and *p* orbitals are lower than the *d* orbitals and can be used in bonding.

CHAPTER 16

16.2

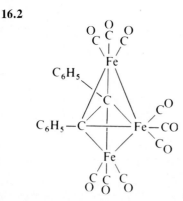

16.5 Diamagnetic $NiCl_4^{2-}$ would be square planar; such a structure is unstable with respect to the tetrahedral form, probably because of repulsions between the chlorines. $Ti(CO)_4^{4+}$ is unstable because there are no $d\pi$ electrons to engage in back bonding. $Cd(CO)_3$ would be expected to be unstable because the d electrons are held so tightly (high nuclear charge of Cd) that they cannot engage in back bonding to the CO groups. AuF_5^{2-} is unstable with respect to AuF_4^- because fluoride ions cannot engage in back bonding and cannot remove the negative charge which builds up on the Au atom when five ligands are coordinated to it.

16.8 The data are consistent with a coordinated NO^+ and a high-spin d^7 Fe(I) atom, with three unpaired electrons.

16.11 (*a*) 13; (*b*) 12; (*c*) 14.

16.15 $(C_4BH_5)Fe(CO)_3$, with a pentagonal pyramid C_4BFe cluster. The number of cluster electrons is 16, corresponding to a nido structure.

CHAPTER 17

17.1 The $Mn(H_2O)_6^{2+}$ ion is a weak-field, high-spin d^5 complex with the spectroscopic symbol $^6A_1$. There is only one combination of quantum numbers corresponding to five unpaired d electrons. Inasmuch as transitions involving a change in multiplicity are forbidden, there are no allowed transitions. The pale color of $Mn(H_2O)_6^{2+}$ is due to weak forbidden transitions.

17.3 CrF_6^{3-}, green; $Cu(NH_3)_4^{2+}$, blue; FeO_4^{2-}, blue-red (purple or magenta); $Co(en)_3^{2+}$, orange.

17.4 CoF_6^{3-}: 16,400 cm^{-1} ($^5T_{2g} \rightarrow ^5E_g$). $NiCl_4^{2-}$: 2400 cm^{-1} [$^3T_1(F) \rightarrow ^3T_2$]; 5400 cm^{-1} [$^3T_1(F) \rightarrow ^3A_2$]; 17,300 cm^{-1} [$^3T_1(F) \rightarrow ^3T_1(P)$]. $Fe(H_2O)_6^{2+}$: 8200 cm^{-1} ($^5T_{2g} \rightarrow ^5E_g$).

17.5 VCl_4, 1; CrO_4^{2-}, 0; FeO_4^{2-}, 3; $Fe(H_2O)_6^{2+}$, 1; $Cr(H_2O)_6^{3+}$, 3; $Ni(NH_3)_6^{2+}$, 3; $Fe(CN)_6^{4-}$, several; $Co(H_2O)_6^{2+}$, 3; $Mn(H_2O)_6^{2+}$, 0; MnO_4^{2-}, 1; $Zn(OH)_4^{2-}$, 0.

17.7 An $O_h d^3$ complex has a nondegenerate quartet ground state and three triply degenerate quartet excited states.

CHAPTER 18

18.2 $CoCl_4^{2-}$ would be expected to have a greater Δ and hence a lower μ. (See Eq. 18.2 and Table 18.4.).

18.4 The NiF_6^{2-} ion in K_2NiF_6 is a low-spin d^6 octahedral complex of Ni(IV). The compound $Ni(NH_3)_2Cl_2$ is probably not a 4-coordinate Ni(II) complex but rather an octahedrally coordinated Ni(II) compound, with bridging Cl$^-$ ions. $Ni(PEt_3)_2Cl_2$ is a square planar complex, and $Ni(Ph_3AsO)_2Cl_2$ is tetrahedral.

18.6 Flattening of the tetrahedron splits the t_2 energy levels into an upper level of energy 2δ and two lower levels each of energy $-\delta$. In a flattened d^8 complex, the ligand-field stabilization energy is $3\delta - 2\delta = \delta$; in a flattened d^9 complex, the stabilization energy is $4\delta - 2\delta = 2\delta$. Elongation of the tetrahedron splits the t_2 levels into two upper levels each of energy δ and a lower level of energy -2δ. In an elongated d^8 complex, the stabilization energy is $4\delta - 2\delta = 2\delta$; in a flattened d^9 complex, the stabilization energy is $4\delta - 3\delta = \delta$.

18.9 The ligand-field stabilization energy of a d^7 tetrahedral complex is greater than that of a d^8 tetrahedral complex, and the LFSE of a d^8 octahedral complex is greater than that of a d^7 octahedral complex.

18.14 Infrared measurement of C—O stretching frequency, XPS measurement of C $1s$ or O $1s$ binding energy. ^{13}C nmr chemical shift, electron or x-ray diffraction measurement of C—O or M—C bond distance, etc.

18.15 $Fe(CO)_2(NO)_2$ tetrahedral; $Fe(CO)_4H_2$ cis-octahedral; $Mn(CO)_4NO$ trigonal bipyramidal with NO equatorial; $Fe(CO)_4CN^-$ trigonal bipyramidal with CN axial; $Fe(CO)_4C_2H_4$ trigonal bipyramidal with ethylene carbons in equatorial plane.

18.17 In neither TiO nor VO does the metal atom have any $e_g{}^*$ electrons that can engage in superexchange interaction.

CHAPTER 19

19.1 $Cr(H_2O)_6{}^{3+} < Al(H_2O)_6{}^{3+} < Fe(H_2O)_6{}^{3+} < Y(H_2O)_6{}^{3+} < Sr(H_2O)_6{}^{2+} < K(H_2O)_6{}^+$.

19.3 A is the cis complex; B is the trans complex. In the cis complex, the ammonia molecules are labilized by the trans chloride ions; replacement of an ammonia molecule by thiourea causes the trans chloride to be labilized. Hence all positions become occupied by thiourea. In the trans complex, the chloride ions labilize each other; only these ligands are displaced by thiourea molecules.

19.5 The $NH_2{}^-$ ions catalyze the displacement of the last coordinated chloride ion by ammonia.

$$Cr(NH_3)_5Cl^{2+} + NH_2{}^- \rightleftharpoons Cr(NH_3)_4(NH_2)Cl^+ + NH_3$$

$$Cr(NH_3)_4(NH_2)Cl^+ \longrightarrow Cr(NH_3)_4(NH_2)^{2+} + Cl^-$$

$$Cr(NH_3)_4(NH_2)^{2+} + 2NH_3 \longrightarrow Cr(NH_3)_6{}^{3+} + NH_2{}^-$$

19.7 If the reaction rate is reduced by the addition of Fe^{2+} to the system, the formation of Sn(III) in a reversible primary step would be strongly indicated:

$$Sn(II) + Fe^{3+} \rightleftharpoons Sn(III) + Fe^{2+}$$

$$Sn(III) + Fe^{3+} \longrightarrow Sn(IV) + Fe^{2+}$$

19.9 The data suggest the following reaction paths:

$$V(H_2O)_6{}^{2+} + V(H_2O)_6{}^{3+} \xrightarrow{k_1} V(H_2O)_6{}^{3+} + V(H_2O)_6{}^{2+}$$

$$V(H_2O)_6{}^{2+} + V(H_2O)_5OH^{2+} \xrightarrow{k_2} V(H_2O)_5OH^{2+} + V(H_2O)_6{}^{2+}$$

If we let K equal the acid ionization constant of $V(H_2O)_6{}^{3+}$, then $k = k_1 + k_2K/(H^+)$.

19.11 A second-order associative reaction of $Ni(CO)_4$ with L implies a very unstable intermediate or activated complex with five ligands and 20 valence electrons on the nickel atom. On the other hand, such an intermediate in the $Co(CO)_3NO + L$ reaction would not be unstable because the NO group can accept a pair of electrons from the Co atom, thus forming a bent NO group and maintaining an 18-electron count.

19.12 The two axial water molecules in $Cu(H_2O)_6{}^{2+}$ are held much more weakly than the four equatorial ones as a consequence of Jahn-Teller distortion.

CHAPTER 20

20.1
$$(OC)_5Cr\!=\!C\!=\!O + LiCH_3 \longrightarrow (OC)_5Cr\!=\!C\!-\!O^-$$

with Li^+ and CH_3 substituents, reacting via $(CH_3)_3O^+$ to give

$$(OC)_5Cr\!=\!C\!-\!OCH_3$$

with CH_3 substituent.

20.4 The activation energies of hydrogen oxidative additions are much lower than the H_2 dissociation energy because the general mechanism involves the exothermic formation of M—H bonds simultaneously with the endothermic H—H cleavage.

20.6 (*a*) $NiL_4 \underset{-H^+}{\overset{H^+}{\rightleftharpoons}} HNiL_4^+ \overset{C_2H_4}{\longrightarrow} L_4Ni\!-\!CH_2CH_3^+$

$CH_3CH_2CH\!=\!CH_2$ ... C_2H_4 ... $L_4Ni\!-\!CH_2CH_2CH_2CH_3^+$

(*b*)

$HNiL_4^+ \overset{C_2H_4}{\longrightarrow} L_4Ni\!-\!CH_2CH_3^+ \rightleftharpoons\ ^+L_nNi\!=\!C\Big\langle{}^{H}_{CH_3}$ with H

$CH_3CH_2CH\!=\!CH_2$ ... C_2H_4

$L_4Ni\!-\!CH_2CH_2CH_2CH_3^+ \longleftarrow$
$$\begin{array}{c} H-\overset{CH_3}{\underset{}{C}}\!-\!CH_2 \\ | \quad | \\ H-\underset{L_n}{Ni}^\pm\!-\!CH_2 \end{array} \longleftarrow \begin{array}{c} H \quad CH_3 \\ \diagdown C \diagup \\ \| \quad CH_2 \\ H-\underset{L_n}{Ni}^+\!\cdots\| \\ CH_2 \end{array}$$

20.8 (*a*) $(\eta^5\text{-}C_5H_5)Fe(CO)_2(COCH_3)$; (*b*) $IrL_2(CO)(CH_2Cl)$; (*c*) $[(CO)_3Co]_2(CR)_2$, tetrahedral cluster compound.

20.10 By using D_2O as a solvent, mass-spectrometric analysis of the product acetaldehyde would indicate whether any of the solvent hydrogen ends up in the acetaldehyde. Such a result seems unlikely, because it would require the effective loss of a proton from a C—H bond.

CHAPTER 21

21.1 The data suggest that CO_2 is absorbed dissociatively, to give adsorbed CO molecules and adsorbed oxygen atoms.

21.2 1.78, 1.54, and 1.09×10^{15} atoms cm^{-2}.

21.4 Platinum is not an electropositive metal; it does not form a stable nitride. Therefore it would not be expected to form a stable intermediate with N_2, as iron does.

CHAPTER 22

22.3 The following is just one possible mechanism:

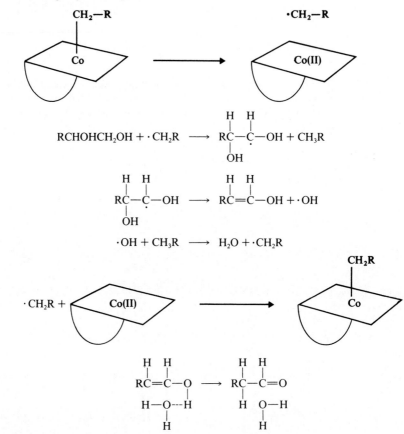

$$RCHOHCH_2OH + \cdot CH_2R \longrightarrow \underset{\underset{OH}{|}}{R\overset{\overset{H}{|}}{C}} - \overset{\overset{H}{|}}{\underset{\cdot}{C}} - OH + CH_3R$$

$$\underset{\underset{OH}{|}}{R\overset{\overset{H}{|}}{C}} - \overset{\overset{H}{|}}{\underset{\cdot}{C}} - OH \longrightarrow R\overset{\overset{H}{|}}{C} = \overset{\overset{H}{|}}{C} - OH + \cdot OH$$

$$\cdot OH + CH_3R \longrightarrow H_2O + \cdot CH_2R$$

$$R\overset{\overset{H}{|}}{C} = \underset{\underset{H-O\text{---}H}{|}}{\overset{\overset{H}{|}}{C}} - O \longrightarrow R\overset{\overset{H}{|}}{C} - \underset{\underset{H}{|}}{\overset{\overset{H}{|}}{C}} = O$$

If the CH_2 group of the diol were labeled with tritium, most of the tritium would end up in the water, and some of it would end up in the Co—CH_2— group of the coenzyme.

22.5 (*a*) From the equilibrium constant expression for the reaction

$$K = \frac{[MbO_2]}{[Mb][O_2]}$$

one can readily derive the expression for the percentage saturation,

$$100\frac{[MbO_2]}{[Mb] + [MbO_2]} = \frac{100 \times K[O_2]}{1 + K[O_2]}$$

By considering the form of the equation under the limiting conditions of very low and very high $[O_2]$, it can be seen that the curve is always concave downward.

(b) In this case, a similar derivation gives

$$100 \frac{[Mb(O_2)_n]}{[Mb] + [Mb(O_2)_n]} = \frac{100 \times K[O_2]^n}{1 + K[O_2]^n}$$

and it can be shown that this equation is concave upward at low $[O_2]$ and concave downward at high $[O_2]$.

FORMULA
INDEX

623

SUBJECT
INDEX